1976 Yearbook of Science and the Future

Encyclopædia Britannica, Inc.

Chicago Toronto London
Geneva Sydney Tokyo Manila
Johannesburg Seoul

1976 Yearbook of Science and the Future

Library of Congress Catalog Number: 69-12349 International Standard Book Number: 0-85229-307-0

THE UNIVERSITY OF CHICAGO
The Yearbook of Science and the Future is published with the editorial advice of the faculties of the University of Chicago

Encyclopædia Britannica, Inc.

Editorial Advisory Board

Science Through the Ages

In our usual preoccupation with contemporary affairs, it is easy to forget the contributions of the past. This is particularly true for science, which is considered by many to be principally a product of the modern world. Scientists tend to reinforce these attitudes because of their orientation toward research and development for present and future benefits.

In an effort to rectify this situation, the *1976 Yearbook of Science and the Future* presents several feature articles that describe achievements in past eras. For example, "Stargazers of the Ancient World" tells of the elaborate structures erected by early man to observe and calculate the motions of the Sun, Moon, stars, and planets. The megaliths of Stonehenge, the Bighorn Medicine Wheel in Wyoming, and the pyramids of ancient Egypt are among the works that exemplify the ingenuity and astronomical skills of these peoples. "Our Architectural Lineage" is a pictorial essay that describes the buildings of man since the cave dwellings of the Stone Age and points out the common elements in their design and construction that have persisted through the centuries.

The bicentennial of the United States provides the occasion for a third feature dealing with the past, "Science in Colonial America." Such men as Benjamin Franklin, David Rittenhouse, John Bartram, and David Bushnell responded to the challenge of the New World with an outlook that stressed the practical benefits of science. The results included the lightning rod, an orrery that could display the positions of planets and their satellites for 5,000 years with less than 1° error, new varieties of hybrid plants, and a forerunner of the modern submarine. These men began a utilitarian tradition in American science that has continued to this day.

The remaining articles in the *Yearbook* focus primarily on the present and future. In several of them, however, the contributions of scientists of the past can again be noted. Thus, in "Periodicity of the Chemical Elements" it is shown how the discoveries of Antoine Lavoisier and Dmitry Mendeleyev in the 18th and 19th centuries laid the groundwork for the development of our present periodic table. "The Recovery of Gold" tells of man's historic craving for this "ultimate metal" and of his efforts to wrest it from the Earth, ranging from the stone mortars of ancient Egypt to the ever increasing sophistication of today's chemical cyanide process. In "The Living Cell" Theodor Schwann, Matthias Schleiden, and Rudolf Virchow are described as chief among the many scientists who contributed to one of the major intellectual achievements of the 19th century, the discovery that life is organized on the basis of cells. And the delightful "Excursions in Topology" recounts the studies in the 18th and 19th centuries of such men as Leonhard Euler, A. F. Möbius, and Felix Klein in seeking solutions to the problems posed by this intriguing branch of mathematics.

Among the other feature articles, the remarkable achievement of U.S. space scientists in probing Jupiter with Pioneers 10 and 11 is described in "Visit to a Large Planet: The Pioneer Missions to Jupiter"; the future of this endeavor also holds promise as Pioneer 11 continues its journey through space toward a rendezvous with Saturn in 1979. The current state of knowledge of the moons in the solar system, a result of space probes and increasingly powerful telescopes, is described in "The Natural Satellites."

The importance of understanding the Earth's climate cannot be overestimated; favorable weather is needed to achieve the level of agricultural production necessary to feed the world's growing population. The variations in climate during past eras and some of the possible reasons for their occurrence are discussed in "The Earth's Changing Climate." Closely related in subject matter is "Dendrochronology: History from Tree Rings," in which the author describes how the annual rings formed in the trunks of certain tree varieties can be studied to reconstruct the climates and environments of past ages.

Can the speed of light be exceeded? The eminent scientist Fred Hoyle deals with this question, one of fundamental importance in physics and astronomy, in "The Speed of Light: A Physical Boundary?" Finally, the phenomena of permafrost and lightning, the world of superhard materials, and the remarkable ways in which honeybees communicate with one another complete the list of feature articles for this volume.

Full-color photographs and drawings abundantly illustrate the articles. As in the past, distinguished authorities have been selected to write these features as well as the Year in Review entries, which concern advances in individual disciplines within science and technology. Year in Review subjects covered for the first time this year include Materials Sciences, Mechanical Engineering, and Optical Engineering.

—THE EDITOR

Contents

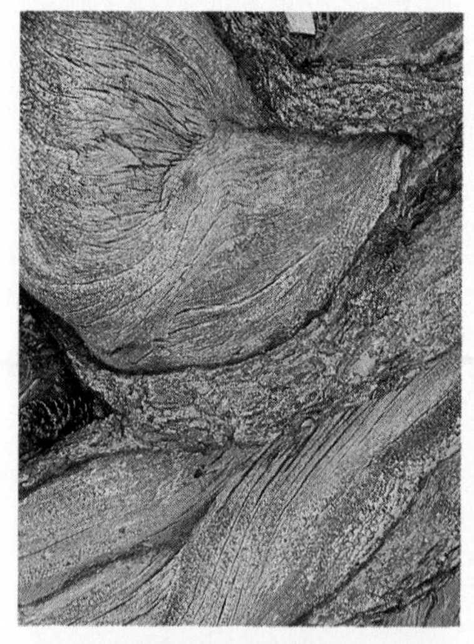

Science in Colonial America

by George Foot

On the edge of a continent that was still largely wilderness, pre-Revolutionary America nurtured a science oriented toward the practical and useful.

John Bartram: He was responsible for the introduction of hundreds of new plants to European gardens.

By the year 1775 the American Colonies comprised about 2.5 million people stretching from Georgia northward to Maine. More than 90% of the population lived on farms and in small towns, set in cleared patches of land scattered through America's primeval forests. Apart from the plantation agriculture of the Southern Colonies, which was based on a single crop, usually tobacco, and the gang labor of thousands of black slaves, American farmers followed the same methods used by English farmers a millennium earlier.

The plow, the sickle, and the flail were essentially unchanged from medieval times, and most men planted and harvested in the ancient rhythms passed down by their fathers. The rotation of crops was virtually unknown, and many a farmer was forced to move his barn because of the enormous piles of rich manure that had been carelessly stacked around it.

Yet it was the American farmer's isolation that proved to be his greatest challenge. In this new wilderness, each family had to supply a multitude of goods ranging from the shirts on their backs to the metal-tipped plows that split the land. Apart from occasional purchases of salt and gunpowder, these people functioned as self-sufficient units, able to carry on for weeks at a time without any outside contact. Despite its obvious disadvantages, this isolation of the American farmers bred a versatility uncommon among their European counterparts. If a farmer was also a self-trained surveyor, so much the better, and if he might practice medicine as well, why, better still! A man had to stretch beyond his reach to master this new land, and every skill, however raw and unpracticed, was an invaluable hedge against an uncertain future.

Set amid these scattered settlements of farmers and tradesmen were the four American communities that dared to call themselves cities. Poised at the very edge of the continent, they drew their strength from the steadily growing trade between Europe and America. While Boston had been the unquestioned leader of the 17th century, New York, Philadelphia, and Charleston had eclipsed her elegance by the end of the colonial period. Life in these metropolises was scarcely more comfortable, however, and usually a good deal less wholesome than in the countryside. Pigs roamed the streets as the only means of sanitation; typhoid and scarlet fever periodically swept away scores of victims. Drinking, as in the countryside, was a common curse, while the dark streets and token police invited violence.

Yet, despite their poverty and isolation, the American Colonies had proved to be an unexpectedly fertile ground for the dissemination of scientific knowledge. Natural philosophy, as it was then regarded, was a field much more suited to the universities of Europe than to the log cabin world of America. But the very nature of the New World was, in itself, a powerful spur to the settlers' curiosity. Full of anticipation, America's first colonists were firmly dedicated to a tradition of

***GEORGE FOOT** is a historian and free-lance writer in Warwick, Massachusetts.*

Paintings by Ron Villani

SEEDS
Flower
SEEDS

exploration and expectation. This was a land of the future where anything, even paradise, might lie over the next ridge.

The emphasis in a society lacking virtually everything was, quite naturally, on the immediately useful. Accordingly, most scientific activity in the early phases of the colonial period was directly related to the practical arts of navigation, surveying, and the investigation of America's cornucopia of new plants, most notably tobacco, for supposed curative powers. Thus began a utilitarian tradition that has since become a hallmark of American society.

The quest for useful science in the New World was complemented by the philosophical theorizing emanating from the Old. Many an American dirt farmer would have agreed with Sir Francis Bacon when he wrote, "Science must be known by its works. It is by the witness of works rather than by logic or even observation that truth is revealed and established. It follows from this that the improvement of man's lot and the improvement of man's mind are one and the same thing." In a land where a man was unprotected by either social class or capital from the consequences of his actions, any knowledge that was worth keeping was worth using.

The merchant naturalists

In 1771 the American Philosophical Society published the first volume of its *Transactions*. After a century of diligent effort by hundreds of Americans, this elegant quarto was solid evidence of America's entry into the international scientific arena. The English *Critical Review* hailed it as "convincing proof of the increasing greatness and prosperity of our American colonies." Composed of some 22 separate essays ranging from the American observations of the transit of Venus to the design of a machine to cut files, it was the high-water mark of colonial science.

The American Philosophical Society itself was an early creation of the brilliant men who settled in the city of Philadelphia during the first third of the 18th century. Set at the confluence of the Delaware and Schuylkill rivers, Philadelphia had become the chief port of the American Colonies, and its rising prosperity and broad religious tolerance had created a favorable environment for the ambitious and the inquisitive. Its recently prosperous men could find no surer sign of gentlemanly status than the cultivation of natural philosophy.

Fortunately, some displayed much the same talent in their avocation as in their profession and, for a few, their investigations began to dominate their lives. Indeed, most of the scientists of 18th-century America sprang from the middle class of merchants and craftsmen who had found success in the New World.

Archibald Spencer: His dramatic performances on the nature of electricity fascinated his audiences, who watched in amazement.

Presiding over the city's scientific circle was James Logan, a wealthy Quaker statesman and merchant who had arrived in Philadelphia in

1699. Trained in mathematics and the classics, he soon assembled the finest nonreligious library in the Colonies. His investigations on the sexuality of plants, which were published in the journal of the Royal Society in 1736, were a genuine contribution to the field of botany and set a standard for those Americans who would follow.

Stimulated by the enormous European fascination with America's plants and animals, the naturalists became the most cohesive scientific group in the American Colonies. Ranging from Dr. Alexander Garden in Charleston, from whom the gardenia received its name, to Cadwallader Colden in New York, this group found its center in John Bartram of Philadelphia. An archetypal Yankee, Bartram was a largely self-educated man who rose to international renown. Working as a farmer to feed his family, as well as practicing medicine among his neighbors, he found little time to mince words and was singularly unimpressed by titles and politicians.

In the late 1730s Bartram found support for his collecting from Peter Collinson, a Quaker merchant based in London, whose untiring curiosity was a considerable resource to colonial America. Bartram was soon forwarding to London seeds collected on expeditions that ranged from Florida to the Great Lakes. He was responsible for the introduction of hundreds of new plants to European gardens. Within a few years he was able to support himself as a professional seedsman and found the time to begin his own botanical garden near Philadelphia. Linnaeus, the foremost naturalist of the 18th century, called Bartram "the greatest natural botanist in the world."

A full decade before Bartram, a young Philadelphia printer was beginning to make his mark on the city. In 1727 Benjamin Franklin set about organizing the Junto, a mutual self-improvement club for such young, middle-class gentlemen as himself. Dominated by craftsmen, it formed the nucleus of scientific activity in the city.

Its success was attested by the creation of the Library Company of Philadelphia four years later. A subscription library designed to serve the pressing needs of Philadelphia's growing intelligentsia, it was able to build, in the words of a European visitor, "a fine collection of excellent works, most of them in English." Indeed, these were men who had little occasion to study scholarly Latin. Their strength lay in their ingenuity rather than in their knowledge.

The concept of electricity

In the spring of 1743 Dr. Archibald Spencer, a traveling Scottish lecturer, began a series of demonstrations along the eastern seaboard of the American Colonies. His dramatic performances on the nature of electricity fascinated his audiences, who sat amazed as he "proceeded to show that [electrical] Fire is Diffus'd through all space, and may be produced from all Bodies, Sparks of Fire Emitted from the Face and

Benjamin Franklin: He concluded that a pointed metal rod set atop a building would draw off disastrous bolts of lightning.

Hands of a Boy Suspended Horizontally, by only rubbing a Glass Tube at his feet."

Among the witnesses to this remarkable event was the ever present printer-philosopher Benjamin Franklin. He later acknowledged that it was these demonstrations which had stirred a curiosity that consumed most of his scientific interest for the next decade. With the help of the Junto and the Library Company, he conducted a series of experiments that culminated in the publication of his *Experiments and Observations on Electricity* in 1751. Soon translated into three other languages, it was undoubtedly the most important American scientific work of the century.

His contribution of the single fluid concept of electricity, as well as the coining of the terms "positive" and "negative" to denote electrical flow, were basic to the scientific study of electricity. Still, it was a spin-off of his work that made his name known throughout Europe. During the course of his experiments, he had noted "the wonderful effect of pointed bodies, both in *drawing off* and *throwing off* the electrical fire." Having already discovered from his famous kite experiment the identity of lightning and electricity, he concluded that a pointed metal rod set atop a building and connected with the ground would draw off the disastrous bolts of lightning that periodically destroyed so many structures. Within a few years the rooftops of Paris and London, as well as those of Philadelphia, were covered with Franklin's newly invented lightning rods.

Electricity was such an unknown field that America's isolation from Europe proved to be only a minor handicap. Indeed, it may have worked to advantage by separating Franklin and his compatriots from the myriad misleading theories being propagated on the Continent.

Early institutions

Though encouraged by such spectacular success, Americans soon realized that their abilities lagged far behind their ambitions. For every Franklin or Logan there were a hundred "philosophers" who were more than willing to propose immediate explanations for all sorts of phenomena. Henry Moyes's proposal that towns be ringed with iron walls sunk deep into the ground to protect them from earthquakes was only one example. Isaac Ledyard's misbegotten *Essay on Matter in Five Chapters* was evidence of the horrors that occurred when a well-meaning American attempted to "clarify" the Newtonian world.

Only time and patience could create the institutions that would one day direct American science. The American Philosophical Society which so proudly published its *Transactions* in 1771 was a dismal failure after its initial organization in 1743. Franklin laid the blame to some "very idle Gentlemen" but, in fact, American society was still too immature to support a full-fledged scientific society. Though Harvard

John Winthrop: American observations of the transit of Venus of 1769 gave new credence to American scientific ability.

had included courses in geometry, physics, and astronomy soon after it opened in 1638, it was not until 1727 that the college endowed a chair of mathematics and natural philosophy. Unfortunately, most of the other eight fledgling colleges were far more reticent.

All but one of these institutions had been conceived primarily as theological schools, a goal that put considerable limitations on their curricula. Their libraries emphasized religious works, and scientific writings were offhandedly included in an attempt to offer a liberal education. Even at Harvard, where the preeminent natural philosopher John Winthrop held the chair, the most attentive students were fortunate if they could grasp basic concepts. One of his students recalled, "He touched on a few matters rapidly: the subjects of course very familiar to him—but to the novitiates, 'it was all Greek.' "

The development of American science had always been predicated on the growing political maturity of the several Colonies and, as tensions between England and America began to build, American scientists eagerly participated in the new wave of cultural nationalism that spread from Charleston to Boston. The decade before the Revolution was a vital, restless period for American science as it struggled to gain its intellectual independence from its European mentors. The moribund American Philosophical Society was stirred to new efforts by the establishment of the American Society in 1766. These two bodies competed with each other for two years, before the Philosophical Society assimilated the American Society's younger and more ambitious membership. It was the fusion of these two groups that led to the successful publication of the *Transactions* in 1771.

Telescopes and timepieces

The American observations of the transit of Venus across the face of the Sun in the summer of 1769 gave new credence to American scientific ability. Widely separate observations were necessary to compile the data needed if scientists were to calculate accurately the distance of the Earth from the Sun; the next transit would not occur for another 105 years. From their New World vantage point, American astronomers eagerly prepared to participate in this worldwide scientific effort. Telescopes and accurate timepieces were commissioned for the observations and, in Philadelphia, three separate observation sites were prepared. All told, some 22 contacts were reported, though many were of doubtful accuracy. Yet the capacity of the American scientific establishment to organize such a widespread effort was a testament to the growing vitality of American society.

David Rittenhouse: This self-educated man desired to fashion a mechanical replica of the solar system that would "astonish the ... curious examiner."

If Americans lacked the formal education for research in the basic sciences, their ingenuity and craftsmanship were clear evidence of their natural genius. Two years before the transit of Venus, one David Rittenhouse, a Philadelphia clockmaker turned astronomer, began

construction of an orrery for the College of New Jersey. Already famous for his exquisite clocks, this self-educated man desired to fashion a mechanical replica of the solar system that would "astonish the skillful and curious examiner."

The result was a device that perfectly expressed the mechanical, Newtonian world of the 18th century. Displaying, with less than a degree of error, the positions of the various planets and their satellites for a period of 5,000 years, it was a visible monument to American skill. Another was soon commissioned for the young College of Philadelphia, which awarded Rittenhouse a master of arts degree in recognition of his accomplishment. Years later Thomas Jefferson would remark, in a moment of pardonable hyperbole, that Rittenhouse stood nearer to God "than any man who has lived from the creation to this day." The Rittenhouse orreries were an important symbol to an entire generation of American patriots who sought to free the Colonies from intellectual as well as political servitude.

From science to revolution

With the outbreak of the war most of this mushrooming scientific activity came to an abrupt halt. Meetings of the American Philosophical Society were disrupted by arguments between Whigs and Tories and were not revived until 1779. By 1780 American science had lost, by death or emigration, some of its brightest lights. Among these was Dr. Garden of Charleston, one of America's greatest botanists, who sailed for Europe a confirmed Tory. John Winthrop, America's greatest natural philosopher, died in 1779.

The cities, which had been the centers of scientific work during the colonial period, were now centers of the struggle for independence. The College of Philadelphia was alternately used as a barracks by American troops and as a hospital by the British. Boston was occupied by British troops for nearly a year, while New York suffered under British military rule for seven years.

America's best minds turned from the love of science to the love of liberty and the very real dangers of the day. Franklin best expressed this concern when he wrote, "Had Newton been Pilot but of a single common Ship, the finest of his Discoveries would scarce have excus'd, or atton'd for his abandoning the Helm one Hour in Time of Danger; how much less if she carried the Fate of the Commonwealth."

The interregnum of the war did encourage some advances in war-connected scientific work. The necessity of examining physicians and surgeons for military service led to the first attempt to regulate the practice of medicine in the Colonies. Prompted by the dysentery and fever that plagued the Revolutionary Army, Dr. Benjamin Rush's 1778 pamphlet, *Directions for Preserving the Health of Soldiers*, was used as an Army medical manual through the American Civil War.

David Bushnell: Though unsuccessful as a weapon of war, the "Turtle" remained an important step in the development of the submarine.

Inventions, such as Thomas Paine's suggestion of incendiary arrows shot from steel crossbows, flourished during the Revolution, but few were ever put to use. The most significant exception was David Bushnell's midget submarine, which he dubbed the "Turtle." Fashioned of oak planks caulked with tar, it was designed to plant a time bomb against the unguarded hulls of British warships. Equipped with hand-driven propellers, a glass conning tower set above the water line, and a barometer to measure the craft's depth, it was a remarkable achievement for the time. Bushnell thought it would be possible to "pulverize the British navy" with it, but the device proved incapable of attaching the bomb to the enemy ships, and after several futile attempts, Bushnell turned to the development of floating mines. Though unsuccessful as a weapon of war, the "Turtle" remained an important step in the development of the modern submarine.

Toward an American science

The euphoria of independence, which prevailed throughout the Colonies in 1783, was nowhere more joyously celebrated than among the devotees of American science. This long-sought freedom brought with it inevitable comparisons with England's grand achievements following its own civil war a century earlier. In a poem written in 1771, Philip Freneau had clearly stated the symbiotic relationship between political freedom and scientific inquiry:

> This is a land of ev'ry joyous sound
> Of liberty and life; sweet liberty!
> Without whose aid the noblest genius fails
> And science irretrievably must die.

Though no American Newton was to be found to proclaim the new American greatness, the energy that had been absorbed by war now poured into the development of a new, purely American culture.

Nearly out of breath with enthusiasm, the American scientist must have been sobered by the intellectual landscape he saw around him when he surveyed his new nation. Eight years of war had separated America from the work of European science and had greatly disrupted the fragile colonial scientific establishment. The horizon was dotted with imposing facades of institutions that had little more substance than movie-lot saloons. In reviewing the unquestionably inferior second volume of the *Transactions* of the American Philosophical Society, published in 1786, the once-enthusiastic *Critical Review* reported that Americans "might be more usefully employed in accumulating facts than in constructing systems."

Peter Carnes: Balloon flight ... was science on a much more engaging level than the transit of Venus ... and Americans eagerly entered the field.

Clearly, the early national period would be spent in rebuilding and strengthening the American scientific base rather than pushing off into new fields of discovery. It was in this spirit that Boston established the American Academy of Arts and Sciences in 1780. Although

quite similar to the Philosophical Society in Philadelphia, it was as much devoted to the dissemination of science as to its advancement.

In medicine, the organization of the College of Physicians of Philadelphia in 1787 gave doctors a national institution similar to the American Philosophical Society to publish their works and help organize their profession. On a more mundane level the Philadelphia Dispensary, established in 1786, was the first medical clinic set up to provide free medical care for the poor. Humane societies organized in Boston and Philadelphia devoted themselves to educating people on methods of resuscitating drowned and suffocated victims—though their methods, ranging from bellows placed in the mouth to the inhalation of tobacco, met with limited success.

While American scientists and educators struggled to import the latest works of European science, popular scientific fads crossed the Atlantic with remarkable speed. In France on June 5, 1783, the Montgolfier brothers offered their first public exhibition of balloon flight, and in October of that year a Frenchman made the first human ascent in a balloon. This was science on a much more engaging level than the transit of Venus, and Americans eagerly entered the field.

A year later Peter Carnes of Baltimore sent a 13-year-old boy aloft in a hot-air balloon. Newspapers and magazines were full of plans for building balloons, and even Benjamin Franklin was caught up in the enthusiasm when he speculated that 5,000 balloons, each manned by two soldiers, would be an effective deterrent to war.

The unity of the 18th-century mind is reflected in the 1771 edition of the *Encyclopaedia Britannica*. Looking under "Philosophy," one is referred to mechanics, optics, astronomy, logic, and morals. That optics and morals should share any congruity is unthinkable to the modern mind. But this was a society still integrally part of the Christian universe, where a scientist could examine God's works via prisms and religious texts with equal dexterity. It was this unique combination of religious-utilitarian enthusiasm that had encouraged scientific activity on both sides of the Atlantic.

Yet, for all its hopes, the 18th century was not a time of great scientific advance in America. Indeed, America made few real contributions to scientific thought during the first two centuries of its existence. Rather, those decades of patient counting and collecting, of fruitless experiments based on mistaken theory, were only the ripples of a more subtle and elusive change in the society as a whole.

Jefferson best explained the thrust of American science when, replying to charges of American ineptitude, he wrote, "In science, the mass of people [of Europe] is two centuries behind ours, their literati, half a dozen years before us." It was in this democratic, educational approach to society that one found the true meaning of all the scientific work attempted in America.

The ambitions of American scientists were clear: to discover knowledge that would ameliorate the human condition. Knowledge without some human utility was useless. These were the children of the future, striving to create a new humanity on a cruel and fertile continent.

FOR ADDITIONAL READING

Bell, Whitfield J., Jr., *Early American Science: Needs and Opportunities for Study* (Russell and Russell, 1971).

Brasch, Frederick E., "The Newtonian Epoch in the American Colonies, 1680–1783," American Antiquarian Society *Proceedings*, New series (1939, pp. 314–332).

Bridenbaugh, Carl and Jessica, *Rebels and Gentlemen: Philadelphia in the Age of Franklin* (Oxford University Press, 1965).

Curti, Merle, *The Growth of American Thought*, 3rd ed. (Harper and Row, 1964).

Hall, A. Rupert, *The Scientific Revolution, 1500–1800*, 2nd ed. (Beacon, 1966).

Hindle, Brooke, *The Pursuit of Science in Revolutionary America, 1735–1789* (The University of North Carolina Press, 1956).

Jaffe, Bernard, *Men of Science in America*, rev. ed. (Simon and Schuster, 1958).

Struik, Dirk J., *Yankee Science in the Making*, rev. ed. (Macmillan, 1962).

Tolles, Frederick B., *James Logan and the Culture of Provincial America* (Little, Brown, 1957).

Wright, Louis B., *The Cultural Life of the American Colonies, 1607–1763* (Harper and Row, 1962).

Visit to a Large Planet: The Pioneer Missions to Jupiter

by Henry T. Simmons

Two U.S. space probes have provided scientists with a wealth of information on the largest planet in the solar system, resolving some questions but raising others.

(Above) Jupiter, the largest planet in the solar system, dwarfs the Earth, shown at the bottom left. (Below) Closest picture of Jupiter's Great Red Spot taken by Pioneer 11 was photographed from a distance of 545,000 km (338,000 mi). View looks northward toward the equator.

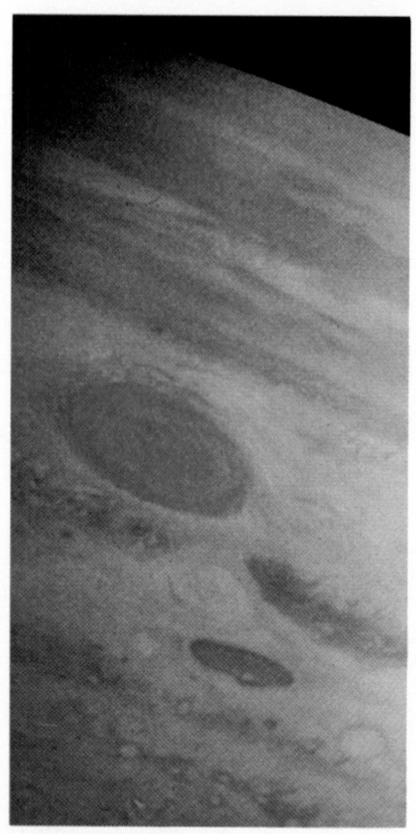

***HENRY T. SIMMONS** has written many articles on the U.S. space program and various aspects of the solar system.*

Illustrations by Dennis Magdich; photos, courtesy, NASA

Within the span of a year, two U.S. spacecraft visited Jupiter to provide mankind's first close look at this strange, vast giant of the solar system. The first of the two 570-lb automated space voyagers flew toward the planet along its equatorial plane, swinging around its right-hand side (as viewed from the Earth, north oriented up) on Dec. 3, 1973. Called Pioneer 10, it passed within 81,000 mi of Jupiter's banded cloud tops. Exactly a year later its twin, Pioneer 11, made a left-hand pass over Jupiter's south polar region, skimming within 27,000 mi of its racing clouds. Both of the spacecraft returned a rich harvest of new findings about the massive planet, including scores of color images that had sharper detail than any produced by the most powerful terrestrial telescopes.

The new data disclose that Jupiter is essentially a huge, fast-spinning mass of liquid hydrogen, totally unlike any of the small, dense, stony planets of the inner solar system (Mercury, Venus, Earth, and Mars). Its weather is convective, driven more by its primordial internal heat than by the radiation it receives from the Sun. Jupiter generates a powerful and complex magnetic field in which swarms of energetic charged particles are trapped. These particles are occasionally accelerated almost to the velocity of light and then squirted great distances into the inner solar system.

Like Jupiter itself, one of its family of 13 satellites—Io—was discovered to possess both an atmosphere and an ionosphere of electrically charged particles. Io also generates an immense cloud of neutral hydrogen, which stretches one-third of the way around the satellite's orbit of Jupiter.

The starlike properties of Jupiter have fascinated many scientists. These include its great mass (318 times that of the Earth), its radiation of more than twice as much heat as it receives from the Sun, and its great output of radio noise as well as charged particles. It even resembles a "mini-solar system" in the sense that its four large moons—Io, Europa, Ganymede, and Callisto—decrease in density with increasing distance from Jupiter. This same pattern occurs in the planets of the solar system, suggesting that Jupiter was hot enough at the time of its formation to drive away the light, volatile gases of its inner satellites, just as the Sun did in the case of the inner planets. Because of these characteristics, according to John A. Simpson, professor of physics at the University of Chicago, "Jupiter is Nature's best gift of what a poor man's star is like."

But Jupiter is definitely not a star. Though the temperature of its central core may reach 30,000° C (54,000° F), about six times the temperature on the surface of the Sun, that figure is hundreds of times too low to ignite the thermonuclear reactions that heat the stars. Large as it is, Jupiter has less than 0.1% of the mass of the Sun. According to theoretical calculations, it would have to be at least 80 times more massive than its present size in order to reach the threshold temperature for the thermonuclear fusion reactions that would allow it to begin to glow like a dim, red-dwarf star.

Visible surface of Jupiter is divided into east–west zones and belts: (1) south polar region; (2) south south temperate zone; (3) south south temperate belt; (4) south temperate zone; (5) south temperate belt; (6) south tropical zone; (7) south component of south equatorial belt; (8) north component of south equatorial belt; (9) equatorial zone; (10) north equatorial belt; (11) north tropical zone; (12) north temperate belt; (13) north temperate zone; (14) north north temperate belt; (15) north north temperate zone; (16) north north north temperate belt; and (17) north polar region. The Great Red Spot can be seen in the south tropical zone.

Magnetic bubble

Jupiter is nonetheless unique and majestic, offering a range of physical conditions to be found nowhere else in the solar system. One of its most striking features is its envelopment in a huge, quivering magnetic bubble at least 10 million mi in diameter. This "magnetosphere" is generated by a Jovian magnetic field 20 to 40 times as strong as that of the Earth. Because Jupiter plies an orbit 480 million mi from the Sun, more than five times the Sun–Earth distance, the pressure of the steady "solar wind" of charged particles from the Sun is only about 4% as great near Jupiter as it is near the Earth. The result is that the Jovian magnetosphere is able to swell into an enormous volume of space. Moreover, because of Jupiter's rotation rate of just under ten hours (the fastest spin of any of the planets) the total magnetic force or moment of its magnetosphere has been estimated to be about 20,000 times that of the Earth.

It has been estimated that if the Jovian magnetosphere were visible from the Earth, at its maximum extent it would fill 2° of space, or about four times the diameter of the solar disk. In this respect, Jupiter is by far the largest object in the solar system, exceeding the Sun's optical disk by about tenfold in total diameter.

Like the Earth's magnetosphere, the vast, tenuous magnetic bubble surrounding Jupiter is heavily populated by charged particles, ranging

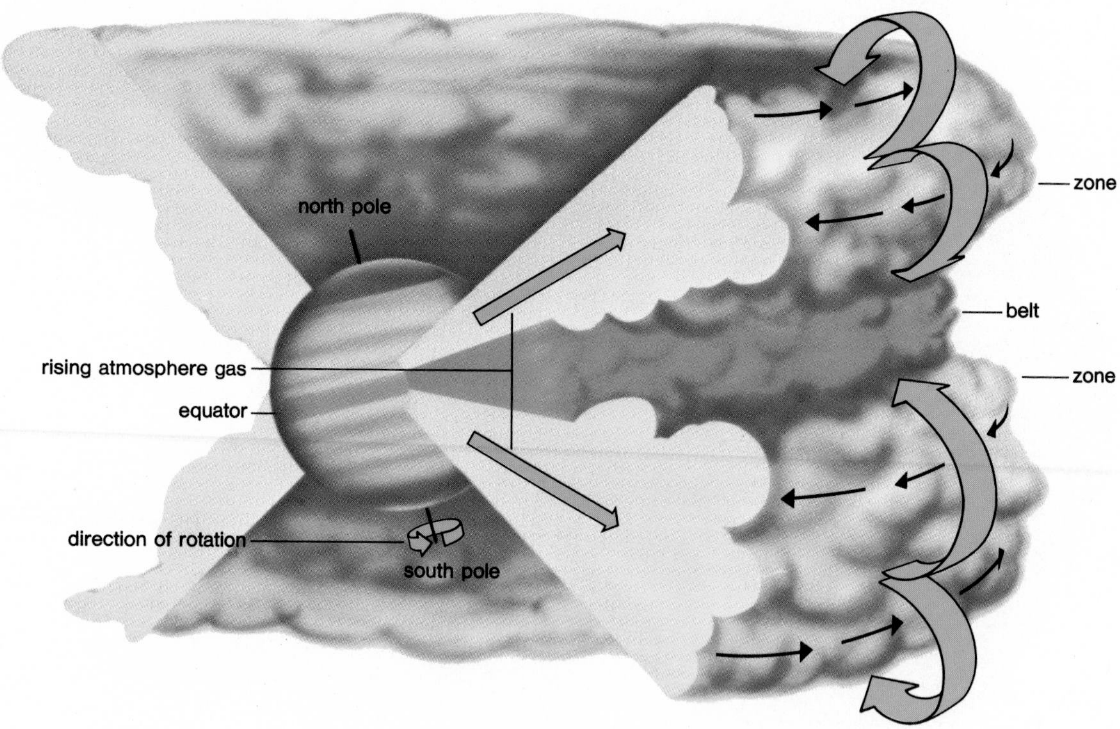

Jupiter's weather is strongly influenced by the Coriolis effect, generated by the planet's fast spin. It causes atmospheric gas that would move toward the equator by means of convection to move instead around the planet against the direction of rotation. Also, gas that would move toward the poles instead moves around Jupiter in the direction of rotation.

from a low-energy plasma of electrons and protons in the outer magnetosphere to extremely high-energy particles gyrating around the strong magnetic field lines in the inner magnetosphere. The geometry of the latter zone is quite similar to the doughnut configuration of the Earth's Van Allen radiation belts, although the intensity of the Jovian radiation belts is far greater.

The outer reaches of the Jovian magnetosphere are readily deformed by strong gusts of charged particles in the solar wind, a phenomenon experienced by both probes. Both Pioneers initially crossed the shock boundary of the magnetosphere at a distance of almost 5 million mi from Jupiter. This boundary is the region where the particles of the solar wind encounter the Jovian magnetic field and are deflected away at a sharp angle, just as the waters of a fast-moving stream are piled up and turned aside by an obstacle. But as the craft flew on toward their rendezvous with Jupiter, they suddenly found themselves once again in the region of interplanetary space as a gust in the solar wind pushed the blunt face of the Jovian magnetosphere inward toward the planet. With the release of this external pressure many hours later, the magnetosphere abruptly snapped outward, and the particle counters and other instruments of the two craft could again map the outer magnetic domain of Jupiter.

Both Pioneer spacecraft found an interesting structure in the outer Jovian magnetosphere, a well-defined "magnetodisk" of relatively strong magnetic field lines containing high fluxes of relatively ener-

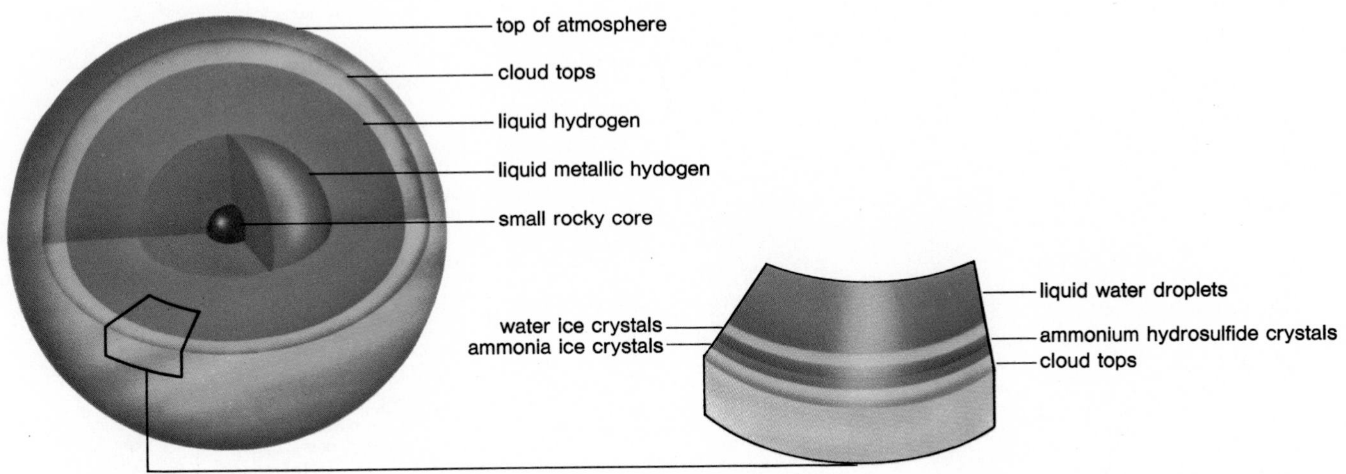

Jupiter's atmosphere, shown in detail at the right, was found by the Pioneer probes to consist of about 82% hydrogen, 17% helium, and a remainder of other elements. The depth of the atmosphere to the liquid zone is approximately 1,000 km (600 mi).

getic particles. Jupiter's fast spin and the consequent great centrifugal force exerted on the outer magnetosphere appear responsible for this structure, but it is not clear whether it is produced by an elongation of the field lines of Jupiter's magnetic equator, or by a concentration in the disk of charged particles that induce a distinct magnetic field of their own. In any case, it appears that an electrical ring current is generated in the disk, and that it flows in a direction opposite to that of Jupiter's rotation.

According to one model of this magnetodisk or "current sheet," it resembles the wide, floppy brim of a fedora hat, with one side cocked up and the other cocked down. Close to Jupiter, the sheet lies in the plane of the magnetic equator. But as the distance from the planet increases, it appears that the centrifugal force generated by Jupiter's rotation bends the sheet into the plane of Jupiter's physical equator; that is, perpendicular to spin axis.

Because Jupiter's magnetic axis is offset about 10°–15° from its rotational axis and because the relatively thin, elongated current sheet co-rotates with the planet, the sheet nods up and down every ten hours. The first clue to this curious situation came from Pioneer 10, which began to detect a rise and fall in particle flux and field strength with a ten-hour variation while still 2.5 million mi from the planet. Because Pioneer 11 approached Jupiter well below its equatorial plane, however, it did not encounter the current sheet until it was within 1.8 million mi of the planet.

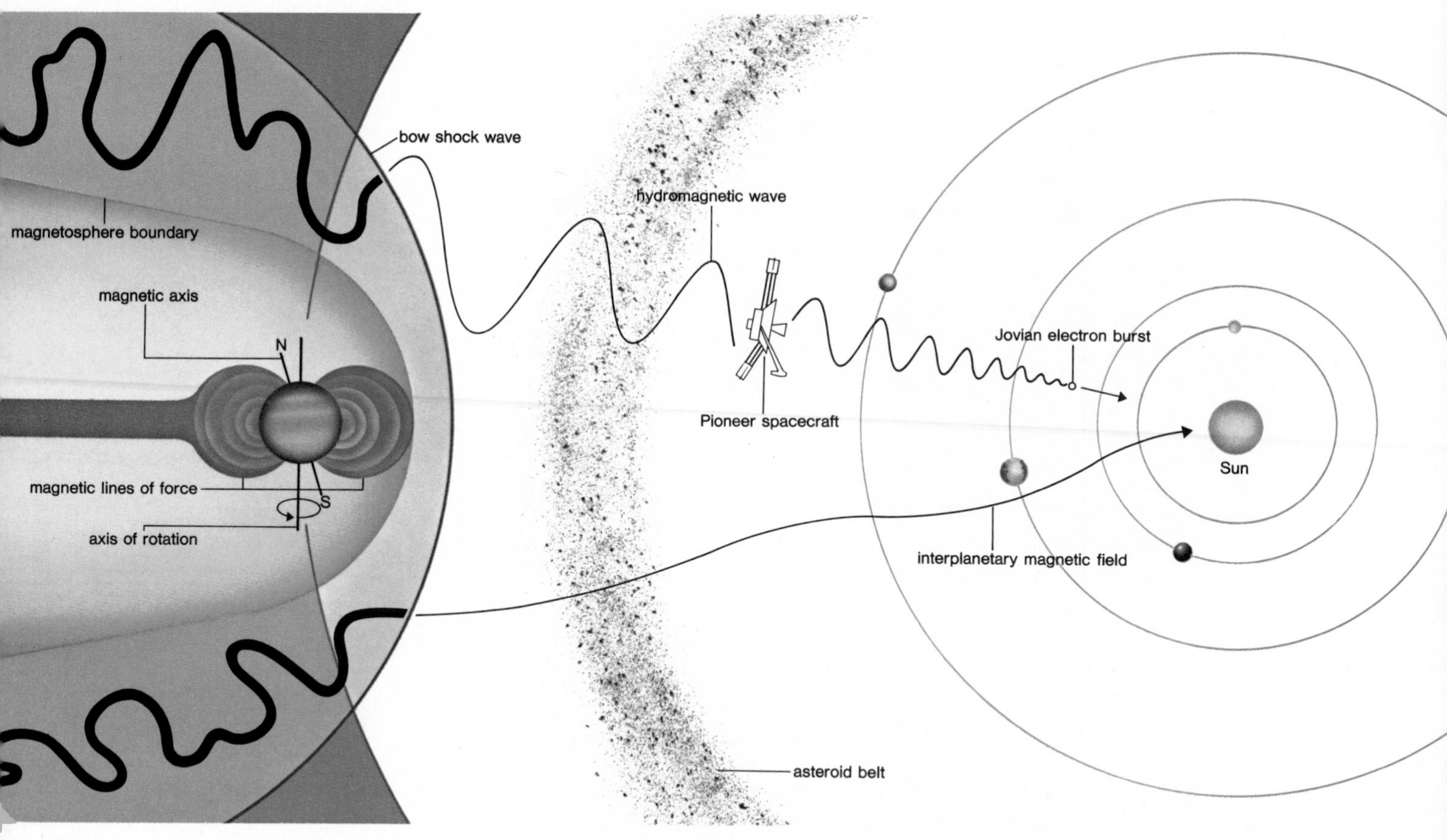

Pioneer probes encountered gusts of electrons originating from Jupiter when they were as much as 140 million mi from the planet. As they crossed the bow shock wave and the boundary into the planet's magnetosphere the radiation became intense. Charged particles are trapped on Jupiter's magnetic field lines of force, bouncing back and forth between one hemisphere and the other.

Radiation and the probes

The inner Jovian magnetosphere is similar to the highly organized but much smaller magnetosphere of the Earth. The field lines arch through space between the north and south magnetic poles. The charged particles are trapped by these lines of force, bouncing back and forth in a mirroring process between one hemisphere and the other but with the maximum concentration of radiation in the equatorial zones. Because of anomalies and instabilities in the Jovian magnetic field, the particles also diffuse inward toward the planet, picking up energy in the process. As a result, the dense, inner radiation belts in the equatorial region represent the greatest hazard for visiting spacecraft.

Pioneer 10 reported a peak bombardment of 13 million electrons per sq cm per sec in an energy range of more than 50 million electron volts (MeV). Of course, the total electron flux on Pioneer 10 reached far higher levels, but the 50 MeV particles, capable of penetrating several centimeters of brass, represented the greatest electron hazard to the spacecraft. Fortunately for Pioneer 10, the flux of damaging protons in excess of 35 MeV peaked at only about four million per sq cm per sec. Even so, the intense storm of radiation disabled or upset several of the probe's scientific instruments and seriously interfered with its operation close to the planet. During the two weeks it spent in the vicinity of Jupiter, Pioneer 10 was subjected to an estimated total radiation dose that was several thousand times the lethal level for human beings.

Because Pioneer 11 was scheduled to fly to within 27,000 mi of Jupiter, three times closer than its predecessor, the engineers at Ames Research Center of the National Aeronautics and Space Administration (NASA) who had built the two craft and the scientists whose instruments were aboard Pioneer 11 were apprehensive that it might be disabled by far more intense radiation than that encountered by Pioneer 10. As one NASA official put it, "With Pioneer 10, we tickled the dragon's tail; with Pioneer 11, we will fly through the jaws of the fiery dragon itself."

Although scientists could not forecast with precision the ultimate radiation intensities that Pioneer 11 might encounter, the probe did have one major advantage in surviving the radiation. This was its high-latitude approach to the planet near the south polar region. In this trajectory, Jupiter's great gravity field would wrench the spacecraft around the planet in a clockwise direction (opposite Jupiter's direction of rotation) and simultaneously bend its path sharply upward so that it would cut across the magnetic equator at a steep angle and at a speed of 107,000 mph. "It's like flicking your hand through a blowtorch," one scientist explained. "If you do it very quickly, you won't get burned."

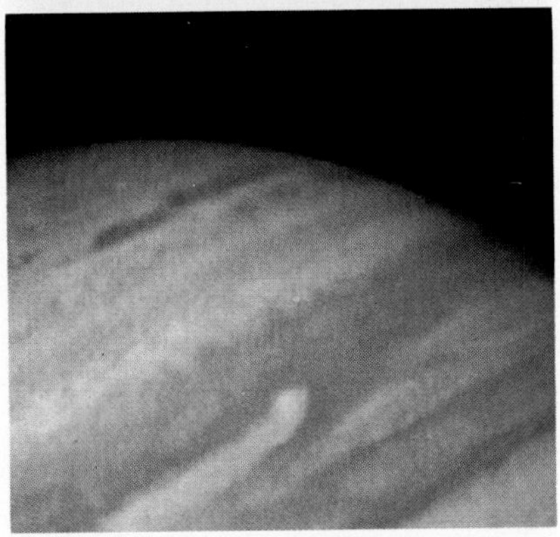

Cloud tops in Jupiter's north temperate region were photographed by Pioneer 10 from distances of (top to bottom) 992,000 km (616,000 mi) and 1.3 million km (808,000 mi). Astronomers determined from these and other pictures that there is an appreciable amount of latitudinal (north–south) motion on Jupiter in addition to the predominant longitudinal (east–west) belts and zones seen in Earth-based photographs. (Bottom) Streaming from right to left above the cloud tops, a cloud plume marks some hidden source of rising and condensing ammonia aerosols.

As it turned out, the tactic succeeded brilliantly. Although the flux of energetic electrons was only a little higher than that found by Pioneer 10, the intensity of 35 MeV protons increased substantially as Pioneer 11 approached its closest point to the planet, peaking at 15 million per sq cm per sec. Oddly enough, this great proton "spike" occurred not at the Jovian magnetic equator but 7° north, and it was one of several radiation peaks close to the planet. These findings plus magnetometer evidence that the Jovian magnetic field is lumpy rather than uniform like that of the Earth suggest that there are "hot" and "cool" zones deep in the inner Jovian magnetosphere, with the highest fluxes confined to different layers or bands.

As the Pioneers crossed the orbits of the major Jovian moons, they detected an interesting phenomenon: a distinct drop in particle flux at each orbital crossing. It appears that the large satellites literally carve holes in the trapped radiation belts, absorbing large quantities of radiation with every ten-hour rotation of Jupiter and of the field lines on which the particles mirror back and forth. The effect appears to be most pronounced for lower energy particles.

Although the two Pioneer spacecraft provided much illuminating data on the Jovian magnetosphere and its trapped radiation, many findings raised new and baffling questions. For example, both probes encountered gusts of electrons of Jovian origin while they were as much as 140 million mi from the planet. In some instances, the particles encountered in interplanetary space even displayed the ten-hour

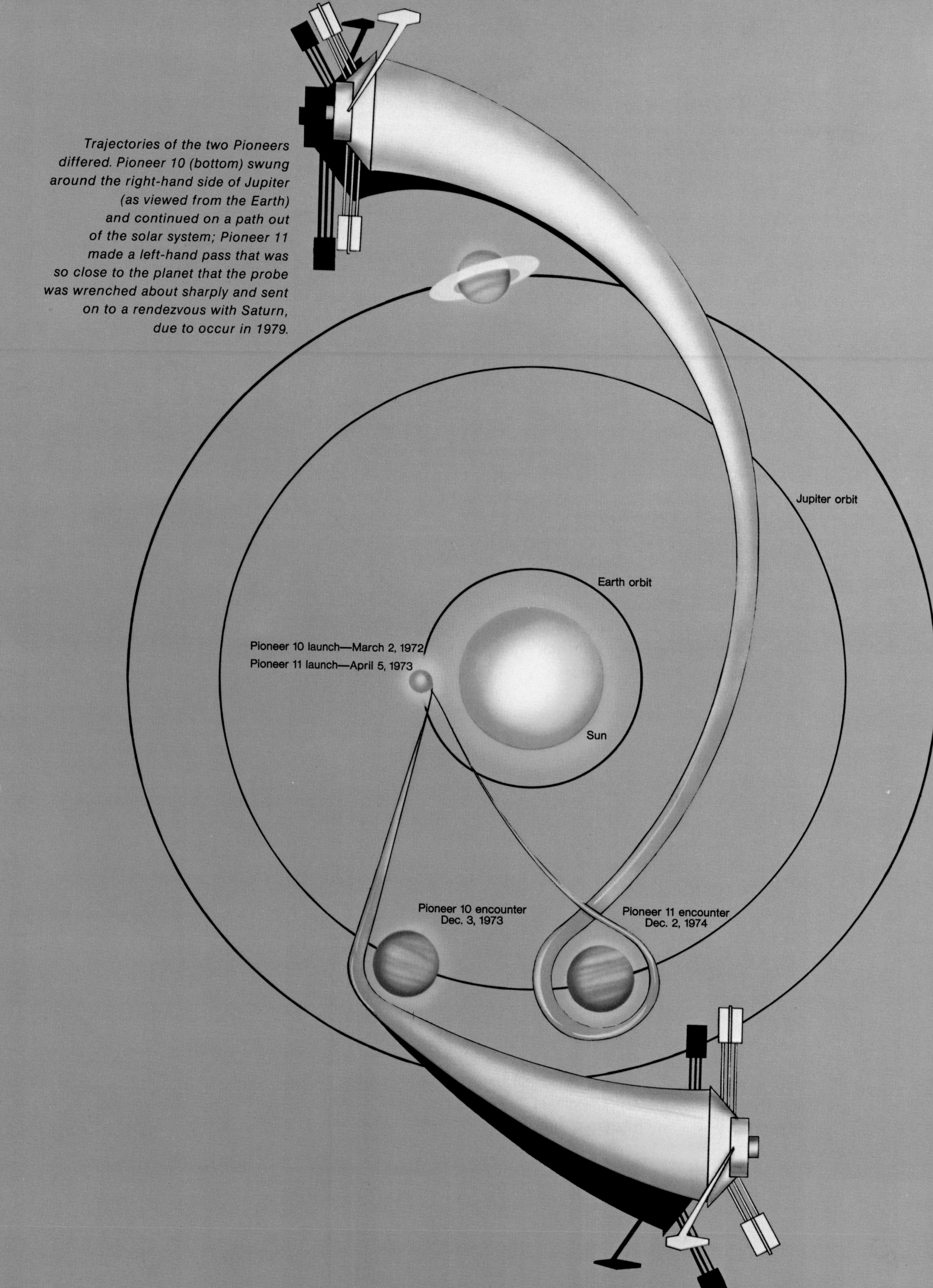

Trajectories of the two Pioneers differed. Pioneer 10 (bottom) swung around the right-hand side of Jupiter (as viewed from the Earth) and continued on a path out of the solar system; Pioneer 11 made a left-hand pass that was so close to the planet that the probe was wrenched about sharply and sent on to a rendezvous with Saturn, due to occur in 1979.

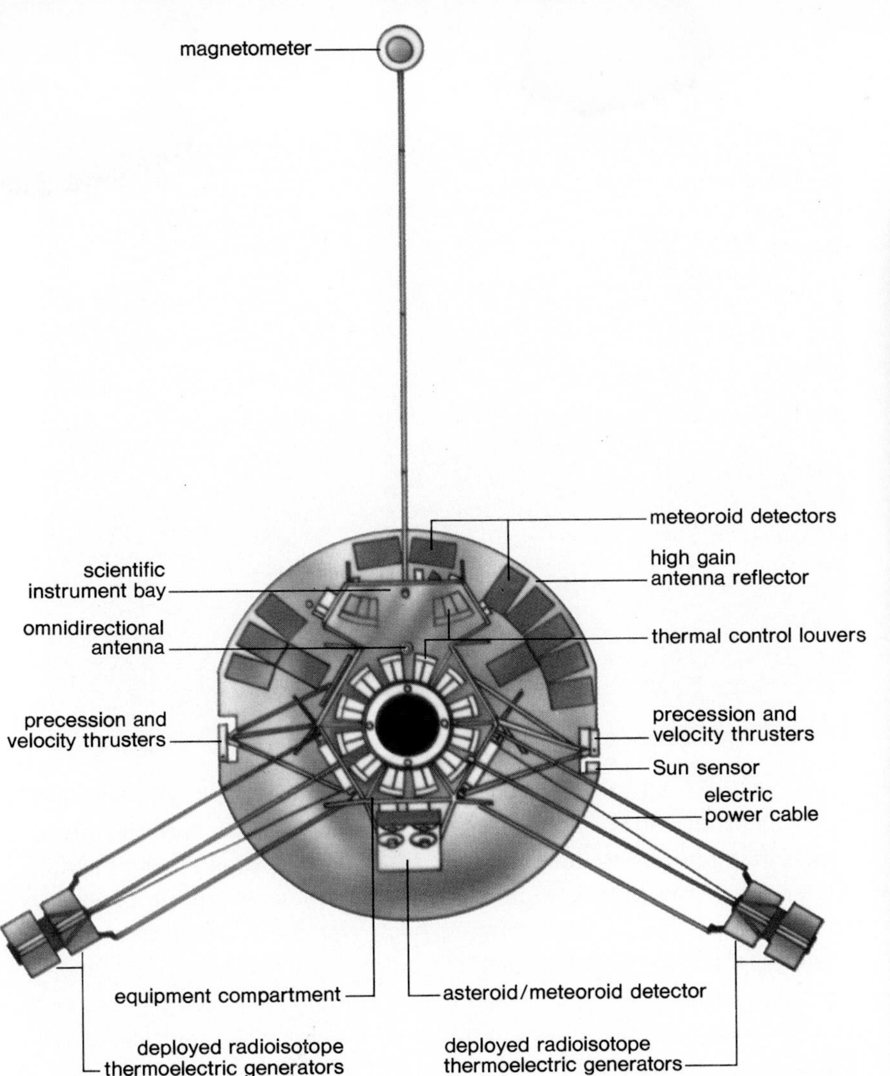

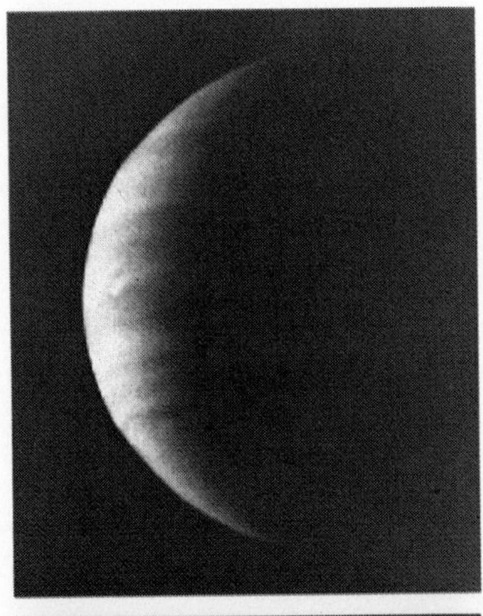

Meteoroid detectors and other equipment are arranged behind the high-gain antennas of Pioneers 10 and 11. Each probe had panels on the back of its antenna for detecting impacts of particles with masses of about 10^{-8} to 10^{-9} g. Photographs taken by Pioneer 10 include (top) crescent showing a bright equatorial nucleus, from 1,863,000 km, and (bottom) face of the planet, revealing clouds in the southern hemisphere that resemble the Great Red Spot, from 2,020,000 km.

periodicity of the current sheet inside the Jovian magnetosphere. Where these particles originate, how they are accelerated, and how they escape the Jovian magnetosphere are all unanswered questions, and some scientists believe they cannot be answered until another spacecraft is placed into a long-lived orbit around Jupiter.

Pioneer observations

More than half the scientific instruments aboard each Pioneer were devoted to particles and fields measurements, but the probes also carried a variety of detectors to map Jupiter's infrared radiation, its ultraviolet characteristics, and the population of micrometeoroids in its vicinity. Each spacecraft carried a small optical scanning device that could record Jupiter in visible light, one scan line at a time like a television camera, and return the digital bits of each picture element to the Earth for computer assembly of complete images of the planet and its large inner satellites.

Operating in wavelengths far longer than those to which the human eye is sensitive, the infrared radiometers of the two spacecraft mapped the temperature over the Jovian cloud tops. They confirmed observa-

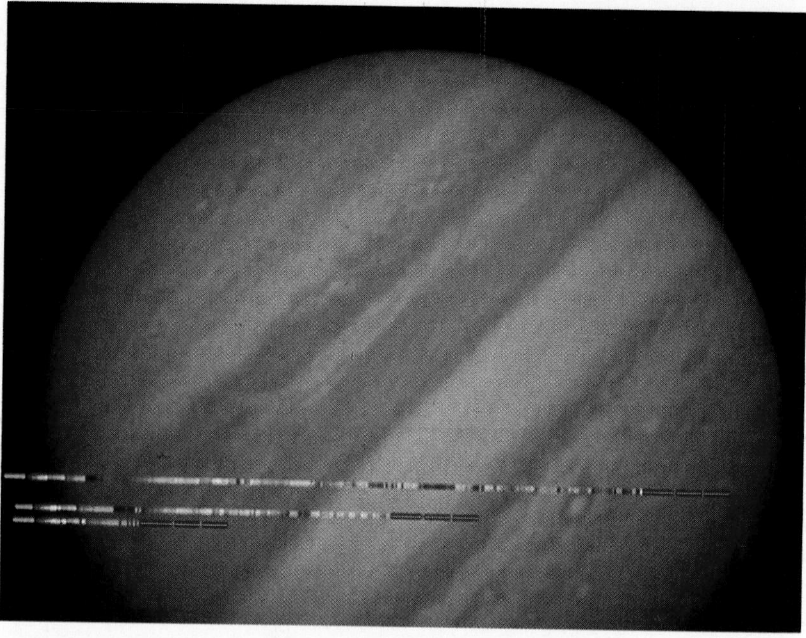

Cloud tops above Jupiter were photographed by Pioneer 10 from a distance of 1.6 million km. In the foreground are scanning lines that were missed in making the image, indicating some of the difficulties that computer processing of Pioneer pictures has to overcome. These gaps were being filled in from the original data records.

tions from the Earth that the planet is radiating more than twice as much heat into space as it is receiving from the Sun, and they also found that there is little temperature difference between the sunlit and dark side of the planet, indicating that Jupiter's dense atmosphere uniformly distributes the incoming heat over the whole planet. The radiometers also reported that the bright, gray-white cloud zones striping the planet are higher and cooler than the dark, reddish-brown belts with which they alternate. Of course, both the bright and dark bands are extremely cold, −144° C (−227° F) in the case of the bright zones and −137° C (−215° F) for the dark cloud belts.

The ultraviolet instruments studied Jupiter and its satellites in wavelengths shorter than those that can be detected by the human eye. Their major contribution was a determination of the amount of helium in the Jovian atmosphere. Scientists had long suspected that Jupiter is a fossil remnant of the original material from which the solar system condensed, and that its composition must closely approximate that of the Sun. The Pioneer observations show that the atmosphere of Jupiter is about 82% hydrogen, 17% helium, and a balance consisting of other elements of the periodic table. (The Sun's atmosphere consists of 88% hydrogen and 11% helium, with most, if not all, of the other naturally occurring elements making up the balance.)

Both Pioneers found a large population of micrometeoroids in the vicinity of Jupiter. The statistics of the strikes rule out the possibility that the dust particles are in orbit around Jupiter. Instead, it appears that Jupiter's great gravity field behaves like a cosmic vacuum sweeper, pulling in everything that wanders into its vicinity.

The most eagerly awaited and dramatic findings were produced by the imaging devices of the two spacecraft. As expected, these returned crisp color imagery of the most prominent features of the

North pole of Jupiter (above) was photographed by Pioneer 11 at approximately latitude 50° above the planet's equator. The pole is roughly on the line of the terminator (boundary between the illuminated and dark areas) across the top of the planet. The Great Red Spot (above left) appears as a giant single eye of Jupiter in a Pioneer 11 photograph taken from a distance of 1.1 million km.

planet, the great banded cloud structures and the mysterious Great Red Spot, which probably has fascinated astronomers since the 17th century. But they also saw detail never before visible from the Earth, in part because their lenses were so close to the planet but also because they viewed it from angles never before available to man.

The major return from the optical data was the important new insight that Jovian weather is convective in character and not baroclinic like the weather on the Earth and Mars. In other words, it is thermally driven by heat from below, like a boiling teakettle, and not, like the weather of the inner planets, by the radiant heat from the Sun. Thanks to the sharp resolution of the Pioneer imagery, it is now possible to identify many convective features in the Jovian atmosphere, including oval convection cells with bright, rising cores and dark, descending outer rims, and bright nuclei that climb high above the cloud bands and trail long plumes. The Great Red Spot itself appears to be a cloud mass towering half a dozen miles above the bright south tropical zone in which it is embedded. The Red Spot is about 25,000 mi in length, large enough to swallow three Earths. Viewed from above, it appears to be a long-lived vortex or super-updraft with a powerful counterclockwise circulation. One of the sharpest of the Pioneer 11 pictures (capable of resolving an object as small as the state of Massachusetts) detected a hint of two ring-shaped structures inside the Red Spot, suggesting that it may have more complex dynamics than does a terrestrial storm system.

The most spectacular overall feature of the Jovian atmosphere, its arrangement into a semipermanent pattern of alternating bright and dark cloud bands, appears to result from two major factors. One is the dominant role of convective heating of the atmosphere from below and the other is the tremendous Coriolis effect exerted by Jupiter's

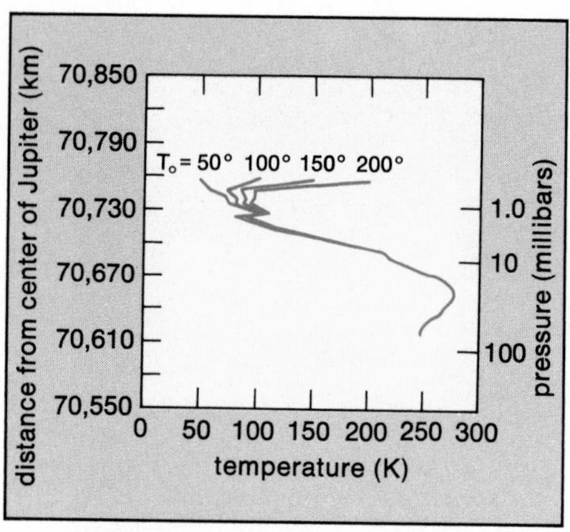

Temperature profile of Jupiter's atmosphere over more than 100 km (60 mi) of altitude is based on a study of the S-band (2,200 MHz) signals of Pioneer 10 as the probe was occulted by the planet. A composition of 85% hydrogen and 15% helium by volume was assumed, and computations were made for four different initial temperatures, T_0. The unexpected temperature inversion at 20 mbar of pressure may be caused by the absorption of solar radiation by dust.

fast spin. Together they prevent the Jovian weather systems from migrating in latitude as do large cyclonic systems on the Earth. As a result, Jupiter's weather systems stretch all the way around the planet. Instead of heat being transported from the Jovian equator to the poles, as is the case for the externally heated atmospheres of the inner planets, on Jupiter the heat is transported and dissipated by means of the strong rising and descending motions of the zones and belts.

Several important findings about Jupiter and its major satellites were produced by two Pioneer experiments that required no special instruments. In both cases, only the radio signal of the spacecraft was necessary to obtain the results.

In the celestial mechanics experiment, precise measurement of the Doppler variation in the signal frequency disclosed spacecraft velocity changes of as little as one millimeter per second relative to NASA's worldwide network of 210-ft steerable dish antennas tracking the Pioneer spacecraft. (The Doppler variation is the change in the observed frequency of a wave because of the relative motion between the observer and the wave source.) Because these velocity changes were brought about by the gravity fields of Jupiter and its satellites, it became possible to determine the mass of Jupiter and its moons with far greater precision than had been possible with Earth-based observations, and it also became possible to reach important conclusions about the internal composition and structure of Jupiter.

One of the most important discoveries in the realm of celestial mechanics made by the Pioneer missions is that Jupiter is even more flattened at the poles and distended in its equatorial plane than ground observations had indicated. Another is the complete absence of any evidence of internal structural rigidity for Jupiter. These data mean that the planet is in hydrostatic equilibrium and, therefore, that almost all of its mass must exist in the liquid state.

Theoretical calculations indicate that Jupiter's atmosphere accounts for only 1% of its total mass and that it extends to a depth of about 450 mi below the cloud tops. Between this point and a depth of 1,800 mi, where the temperature reaches 5,500° C (10,000° F) and the pressure climbs to 90,000 Earth atmospheres, there is a transition zone of gradually increasing viscosity in which the gaseous hydrogen is compressed into a liquid state that has about one-fourth the density of water. At a depth of 15,000 mi, the temperature increases to 11,000° C and the pressure to three million Earth atmospheres. At this level, the liquid hydrogen changes to a liquid metallic phase that is capable of conducting electrical currents. While there may be a small rocky core at the very center of Jupiter, equivalent to a few Earth masses, it is the great interior ball of liquid metallic hydrogen, some 60,000 mi in diameter, that contains the bulk of Jupiter's mass.

A substantial amount of primordial heat accumulated by Jupiter during its condensation into a planetary body about 4.6 billion years ago remains stored in this metallic hydrogen core. Because of the planet's powerful magnetic field and its slow radiation of heat into space, a large system of eddy currents must be circulating deep within its liquid interior. Moving at a rate of 1,500 mi per year, these currents transport the heat of the central core to the upper layers of Jupiter as well as generate the planet's magnetic field by means of a dynamo process, similar to that believed to be at work in the Earth's core.

The Jovian satellites

The celestial mechanics experiment also allowed a much more precise calculation of the total masses of Jupiter and its four major satellites. The system as a whole is one or two lunar masses larger than previously suspected, but the biggest surprise involved a new mass determination for Io. Tracking by the spacecraft revealed that this innermost of Jupiter's four "Galilean" moons (after Galileo Galilei, who discovered them in 1610) is about 20% more massive than expected. Its new density value is 3.5 times that of water, or slightly greater than the density of the Earth's own Moon. The new density values for the remaining Galilean satellites, progressing outward from Jupiter, are Europa, 3.1; Ganymede, 1.9; and Callisto, 1.6. This systematic decrease in density with distance from Jupiter supports the idea that the planet radiated great heat during its condensation. This would have caused such volatile substances as hydrogen to be driven away from the inner satellites, while temperatures were remaining cool enough on the outer satellites to allow water vapors and other volatiles to condense into ice. Such a temperature history would account for the greater diameters and significantly lower densities of the outer Galilean satellites.

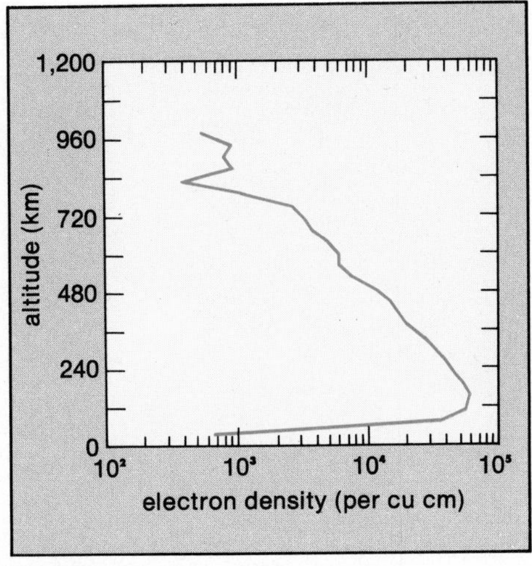

Electron density of the ionosphere of the Jupiter moon Io was deduced by analyzing an occultation of Pioneer 10 by the satellite. The peak density was found to occur at an altitude ranging from 100 to 160 km (60 to 100 mi).

The other major experiment involving the Pioneer radio signal used the spacecraft radio beam to probe the atmosphere of Jupiter and its satellites. By arranging the trajectories of the Pioneers so that they would swing behind these bodies relative to the Earth line of sight, scientists allowed the microwave radio signal to cut into the planet's outer ionospheric and atmospheric layers. Measuring the change in the signal's characteristics during its extinction and subsequent restoration as the spacecraft moved back into view made it possible to derive important information about these outer layers. In the case of Jupiter, the interpretation of the occultation data is difficult because of the planet's complex, multilayered ionosphere and problems of multiple-path signal propagation in the layers. However, the technique worked in a straightforward fashion with Io, showing that it has a distinct ionosphere and, by inference, an extremely thin atmosphere. It now appears likely that the other Galilean satellites will be found to have a trace atmosphere.

Io's relationship with Jupiter has fascinated radio astronomers for more than a decade because of its apparent role in modulating the planet's radio outbursts in the decametric (10–100 m) wavelengths. These explosions of radio noise have the power of several hydrogen bombs and make Jupiter the "brightest" radio source in the heavens after the Sun. Io's orbital position with respect to both Jupiter and the

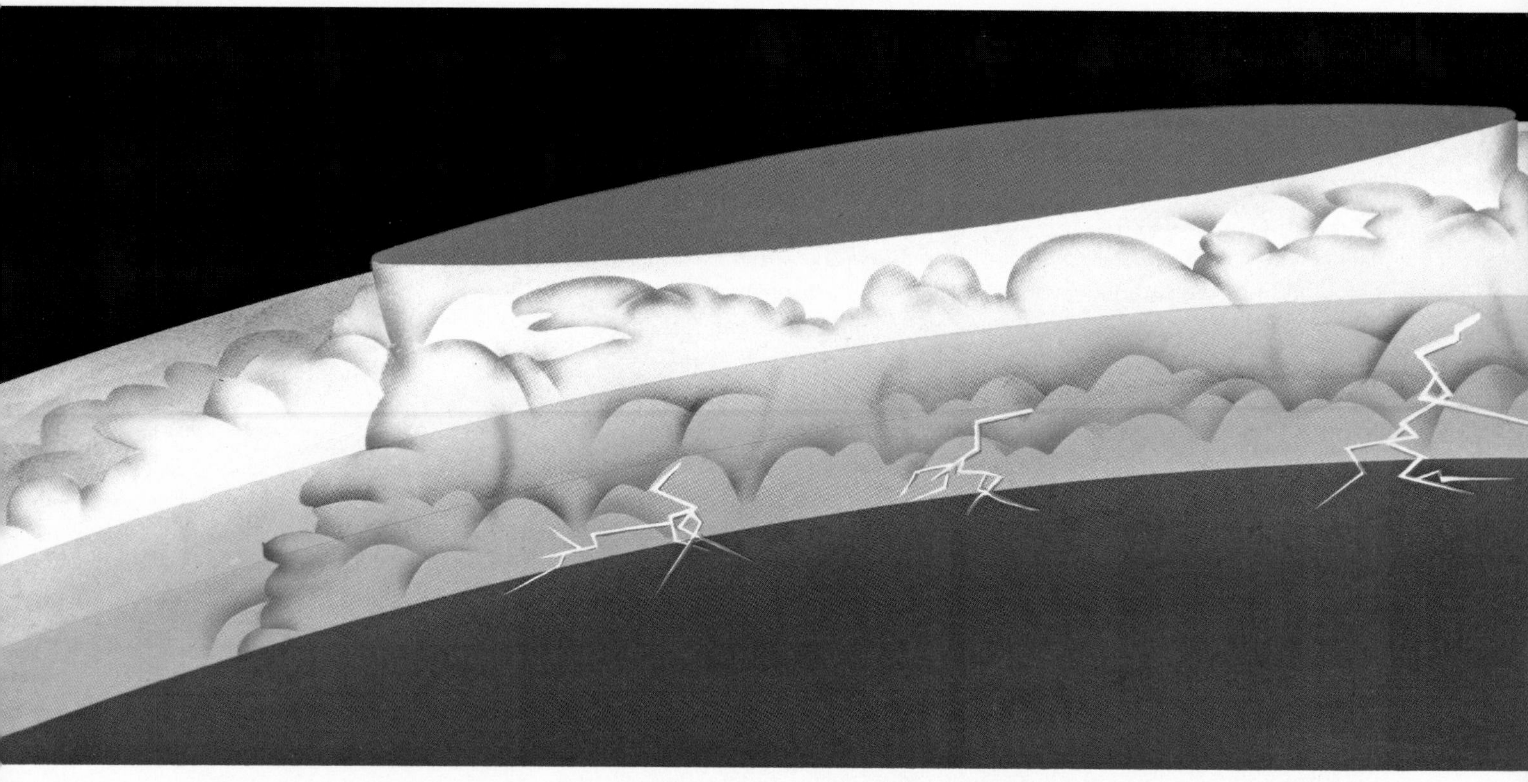

Great Red Spot on Jupiter appears to be a swirling cloud mass that towers half a dozen miles above the south tropical zone. About 40,000 km (25,000 mi) in length, it is large enough to encompass three Earths. Astronomers speculate that the reddish hue may be caused by phosphorus compounds. Lightning flashes in the water clouds below.

Earth appears to be an important factor in the detection of these radio bursts on the Earth, but the exact nature of the electromagnetic interaction is not clear.

According to one model, Io is electrically conducting. As the Jovian magnetic field lines sweep through it at 34 mi per sec, Io behaves like the armature of a direct-current generator spinning between the pole pieces of a magnet. In this analogy, the "brushes" that collect and transport the current (as much as 10,000 Mw) are the charged particles in the magnetic flux "tubes" that connect Io at any given time to the north and south magnetic poles of Jupiter. This model suggests that Io discharges relatively continuously to those regions deep in the Jovian atmosphere but that the phenomenon can be detected on the Earth only when the geometry is favorable. Other models propose that Io behaves more like an electrical "short," making a circuit with Jupiter only when it accumulates a sufficient charge. It was hoped that Pioneer 11 might fly close enough to one of these narrow flux tubes to detect Io's electromagnetic activity, but the distance proved too great for the experiment to work.

Earth-based observations

Although the Pioneer flights have provided the most dramatic new data on Jupiter and its major satellites, ground observations have also provided some important new insights. One of the most interesting was the discovery that Io is accompanied in its orbit by a huge cloud of

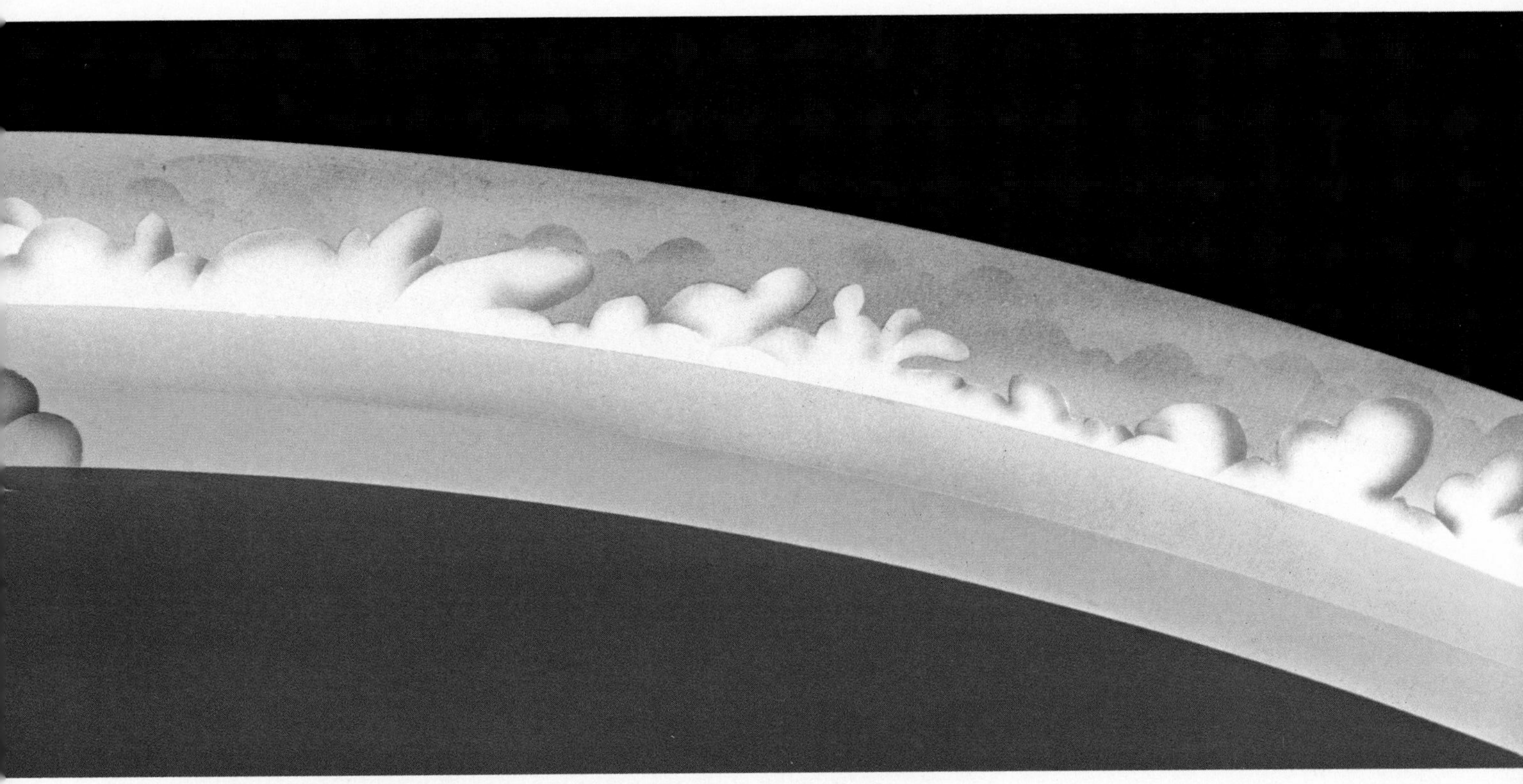

excited sodium atoms as well as by the cloud of hydrogen atoms discovered by Pioneer 10 and confirmed by Pioneer 11. It has been suggested that Io's surface is rich in sodium evaporite salts as well as frozen ammonia. The hydrogen, sodium, and probably nitrogen as well may be "sputtered" off the surface of Io by the constant bombardment of the Jovian radiation belts. Other theories propose that the surface material is dissociated by solar ultraviolet radiation and excited to thermal escape energies, or raised to these energies by localized electrical discharges between Io and its primary. Whatever the mechanism for the production of the hydrogen and sodium clouds by Io, it appears that these are major constituents of both its thin atmosphere and its ionosphere. It is certain that unique physical processes are at work on Io and that it is participating in a variety of significant interactions with Jupiter and its vast belts of trapped radiation.

With respect to Jupiter itself, an important recent finding was the detection of water vapor deep in the atmosphere by astronomers of the University of Arizona. Detected through holes in the cloud deck of the dark Jovian belts, the vapor is quite hot relative to the Jovian cloud tops and is subject to a pressure of up to 20 atmospheres. But the amount discovered is quite low in relation to the hydrogen on Jupiter. The implication of this finding is that oxygen is a thousand times less abundant on Jupiter than in the Sun, an important disparity in elemental abundance that could have significant cosmological implications for the evolution of Jupiter and the other outer planets.

In addition to water vapor, ground-based studies have recently discovered acetylene (C_2H_2), ethane (C_2H_6), and phosphine (PH_3) in the Jovian atmosphere. Red-hued phosphorus compounds could be responsible for the coloration of the Great Red Spot. However, the discovery of water vapor and the other hydrocarbon molecules in the upper atmosphere is of greater importance because it virtually assures that these upper levels are manufacturing complex organic compounds on a large scale. Few scientists are bold enough to suggest that biological life forms may exist in the Jovian atmosphere, mainly because the turbulent vertical motions of the atmosphere would subject even primitive life forms to destructive variations in heat and pressure. The relatively stagnant polar regions of Jupiter may prove more hospitable to life than the equatorial latitudes, however, and in any case the large-scale production of organic compounds and perhaps amino acids (the building blocks of cellular protein) suggests that the appearance of life is a natural, even widespread, phenomenon in the universe, rather than a unique event confined to the Earth.

On to Saturn

In its 1973 swing around Jupiter, Pioneer 10 was accelerated to a speed of 82,000 mph in the great Jovian gravity field. As a result of this extra gravitational kick, the probe will become the first human artifact to escape from the solar system. The craft's radioisotope thermoelectric generators are slowly losing power as their plutonium fuel runs through its decay half-life. Its signal is expected to fail by the time it crosses the orbit of Uranus in 1979. Eight years later, it will cross the orbit of Pluto, the outermost planet, and then drift through interstellar space at a speed of about 25,000 mph in the direction of the constellation Taurus.

As for Pioneer 11, it has another major task to perform. Its close pass of Jupiter wrenched its trajectory around sharply and accelerated the spacecraft into a path that will carry it high above the plane of the ecliptic and far across the solar system to a rendezvous with Saturn in September 1979.

Orbiting the Sun at a mean distance of 886 million mi, Saturn is the sixth planet of the solar system. With a mass of 95 Earths, it is second only to Jupiter in size. It is also second only to Jupiter in rotational velocity, with a day of only 10 hours and 14 minutes. Its most spectacular feature is a system of flattened rings that encircle the planet in its equatorial plane. The rings have been thought to consist of an immense swarm of tiny ice crystals of ammonia or water, although recent radar studies suggest that some of the particles may be the size of rocks and even boulders. The span of the outermost ring is 169,000 mi from edge to edge, while the inner edge of the closest ring orbits the planet only 9,000 mi above the cloud tops. Although the rings have great width, it appears that they are confined to a razor-thin plane at Saturn's equator with a thickness of less than ten miles and perhaps as little as a few inches.

Like those of Jupiter, the cloud tops of Saturn occur in alternating light and dark bands, and it also appears that the planet emits more heat than it receives from the Sun. Scientists recently found that Saturn generates bursts in the one-megahertz frequency range. This suggests that it has an internal magnetic field as well as radiation belts of trapped electrons like Jupiter. (It is the electron component of the radiation belts that is responsible for the radio noise.) Because of the absorbing effects of the material in Saturn's rings, it is doubtful that the radiation intensity of Saturn can approach that of Jupiter.

Pioneer 11 may be able to shed light on many of the questions concerning the Saturn system. Though a final decision on the probe's trajectory had not been made by mid-1975, NASA was considering a route that would send the spacecraft between the innermost ring and the planet. Such a pass would bring Pioneer 11 within 2,300 mi of the cloud tops. Provided the spacecraft is in good working order after its additional five-year mission, it should be able to obtain good optical imagery and infrared and ultraviolet data concerning the sunlit side of Saturn; it was also expected to transmit radio occultation data on the nature of the rings and of Saturn's atmosphere, authoritative data on the mass distribution in the Saturn system, and detailed magnetic field and particle data like that gathered at Jupiter. If NASA should decide in favor of such a flyby of Saturn, the planet's gravitational field will swing the spacecraft into a trajectory carrying it outside the solar system. As with Pioneer 10, the probe's power source will fail long before it enters interstellar space.

But the two doughty little spacecraft may still have another function to perform. Both carry identical gold-anodized plaques bearing the engraved figures of a human male and female. The figures are scaled against an outline of Pioneer's antenna to show their size, and the man's right hand is upraised in a gesture of peace. The Sun's location in the universe is expressed in terms of the present frequency and position of 14 pulsars visible at the Earth, with the units expressed as binary multiples of the 1,420-MHz radiation frequency emitted during a characteristic transition of the hydrogen atom. The planets are depicted symbolically, showing that the Pioneers originated from the third planet from the Sun to visit the giant fifth planet.

The odds against an alien intelligence ever finding and interpreting these messages from our own civilization are quite large. But the rapid advance in our understanding of the Earth's planetary neighbors and their remarkable physical and chemical processes has strengthened the notion that life may not be unique to the Earth and has rendered less fanciful the idea that there may be other civilizations with which we might someday communicate.

Our Architectural Lineage

by Lawrence K. Lustig

Architecture has been variously described by persons of differing persuasion and perception. Some have termed it the organization of space in an efficient and pleasing manner, whereas others prefer, simply, the art and technique of building. Both definitions are somewhat limiting, and most students of the subject today tend to emphasize that architecture is a structural art form which is expressive of the total culture of the builder and the environment of the building. Indeed, it is all this and more, for architecture represents the physical creation of an edifice whose plan first took form in the mind of man. It is therefore no extravagance to argue that architecture is one of a mere handful of truly creative human endeavors.

This common element of creativity is apparent to any thoughtful visitor to one of the world's monuments of civilization, ancient or modern. Who would not stand in awe and contemplation before such silent splendor as the Great Pyramid of Khufu, or the walled Andean city of Machu Picchu; a calligraphically adorned Eastern mosque, or a buttressed Gothic cathedral; the symmetrical Taj Mahal, or the twin-towered World Trade Center? But the casual observer may search in vain for binding threads beyond this, particularly when more primitive structures are considered. What common factors relate such disparate human accomplishments as the simple shelter of the Eskimo or desert Bedouin on the one hand, and the soaring, steel-ribbed piercements of the sky that exemplify a modern city on the other? Setting aside for the moment the concept that architecture is reflective of culture, or that it represents a triumphant blending of form and function, the rather pervasive influence of the kinds of available building materials and techniques, and of climate and the general environment, becomes apparent. Consideration of these factors suggests the existence of an architectural lineage through time—which is the subject of this pictorial essay.

Roberto Bunge from Black Star

(Above) Machu Picchu, walled city of the ancient Incas, high in the Peruvian Andes. (Opposite, top) Bedouin encampment in the Mauritanian Sahara. Rock churches (bottom left) of the early Christian hermits, at Göreme, Turkey. Traditional Mongolian tent (bottom right) with handsome fitted doors.

LAWRENCE K. LUSTIG *is managing editor of* Encyclopædia Britannica Yearbooks.

Cavern beginnings

In the beginning, man was principally a cave dweller and wherever naturally formed caves were to be found our earliest ancestors put them to good use. Caves are admirable shelters against the elements, albeit somewhat uncomfortable and inhospitable from our vantage. But they sufficed to nurture life and the very race of man to which we all belong. And they also nurtured art, for in such primitive habitations as the Magdalenian cave of Altamira in northern Spain are found depictions of most marvelous, but now extinct, Pleistocene animals of several kinds.

Our architectural lineage, then, begins in these caves of Early Man. Lest it be thought that no counterpart more recent than the well-known European Neanderthals can be cited, examples from Ethiopia and Turkey are instructive. In the valley of Göreme, in central Anatolia, early Christian hermits practiced and preserved Byzantine customs until the 14th century. They did so in cells and churches which were hewn by hand from the porous volcanic rock of the region, thus lending an austere and alien mien to the hillsides by virtue of their substantial abundance.

The Christian hermits used to full advantage the available building materials of the site. The rock was so porous and friable that construction of caves provided an architectural solution to the problems of shelter and worship which was superior to the erection of buildings from the same stone. In more recent time the Chinese Communists have illustrated this cavern heritage; they persevered for years in the caves of northern China against then-extant Nationalist forces.

Folk architecture

Beyond the cave, there exists an immense variety of simple structures in all parts of the world which also utilize available building materials to full advantage. Designated as folk architecture, each such structure answers the need for protection from the elements, privacy for the family or communal group, and, in the case of nomadic peoples, the need for portability and ease of assembly as well. Included in this category are tents of every description; rudimentary windbreaks and sunscreens of reed, palm, or branches; and houses of mud, stone, sod, thatch, log, and ice.

The tent, for example, dates at least from Assyrian times and may well be even older. It exists in two basic forms, conical and domed, but these give rise to many varieties. The simplest conical tent has a central pole or upright, about which fabric is stretched outward and downward in circular fashion and then staked to the ground. The Arab tent, or *beyt es-shaam*, is basically of this kind; black worsted or cloth of hair is the usual fabric, although brown tents are commonplace among the Bedouin in some parts of the Sahara. Contemporary military and circus tents utilize the same principles of construction, but they generally exhibit two, three, or more centrally located and aligned uprights beneath canvas.

N. deVore III from Bruce Coleman Inc.

Stephen Deutch

Jørgen Bitsch from Black Star

Bill Strode from Black Star

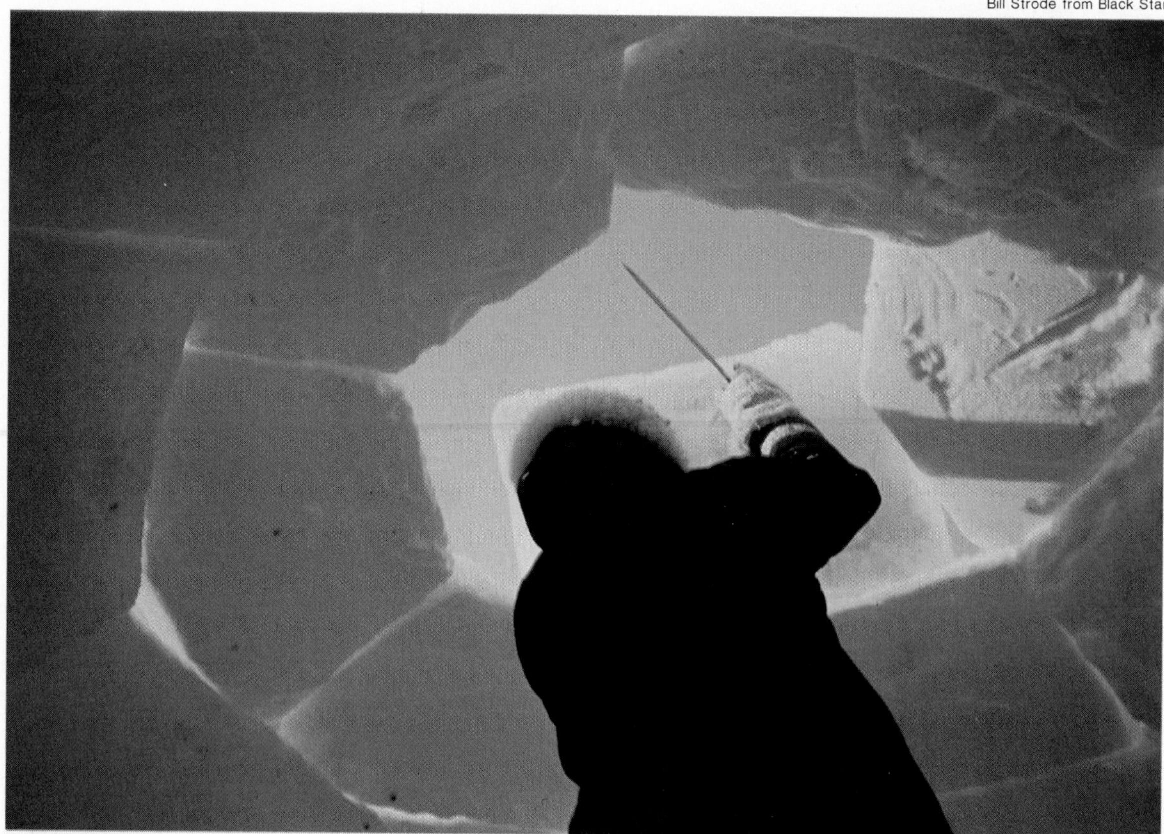

L. Lyon from Bruce Coleman Inc.

The traditional Mongolian tent, the yurt or gher, is also conical, invariably white, and is often fitted with doors which are taken with the dwelling when the herdsmen migrate from one pasture to another.

In North America, the tepee of the Plains Indians provides still another variation. Approximately 25 poles are used to form the framework of a cone, about which buffalo hide is stretched, fastened, and decorated to complete the tent. Much more abundant, of course, prior to western settlement by the white man, the conical tepee is still a standard dwelling of some Indians in the U.S. and Canada.

Domed tents, on the other hand, tend to be the preferred form among more sedentary peoples, or when the intent is not to carry the dwelling from place to place during periodic migrations. The standard wigwam, for instance, was the permanent home for many North American Indians. It was constructed by placing saplings in the ground at the corners of a square or rectangle, bending these inward, and then tying the ends together to form a rude dome. This framework was then covered with bark, perhaps rushes and mud, or sometimes animal skins. The huts of certain peoples in the Afar and Danakil areas of East Africa reflect a similar mode of construction.

Still another tent worthy of mention here is that indigenous to Tibet, where yak hair is woven into the fabric used. Tibetan tents are rectangular dwellings with lengths ranging to 50 feet.

It should be noted that tents have served well in cold climates as well as in the world's hot deserts. At either climatic extreme, where building materials ultimately become so scarce that the vanishing point seemingly is reached, folk architecture again provides solutions.

In extremely arid climates, the early inhabitants required no overhead shelter, so they simply assembled the sparse brush, reeds, or branches that could be found in the area and made circular windbreaks and sunscreens. The Indians who originally lived in Death Valley and adjacent interior basins followed this practice. In winter they stayed on the valley floor and when summer came, they moved to cooler elevations in the surrounding mountains, leaving their windbreaks behind. Remnants of these circular forms occur in archaeological sites that date back to the beginning of the Christian era.

The shelter provided by palm screens at oases in central Arabia today is not truly different in concept. The cooling effect when one passes from open desert to its shaded confines is astonishing.

But it is the Eskimo igloo that serves as the most outstanding example of folk architecture in a forbidding or prohibitive environment. Constructed by arranging blocks of dense snow in a spiral-circular form, the resulting dome becomes tightly sealed and marvelously well insulated against the rigors of the Arctic winter. The body heat of the inhabitants and their burning oil lamps cause slight interior melting and consequent glaze formation. The interior of an igloo attains temperatures of about 68° F without benefit of the modern furnace. It need scarcely be pointed out that the Eskimo utilizes the *only* available building material in the construction of his home.

N. deVore III from Bruce Coleman Inc.

(Above) Tied poles emerge from the smoke hole of a Sioux tepee in Montana. (Opposite, top) An Eskimo completes his winter igloo made of hand-hewn blocks of snow. Domed hut (bottom) in the Danakil Desert, East Africa.

Elliott Erwitt from Magnum

Toni Schneiders from Bruce Coleman Inc.

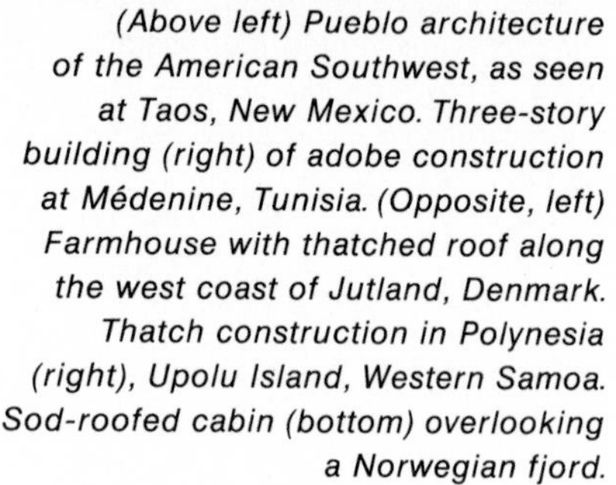

(Above left) Pueblo architecture of the American Southwest, as seen at Taos, New Mexico. Three-story building (right) of adobe construction at Médenine, Tunisia. (Opposite, left) Farmhouse with thatched roof along the west coast of Jutland, Denmark. Thatch construction in Polynesia (right), Upolu Island, Western Samoa. Sod-roofed cabin (bottom) overlooking a Norwegian fjord.

Available materials also provide good thermal insulation in a variety of structures other than igloos. Mud and brick buildings plastered with adobe were and are nearly ubiquitous in our architectural lineage. Across Africa, from the distinctive Nigerian dwellings at Dogon, through the arid cities and towns of the Sahara, to the mud-walled dwellings and minarets of Arabia and the villages of Iran, Afghanistan, and India—everywhere can be found structures not too unlike the pueblo, so typical of the American Southwest. Where there is not sufficient wood or stone, dried mud must do, and in adequate thickness it absorbs the heat of the day without permitting interior penetration and then reradiates that heat at night.

The same can be said of stone, of course. In desert regions the daytime ground temperatures—where they have been measured—commonly attain values as great as 170° F. Exposed rock surfaces thus become extremely hot during the day, but the thermal conductivity of rock is low, their albedo (reflectivity) is relatively high, and cooling thus occurs rapidly as soon as the sun goes down.

Thatch and sod also exhibit good insulating qualities, in part because the former contains air spaces when arranged in layers, and because sod or soil is essentially equivalent to adobe in undried form. Roofs of these materials can be found in such diverse locations as Iceland, Ireland, Scandinavia, and the islands of Polynesia; and houses entirely of sod were erected in the western plains of the U.S.

Thus folk architecture serves everywhere to illustrate the adaptation of man to his environment. Structures of enormous diversity characterize such architecture, but all reflect the utilization of materials at hand to provide portable or permanent shelter.

Toni Schneiders from Bruce Coleman Inc.

N. deVore III from Bruce Coleman Inc.

Rock strength, pillars, and the arch

Aside from thermal properties of building materials, their compressive and tensile strengths loom large in influencing our architectural lineage. Compressive strength, or resistance to failure by crushing, obviously governs the heights that buildings can attain. To cite an absurd extreme, skyscrapers cannot be erected from adobe because the compressive strength of mud is too weak and the walls would fail under the burden of their own weight.

Beyond this, however, it should be noted that a considerable range in strength exists for natural materials. The compressive strength of chalks and siltstones is less than 5,000 pounds per square inch (psi), for example, whereas sandstones and limestones are in the 10,000–25,000 psi range, and many igneous rocks and some quartzites exhibit strengths of 25,000–40,000 psi or more. Fine-grained rocks are generally stronger than coarse equivalents. Thus, three- and four-story cliff dwellings, as at Mesa Verde, Colorado, and the massive pillars of the Temple of Poseidon and other Greek places of worship easily bore the weight required because fairly strong sandstones or limestones were used in their construction. In the case of pillars, they were always cut in such a way as to ensure that their vertical axes were perpendicular to the bedding (horizontal layering) of the stone. This orientation is associated with the greatest compressive strength; that parallel to the bedding would exhibit least strength. Elsewhere, buildings of cobbles and boulders, or of cut stone, could be erected to almost any desired height if igneous rocks, of still greater strength, were available.

The tensile strength of rocks—their ability to resist failure by stretching while under a load—is nearly negligible, however, and this

Toni Schneiders from Bruce Coleman Inc.

Photos, Toni Schneiders from Bruce Coleman Inc.

(Above) Pillars of the Temple of Poseidon. Shrine of the oracle of Delphi (right) on Mount Parnassus, Greece. (Opposite, top) Pre-Columbian cliff dwellings at Mesa Verde National Park, Colorado. Medieval abbey of Saint-Jean-des-Vignes (bottom left), Soissons, France. The Roman aqueduct at Segovia, Spain (bottom right).

fact has had important architectural consequences. A long granite column can sustain a great weight indeed when placed upright. But when laid on its side between two supports, a column of the same diameter will readily fail beneath a much lesser load. This is because tensile, rather than compressive, strength is involved in the latter situation. Accordingly, one finds abundant evidence of the prosaic constraints imposed by the tensile strength of rocks amidst the structural grandeur of Greece and the ancient world. The pillars of Poseidon, like those of the exquisite oval on Mount Parnassus, shrine of the oracle of Delphi, had to be quite closely spaced. Had they not been so spaced, the overlying stone beams supporting the pediment would surely have failed. It is for this reason that the interiors as well as exteriors of Grecian structures were characterized by ever present columns or pillars. So too, in mute testimony, stand ruins in North Africa which were modeled according to Greek design and much older works as well, such as the Egyptian Temple of Karnak. These ruins again illustrate the basic point: stone is strong but it cannot be used in the form of beams to span any substantial horizontal distance, regardless of the stonemason's fitting skills.

Solution to the problem was accomplished by use of the arch, a monumental invention of deceptive simplicity, whose origins are basically uncertain. Some have even suggested that discovery of the arch was accidental. Involving the use of wedge-shaped stones to form an arc, in which stress is directed outward, from keystone to the supports, the arch certainly was not unknown to the Egyptians and Greeks. As indicated above, however, it was apparently not deemed suitable for use in their greatest works. It was left to the Romans, beginning around 750 B.C., to immortalize the glory of their civilization in the symmetrical beauty of repeated arcuate apertures in buildings,

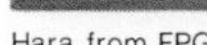

Hara from FPG

(Top) Hallian from FPG; (bottom right) Toni Schneiders from Bruce Coleman Inc.

Eric Lessing from Magnum

Herb Archer from Alpha

John Padour

(Above left) Rustic wooden mill in the Virginia countryside. Typically gabled stave church (right), near Borgund, Norway. (Opposite) Onion-dome turrets of the Church of the Transfiguration, in the lake country northeast of Leningrad.

amphitheaters, bridges, and aqueducts. The repetitive grace of these structures is such that the uninitiated can scarcely believe that the splendid aqueduct of Segovia, for example, was constructed for so utilitarian a purpose as the transport of water! It is a waterway of immense beauty. Significantly, the widely used Roman arch provided another link in the architectural continuum, one which led directly to the great wonder of the medieval world, the Gothic cathedral. Indeed, the churches and abbeys of Europe that were built between approximately 1100 and 1600 seem veritable advertisements for the splendor of arch upon arch upon arch. The abbey at Soissons in France is perhaps not the most remarkable among these, but it is clearly indicative of the distance traversed by man between his early cavern heritage and the advent of these gracious vaults. It is testimony to the effective merger of the spiritual and engineering worlds.

Finally, in this vein, mention must be made of wood. The beauty of wood grain has always been appealing to man's aesthetic sense, but the fact that the tensile strength of wood is approximately twice its compressive strength has more to do with its architectural use. The strength difference means that where rock in the form of long beams between distant supports will fail, wood will yield slightly but will hold. Phrased differently, wood can bend without breaking, whereas rock

The Recovery of Gold

by Michael I. Brittan

Gold is the one commodity of intrinsic value that commands virtually universal trust. The recent rise in price has led scientists and engineers to renew their efforts to extract the maximum amount of gold from its ores.

The properties of gold have made it a prized and useful commodity for many centuries. A 7-in. raft (opposite, top left), crafted in pre-Columbian times by the Chibcha Indians of South America, demonstrates the malleability and ductility of gold, while its luster can be seen in the butterfly poporo *(top right), also made by pre-Columbian Indians in South America. A contemporary application (opposite bottom) is the thin gold film on the plastic visor of U.S. Apollo 9 astronaut Russell Schweickart, used because gold reduces the glare from sunlight but permits good visibility.*

MICHAEL I. BRITTAN *is Gold Project Coordinator and Head, Computer Applications, of the Research Division, Anglo American Corporation of South Africa Ltd.*

(Overleaf) Courtesy, Chamber of Mines of South Africa

Gold, the most illustrious of the metals, has been treasured since antiquity. The almost obsessive desire for gold transcends its utilitarian applications and even the psychology of rarity. The key to its interaction with human aspiration lies enmeshed in the remarkable properties of this yellow metal.

Perhaps the most significant property of gold is its nobility, its chemical inactivity. It may be observed that, in nature, this factor kept it in regal isolation from combination with such common elements as oxygen and sulfur; its persistence almost exclusively as a native metal of distinctive luster must first have attracted early man. The first references to gold date back to Egypt some 5,000 years before the Christian era. The sheen, malleability, and ductility of the metal encouraged craftsmen to enrich early civilizations by fashioning it into ornaments, decorations, and jewelry. Dramatic proof of the imperishability, immutability, and nontarnishing qualities of gold has been provided by the unearthing in recent times of some of these artistic treasures. For example, those recovered from the Mesopotamian tomb of Queen Shub-ad were as pristine as when first buried about 4,600 years ago.

As early as its association with the creative drive must have been the alliance of gold with religion. Its enduring qualities established it as a symbol of immortality. The Theban Egyptians linked gold to Re (Ra), the sun god. The pharaohs, believed to be descendants of the gods, amassed the metal as one of the trappings of power and to ensure their divine afterlife. The Incas linked gold to the Sun, and the Aztecs adorned their temples with it. Gold has sacred spiritual connotations for the Hindus of India and the Ashantis of Ghana.

Gold also left its legacy in scientific endeavor. The craving for this "ultimate metal" caused the alchemists of the Middle Ages to try to create it from lesser materials. Their relentless torture and examination of the common elements spawned the art (subsequently the science) of chemistry. In 1489 Leonardo da Vinci wrote thus of their vain struggles:

"By much study and experiments the old alchemists are seeking to create not the meanest of Nature's products, but the most excellent, namely gold, which is begotten of the Sun, inasmuch as it has more resemblance to it than anything else that is and no created thing is more enduring than this gold. It is immune from destruction by fire, which has the power over all the rest of created things, reducing them to ashes, glass or smoke."

The unique blend of physicochemical attributes of gold comes into play in providing the key to its modern technological uses. Being rare and precious, it is fortunate indeed that gold can be used sparingly while still permitting exploitation of its immunity to corrosive environments, its prowess as a reflector of heat and light, and its efficiency as a conductor of heat and electricity. It can be effective for these purposes when reduced to extremely thin foil or even translucent films only millionths of a centimeter thick; one troy ounce can be drawn into a thread 80 km (50 mi) long.

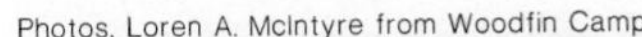
Photos, Loren A. McIntyre from Woodfin Camp

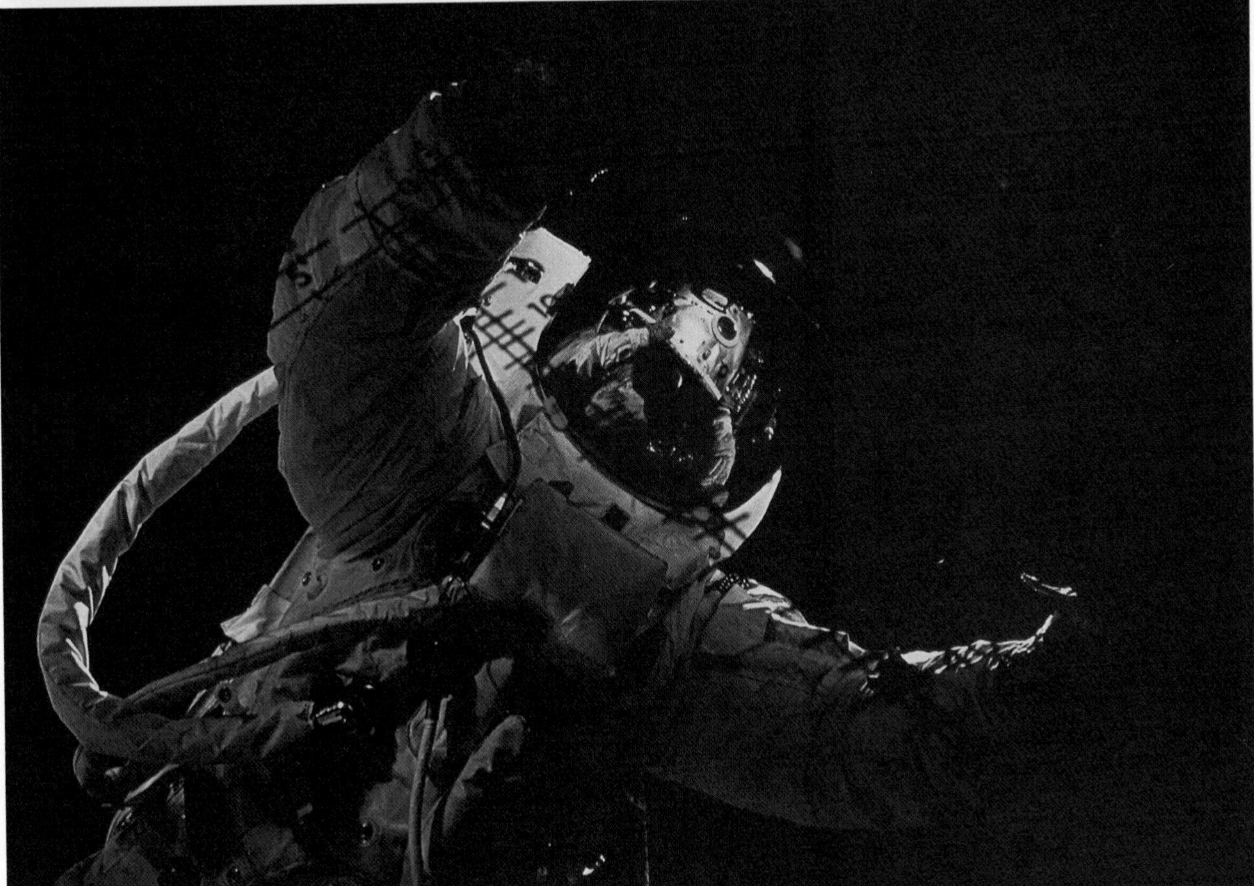

NASA from Black Star

Extracting gold from Witwatersrand ores proceeds as follows: (1) Elevator loads of ore fragmented into rocks are hauled to the surface. The gold content of the rocks is about ten parts per million. (2) Jaw crushers and cone crushers break down the rocks, and waste pieces containing no gold are discarded. (3)The crushed ore and water are fed to rotating cylindrical mills containing steel balls, rods, or pebbles. The ore is crushed to a powdery fineness sufficient to liberate the gold. (4) During the milling operation the ore-water mixture may be passed over endless riffle belt concentrators, where gravity separation causes the heavier gold particles to be trapped in the riffles. (5)The gold concentrate, containing up to 20 kg of gold per ton, is mixed with mercury and agitated in amalgamation barrels. The gold is then extracted as amalgam. (6) After excess mercury is removed from the amalgam, it is heated in a furnace to 600° C. The mercury is distilled, leaving the impure "sponge" gold. (7) The pulp leaving the mills is dewatered in large tanks. The ore particles settle, and the dense pulp is withdrawn. (8) The thickened pulp, to which small amounts of cyanide and lime are added, is agitated and oxygenated by compressed air injection. The gold dissolves into the cyanide solution. (9) Rotary vacuum filters separate the gold-bearing solution from the barren solids. (10) after the solution is clarified and de-aerated, zinc dust is added to it to precipitate the gold. The zinc is then removed by sulfuric acid, and the gold is calcined at 600–700° C to oxidize residual impurities. (11) The impure gold from both processes is smelted with a borax/silica flux and poured into molds. Most residual contaminants are removed as slag. (12) The final product is a 32-kg bar containing about 10% silver and trace amounts of other metals.

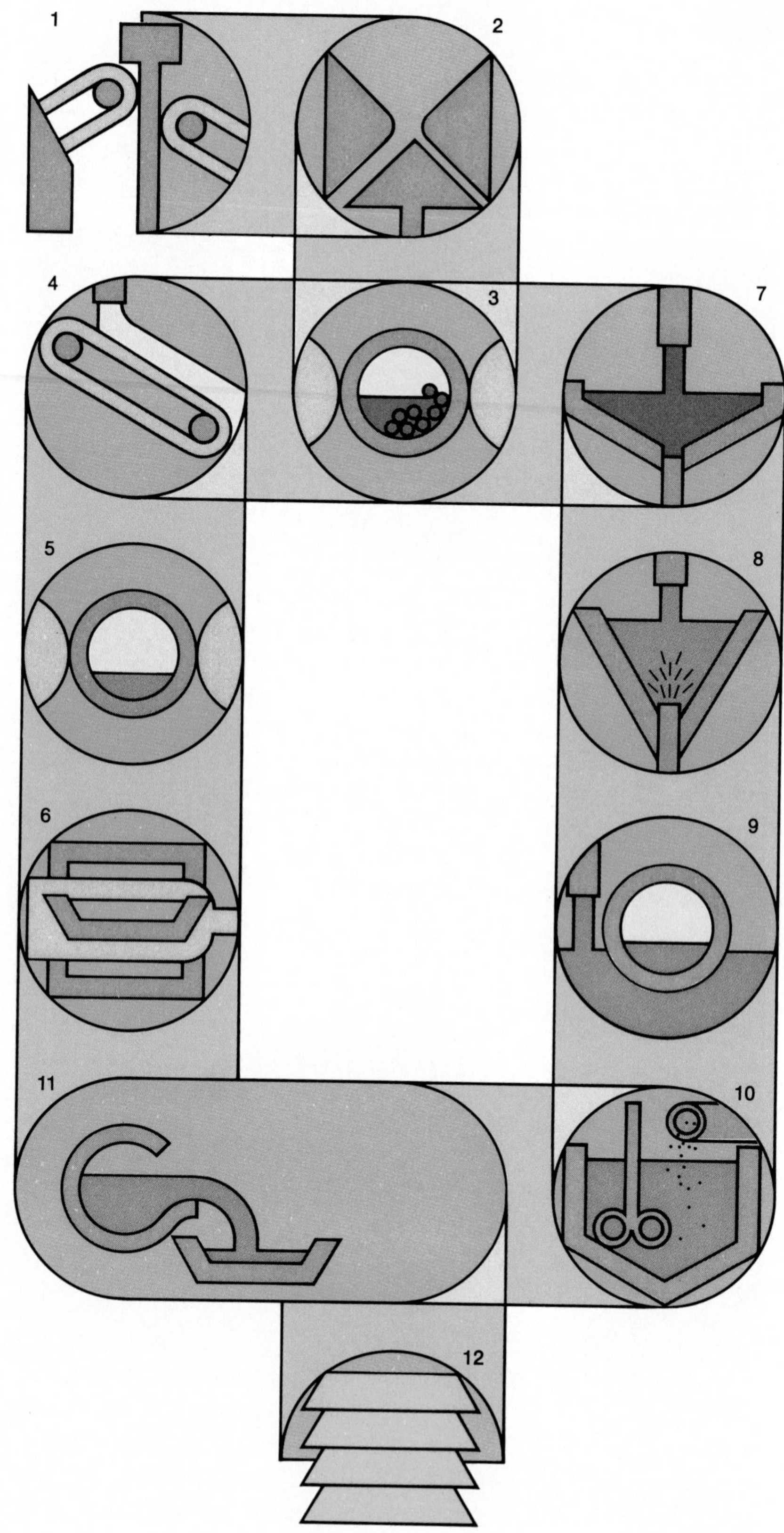

Illustrations by Dave Beckes

The inherent softness of gold endeared it to early craftsmen but might have inhibited its industrial inroads into dentistry, jewelry, and electronics were it not for its capacity to alloy with a wide range of metals. This can bestow the required tensile strength and resistance to wear while retaining the special properties of the native metal. As an artistic bonus, alloying can impart an infinite variation in hue, including shades of white, red, green, and blue. Exploitation of alloying is far from a recent innovation: the goldsmiths of ancient Egypt and Sumeria were skilled in the art of brazing and coloration using gold-based alloys. The first white gold containing up to 75% platinum was made by the Incas. This represents no mean feat since platinum has a high melting point, 1,769° C, compared with that of gold, 1,063° C. A sintering technique was used, the forerunner of a metallurgical process that was independently "reinvented" in the 19th century.

Marvin E. Newman from Woodfin Camp

Lee Boltin

Gold occurs in nature in various forms. (Top) Miners work the underground veins of gold-bearing ore at the Western Reef mines in South Africa; only a small fraction of these ores is gold. (Bottom) gold crystals measuring an inch or more across were found in alluvial deposits in California.

Measure of value

The conception of gold as a measure of value has long been branded on the human psyche. Expansion of trade and the evolution of more sophisticated economic systems probably provided the basic incentive for the use of an indestructible and universally esteemed commodity such as gold as a medium of exchange. Gold, as rings or small lumps, was in use as currency by 2000 B.C., and gold coins have been minted since the 7th century B.C.

The modern counterpart of the metal's monetary association, the gold-exchange standard, relied for its workability on stable currency exchange parities and the ultimate convertibility into gold of the primary reserve currency, the United States dollar. In recent years, however, the system has broken down, culminating in the official closing of the gold "window" by the U.S. Treasury in August 1971. Scarcity in the supply of gold in the face of an increase in private demand had already, in March 1968, forced international monetary authorities to abandon their efforts to contain the gold price at the level of $35 per ounce established in 1934 by U.S. Pres. Franklin Roosevelt. (That relatively attractive price had stimulated demand, initially for industrial use and later as a psychological haven and bulwark against political and economic uncertainty.) These and other milestones in the financial role of gold were symptomatic of growing disarray in the international monetary system. The basic cause was an erosion of confidence in major currencies accompanying a declining gold content of total reserves (which, in turn, had declined in relation to the value of world trade), rampant inflation, and mountainous trading deficits.

Will gold be recalled to play a disciplinary role in a permissive monetary system racked with international payments difficulties? Ultimately, the main prerequisite of any monetary base is universal confidence, and, against a backdrop of international discord and paper currency excesses subject to the whims of politicians, gold stands as the one tangible commodity of intrinsic worth able to command universal human trust.

Courtesy, Chamber of Mines of South Africa

A city built by gold, Johannesburg is a metropolis of more than 1.5 million people and is South Africa's largest urban center. In the background are tailing dumps produced by gold mining operations. Western Deep Levels, the world's deepest mine (opposite page), is about 80 km (50 mi) west of Johannesburg. Two shaft complexes provide access to the two gold-bearing reefs, the Ventersdorp Contact and the Carbon Leader, and these allow the fractured gold-bearing rocks to be hauled to the surface. Crosscut tunnels are driven from the shafts to intersect the reefs at different levels. Other tunnels called drives then follow laterally the plane of the reef, and from these more tunnels are driven, both up (raises) and down (winzes) the reef. Mining takes place at the ore faces along the walls of these tunnels, creating narrow excavations called stopes.

Extraction of gold

Gold is found in minute quantities in most common rocks and in solution in seawater, but commercial exploitation is limited to isolated deposits where natural processes have concentrated it during the course of geological time. Even in these places the gold content is incredibly low, generally only a few parts per million. This suggests the magnitude of the task facing the extractive metallurgist in his attempt to win this minute quantity of gold from such an overwhelming proportion of host rock. Extraction of the maximum amount of gold from its ores continues to pose a formidable challenge to scientists and engineers. In some instances, base-metal and other polymetallic ore bodies contain sufficient traces of gold to warrant its recovery as a by-product.

Rich nuggets or veins of gold never presented any extractive obstacle through the ages because the prize always yielded to a little hammering and handpicking. Such rich occurrences, however, have never been as extensive as fortune hunters might have wished, and attention had to be turned to the more abundant alluvial deposits. With these, ingenuity had to be exercised to induce them to give up their fine grains of gold. This could be accomplished by exploiting the density of the metal (specific gravity of about 19.3 for pure gold; that is, 19.3 times the weight of an equivalent volume of water) to enable the gold particles to be washed free of the lighter valueless gangue (specific gravity usually less than 3).

Such practices of gold recovery by gravity concentration began at least as early as recorded history and constitute, in effect, an emulation of nature's handiwork in laying down alluvial sediments in the first instance. In primitive operations, the material was washed down a sluice equipped with small transverse riffles, which retained the

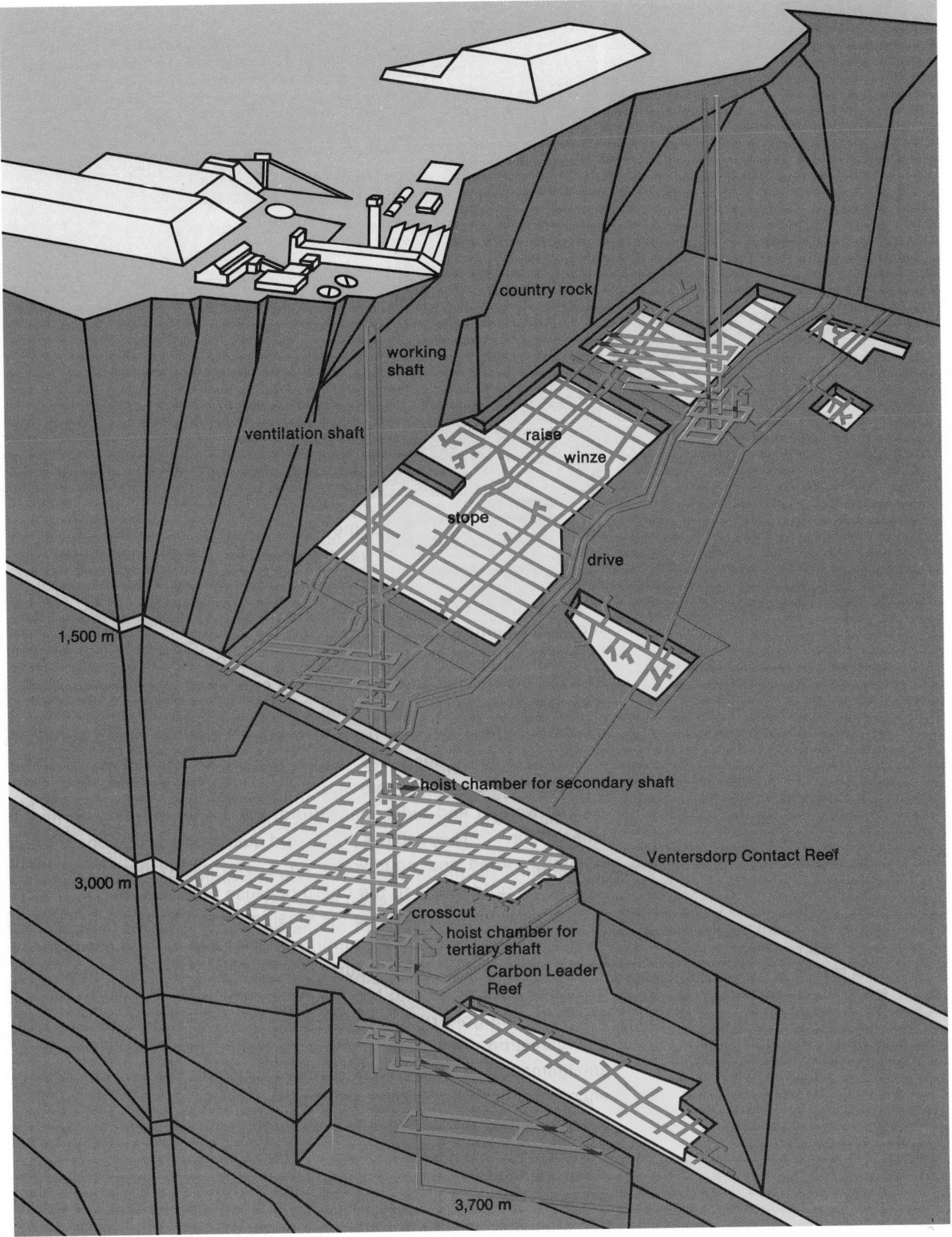

country rock
working shaft
ventilation shaft
raise
winze
stope
drive
1,500 m
hoist chamber for secondary shaft
Ventersdorp Contact Reef
3,000 m
crosscut
hoist chamber for tertiary shaft
Carbon Leader Reef
3,700 m

Permafrost: Challenge of the Arctic

by Troy L. Péwé

As man increases his activities in the Arctic, he must learn to deal with the problems posed by permafrost, the layer of permanently frozen ground that underlies the Earth's surface in the polar regions.

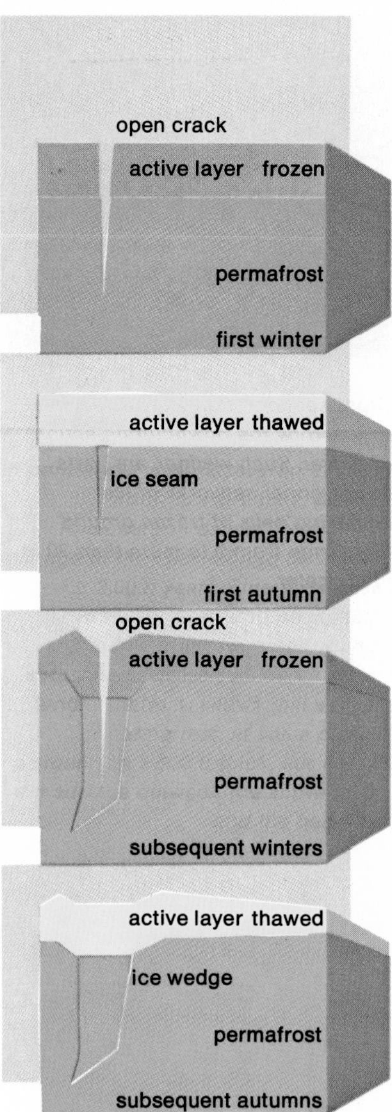

Ice wedge formation begins with a polygonal thermal-contraction crack in frozen ground. Water from melting snow runs down the crack in the spring and freezes, producing a vein of ice in the permafrost. This process is repeated during succeeding years, eventually creating an ice wedge. Polygonal ground (above right) in the Atigun River valley of northern Alaska indicates the presence of ice wedges.

Steve McCutcheon

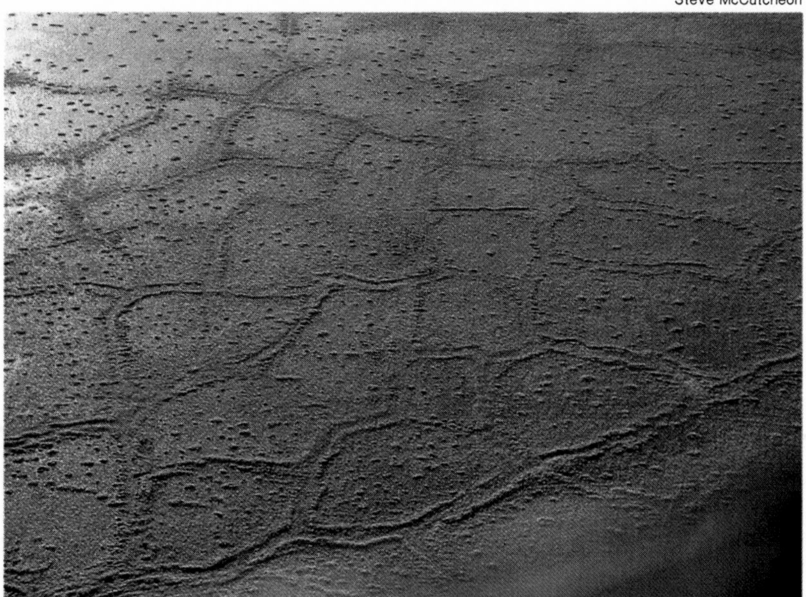

the main evidence that a belt of permafrost existed south of the continental glaciers that covered northern Europe and North America about 15,000 to 22,000 years ago. Ice-wedge polygons and perhaps pingos were characteristic parts of the landscape at that time in such areas as present-day Belgium, Germany, and Poland, and ice wedges also existed in front of the ice sheet in the United States in the region from Montana to Ohio.

Age of permafrost

Perhaps the earliest evidence of permafrost in North America is indicated from central and western Alaska. In central Alaska on the campus of the University of Alaska are ice-wedge casts thought to be at least one million years old. Ice-wedge forms thought perhaps to be 1.5 million years old have been reported from Cape Deceit on the south shore of Kotzebue Sound. This early permafrost disappeared during interglacial times. Permafrost undoubtedly formed, thawed, and reformed at various times during periods of refrigeration in the last one or two million years.

Most of the existing permafrost and perhaps all of the existing ice wedges formed in the latest of glacial times. All radiocarbon dates on ice wedges are less than 37,000 years old, and dates on frozen carcasses of ice-age animals also are in that range.

An interesting aspect of permafrost is the preservation of carcasses and partial carcasses of modern and extinct Ice Age mammals within it. Frozen carcasses have been known in the north for centuries. The first direct report of a carcass of a mammoth in frozen ground in Siberia was by E. Yssbrants Ides in 1692. At least one or two almost complete frozen mammoths have been reported from Siberia. In North America partial carcasses of mammoth, bison, musk ox, moose, horse, lynx, caribou, and ground squirrel have been found in Alaska.

Perhaps the most celebrated find of an Alaskan frozen carcass was the partial forequarters of a baby mammoth, collected in 1948 near Fairbanks. The well-preserved hide of the head, neck, trunk, and one front leg was about one-quarter of an inch thick and almost hairless. An important discovery of a partial carcass of an extinct superbison 31,400 years old was made in 1951 in central Alaska.

Stories have been reported for many years of mammoths dying with buttercups in their mouths, perishing as cataclysmic climatic changes took place that suddenly turned tropical climates into frigid Arctic conditions. It is not necessary, however, to entertain fantasies about the distribution, preservation, and extinction of Ice Age mammals. Work in Alaska demonstrates that all the carcasses are between 10,000 and 70,000 years old and could not have been older because they would not have survived the preceding interglacial warm period when the ground and its enclosed carcasses thawed. In order to preserve the carcasses in nature's deep freeze, it is evident that most of the permafrost present today, in central Alaska at least, has existed since latest glacial times. Geological relationships and radiocarbon dates indicate that the carcasses are remains from the last glacial cold period and are not "a million years old" as casually announced on occasion. All evidence suggests that most carcasses represent natural deaths in cold, harsh climates.

From Péwé, Church, and Andresen, The Geological Society of America, Special Papers #103 (1969), p. 23

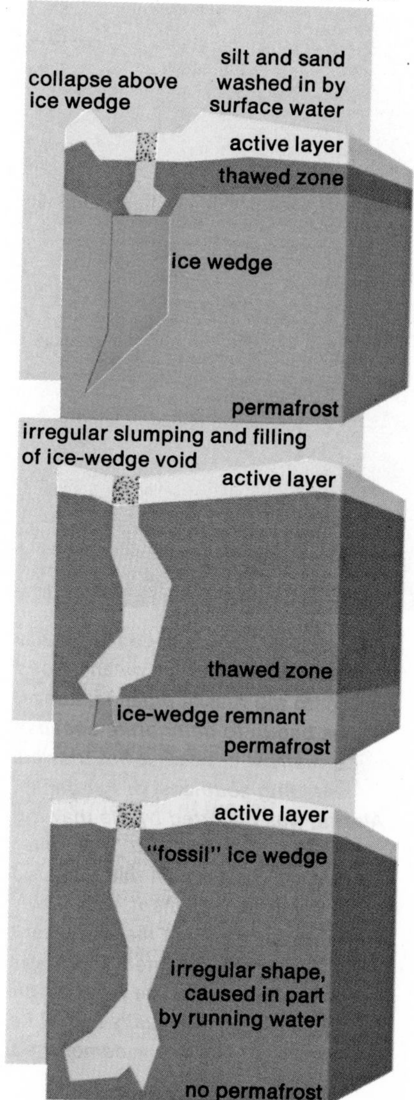

"Fossil" ice wedges (ice-wedge casts) form when ice wedges slowly melt. This occurs when the permafrost table is lowered during a period of warming climate.

Permafrost and man

Development of the polar regions demands that man understand and be able to cope with the problems caused by permafrost. The most dramatic, widespread, and economically important examples of the influence of permafrost on life in the North deal with construction and maintenance of roads, railroads, airfields, bridges, buildings, dams, sewers, pipelines, and communications lines. Engineering problems are of four fundamental types: (1) those involving thawing of ice-rich permafrost and subsequent subsidence of the surface under unheated structures such as roads and airfields; (2) those involving subsidence under heated structures; (3) those resulting from frost action, generally intensified by poor drainage caused by permafrost; and (4) those involved only with the temperature of permafrost, causing buried sewer, water, and oil lines to freeze.

A thorough study of the frozen ground should be part of the planning of any engineering project in the North. It is generally best to attempt to disturb the permafrost as little as possible in order to maintain a stable foundation for engineering structures, unless the permafrost is thin; in that case, it may be possible to destroy the permafrost.

Because thawing of permafrost and frost action are involved with almost all engineering problems in polar areas, whether dealing with highways, buildings, or sewer lines, it is well to consider the general principles of these phenomena. The delicate thermal equilibrium of permafrost is disrupted when the vegetation, snow cover, or active layer is compacted. The permafrost table is lowered, the active layer is

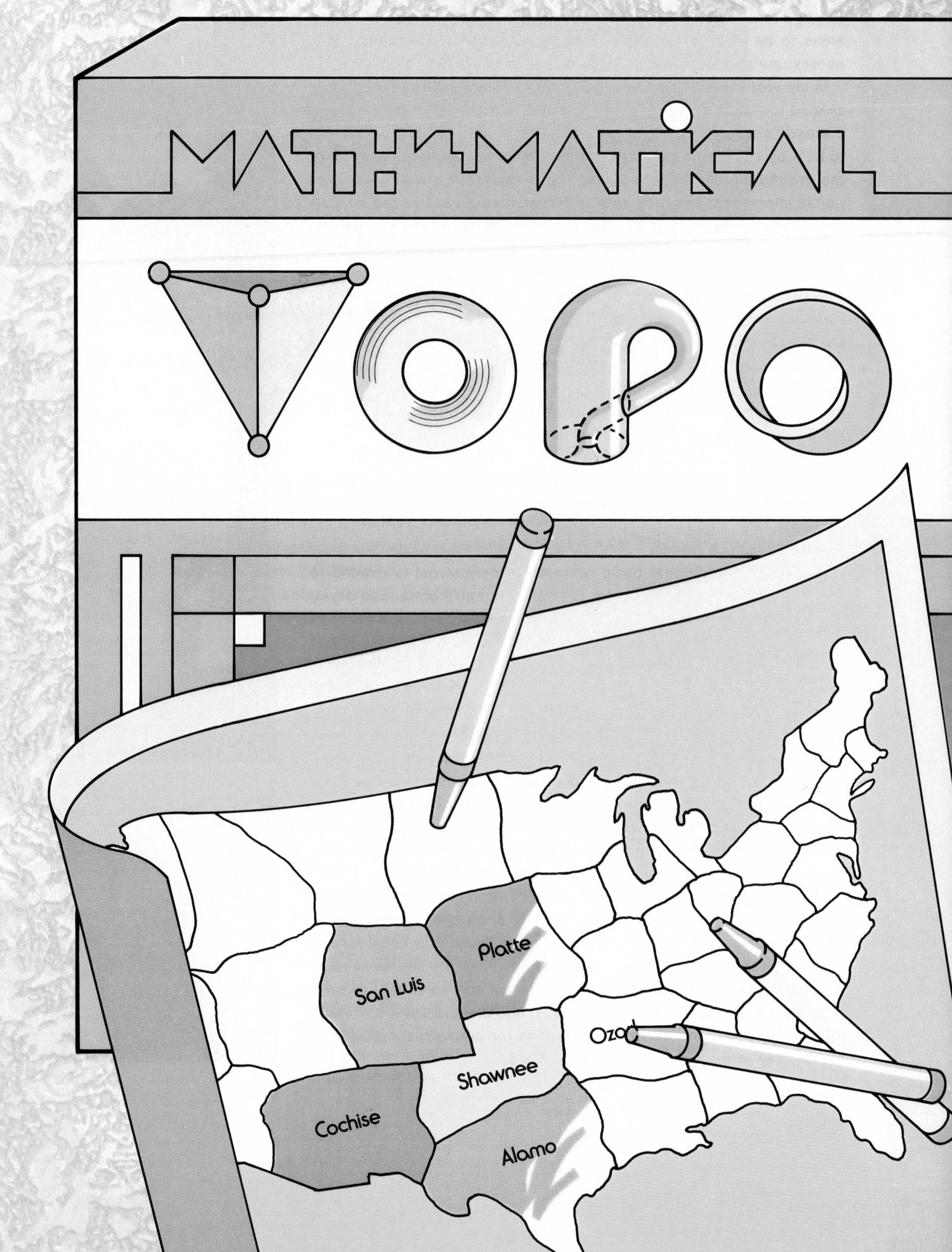

MATHEMATICAL
TOPO
Platte
San Luis
Ozar
Shawnee
Cochise
Alamo

Excursions in Topology
by Harold R. Jacobs
A map with four colors and a bottle with no inside or outside are two of the toys in this most entertaining branch of higher mathematics.
CRAIG

Figure 1. Seven bridges (1–7) in the city of Königsberg, Germany, joined two islands and the mainland. The townspeople found it impossible to cross all seven in one continuous trip without traveling over any one of them more than once. A, B, C, and D represent the vertexes devised by Swiss mathematician Leonhard Euler in his analysis of the problem (see figure 2:1 and 2).

Recently a geographer named G. Etzel Pearcy suggested that, in the interest of more efficient and economical government, the United States should be reorganized into 38 states. The only state that would remain intact after his proposed rearrangement is Hawaii. Alaska would become two states, and the rest of the country would be divided as shown in the map on page 106.

If this map were colored in such a way that no two states that share part of their borders were the same color, how many colors would be required? Although the answer to this question is not immediately obvious, it is easy to show that three colors are not enough. Consider the state of Shawnee and the five surrounding states. Shawnee, Cochise, and San Luis must be colored differently because each of them touches the other two; suppose they are yellow, blue, and green, respectively, and that we try to color Platte, Ozark, and Alamo without introducing a fourth color. It is clear that Platte must be blue and that Alamo must be green—hence, Ozark can be neither blue nor green. Because it cannot be yellow, a fourth color must be introduced.

We have shown that four colors are necessary to color Pearcy's map; it is easy to verify, by actually coloring the map, that four colors are also sufficient. Is it possible to divide the country into a different arrangement of states so that five colors are necessary? Although mathematicians have spent much time trying to answer this question conclusively, no one has yet succeeded. The problem of coloring a map first gained wide attention when the brilliant British mathematician Arthur Cayley admitted in 1878 that he had attempted to prove that no more than four colors are ever needed but could not do so. Since then, it has been rigorously proved that every map in a plane can be colored with five colors. This is not very satisfying, however, because all known maps require only four.

The "four-color map problem" is a seemingly simple, yet remarkably elusive puzzle in the branch of mathematics known as topology. Originally known as analysis situs, topology might be considered an exceedingly basic sort of geometry that deals with those properties of figures that remain unchanged when the figures undergo certain transformations. Among these transformations are bending, stretching, and twisting—any deformation, in fact, that does not result in disconnecting or connecting parts of the figure. Such transformations, in which neighboring points of the figure remain neighboring points, are said to be "continuous." The properties of figures that are not altered by continuous transformations are called topological and are of great importance in many mathematical investigations. Some of these properties will be considered in this article.

__HAROLD R. JACOBS__ is Chairman of the Mathematics Department, U. S. Grant High School, Van Nuys, California.

Illustrations by John Craig

The Königsberg bridges

One of the oldest problems in topology concerns seven bridges in the city of Königsberg in old Germany (now Kaliningrad, U.S.S.R.). The center of the city was on an island in the Pregel River. This island was linked to a neighboring island by a bridge, and the two islands

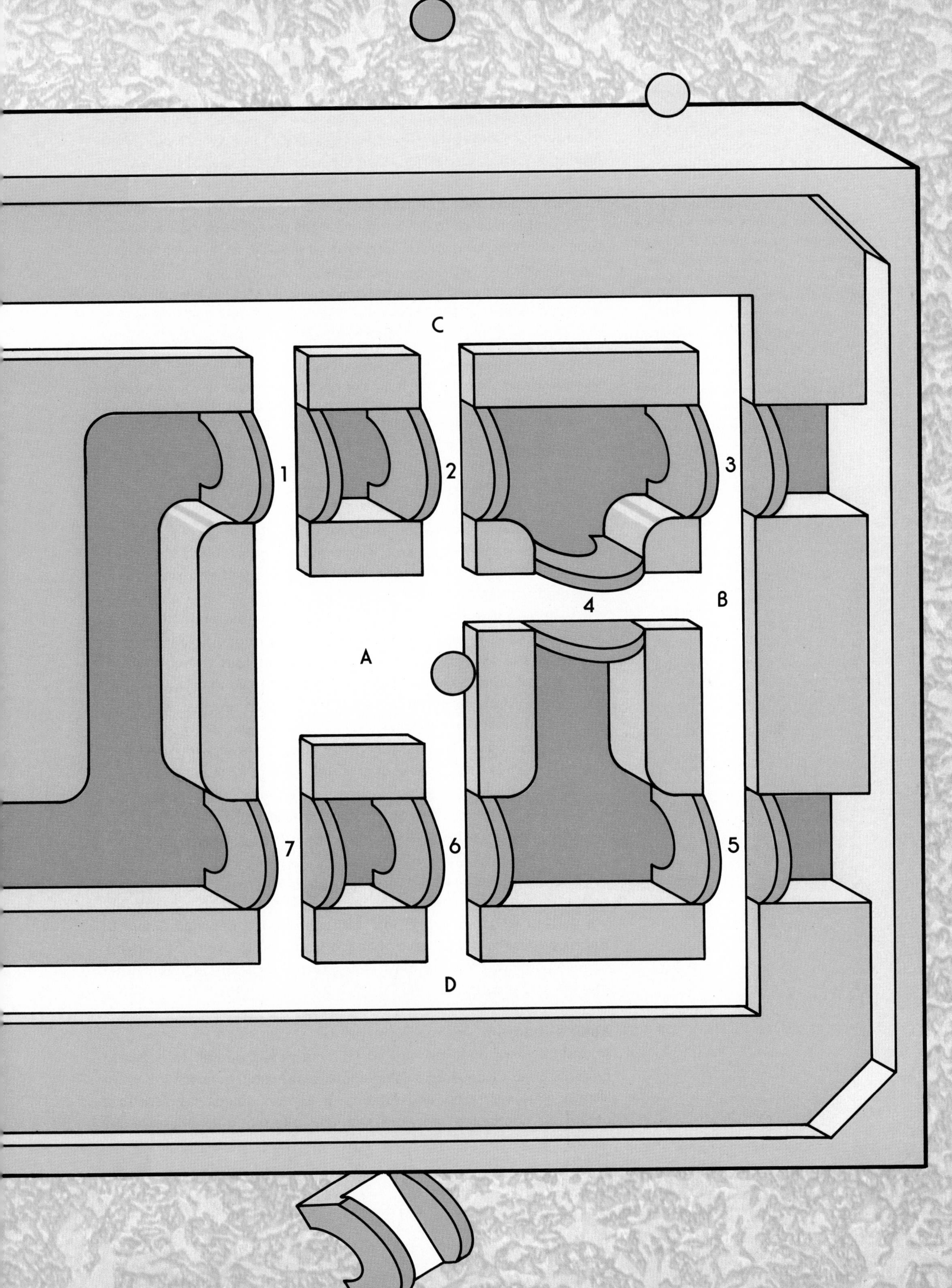
C
1
2
3
4
B
A
7
6
5
D

Figure 2. Swiss mathematician Leonhard Euler analyzed the Königsberg bridge problem by devising a diagram (network), 1, in which the four land areas involved became vertexes A–D and the seven bridges were represented by arcs. When an eighth bridge was added, 2, it became possible to cross all bridges in one continuous trip without traveling over any one of them more than once. In 3a and 3b and 4a–4g Euler discovered that in a connected network in a plane the number of vertexes plus the number of regions (one of them exterior to the network) minus the number of arcs always equals 2.

were connected to the rest of the city by six additional bridges. The townspeople wondered if they could cross all seven bridges in a continuous trip without traveling over any one of them more than once. After many attempts it was generally agreed that such a trip was impossible, but no one knew why. A schematic representation of the problem can be seen in figure 1.

The problem came to the attention of the great Swiss mathematician Leonhard Euler, who wrote a paper on it in 1736. In it, he observed:

> The branch of geometry that deals with magnitudes has been zealously studied throughout the past, but there is another branch that has been almost unknown up to now; Leibnitz spoke of it first, calling it "the geometry of position." This branch of geometry deals with relations dependent on position alone, and investigates the properties of position. . . . Recently there was announced a problem that, while it certainly seemed to belong to geometry, was nevertheless so designed that it did not call for the determination of a magnitude, nor could it be solved by quantitative calculation; consequently I did not hesitate to assign it to the geometry of position. . . . In this paper I shall give an account of the method that I discovered for solving this type of problem.

In essence, what Euler did was to replace the map of the city of Königsberg with a diagram in which the four land areas were represented by points and the seven bridges by lines connecting them (figure 2:1). We will refer to this diagram as a *network* and to its points and lines as *vertexes* and *arcs*, respectively. The problem of crossing the seven bridges of Königsberg is equivalent to that of traversing the seven arcs in this network.

It is apparent that vertex A of the network is the endpoint of five arcs, whereas vertexes B, C, and D are each the endpoint of three arcs. The key to the solution of the problem is that all four vertexes are the endpoints of an odd number of arcs. If it is possible to travel each arc exactly once in a continuous path, then to each arc that is used to reach a vertex there must correspond another arc along which the path may continue. Hence, except for the points at which the path begins and ends, the number of arcs that meet at each vertex must be even. In other words, a network cannot be traveled if it contains *more than two vertexes that are the endpoints of an odd number of arcs.* Because the Königsberg network contains four, we have proved that no solution for it is possible.

Long after Euler's analysis of the problem, an eighth bridge was built in Königsberg. This change is shown in the second network (figure 2:2), in which an eighth arc has been added between B and D. Because these vertexes have changed from odd to even, the network contains only two odd vertexes, A and C; the path can begin on either of them and end on the other.

Euler's theorem

About 15 years after he had written his paper on the Königsberg bridges, Euler discovered a remarkable relationship between the numbers of vertexes, arcs, and regions of a "connected" network in a plane. By "connected" network, we mean one that does not have any

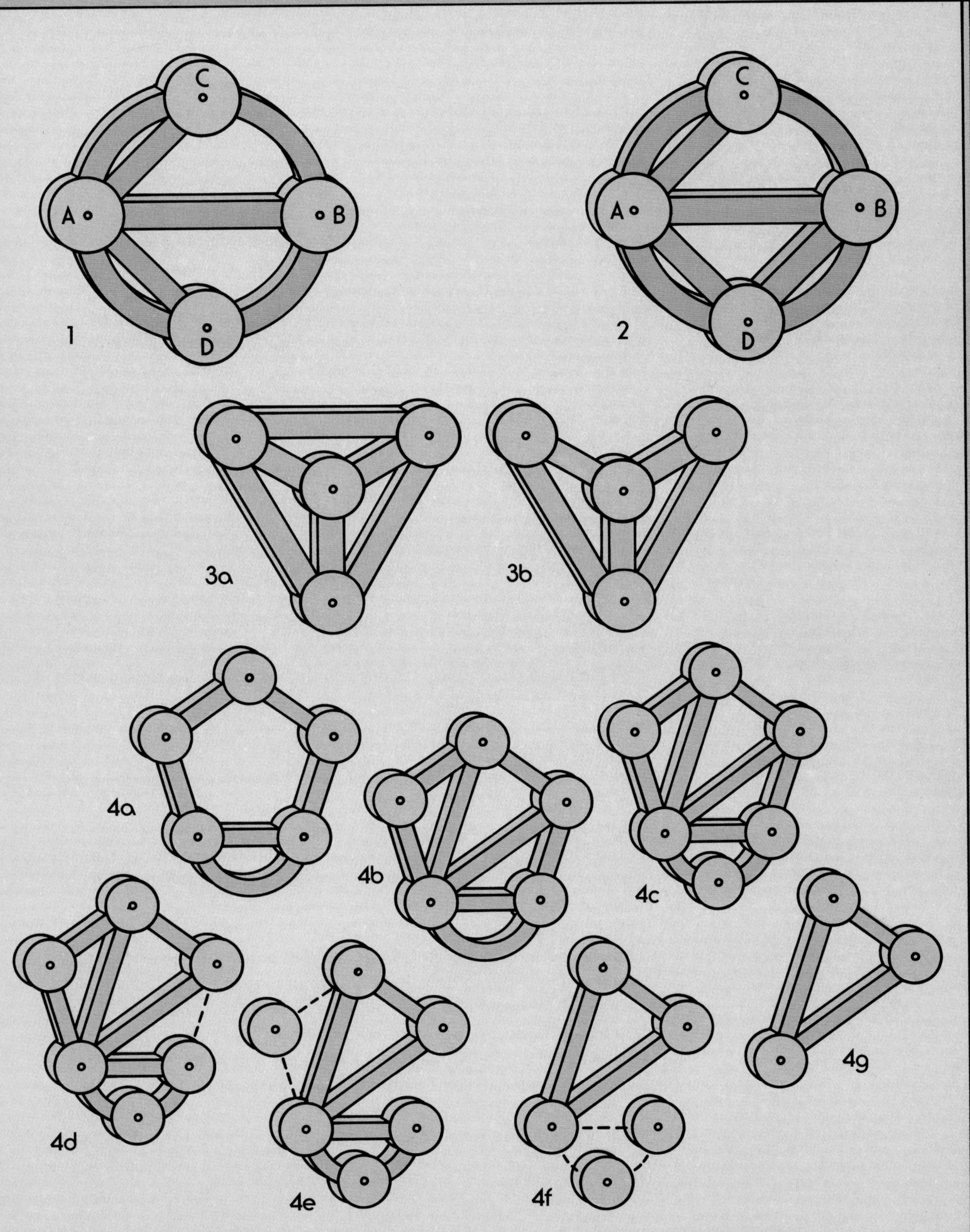
C
A
B
D
1
C
A
B
D
2
3a
3b
4a
4b
4c
4d
4e
4f
4g

Figure 3. When a regular polyhedron such as a tetrahedron is projected onto the surface of a sphere and its edges are pulled around to one side, the vertexes (V), *faces* (F), *and edges* (E) *of the tetrahedron become vertexes, regions, and arcs. During the transformation the number* V+F−E *(or* V+R−A*) remains unchanged, suggesting that its value is 2 not only for networks in a plane but also for those on the surface of a sphere.*

unattached parts. Euler's discovery is now considered one of the fundamental theorems of topology. The first network (figure 2:3a) contains four vertexes and six arcs; it divides the plane of the paper into four regions (one region is the portion of the plane that is exterior to the network). Letting V, A, and R represent the numbers of vertexes, arcs, and regions, respectively, it is evident that $V+R-A=2$. Now suppose that we remove one of the arcs of the network, as shown in figure 2:3b. Although A has decreased by one, so has R. Therefore, it is still true that $V+R-A=2$. In fact, by continuing to reason in this manner, it is easy to prove that this formula is valid for all connected networks that lie in a plane.

In the example just considered, we began with a network all of whose internal regions were triangular. Consider figure 2:4a, in which one of the internal regions is bounded by five arcs and the other by two. We can convert it to a network all of whose regions are triangular (bounded by three arcs) by making some simple additions. By adding two arcs from one vertex of the upper region, we can divide it into three triangular regions (4b). In general, by adding $(n-3)$ arcs from one vertex of a region bounded by n arcs, $(n-3)$ additional regions are created. This does not change the value of $V+R-A$ for the figure since $V+(R+n-3)-(A+n-3)=V+R-A$. Also, by adding one vertex to the boundary of the lower region, we can change it into a triangular region (4c). In general, by adding n vertexes to an arc, n additional arcs are created. Again $V+R-A$ remains unchanged because $(V+n)+R-(A+n)=V+R-A$.

After a given network has been "triangulated" in this way, triangles can be removed from the border of the network in one of the following three ways. One arc can be removed (4d), resulting in the loss of one region: $V+(R-1)-(A-1)=V+R-A$. Two arcs can be removed from the same region (4e), resulting in the loss of both a region and a vertex: $(V-1)+(R-1)-(A-2)=V+R-A$. And three arcs can be removed from the same region (4f), resulting in the loss of a region and two vertexes: $(V-2)+(R-1)-(A-3)=V+R-A$. In all three cases, $V+R-A$ remains unchanged. After this process has been carried out to the point at which only one triangle remains (4g), it is apparent that $V=3$, $R=2$, and $A=3$, so that $V+R-A=2$. Because we have established that $V+R-A$ remains unchanged throughout, it is clear that $V+R-A$ must equal 2 regardless of the complexity or shape of the original network.

Euler first conceived of his theorem in terms of the five geometric solids called the regular polyhedrons. Well known to the early Greek mathematicians (Plato described all five in his *Timaeus*), these solids are named according to the numbers of faces they possess. Three have triangular faces: the tetrahedron, with 4; the octahedron, with 8; and the icosahedron, with 20. One has square faces: the cube, or hexahedron, because it has 6. And one has pentagonal faces: the dodecahedron, with 12.

Letting V, F, and E represent the numbers of vertexes, faces, and

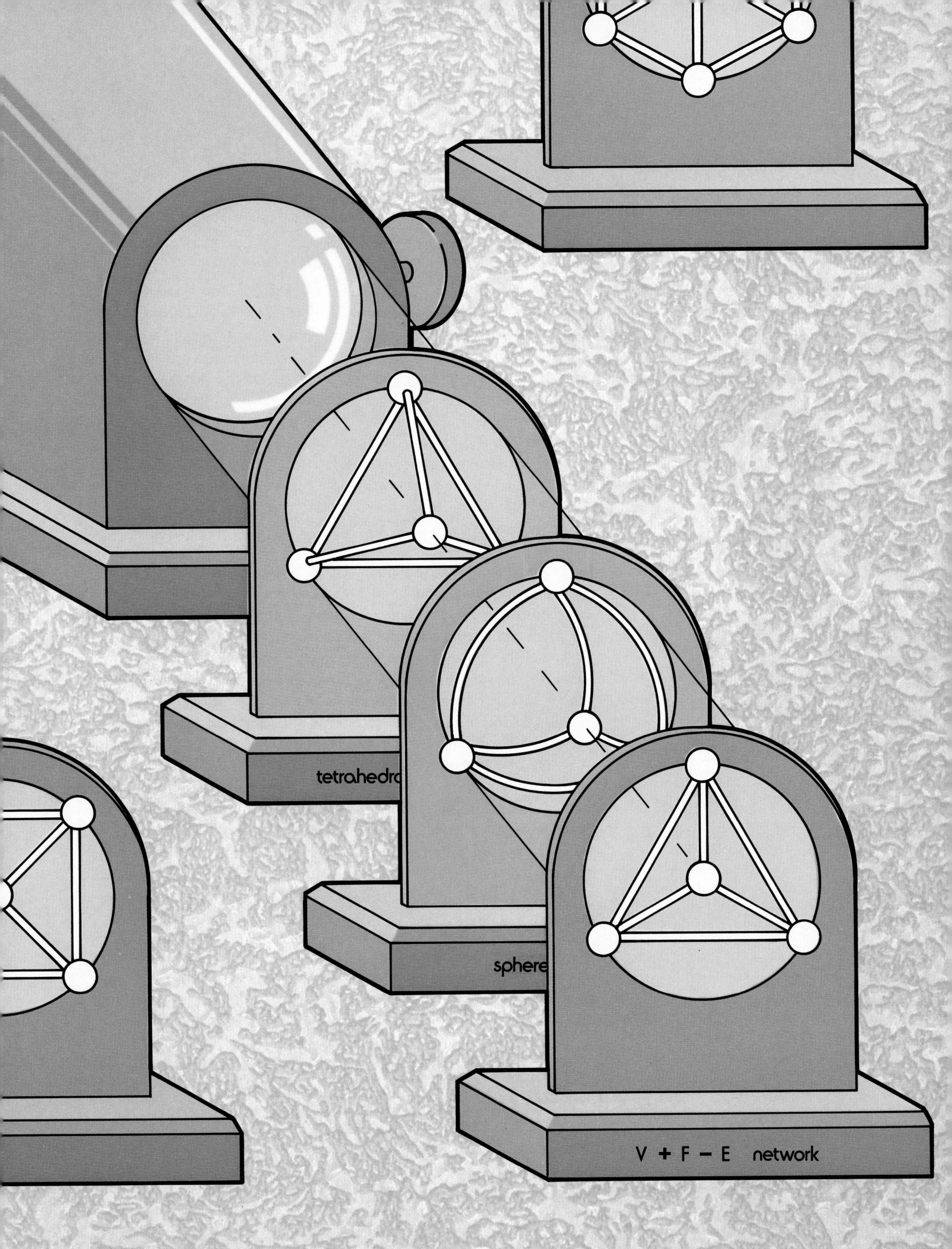
tetrahedra
sphere
V + F − E network

Figure 4. Jordan curve theorem states that a simple closed curve divides the surface of a plane or sphere into two distinct regions, an "inside" and an "outside." In regard to the torus, however, the theorem is not always applicable. It does apply in (a), but does not in (b), where the curve (green circle) encompasses the hole of the torus, or in (c), where the curve passes through the hole. In (d), where one set of curves goes around the hole and another set passes through it, V+R−A *no longer equals 2 but is now 0.*

edges, respectively, the following table reveals that, for each polyhedron, $V + F - E = 2$.

polyhedron	V	F	E
tetrahedron	4	4	6
cube	8	6	12
octahedron	6	8	12
dodecahedron	20	12	30
icosahedron	12	20	30

That this formula is the same as the one we derived for networks in a plane is easy to see. Suppose we project one of the polyhedrons, say the tetrahedron, onto the surface of a sphere. If the edges are pulled around to one side, the result is a network in which the vertexes, faces, and edges of the original polyhedron have become vertexes, regions (one face has become the region outside), and arcs (figure 3). The fact that the number $V + F - E$ (or $V + R - A$) remains unchanged during this transformation suggests that its value is 2 not only for networks in a plane but also for networks in the surface of a sphere. This number, said to be a "topological invariant," is called the *Euler characteristic* of the surface. The Euler characteristic of both plane and spherical surfaces is 2.

The Jordan curve theorem

A curve is termed closed when it has no endpoints, and simple when it does not intersect itself; circles, ellipses, and the circumferences of polygons are examples of simple closed curves. Although it is seemingly self-evident that a simple closed curve divides the surface of a plane or sphere into two distinct regions (an "inside" and an "outside"), this fact, called the Jordan curve theorem, is surprisingly difficult to prove. In fact, the mathematician after whom the theorem is named gave a proof of it that was eventually recognized to be invalid! A variety of rigorous proofs have been discovered since.

The Jordan curve theorem does not apply to all simple closed curves that might be drawn on the surface of the doughnut-shaped solid known as a *torus*. The theorem does apply to the closed curve on the torus shown in figure 4a. However, a curve that encompasses the hole of the torus (b), or that passes through the hole (c), does not divide the surface into two separate regions because any pair of points not on the curve can be joined together without crossing it.

One consequence of this is the fact that the Euler characteristic of a torus differs from that of a plane or sphere. If a network is drawn on the surface of a torus so that at least one set of arcs goes around the hole and another set of arcs goes through it (d), the value of the number $V + R - A$ is not 2, but 0. In the simple example shown here, the network consists of 1 vertex, 1 region, and 2 arcs: $1 + 1 - 2 = 0$. Although there are networks on a torus for which $V + R - A = 2$, the Euler characteristic of a given surface is taken to be the minimum value of $V + R - A$ that a network in the surface may have.

a
b
c
V + R − A = ?
d

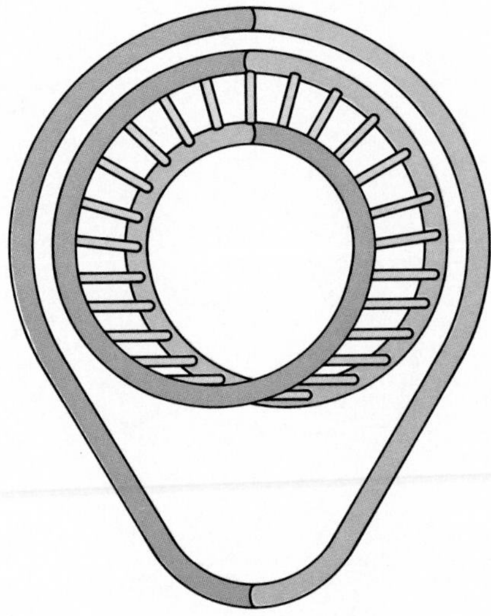

In general, the Euler characteristic of the surface of a solid depends upon the number of holes that are present in the solid. It decreases by two for each additional hole, so that a slice of Swiss cheese containing seven holes, for example, would have an Euler characteristic of −12. Although this seems to indicate that there are no surfaces for which the Euler characteristic is odd, this is not the case. The Möbius strip is such a surface.

The Möbius strip

In 1858 a German astronomer and mathematician named A. F. Möbius discovered a remarkable surface that has since become one of the most delightful playthings of topology. Called a Möbius strip, it can be made by taking a rectangular strip of paper and taping a pair of opposite edges together after twisting one of them through 180°.

The most surprising property of the Möbius strip is its behavior upon beng cut along the line midway between its edges: it remains in one piece! The result is a single loop that contains a 720° twist. (If this loop is cut along a line perpendicular to its edge, one end would have to be turned over four times to remove the twist.) If this loop also is cut along a line midway between its edges, it becomes two interlocking loops that are equal in length.

It is also interesting to discover that a Möbius strip has just one side and one edge. If a point is chosen somewhere on the strip and a corresponding point is found by poking a small hole through it to the "other side," then the two points can be connected by a continuous curve that does not cross over the edge of the strip. An amusing demonstration of this fact is sometimes made by laying track for a toy train around a model of a Möbius strip (figure 6). As the train travels in a continuous circuit about the strip, it travels on "both sides."

Figure 6. Track for a toy train is laid around a model of a Möbius strip (opposite). As the train moves continuously about the strip, it travels on "both sides" of it. Picture of a left foot on a plane (top left) or on a spherical surface (center) always remains that of a left foot no matter how it is slid about the surface. But on a Möbius strip (bottom right) a picture of a left foot can be moved about and return to its starting position as a picture of a right foot. In figure 5 (above) a "molecular Möbius strip" is made with a ring-shaped molecule having a double strand of atoms (inner loop); breaking the bonds within this molecule along a line midway between its edges would produce a ring twice as long (outer loop).

Because the Möbius strip is one-sided, it is a *nonorientable* surface. To understand what this means, think of the strip as having no thickness, just as a student of geometry imagines a plane as having no thickness. The pair of points described in the previous paragraph as being on a physical model of a Möbius strip, then, become one point which is "in" the surface that is the strip, rather than on either side. With this idea in mind, consider a picture in a plane of somebody's left foot (figure 6). No matter how the picture is moved about on the plane, it remains that of a left foot. If the picture slides over the surface of a sphere, the result is the same. Because left and right always remain distinct in both plane and spherical surfaces, they are said to be *orientable*. In contrast, if a picture of a left foot is moved around a Möbius strip, it can return to its starting position as a picture of a right foot. Hence, the Möbius strip is nonorientable. Other surfaces of this type have been discovered by topologists; in three-dimensional space, such surfaces are always one-sided.

The Euler characteristic of a Möbius strip can be determined by considering the network corresponding to the cut midway between its edges described above. Letting an arc represent the cut and a vertex

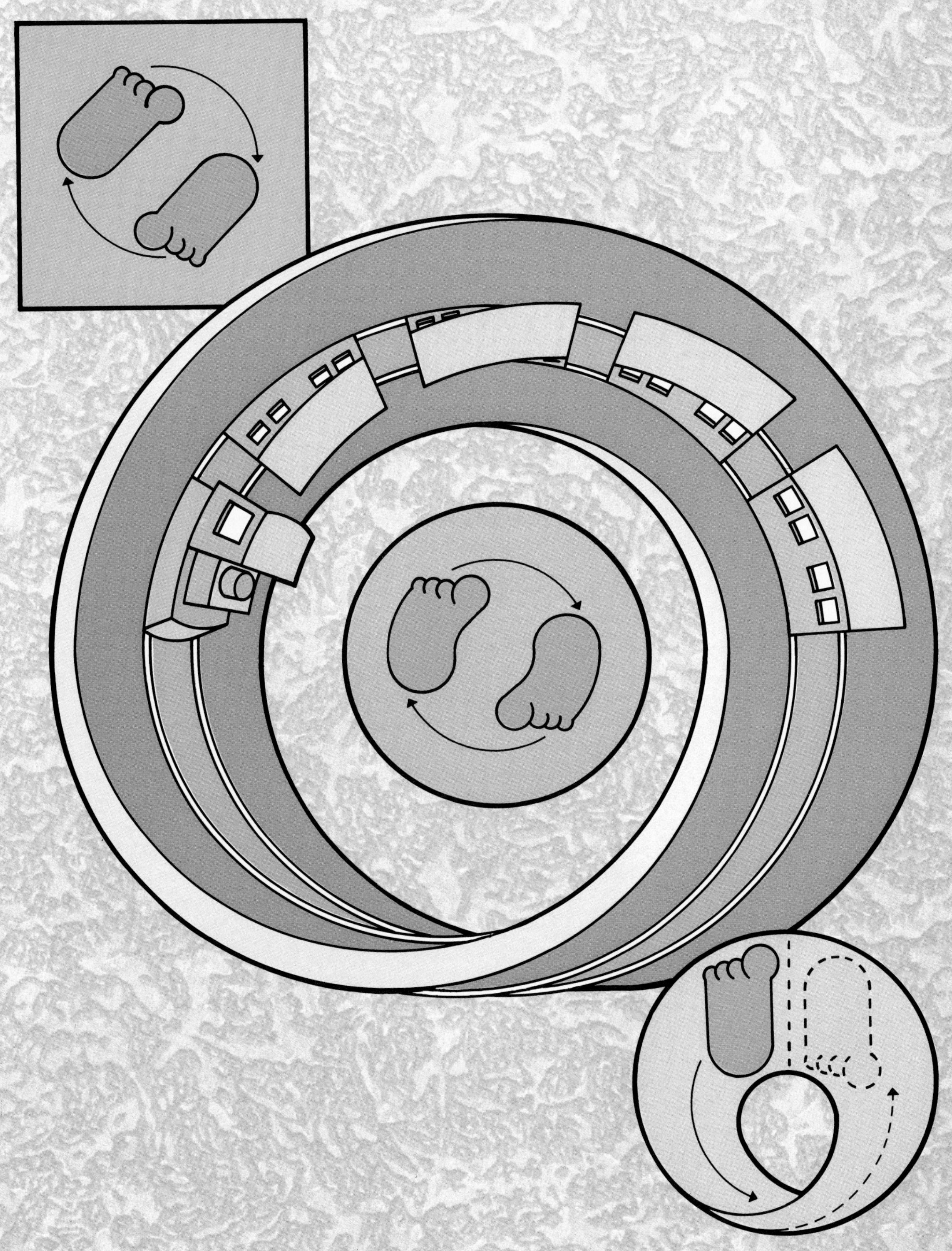

Figure 7. Klein bottles (top) are closed, one-sided surfaces. To understand their construction one can compare making a Möbius strip from a rectangular piece of paper (bottom left) with joining both ends of a twisted cylindrical tube (bottom right). In making a Klein bottle from the tube, however, it is not enough merely to "turn over" one end; one end must be put through the wall of the tube so that it can be attached from the "other side." This requires making a hole in the wall, but in theory Klein bottles have no such holes. They exist, therefore, only in the imagination. If a Klein bottle is split in a certain way (center), it falls apart into two Möbius strips.

on it represent the point in which the scissors penetrate the paper, we have a network consisting of one arc and one vertex. Because of the one-sidedness of the Möbius strip, this network does not divide it into two regions; hence there is only one. The value of $V + R - A$, consequently, is $1 + 1 - 1 = 1$.

A variety of practical uses have been discovered for the Möbius strip. The B. F. Goodrich Co. has patented a rubber conveyor belt in its shape; the belt lasts longer since both "sides" are actually one and receive equal wear. A continuous-loop recording tape sealed in a cartridge will play twice as long if it has a twist in it. An especially remarkable application is based on the discovery that if an electrical resistor is twisted and its ends are joined together to form a Möbius strip, properties that would ordinarily adversely affect the performance of the resistor are virtually eliminated. Chemists exploring ways of creating ring-shaped molecules have speculated about the possibility of synthesizing a "molecular Möbius strip." If such a molecule, which would consist of a "double strand" of atoms, were split in the same way as described for a Möbius strip made from paper, it would produce a ring twice as long (figure 5); a substance made up of such molecules might undergo a startling physical change were this to occur. On a less profound level, a new acrobatic trick performed by freestyle skiers is now commonly known as the Möbius flip.

The Klein bottle

In his book *Sylvie and Bruno Concluded*, Lewis Carroll refers to a "Purse of Fortunatus." The purse, like a Möbius strip, has only one surface. Because of this, its "inside" and "outside" are the same, so that whatever is inside the purse is outside it and whatever is outside it is inside. It is called the Purse of Fortunatus because the entire wealth of the world is inside it.

Just a few years before Carroll wrote about this enviable container, the great German mathematician Felix Klein discovered the surface now known as the Klein bottle. This surface, like the surfaces of a sphere and torus, is closed; like a Möbius strip, it has just one side. The basis for the construction of a Klein bottle is best understood by first comparing the construction of an ordinary belt-shaped loop with that of a Möbius strip (figure 7). In each case, the model of the surface can be made by taking a rectangular strip of paper and connecting a pair of opposite edges; the difference between the two results from the turning over of one end before the connection is made.

Now instead of starting with a flat strip, suppose we take a cylindrical tube and connect its ends together. If this is done in the same way that the belt loop is made, the result is a torus. If one end is "turned over" in the process, the result is still a torus. To make a Klein bottle, the orientation of one end of the cylinder has to be reversed with respect to the other before the connection is made. In other words, if we assign a rotational direction to the cylinder before it is bent, both ends will have the same direction when they are connected, even

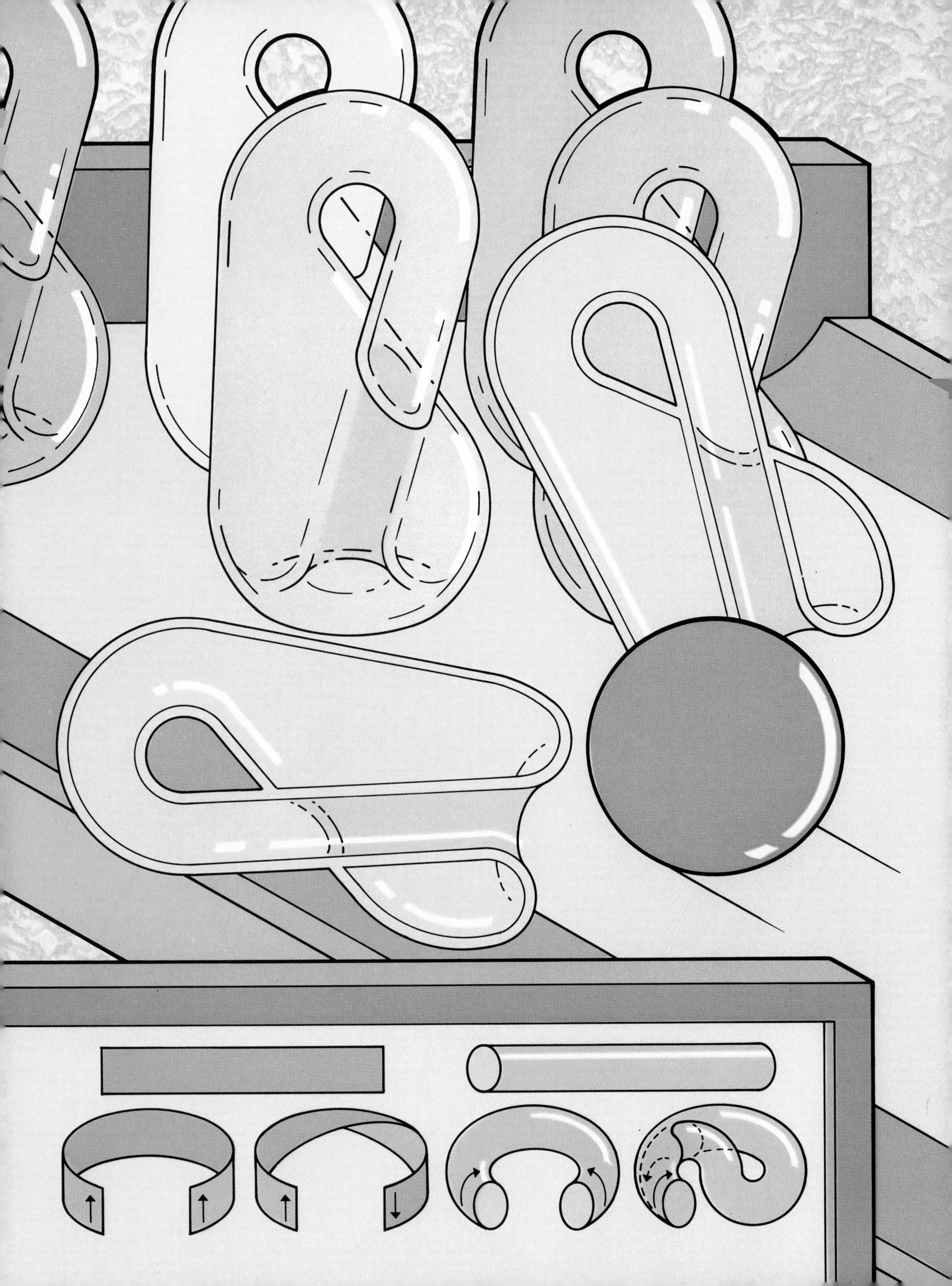

C
B
A
E
D
C
B
A
E
D

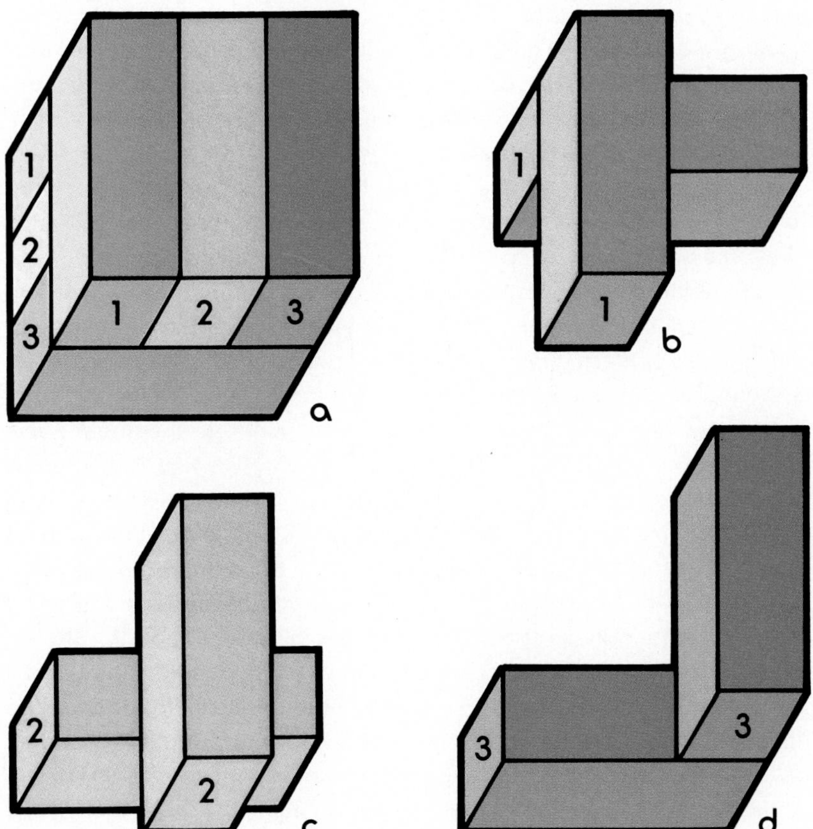

Figure 8. Five regions (opposite, top) cannot be laid out so that each shares part of its border with each of the others. Networks of five vertexes above the regions show that such an arrangement is not possible on a plane surface. A bridge connecting B and D (bottom) solves the problem; considered on a sphere, the bridge becomes a handle and the figure becomes topologically equivalent to a torus, on which a map containing five mutually neighboring regions can exist. In figure 9 (above) the neighboring regions problem is extended to three dimensions. By pairing bricks set crosswise to one another (b, c, and d), it can be demonstrated that the number of neighboring regions that can exist in space is infinitely large.

though one has been twisted beforehand. In order to connect the ends so that their directions are reversed, it is necessary to take one end and put it through the wall of the tube so that it can be attached from the "other side." Unfortunately, this procedure requires making a hole in the wall of the tube, which is what a glassblower must do in constructing a physical model of a Klein bottle. In theory, a true Klein bottle has no such hole; consequently, it exists only in the imagination. Even so, this does not prevent the topologist from studying its properties. Among them is the interesting fact that if a Klein bottle is cut in a certain way, it falls apart into two Möbius strips.

Neighboring regions

In a lecture given in 1840, Möbius presented a fundamental topological problem in the form of the following story. A prince with five sons specified in his will that his kingdom was to be divided into five regions so that each region would share part of its border with each of the others. Furthermore, the sons were to build roads connecting each pair of regions so that the roads would neither cross each other nor pass through the regions they do not connect. How should the kingdom be divided?

The network of roads in figure 8 (top) comes close to solving the problem. The only flaw in it is that the regions labeled B and D do not

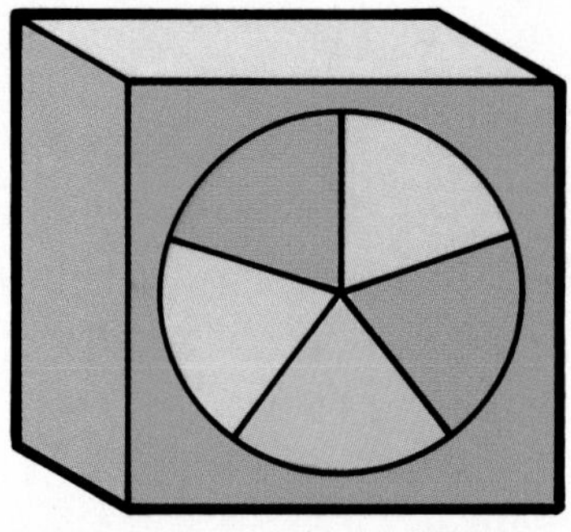

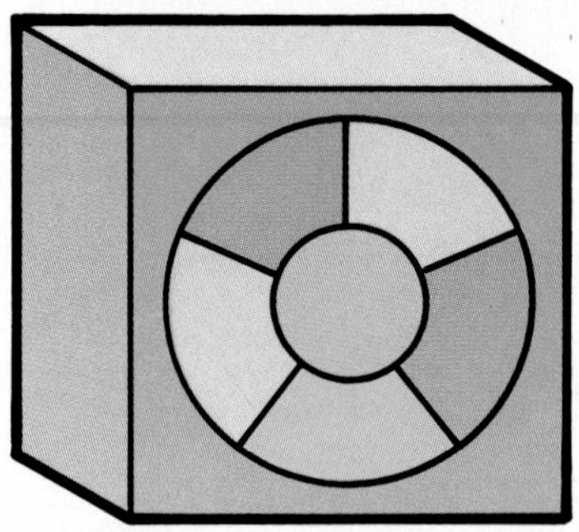

Figure 10. Maps on a plane surface cannot contain more than four mutually neighboring regions, but this does not prove that four colors are sufficient to color them. For example, the map at the top does not contain three mutually neighboring regions but does require three colors. The second map does not contain four mutually neighboring regions yet requires four colors.

share any portion of their borders, so that a road cannot be built connecting them. It is not difficult to prove that the problem has, in fact, no solution. The regions can be represented as vertexes of a network and the roads connecting them as arcs. A network of five vertexes in which each pair of vertexes is joined by an arc so that no arcs intersect cannot be drawn on a plane surface. The circular network at the top left has four vertexes and satisfies the given requirements; all such networks are equivalent to it. It is apparent that the network separates the plane into four regions: three neighboring regions within and the region outside. If the fifth vertex is put inside one of the three interior regions (top center), it cannot be joined to the vertex outside its region without crossing one of the three arcs bounding it. If the vertex is put outside the network (top right), it cannot be joined to the interior vertex without crossing an arc. Hence there is no network of five vertexes satisfying the requirements of the problem. Not only is a planar map which contains five mutually neighboring regions impossible, such a map on the surface of a sphere can be shown to be impossible as well.

If the map is not restricted to a plane or spherical surface, the problem has a solution. As the roadway network in figure 8 (bottom) shows, a bridge might be built connecting B and D, the two regions that previously did not share a border. Thinking of the prince's kingdom on the spherical Earth, the added bridge is equivalent to putting a "handle" on a sphere. This makes it topologically equivalent to a torus, as the sequence of figures above the network indicates. A map containing five mutually neighboring regions, then, can exist on a torus. In fact, a map having as many as seven neighboring regions can be constructed on a torus. On a Möbius strip, the maximum number is six.

The problem of neighboring regions can be extended from a surface to three-dimensional space. The maximum numbers of neighboring regions on a plane or spherical surface, Möbius strip, and torus, are four, six, and seven, respectively. What is the maximum number of neighboring regions that can exist in space? The answer to this question can be found by considering a structure made from rectangular bricks, as shown in figure 9. Six identical bricks are arranged in two layers; the three in the top layer are set crosswise with respect to the three on the bottom (a). Now suppose that the two bricks numbered 1 are glued together so that they form a single region (b), and that the bricks numbered 2 and 3 are paired in the same way (c and d). It is evident that each of the three resulting regions shares part of its surface with the other two. Since this sort of structure can be built from any number of pairs of bricks (assuming that they are made sufficiently long), the number of neighboring regions that can exist in space is infinitely large!

The problem of neighboring regions in a surface is often confused with the map-coloring problem discussed at the beginning of this article. The fact that a planar map cannot contain more than four mutually neighboring regions does not prove that four colors are sufficient to color it. Consider the two simple maps in figure 10. Although the first map does not contain three mutually neighboring regions, it requires three colors. The second map does not contain four mutually neighboring regions, yet four colors are necessary to color it. So it is certainly possible that although a map does not contain five neighboring regions, it might require five colors. It has been proved that the maximum numbers of neighboring regions that can exist on a Möbius strip and torus—six and seven, respectively—are indeed the numbers of colors sufficient to color any map that can be drawn on them. Why no one has succeeded in proving that four colors are sufficient for any map in a plane is one of the most puzzling mysteries of mathematics.

In conclusion

When the map-coloring problem was first proposed in the mid-19th century, topology did not yet exist as a distinct subject. Although its beginnings can be traced back even further, the major development of topology has taken place in the last few decades. Its influence on other areas of mathematics has been immense. As a unifying force in abstract mathematics, topology has been described as "the mathematics of the possible" because of its ability to determine whether or not certain conditions in other branches of mathematics are possible and whether or not solutions to certain problems within them exist. It has also proved to be a powerful tool for solving a wide variety of practical applications and, because of its apparent paradoxes and often astounding surprises, an abundant source of intellectual pleasure.

FOR ADDITIONAL READING

Barr, Stephen, *Experiments in Topology* (Crowell, 1964).

Cadwell, J. H., chap. 8, "The Four-Color Problem," in *Topics in Recreational Mathematics* (Cambridge, 1966).

Gardner, Martin, "Mathematical Games: Curious Topological Models," *Scientific American* (June 1957, pp. 166–172); "Recreational Topology," *Scientific American* (October 1958, p. 124 ff.); "The Four-color Map Theorem," *Scientific American* (September 1960, p. 218 ff.); "Topological Diversions [Klein Bottles and Other Surfaces]," *Scientific American* (July 1963, pp. 134–145); "The World of the Möbius Strip," *Scientific American* (December 1968, pp. 112–115).

Hilbert, David, and Cohn-Vossen, Stephan, chap. 6, "Topology," in *Geometry and the Imagination* (Chelsea, 1952).

Lietzmann, Walther, *Visual Topology* (Chatto & Windus, 1965).

Tietze, Heinrich, chap. 4, "On Neighboring Domains," and chap. 11, "The Four-Color Problem," in *Famous Problems of Mathematics* (Graylock Press, 1965).

Tucker, Albert W., and Bailey, Herbert S., Jr., "Topology," *Scientific American* (January 1950, pp. 18–24) and in *Mathematics in the Modern World*, Morris Kline, ed. (W. H. Freeman, 1968).

Stargazers of the Ancient World

by Gerald S. Hawkins

Ancient man knew the stars in their courses and built many of his structures to align with astronomical events. Recently rediscovered, these alignments reveal a hitherto unknown aspect of primitive culture.

The astronomical expertise developed by the ancient civilizations of Mesopotamia has long been known to archaeologists and historians. In recent years, however, surveys of ancient structures and calculations involving their alignments have demonstrated that a basic concern with what we narrowly call astronomy existed in most prehistoric cultures—in Europe, Asia, Africa, the New World, and the Pacific. In instance after instance, it has been found that the line of a wall or the position of two mounds pointed, at the time of construction, to the rising or setting of some heavenly body—Sun, Moon, planet, or star. Reports of several newly discovered alignments were published during 1974. These results are of importance not so much for the quantitative science involved as for the insight they provide into the mind of ancient man.

To the earthbound observer, the heavenly bodies appear to move in regular patterns, marking the seasons and longer-term cycles spanning many years. The alignments, therefore, show that prehistoric man was observing the motion of celestial objects to an extent hitherto unsuspected, and that he was cognizant of their periodicities. Despite the lack of written evidence, it is apparent that there was a well-developed "pre-science" in prehistory, and this basic intellectual involvement with the Earth-sky environment was an important factor in the early development of religion and philosophy.

Although the alignments would have had some practical value, as in the prediction of the seasons or putative divination and the prediction of future events, one must, in general, regard these efforts as nonutilitarian, a quantitative counterpart to the artistry of cave paintings. For example, the archways and stones at Stonehenge, in southern England, pointed at the time of their construction to the rising and setting of the Sun and Moon over a periodic cycle of 18.6 and possibly 56 years. Yet making these alignments did not demand the prodigious effort of cutting and hauling tons of stone, or the architectural marvel of the trilithon archways. Posts in the ground would have sufficed. The involvement of this particular culture with the remote patterns in the sky clearly transcended the utilitarian in a way the modern mind finds difficult to comprehend.

The Bighorn Medicine Wheel

As late as the 18th century, astronomical observations were being made by the American Indians in Wyoming to a much deeper extent than local oral tradition would indicate. This was revealed by a combined astronomical and archaeological survey of the stone-scribed Medicine Wheel, which stands on an exposed shoulder of the Bighorn Mountains. Located at longitude 107° 55′ W and latitude 44° 50′ N, at an altitude of 9,640 ft just above the timberline in the Bighorn National Forest, this structure consists of 28 lines of stones set out like the spokes of a wheel. A central cairn of stones about four meters in diameter forms the hub, and there is a flattened circle 25 m in diameter around the edge. Five large cairns and a sixth smaller cairn are placed just outside the rim.

The Indian term "medicine" can be translated more correctly as "magic," though even this does not convey the full burden of its meaning from prehistoric to modern times. In the traditions of the Crow tribe, the Bighorn Wheel was "the Sun's Tipi" and "it was built before the light came." Another Crow legend claims "the Sun built it to show us how to build a tepee." Superficially the wheel does resemble the floor plan of a 28-post, ceremonial Sun Dance lodge, but the legend might have been suggested during later visits by individuals who had no part in the design or operation of the structure. Certainly the site is not a practical place for a tepee or for an architectural exercise. It is uninhabitable and even as late as the summer solstice, about June 21, it can be struck by immobilizing snowstorms.

A portion of a tree limb was found embedded between stones at the base of cairn F. The tree rings matched those of other trees in the area, and by cross-referring the patterns it was deduced that growth stopped in A.D. 1760. This is, therefore, the earliest possible date for the placing of the stones in the cairn. It was broadly confirmed by potsherds, beads, and arrowpoints found sparsely scattered in the soil between the spokes and in the interiors of the tower-shaped cairns.

A survey published by J. A. Eddy in 1974 showed that the sight lines connecting the cairns pointed to the rising of the Sun and a bright star

GERALD S. HAWKINS, *formerly Astronomer at the Smithsonian Astrophysical Observatory, is currently a Senior Scientist at the Air Force Systems Command in Massachusetts. He was the discoverer of the fundamental astronomical alignments present at Stonehenge.*

(Overleaf) Photograph by Gerald Brimacomb from Black Star

Tony Howarth © Daily Telegraph Magazine from Woodfin Camp

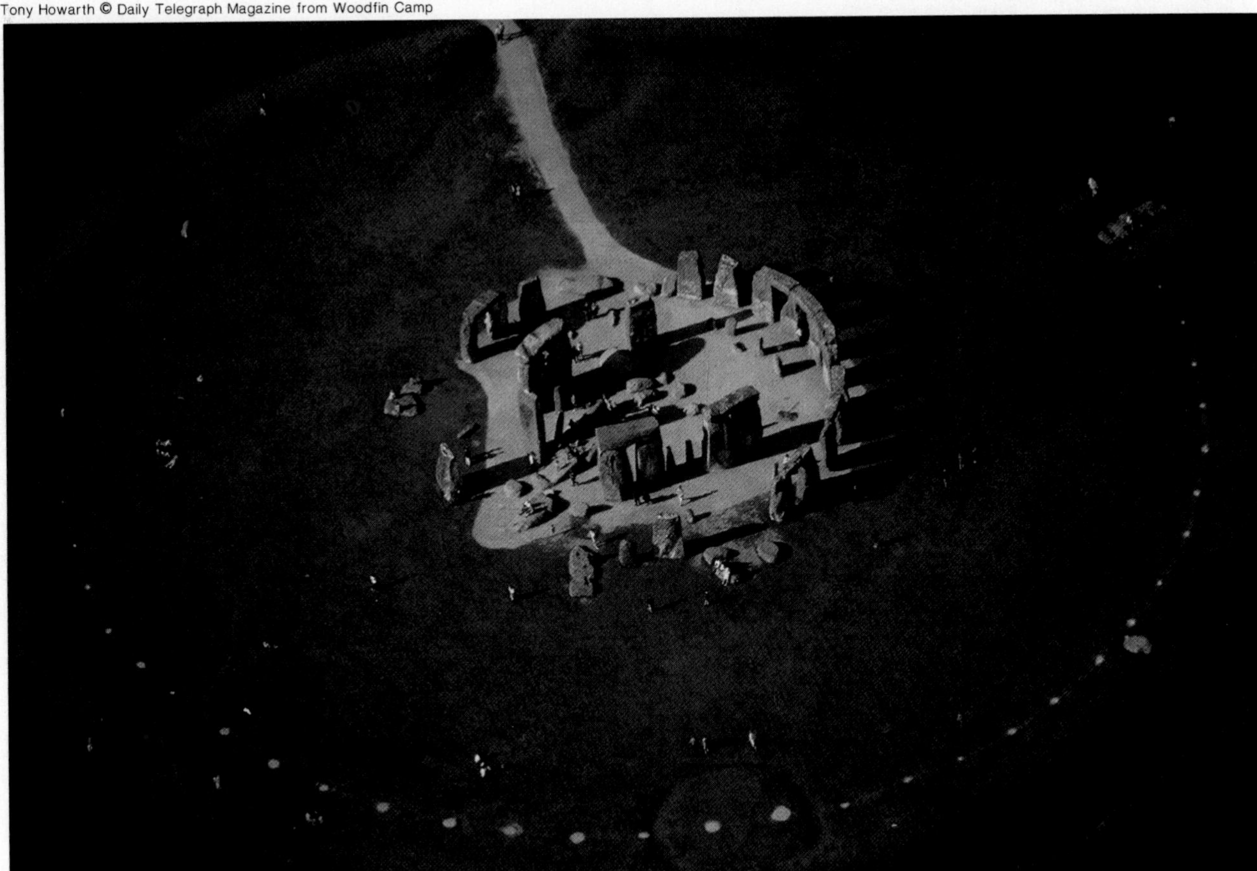

David Moore from Photo Researchers

George Gerster from Rapho/Photo Researchers

The astronomical alignments of Stonehenge, near Salisbury in southern England, bear mute witness to the skill of Neolithic man in observing and recording the cycles of the heavenly bodies.

From John A. Eddy, "Science," Vol. 184, p. 1037, June 7, 1974 © American Association for the Advancement of Science

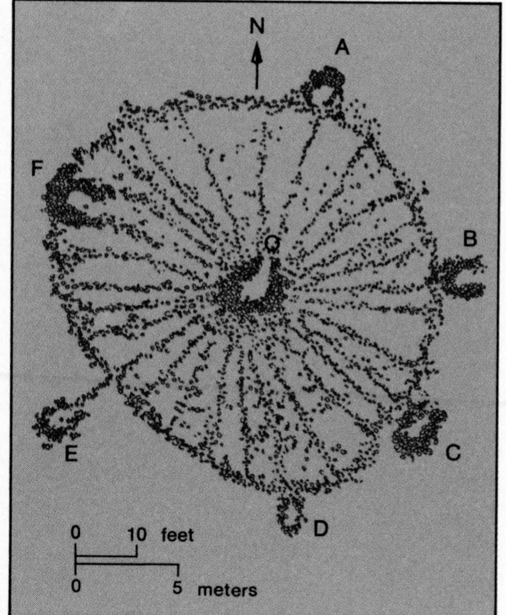

at the date of the summer solstice (midsummer) during A.D. 1500–1900, and of two other stars later in the season. An additional, less accurate alignment was to midsummer sunset. An observer standing at E would see the Sun rising over the central cairn, O, on June 21 and for a few days before and after. This marks the northernmost extrema of solar azimuth and declination (+23.4°). The error in the alignment is 0.3°, about one solar diameter, but the true amount of misalignment cannot be determined precisely because of the irregular nature of the stone piles and the fact that they may have been disturbed. At the central cairn there is a conical hole, cut into bedrock to a depth of one meter, which might originally have held a post that provided a more precise marker. The tree limb found at F indicates that wood was indeed jammed between the stones during construction.

A solstice marker might be of significance as a permanent record of the recognized pattern of solar movement, as with the Heel stone at Stonehenge, but it is not an accurate marker for calendric purposes. The movement in azimuth on the horizon from sunrise to sunrise is imperceptibly slow at the time of the solstice, and the precision can be no better than a week. The star alignments, on the other hand, were potentially accurate to within one day. The line from F to A marked declination + 16.6°, the rise of Aldebaran, a reddish star of magnitude 0.8 in the constellation Taurus. Between 1500 and 1900 this star, obscured by the glare of the Sun through the month of May, first appeared on the eastern horizon on June 21 as the Sun shifted to the left around the zodiac. On that day Aldebaran rose heliacally, that is to say, it was just visible, low on the skyline, as dawn filled the sky. No other first-magnitude star was visible with a dawn rising on that date. Thus, during a 400-year period in agreement with the most probable archaeological date, the slowly changing solstice over cairn O would be more precisely calibrated by the appearance of Aldebaran over A.

This star-Sun coincidence has lapsed owing to the gradual change in the Earth's axis of rotation, and Aldebaran, in 1975, did not appear at dawn until after June 23. There will not be a replacement star for several centuries to come, and there was no first-magnitude heliacal solstice star in the millennium before the building of the Bighorn Wheel. Thus the use of a bright star as a midsummer warning could not have been practiced before about 1500, and the method became obsolete around 1900.

Cairn B as seen from F marks the rise of Rigel, a bright blue-white star which is next to appear at dawn as the Sun proceeds eastward. It is heliacal in July. Sirius, at its first dawn appearance in August, is seen from F at O. By the end of August the weather begins to close in on the mountain, and Sirius would have provided a convenient end-of-the-summer marker. Alternatively, the lines might memorialize those times in previous millennia when Sirius, and later Rigel, had been solstice stars.

The spokes of the stone "wheel" are somewhat irregular and have no recognized astronomical alignments. It has been suggested that

Courtesy, U.S. Forest Service

The Bighorn Medicine Wheel in northern Wyoming (above). The cairns, identified by letters in the plan of the wheel (opposite page) were aligned with midsummer sunrise and the heliacal rising of three bright stars as they occurred between about A.D. 1500 and 1900.

they are cabalistic, representing a mystic number sacred to the Sun, or the days of the month. If the latter is correct, 28 is, more accurately, the number of nights when the Moon is visible; mean lunation is 29.53 days, and there is a period of one day when the Moon is invisible.

But beyond the surveys and calculations there remains the question of motive. The general area was inhabited by nomadic Plains Indians —Crow, Sioux, Arapaho, and others—who might have needed an approximate calendar to guide their migrations but who had no need of a precise agricultural or legal calendar. It is possible that the famous Sun Dance ceremony of the North American Indians, which took place when "the growing power of the world is strongest," was timed as precisely as possible to coincide with the solstice, but to modern minds such precision appears unnecessary or even esoteric. Nor is there a practical explanation for the choice of site. In a world still free of atmospheric pollution, a mountaintop would have given only a marginal increase in transparency. Further, in the Bighorn Mountains this slight advantage was more than offset by the high frequency of mountain storms. Perhaps the elevation was chosen as a secluded spiritual refuge or was the point of contact between the "medicine man" and the great sun-god of the sky.

The Newgrange tomb

Also published in 1974 was an analysis of the alignment at the Neolithic tomb at Newgrange, Ireland, some 30 mi NNW of Dublin. This structure is architecturally composite, the first construction having been started as early as 3100 B.C. The tomb is a narrow passage, 19 m long, lined with vertical slabs and roofed over with lintels and corbeled stones. The ground plan of the end of the passages is cruciform,

From "The Sacred Pipe: Black Elk's Account of the Seven Rites of the Oglala Sioux," J. E. Brown, ed. © 1953, University of Oklahoma Press

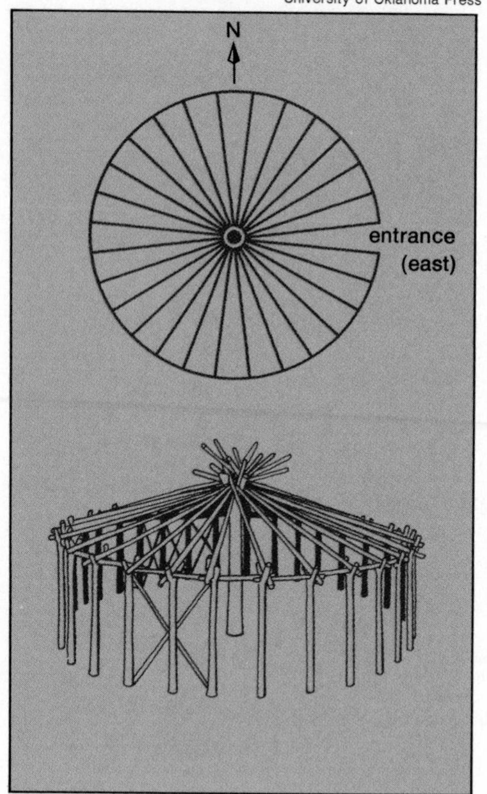

with an end chamber and two side chambers. Excavations in 1967 uncovered stone pendants, beads, and burnt bone fragments. The tomb is covered over by a huge cairn of stones and surrounded by a curb of stone slabs. It is believed that the curb was surmounted by a retaining wall of white quartz, 3 m high, which ran for 30 m on each side of the entrance. The cairn is set within a ring of standing stones which may predate the tomb structure.

Newgrange has a feature that is apparently unique. There is a roof box about 1 m wide and 25 cm high over the entrance lintel. Thus, even if the entrance were closed with a large slab, light could penetrate through the oblong window over the door. The ground on which the passage was built slopes upward, and the design is such that a line from the distant horizon passes through the roof box to the floor of the burial chamber and the back wall of the end chamber. The passage was built with a slight sinuosity, and the double curve somewhat restricts a beam of light entering the window. A survey by Jon Patrick shows that, taking the curvature into account, the most direct line through the passage runs at an azimuth of approximately 136° 03′. This meets the distant skyline (elevation 51′) at a solar declination of −24° 25′, or within 0.5° of the southern extremum of the Sun around the year 3000 B.C. Patrick concluded that the axis and open roof box were designed and engineered to permit the sunlight to enter the far tomb momentarily on the morning of the winter solstice (midwinter). The tolerance in the design was such that the phenomenon still occurs today, even though the declination is now −23° 24′.

Newgrange is also remarkable for its carvings. Many of the slabs are decorated over almost their entire surface with double spirals and lozenges. There are lozenges, and a cut presumably made for rainwater drainage, on the lintel of the roof box. The far wall in the deep recess of the tomb is decorated with three interlocking double spirals that wind clockwise and unwind counterclockwise. This motif, so far undeciphered, may have astronomical significance since it is illuminated briefly by the Sun at dawn at the winter solstice. The line of each spiral curls clockwise toward the center, which perhaps represents the lowest point of the solstice, and then reverses itself and unwinds. This could be a highly abstract representation of solar motion. There is no ready explanation for the triplication of the pattern, or for the four-sided lozenges. It may be that there is a unity in the tomb, the artwork, and the alignment that presently escapes us.

Stonehenge

In 1974 Richard Brinckerhoff, with the permission of the U.K. Department of the Environment, made an examination of the tops of the lintels at Stonehenge. Previous work by Gerald S. Hawkins had established an almost complete pattern of ground-level alignments to the four extrema of the solstice Sun and the eight of the Moon. Cup-mark depressions found on the lintels confirmed the ground-level alignments and indicated an attempt toward greater exactness on the part

Photos, Nicholas deVore III from Bruce Coleman Inc.

The superficial resemblance between the Bighorn Medicine Wheel and the medicine lodge of the Plains Indians, apparent when the plan of the wheel is compared with that of the Sun Dance lodge of the Oglala Sioux (opposite page), has led some authorities to associate the wheel with the Sun Dance ceremony performed at midsummer. A similar connection occurs in a Crow legend in which the Sun built the wheel to show the Crow how to make their tepees (above). Neither explanation accounts for the remote site of the wheel or its astronomical relationships.

of the builders. There are two cup depressions on the lintel of the central sarsen arch which make a line toward the midsummer sunrise. A deeper cup, found on the next lintel to the northwest, may mark the extreme northerly rising point of the winter full Moon. Unfortunately only 9 of the original 35 lintels are now in position. Patrick suggests that, although they were separated by almost a millennium, there may have been a transfer of interest and information between the Newgrange and Stonehenge cultures.

Postholes some four feet in diameter were found by the Department of the Environment during excavation for a parking lot extension at Stonehenge. Tree bark at their bottom edges indicated that they had originally held large, crude posts. Calculations made in 1972 showed that these holes aligned precisely with the extrema of the Moon when viewed from station stone 94. It would seem that astronomical observations and recording were in progress at the earliest phase of Stonehenge, which recent radiocarbon dating on the revised scale places around 2500 B.C.

The temples of Egypt

Ancient Egypt, which possessed hieroglyphic writing, cannot be classified as prehistoric, but it is generally conceded that the inscriptions and papyrus documents do not give an exact and complete description of the state of knowledge during the pharaonic period and that further information may be obtained from architectural studies. For example, the base of the Great Pyramid at Giza was aligned to the four points of the compass within about three minutes of arc, which is the limit of accuracy of the unaided eye. This is of both geographic and astronomical importance. The rectangular grid aligns to the geographic meridians and parallels, and the pairs of sides point to the prime vertical, the great circle intersecting the celestial poles and the

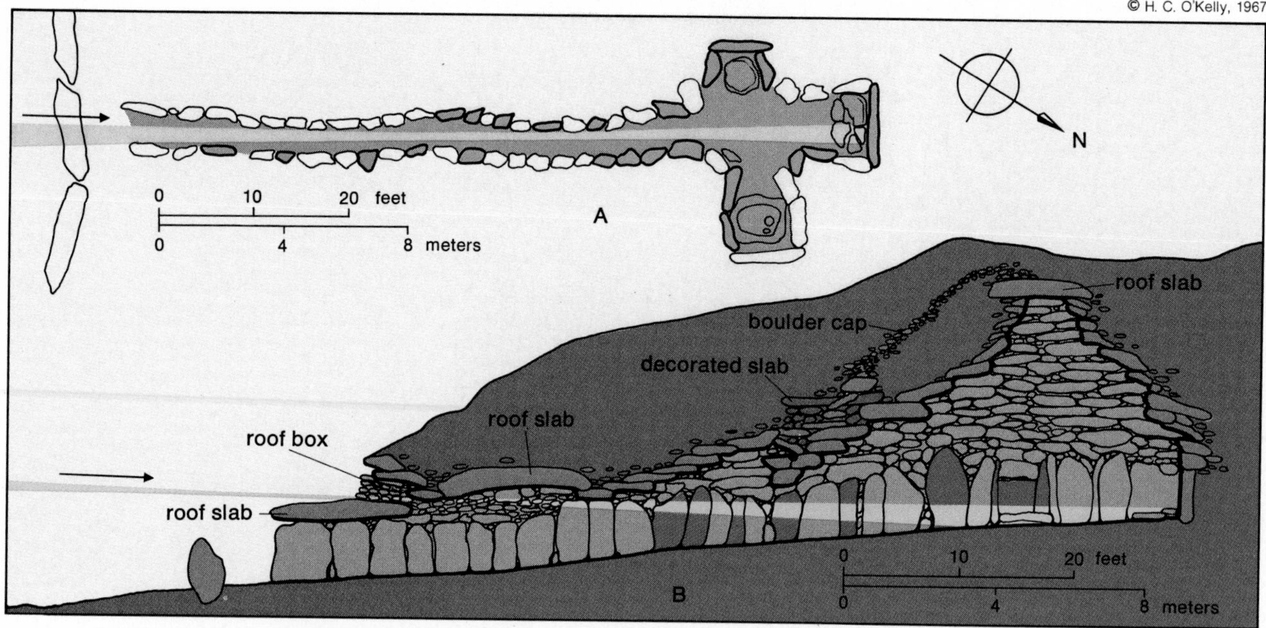

On midwinter morning the Sun's rays, indicated by the yellow lines, enter the roof box above the entrance of the Neolithic tomb at Newgrange, Ireland, and strike the burial chamber at the far end. The bird's-eye view (A) reveals the slight double curve of the passage. Decorated stones are shown in blue. The upward slope of the floor is apparent in the cross section (B).

zenith, and the equinoctial colure, the great circle at right angles to it. It seems apparent that this alignment must have been obtained by astronomical observation of the highest precision, either of stars on the horizon or of solstitial rising and setting. Yet there is no written description of any such methodology, rationale, or even plain recognition of the fact that observations of this type were carried on.

The Sphinx conforms to the grid of the Giza pyramids, and the axis is along the east-west parallel. Thus this sun-god effigy, in the likeness of the pharaoh Khafre, was accurately sculptured from the living rock to point to the vernal equinox sunrise. This particular azimuth was presumably of importance to the culture, but for reasons that are still unknown.

Similarly, the Nile temples were built to conform to various rectangular grid systems. The outer walls are oblong, and there are side chapels which are parallel or orthogonal to the main axis. Often a secondary temple was erected in the vicinity to a god subordinate or related to the main god, and this structure conforms to the grid to within a fraction of a degree. These Nile temple alignments have been tested for astronomical significance with positive results.

Hawkins in 1974 showed that the Great Temple of Amon-Ra at Karnak was aligned to the midwinter sunrise during the epoch of Thutmose III, around 1480 B.C. The alignment for the Sun's disk tangent on the skyline was at least equal in precision to the modern survey, which was carried out to an accuracy of $\pm 0.1°$. This temple was extensively rebuilt and extended by Thutmose III on the site of previous Middle Kingdom structures. Other temples of the citadel, such as those of Mont and Amon-who-hears-the-prayers, and the structures of Seti II and Ramses III, are built to the same grid and therefore share the principal astronomical alignment, though this may have no greater

Green Studio Ltd., Dublin—EB Inc.

The roof box above the lintel of the Newgrange tomb. Covering the slab in front of the entrance are the distinctive double spirals and lozenges used to decorate many of the stones within the structure.

significance than the fact that Eighth Avenue in New York City is parallel to Fifth Avenue.

One small chapel on the northeast corner of the Great Temple, now known as the High Room of the Sun, is significant, however. It is a roof temple dedicated to Ra-Hor-akhty, the sun-god on the horizon. The chapel contains an alabaster altar, and there are remains of what appears to have been a window in the east wall. Most probably, observations of the direction of sunrise were made routinely from this position, and the approach of the winter solstice was noted. The roof temple surmounts the long outer wall of the Thutmose temple, which carries many reliefs showing the temple's surveying, dedication, and building. It has been argued that the pylons of the temple deviate from a straight line by about 3 ft in the 1,860-ft length of the axis, and that the builders were not precise or astro-oriented in intent. However, these pylons were added by later dynasties; the Thutmose wall is essentially straight.

Near the temple of Amon-Ra is the temple of Khons, the moon-god. This was built to a rectangular grid rotated by 1.9° to that of Amon-Ra, and subsidiary chapels also fit this grid. Calculations show that the transverse axis was aligned to a lunar declination of +27.4° at its intersection with the distant hills of Thebes, the northernmost extremum of the setting of the new-Moon crescent at the time of the summer solstice. This crescent phase was of calendric importance in ancient Egypt, particularly at the month following the solstice, and it is not surprising to find the moon-god's temple so oriented. The hieroglyphs do not state this explicitly, but there are indirect references on a window of a chapel on the roof through which the crescent would have been visible. Significantly, Khons is represented as a horizontal crescent similar to the new Moon as it sets on the western horizon.

(Below) Brian Brake from Rapho/Photo Researchers; (opposite) FPG

Though there is little in the written record that casts light on the state of astronomical knowledge in ancient Egypt, the alignments of Egyptian structures indicate a high degree of interest and comprehension. This is evident even in the pyramids of Giza (shown, above, at sunrise), built during the Old Kingdom (c. 2686–c. 2160 B.C.). The diagram (opposite page) indicates the highly sophisticated alignments of the temple complex at Karnak, constructed during the New Kingdom (1567–1085 B.C.). Like the Great Temple of Amon-Ra at Karnak, the Colossi of Memnon (upper right), which once flanked the mortuary temple of the New Kingdom pharaoh Amenhotep III in western Thebes, face the midwinter sunrise.

South of the temple of Khons, reached by an avenue of sphinxes, is a complex of three or more structures dedicated to Mut (the wife of Amon-Ra), Khons, and Amon. This complex was rotated by 9.9° to the Khons grid—an amount which picks up the southern extremum of the setting of the summer crescent Moon. Within the Mut complex, with its major axis directed toward the crescent, is a temple dedicated to Khons the child. Thus it would seem that the 18.6-year cycle of movement of the Moon along the outline of the hills of Thebes is delineated by the orientation of temple structures. Possibly this astronomical fact may be of value in interpreting some of the recorded myths. It is also of interest to note that the southern end of the moonset cycle as seen from the Mut temple is over the Valley of the Tombs of the Queens while the northern end as seen from the Khons temple is over the Valley of the Tombs of the Kings behind the ridge.

Other recent surveys in Egypt have confirmed and extended the astro-orientation discoveries. (*See* table.) The Colossi of Memnon, the pair of statues built by Amenhotep III in western Thebes, face the direction of winter solstice sunrise, as does the Karnak temple on the opposite bank of the Nile. The Colossi originally flanked the entrance of a large temple which was similarly oriented.

The main temple at Abu Simbel, cut 185 ft into a solid sandstone cliff on the west bank of the Nile, was aligned to the sunrise on I Peret 1, a critical day in the Egyptian civil calendar, during the epoch of Ramses II, *c.* 1304–1237 B.C. Possibly the entire structure was designed to permit sunlight to penetrate to the effigy of Ramses II on that date, when the pharaoh celebrated one of his jubilees and supposedly was reborn. A side chapel at the northern entrance to the temple was rotated 15° to the main structure. At azimuth 116° it pointed, at the time of building, to the position of midwinter sunrise, paralleling the celestial orientation of the Karnak complex and the

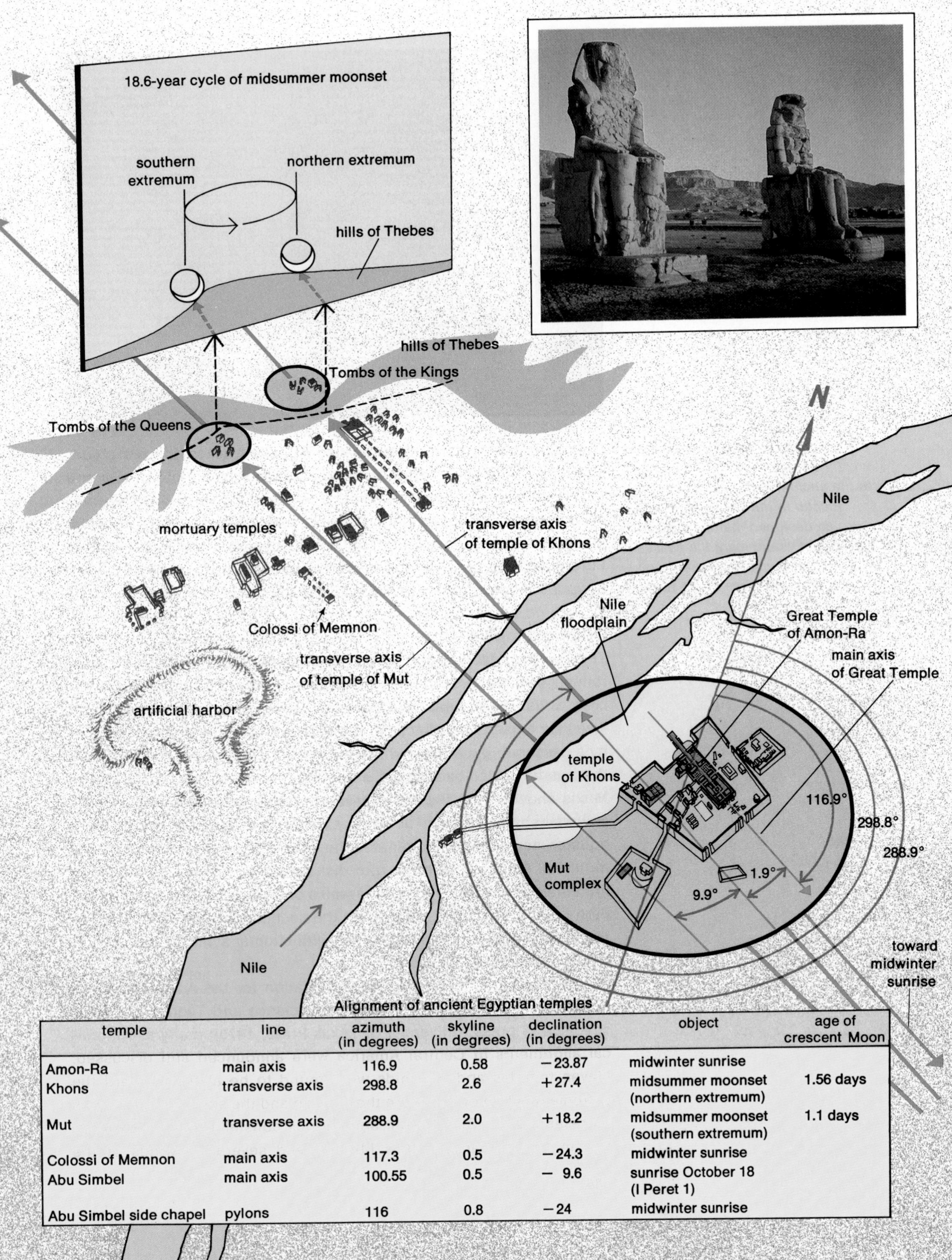

Alignment of ancient Egyptian temples

temple	line	azimuth (in degrees)	skyline (in degrees)	declination (in degrees)	object	age of crescent Moon
Amon-Ra	main axis	116.9	0.58	−23.87	midwinter sunrise	
Khons	transverse axis	298.8	2.6	+27.4	midsummer moonset (northern extremum)	1.56 days
Mut	transverse axis	288.9	2.0	+18.2	midsummer moonset (southern extremum)	1.1 days
Colossi of Memnon	main axis	117.3	0.5	−24.3	midwinter sunrise	
Abu Simbel	main axis	100.55	0.5	− 9.6	sunrise October 18 (I Peret 1)	
Abu Simbel side chapel	pylons	116	0.8	−24	midwinter sunrise	

Language of the Bees

by Donald R. Griffin

Honeybees communicate with one another by means of intricate "dances" on the surface of their honeycombs.

Adapted from Karl von Frisch, "The Dance Language and Orientation of Bees" (1967), The Belknap Press of Harvard University Press

When the feeding dish was within a relatively short distance of the hive, only round dances were observed, but as it was moved farther and farther away the round dances changed progressively to waggle dances. The first modification was a very short series of waggles interposed between clockwise and counterclockwise circles of a round dance. When the distance was somewhat greater, the waggles were extended into short straight runs, and when it reached several hundred yards the straight run with vigorous lateral movement of the abdomen became quite prominent. Frisch then realized that waggle dances were used when the food source was at a considerable distance, regardless of the nature of the food itself.

Further study showed that the tempo of the dance slowed as the distance to the food increased. In the particular strain of bees used in these experiments, the waggle dances first become clearly distinct at about 100 yd, and each figure 8-shaped cycle lasts about 1.25 sec. When the food is about 1,000 yd away, each dance cycle lasts about three seconds. On rare occasions these bees have been induced to gather food from sources as far away as five or six miles, and in these cases each dance lasts about eight seconds.

Furthermore, Frisch discovered, the dances indicate not only distance but direction. When bees gather food from the same source over a period of several hours, the direction of the straight portion of the waggle run changes gradually. Between dawn and dusk it shifts about 180°. Waggle dances are carried out almost exclusively in clear weather, and their direction is related to the position of the Sun. When the food is located directly toward the Sun, the waggle run points straight upward over the vertical surface of the honeycomb. If food is roughly 90° to the right of the Sun, the waggle run points about 90° to the right of the vertical. At other positions and at different times of day, the same relationship holds. Thus the "code" of the bee dances can be stated as follows: The waggle run points at the same angle relative to the vertical as the food is located relative to the Sun.

To enable observers to follow the activities of individual bees, Frisch devised a method of color-coding his bees with droplets of quick-drying paint (opposite page). A white, red, blue, yellow, or green spot on the anterior thoracic margin means 1, 2, 3, 4, or 5, respectively, whereas the same colors on the posterior thoracic margin indicate 6, 7, 8, 9, or 0. Double figures are signified by two spots on the thorax and hundreds by an additional spot on the abdomen. With variations, bees can be numbered in this way up to several thousand. Glass windows in the observation hive (above) permit the observer to watch the dances performed by bees after they return from foraging.

Can insects really do this?

The systematic relationship between the location of a food source and the dance pattern was a most surprising discovery. As Frisch himself put it, "No competent scientist ought to believe these things on first hearing." During the 20th century biologists and psychologists have become extremely cautious in interpreting the behavior of animals. For excellent reasons, they are extraordinarily reluctant to ascribe even to the higher animals anything comparable to human language or even simple signaling. Thus, when these honeybee dances were first discovered, all scientists, including Frisch, were both excited and, at the same time, anxious to ascertain as accurately as possible what information the dances actually conveyed.

The first step was to analyze carefully the correlation between dance pattern and food location. Frisch found that, while this correlation was not absolutely perfect, he could deduce where the honeybees were gathering food by watching the dances, even when the food source had been set out by another experimenter. The directional indication appeared to be accurate within about 2°–3°, and the distance of the food source from the hive could be determined by an experienced human observer to within roughly ±10%.

With such an important point at issue as the discovery of a complex signaling system in an insect, it was of great scientific importance to ascertain whether the information available in the dances was actually used. To study this question, Frisch conducted two types of experiments in which several food dishes, as nearly identical as possible, were set out and observers watched to see whether bees came selectively to some but not to others. In what he called fan experiments, an unscented feeding dish was set out and bees were led to it with a sugar solution diluted to prevent them from dancing at the hive and causing a premature arousal. Other dishes without food but scented were positioned in a fanlike array on either side of the original feeding dish, usually 15° apart but slightly nearer to the hive. Then the feeding dish was refilled with a much sweeter, scented solution and the bees

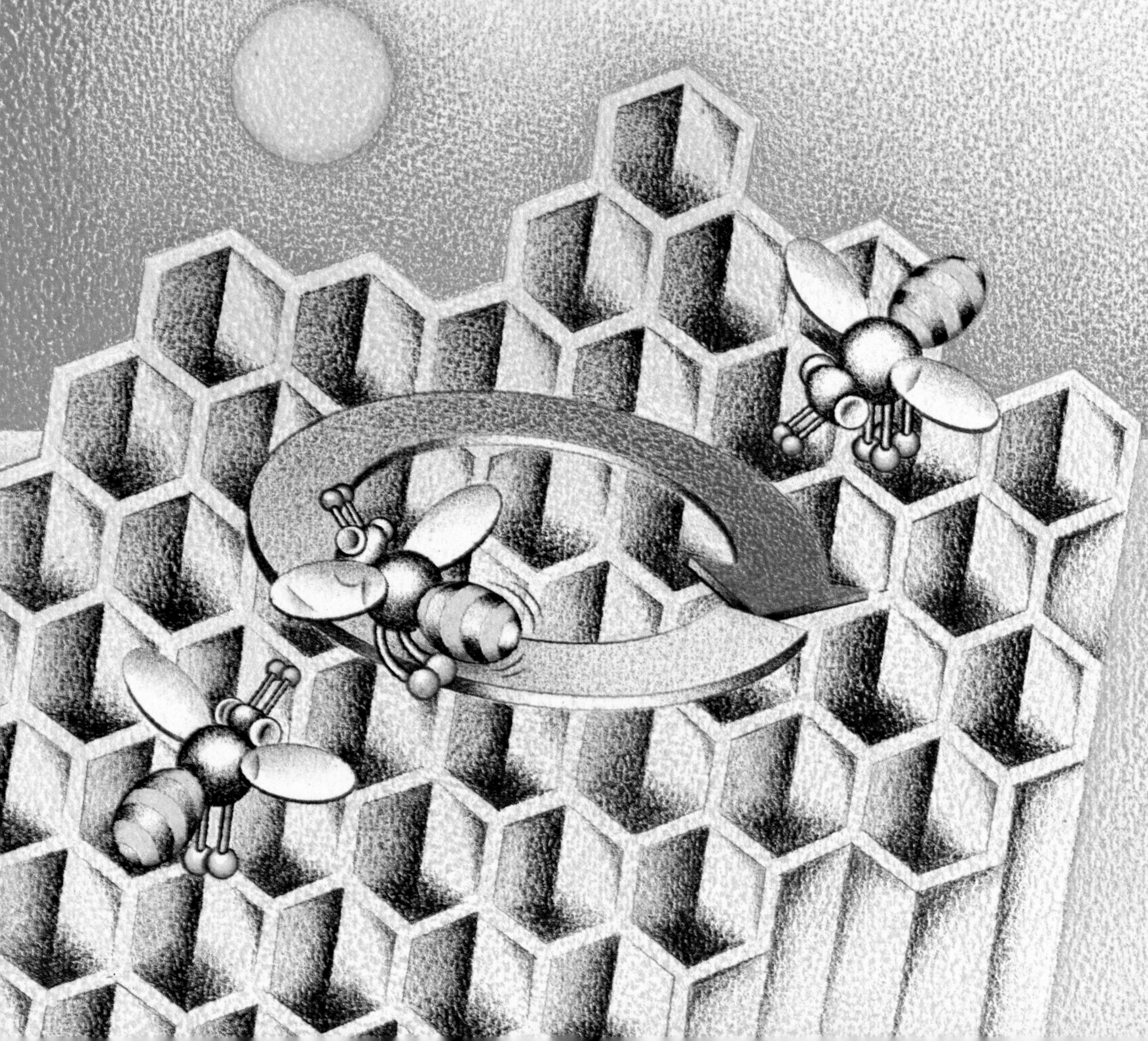

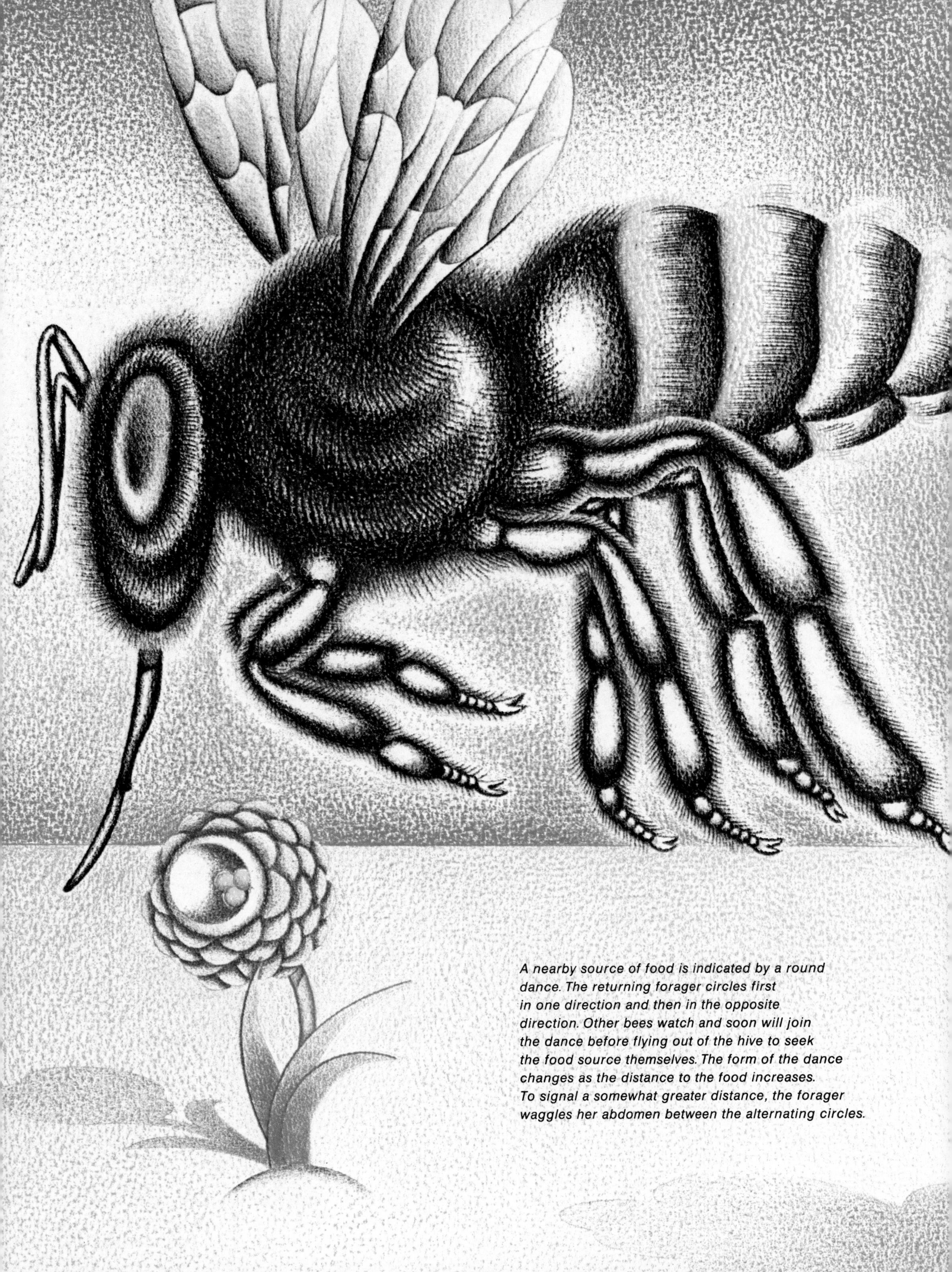

A nearby source of food is indicated by a round dance. The returning forager circles first in one direction and then in the opposite direction. Other bees watch and soon will join the dance before flying out of the hive to seek the food source themselves. The form of the dance changes as the distance to the food increases. To signal a somewhat greater distance, the forager waggles her abdomen between the alternating circles.

Worker bee returning from a distant food source signals not merely the existence of food but also its distance and direction from the hive. Now the dancer traces clockwise and counterclockwise circles, forming a figure 8, and between circles she waggles her abdomen vigorously and makes a straight run over the surface of the honeycomb. The dance tempo slows in proportion to the distance to the food source, while the straight run has the same relation to the vertical as the location of the food source does to the Sun.

at the dish were allowed to return to the hive. During the next hour or so bees were seen to come in considerable numbers to test feeders near the central position where food had first been available, while only a few went to dishes at the outer edge of the arc.

In the second type of experiment, Frisch used the same basic plan, but in this case the test feeders were all aligned in the same direction, with some closer to the hive than the original food source and others farther away. Again, when the original source was enriched and scented, observers counted many more bees arriving at nearby dishes than at identical feeders that were closer or more distant. This held true even when the distance from the hive to the original source was as great as three miles. It seemed clear that information obtained from the waggle dances brought the bees close to the correct location, although locating the actual feeding dish (or, under natural conditions, a particular group of flowers) involved searching for the appropriate odor.

These experiments have been criticized by Adrian Wenner of the University of California, Santa Barbara, who pointed out that bees can

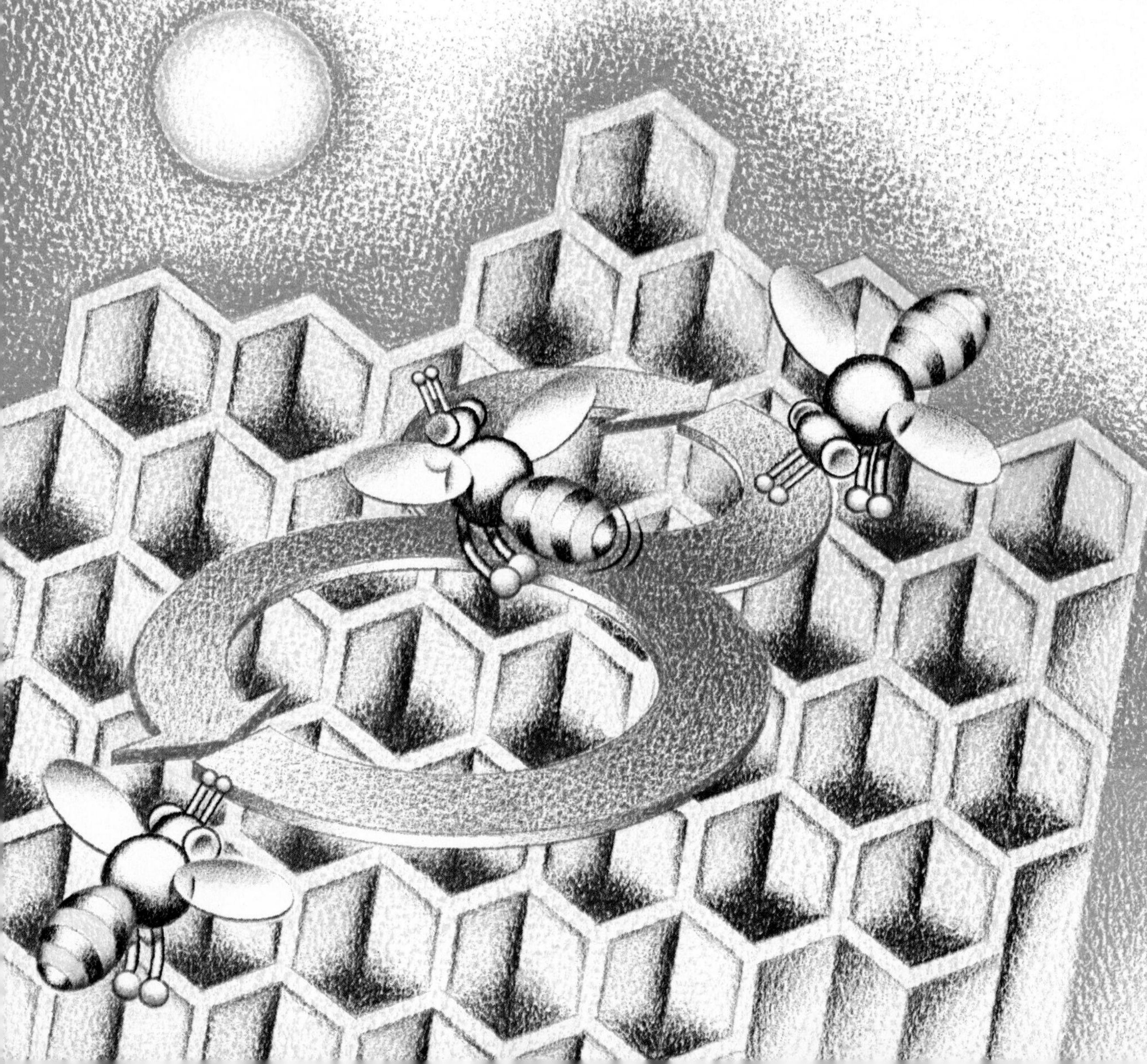

find food sources whose location has not been predicted by the dances. Therefore, he and several colleagues have argued that the correlation between food location and the dances is accidental, a by-product of no biological significance. Few biologists accept Wenner's interpretation, but his criticisms stimulated several of them to design new and more rigorous experiments.

The question of whether bees can indeed find a food source they have never visited before solely by means of information conveyed through the dances is not as simple as it originally had seemed. Wenner noted that not only the scents of flowers but other odors in the vicinity of a food source might well be transferred to other bees by the vigorous motions of the dance. Furthermore, bees often mark food sources by emitting scented secretions from specialized glands. Odors from the paints used to mark the bees or even from the experimenters or observers might have guided bees to the correct location. No matter how careful the controls in experiments of this type, it could always be argued that the returning forager brought back odors characteristic of the site of the food source, and that the bees stimulated by the dance simply searched for those odors.

Recently, James Gould of the Rockefeller University developed a new and improved type of experimental procedure free from the problems noted in Wenner's criticisms. To understand Gould's experiments, it is necessary to appreciate certain details of the bee dances and certain ways in which they can be influenced experimentally. As Frisch showed many years ago, if an observation hive is located inside a dark chamber, a bright light bulb placed a few feet from the window of the observation hive will often be accepted by the bees as the Sun. This behavior is doubtlessly related to the fact that under special circumstances, honeybees perform waggle dances on a horizontal surface at the hive entrance, pointing their waggle runs directly toward the food and making an angle with the Sun appropriate for a bee flying directly to the food. With an artificial light replacing the Sun, the dance is oriented at the same angle relative to the light as it would normally be to the vertical. Thus, if the food is located directly toward the Sun, both dancer and attending bees point their waggle runs toward the artificial light.

While the bee uses its large compound eyes to determine the Sun's direction, smaller and much simpler eyes called ocelli inform it about the general brightness of the surrounding illumination. An "artificial sun" used in these experiments does not reorient dances unless it is quite bright. If it is dim, the bees still dance relative to the vertical. Gould took advantage of this. While bees were gathering food at a test feeder, he covered their ocelli with opaque paint. On returning to the hive, they did not sense a light bright enough to cause their dances to be deflected. But the bees clustering around had normal ocelli, and the artificial light caused them to act differently. While the dancers oriented their dances relative to gravity, information received by the other bees from these dances was interpreted relative to the light.

Gould took great care to provide no information to the bees except through the dances of the foragers whose ocelli had been covered. He set out identical feeding stations in various directions and used an automatic recording system instead of human observers to count the number of bees arriving at each. Bees were anesthetized as they came to the test feeders so they could not return to the hive.

Almost every bee stimulated by the experimentally altered dances flew not to the feeder where the dancer had collected food but to the place indicated by the dances. Thus these experiments established more conclusively than ever before that bees really do communicate to one another the distance and direction of a food source.

One striking aspect of the honeybee waggle dances indicates both their appropriateness for communicating distances and the degree to which they are related to events somewhat removed in time from the dance itself. When bees must fly against the wind to reach a food source, the resulting dances indicate a correspondingly greater distance. If the bees have a tailwind when they are flying to the food, a shorter distance is indicated. When there are crosswinds, the duration of the waggle run is approximately proportional to the amount of energy a bee must use to reach the food. It is especially noteworthy that this dance pattern corresponds to a flight from the hive the dancer made some time previously, not to the flight back from a food source which she has just completed.

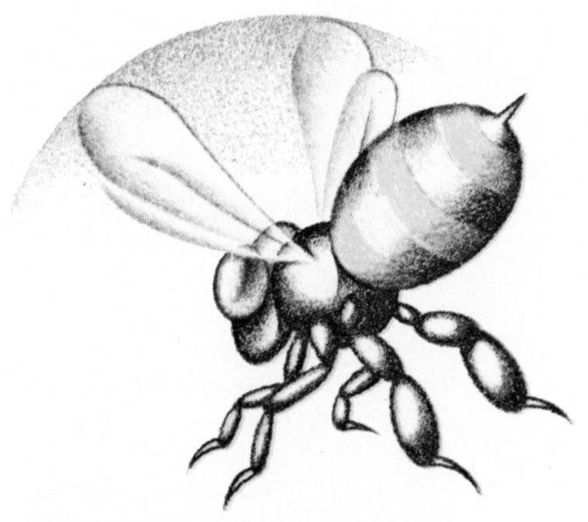

Bee signals the presence of danger by emitting an alarm substance and fanning its odor toward her hivemates (above). Similarly, bees evert their scent organs and fan the odor backward to assist inexperienced bees to find the hive or to guide their hivemates to a new dwelling during swarming. Fanning is also used to circulate air, thus regulating the temperature in the hive in hot weather. Specialized hairs on the hind leg form the pollen basket (below), in which workers transport pollen.

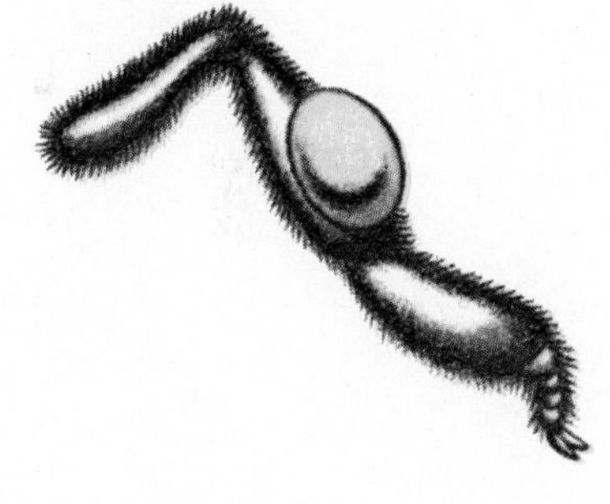

Similar behavior in other insects

The familiar honeybee (*Apis mellifera*) is one of the most specialized of social insects. The colonies are large, with workers numbering up to 50,000 or more, and there is elaborate division of labor among various groups of workers, differentiated primarily on the basis of age. No other insect—indeed, no other animal to the best of our present knowledge—normally uses as complex and versatile a symbolic communication system as the honeybee. Closely related species of bees have similar but simpler communication systems. For example, the dwarf honeybee *Apis florea* has a dance system akin to that of the honeybee, except that the waggle dances are performed only on horizontal surfaces rather than on vertical surfaces inside the dark hive. Thus these bees lack the ability to transform a directional reference based on the Sun's position to a reference based on the vertical.

Other bees have dances somewhat like the round dances of honeybees, but these appear to do nothing more than convey a general arousal and transfer odors of newly found food sources. For example, some species of *Trigona*, a genus of South American stingless bees, form into relatively small colonies of several dozen to a few hundred workers together with a single queen. Scouts from these colonies often locate new food sources at distances of a few hundred yards, and on returning to the colony stimulate recruits by running excitedly about and jostling them. But no information about direction or distance is included in the "dance," and the bees rely entirely upon

The dermal scent gland, which emits sexual attractant in many female insects, serves a social purpose in the sterile worker bee. The abdominal section (a) shows the scent gland (2) everted; in (b) it is in the rest position. The queen (c) lacks the scent gland but other glands (1) produce an odor that is noticeable during nuptial flight. (II–VII) are dorsal abdominal plates. The large compound eye (d) is used to determine the Sun's direction. (Opposite page) Segment of the antenna of a worker bee (top) and cross sections of the antenna's sensory organs.

From Karl von Frisch, "The Dance Language and Orientation of Bees" (1967), The Belknap Press of Harvard University Press

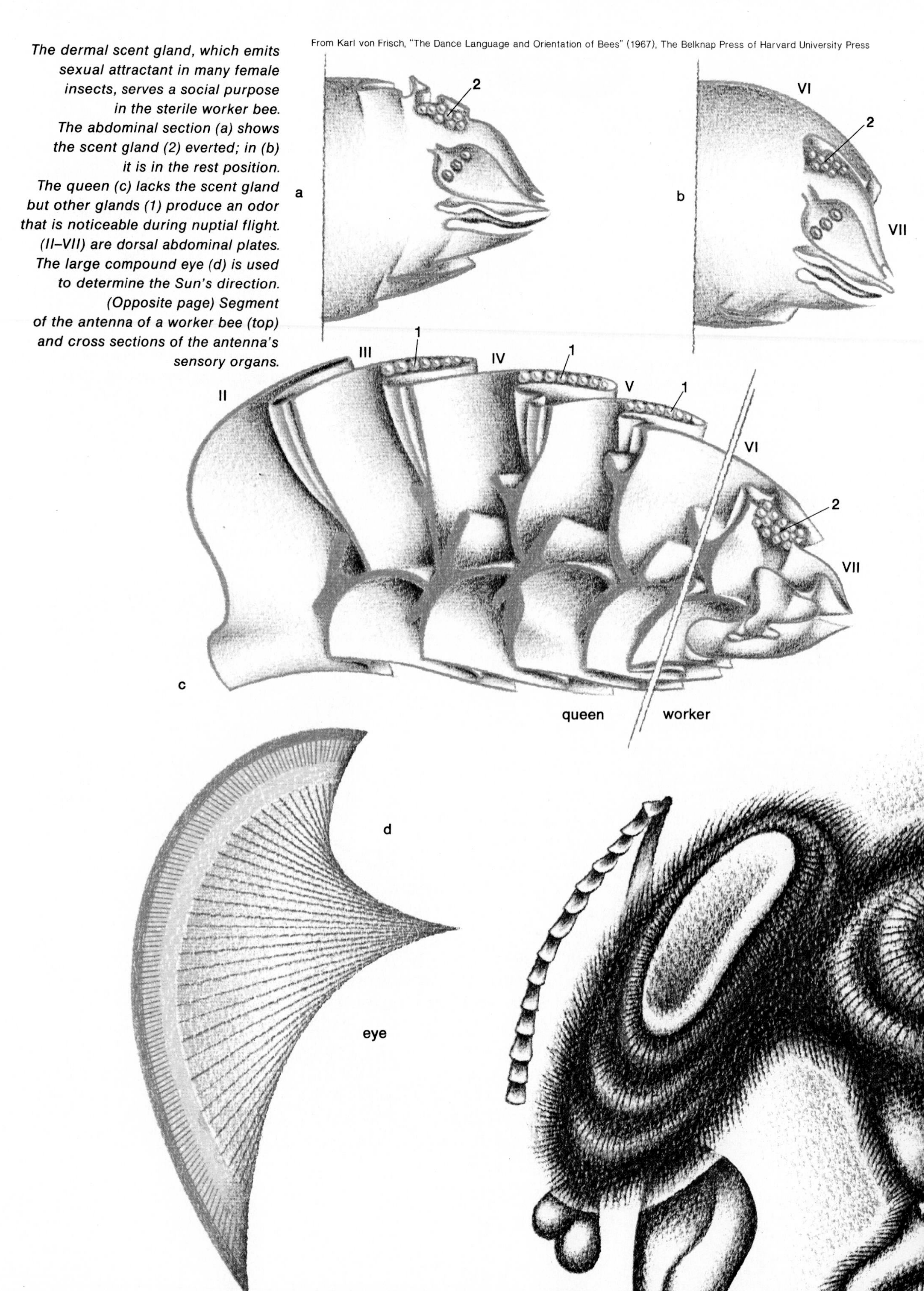

odors. As with honeybees, the characteristic odor of the flowers is conveyed mechanically or by regurgitation. In addition, special glands are used to deposit scent marks along the ground every few yards all the way from the food source back to the colony, and newly aroused recruits use these as guides.

Other insects, only distantly related to bees, display some elements of the honeybee communication system. Many ants leave scent trails by which other members of the colony find their way to new food sources. Blowflies (genus *Phormia*) are nonsocial insects, but when many hungry flies are confined in a small space, a fly that has located a newly provided drop of sugar solution carries out circular or elliptical motions and transfers food by regurgitation to other flies. This behavior stimulates previously inactive flies to search for food. The searching, however, does not appear to be directed.

The lateral abdominal movements of honeybees during the waggle dance are similar in some ways to side-to-side rocking motions performed by certain moths after they have completed moderately long flights. Furthermore, the longer the flight just completed, the longer the lateral rocking continues. These motions are not performed in the vicinity of other moths and have no communicatory function, but they do show that lateral motions proportional in duration to the length of an immediately preceding flight occur in a distantly related group of insects. Another parallel to one part of the honeybee dance pattern can sometimes be observed in the agitated running of ants when they are exposed to a bright light. At such times, the ants tend to move toward or away from the light in preference to other directions. If the light is turned off and the horizontal surface on which the ants have been moving is tilted, they often continue similar motions but with the movements oriented up and down over the inclined surface. This is analogous in some ways to the transformation of honeybee dances from toward the Sun on a horizontal surface to upward on a vertical surface. Nevertheless, although these basic behavioral elements can be found in other insects, only in bees of the genus *Apis* are they combined into a symbolic communication system.

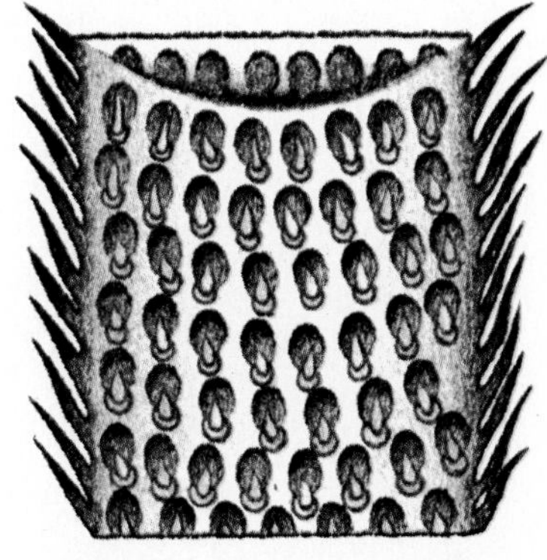

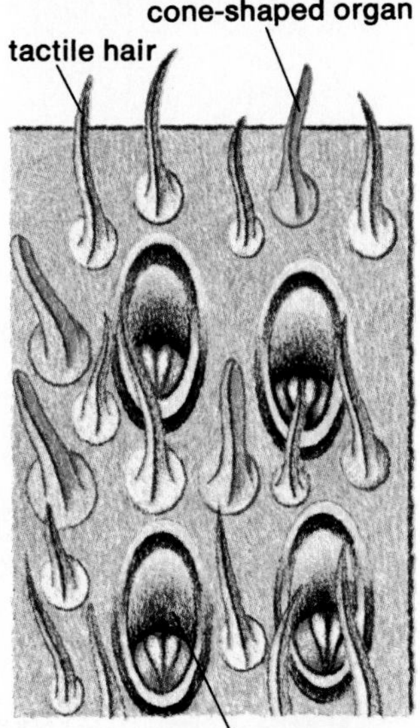

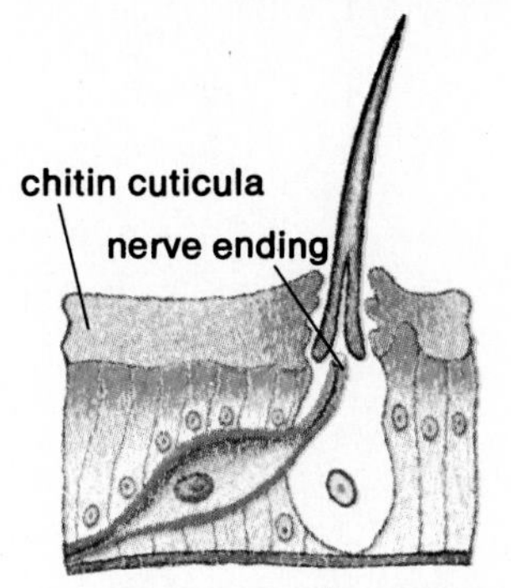

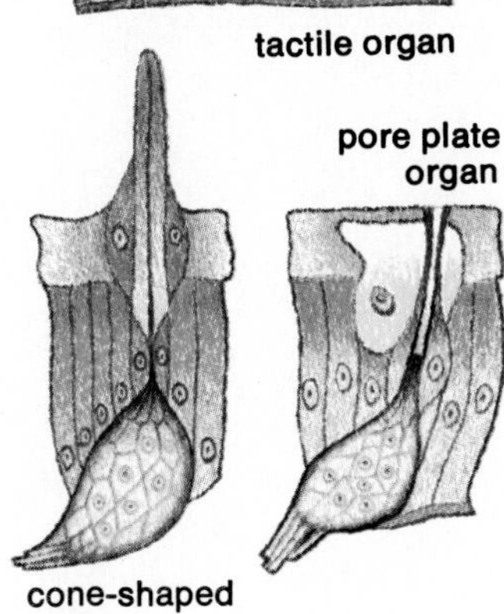

Use of the dances for other purposes

The round and waggle dances are often thought to be rigidly linked to food gathering, because they were first discovered in this context and because this is their primary use. But when a honeybee colony is in serious need of something other than food, the dances are used in the same way to transmit information about its location.

One common situation occurs in hot weather. Honeybees have a well-developed system of regulating the temperature inside the hive by fanning with their wings to circulate air. They also bring in water, which they regurgitate in small drops, and this cools the air as it evaporates. When the temperature in the hive rises above approximately 95° F, many workers move near the entrance and fan actively. This sets up air currents that circulate into and out of the hive, and a

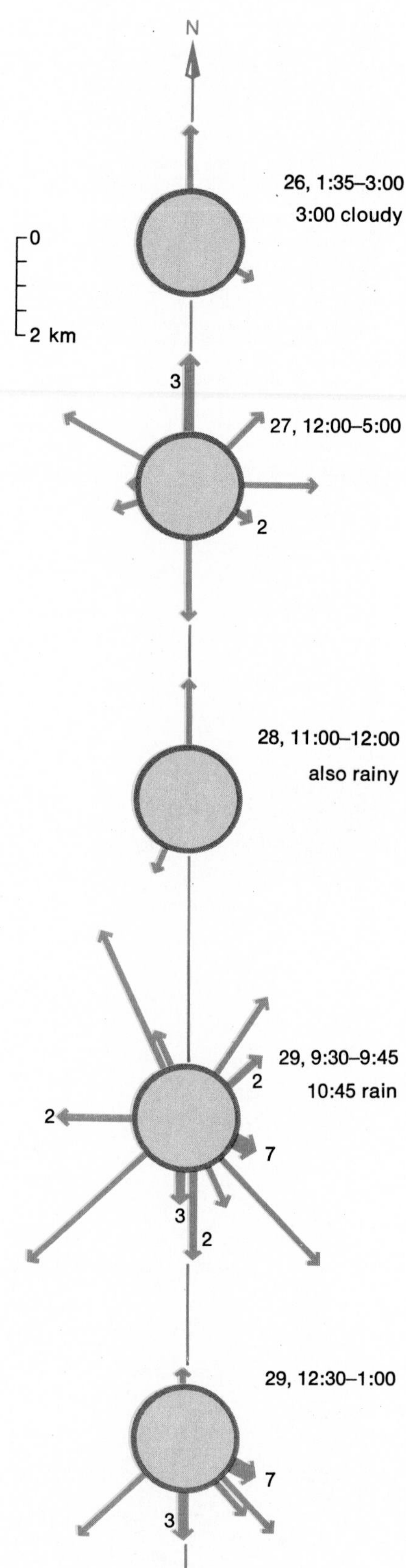

considerable amount of water is evaporated in the process. After some time the colony as a whole comes to need water for this purpose. At such times foragers returning with pollen or nectar are not received enthusiastically, and they have difficulty finding other bees that will accept liquid material from their stomachs. Foragers returning with water are greeted enthusiastically, however, and their regurgitated stomach contents are eagerly accepted. Under these conditions, foragers that have discovered a good source of water dance in a way that communicates its distance and direction, just as the dances normally do for food.

In another special situation, bees gather resinous material from plants, chiefly from buds of trees, in order to construct parts of the honeycomb or to close openings in the cavity where the hive is located. When this material, commonly called propolis, is badly needed, foraging bees that have located a good source dance to indicate its location.

When the colony has been very successful, has collected abundant food, and has increased in numbers until the hive is too small, a large number of bees—roughly half the population including the queen—move outside and cluster together in a swarm up to one or two feet across. Ordinarily the beekeeper intervenes at this point by providing an empty hive, and the bees occupy it at once. But if the beekeeper has not made such a provision, or if the colony is in the wild, some time elapses before the bees fly off together to a new location. In the meantime, scout bees search over a wide area, if necessary, and inspect cavities such as hollow trees and rock crevices. Martin Lindauer, a student of Frisch, found that when scout bees have discovered a suitable cavity they return and dance about it on the upper surface of the swarm itself. The same waggle dances are used to indicate distance and direction.

The discovery that these dances are used by swarming bees was a remarkable one. An individual worker bee lives no more than a few weeks, but many months or even years may elapse between the establishment of a colony and the time it grows large enough to require swarming. Thus the scout bees have never engaged in such behavior

Dances performed by scout bees on a cluster observed by Lindauer between June 26 and June 30 are represented in the diagrams by arrows that indicate direction and distance of the potential nesting sites. Only new dancers are recorded, although many bees continued to dance actively for their proposed sites throughout the period. The presence of more than one scout dancing for the same site is indicated by the thickness of the arrows and, where appropriate, by the number. On the first day (top left), only two possibilities were discovered. Of the many added in the following days, some were abandoned quickly whereas others received continuing attention. Gradually, however, more and more scouts were won over to a hole in a wall 300 m ESE of the swarm. When unanimity had been achieved (opposite page, bottom), the swarm departed for that site.

before. When the time for swarming has come, something in their genetic makeup not only induces them to move out of the established colony but, if they are workers of the appropriate age to be foragers, causes them to search for something they and their fellow workers have never sought before.

Furthermore, scouts that have located a new cavity indicate by their dances not only its distance and direction but also its suitability. They do this by the vigor and persistence of their dancing, just as when food is scarce they dance more vigorously when they have returned from a very rich source of nectar. As one would expect, different scouts locate and dance about various cavities. Many bees cluster around these dancers and, if the dances are vigorous, several visit the cavity thus indicated. The entire process continues for many hours or even a few days, and ordinarily the dancers come more and more to concentrate on one of the several cavities first announced by various scouts. Only when a great majority of the returning scouts are dancing about the location and suitability of one particular cavity does the swarm fly off to it. Scented secretions are almost certainly used to mark these cavities, and this undoubtedly helps the swarm to locate the precise spot when, on the basis of information obtained from the dances, they have reached its general vicinity.

Significance of the dances

Frisch speaks of "the language of bees," but most behavioral scientists prefer the term communication because language has generally been considered peculiar to our own species. Certainly honeybee dances do not achieve anything remotely resembling the large vocabulary, the flexible recombinations of a few basic signals, or the adaptability to new situations that characterize human speech. But the dances are symbolic gestures used to signal important information. They are varied appropriately according to circumstances, and, at least in the case of dances about cavities, the same bee may serve first as a transmitter of signals and then as a receiver. According to the signals she receives from other bees, she may stop dancing about a mediocre cavity and join her sisters in dancing about a superior one that they have located.

Clearly, the honeybee dances have many of the basic properties of human language, but in a much simpler form and one that is far more limited. Since the nervous system of a honeybee is capable of flexible and symbolic communication, it is evident that large brains are not necessary for carrying out, at least in simple form, the basic processes of language use, which previously had seemed to be uniquely human achievements.

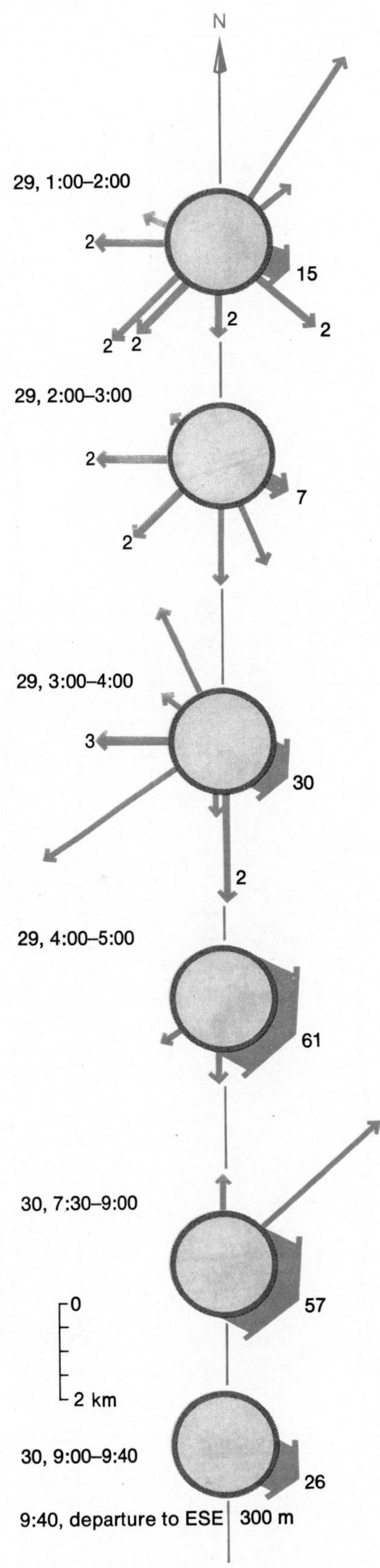

From Karl von Frisch, "The Dance Language and Orientation of Bees" (1967), The Belknap Press of Harvard University Press

Electricity in the Sky

by Richard E. Orville

One of the most spectacular and destructive displays of nature is the lightning flash. Scientists study its properties in order to predict, and perhaps control, its occurrence and intensity.

Lightning strikes the earth several million times each day. Each year in the United States alone, this is sufficient to kill more people than are killed by tornadoes and to cause property damage in excess of $100 million. Lightning strikes to the world's forests and grasslands typically occur 180 million times per year. A fraction of these strikes produces 10,000 wilderness fires in the U.S. that cost approximately $100 million to control, about one-third of the nation's total annual fire control expense. Not all lightning damage occurs on the ground. Commercial airplanes are involved in lightning strikes about once in 5,000 flying hours. The damage is usually slight, and only rarely is a plane lost. The exceptions range from a Boeing 707 that was downed by lightning on Dec. 8, 1963, to a Boeing 727 that successfully sustained five separate lightning hits in 20 minutes while in a holding pattern over Chicago on Sept. 26, 1964. The variety of damage reports matches the natural variability of the lightning flash. To understand the effects of lightning strikes on the ground and in the air, the physical processes that occur in a lightning flash must be examined.

Anatomy of a lightning discharge

The lightning flash is a transient, high-current electrical discharge in the atmosphere. It usually occurs within a cloud (intracloud discharge) or between the cloud and ground (cloud-to-ground discharge). Only infrequently does a flash occur between clouds (intercloud discharge) or from the cloud terminating in the surrounding air (air discharge). Intracloud discharges are the most frequent, but those from cloud to ground cause the most damage and deaths.

A lesser known type of lightning is the discharge that originates at the ground from a tall conducting object, such as a tower or skyscraper, and propagates to the cloud above. This discharge is caused by the presence of the conductor and would not occur in the absence of the man-made object. It is called "triggered lightning" because it is initiated or "triggered" by the conductor. To understand the danger of lightning and the amount of damage it can cause, it is instructive to examine the physical events that occur in a lightning discharge involving the ground.

The source of the lightning discharge is linked to the accumulation of net charge in regions of the cumulonimbus cloud (the thundercloud). Just how this occurs is one of the unsolved mysteries of the atmosphere. There is little doubt, however, that the processes of nature typically generate a region of net positive charge in the top of a cumulonimbus cloud and a net negative charge in the bottom, both perhaps with magnitudes of 40 or 50 coulombs. Frequently, a small positively charged region with a magnitude of a few coulombs is detected near the cloud base. The charge resides on ice crystals and water drops. It is thought that where the electric field in the cloud exceeds a few hundred thousand volts per meter, charge will drain from droplets and ice crystals to initiate a lightning flash. If this occurs between the large negatively charged region and the small positively

RICHARD E. ORVILLE *is Associate Professor of Atmospheric Science at the State University of New York, Albany, New York.*

Illustrations by Kerig Pope

Courtesy, Richard E. Orville, SUNYA

"Triggered lightning" (above) is generated by the tall tower on top of Mt. San Salvatore near Lugano, Switzerland. At the left is the spectrum of a lightning flash as photographed with a transmission diffraction grating and color film. The most intense lines are from singly ionized nitrogen atoms (atoms that have lost one electron).

William S. Bickel, University of Arizona, Tucson

Adapted from M. A. Uman, "Understanding Lightning" (1971); BEK Technical Publications, Inc., Pittsburgh, Pa.

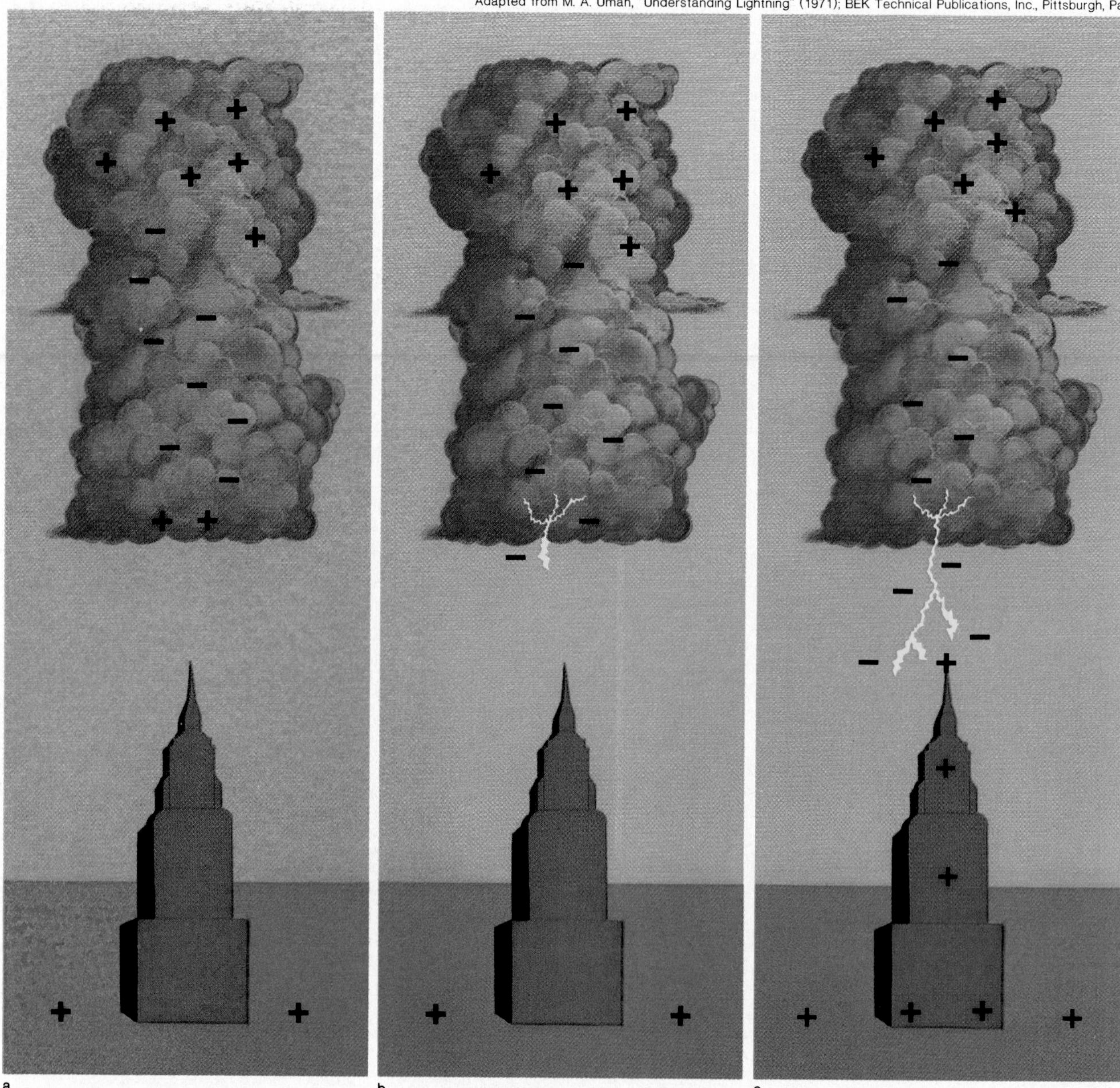

Cloud-to-ground lightning flash begins (a) when a stepped leader is generated between the positive and negative charge regions in a cumulonimbus cloud, (b) emerges from the base of the cloud, and (c) is propagated toward the ground in 50-m steps.

charged region in the base of the cloud, a cloud-to-ground discharge begins in the form of a stepped leader.

The first view of the leader process is obtained when the luminous channel emerges from the cloud base. It is very faint and is composed of an electrically charged column with a luminous core a few centimeters or less in diameter. The leader moves in a steplike manner toward the ground; each step is typically 50 m long and is traversed in less than a microsecond, followed by a pause before another 50-m step. Several channels may develop toward the ground to give the appearance of downward branching. The stepped leader is characterized by a two-dimensional velocity of 2×10^5 m per sec (about 1/1,500 the speed of light) and a current of, typically, 100 amperes which deposits

The return stroke (d–g) begins with upward discharges from the ground to meet the stepped leader. Electrons flow toward the ground, while the visible effect of this flow, the luminosity, appears to move up the channel. A typical return stroke reaches the cloud base from the ground in only 70 microseconds. Dart leader (h–j) follows the return stroke. No stepping or branching occurs, and the dart usually contacts the same point as the previous leader. Each leader is followed by a return stroke (k).

Example of cloud-to-ground lightning (opposite top) shows a flash striking a tree. A 13-m section of a lightning channel is reconstructed in color from a black and white negative by a pseudocolor video densitometer (opposite bottom). Regions of the channel that have the same color had the same density (transmitted the same amount of light) on the negative.

5 coulombs on the channel in the 20 milliseconds it takes to reach the ground. When one of the leader branches comes to within about 50 m of the ground, one or more discharges occur from the ground. The upward discharge that connects with the leader short-circuits the cloud to the ground, and the return stroke phase then begins.

The return stroke can be thought of as the high current flow that drains the negative charges deposited on the channel. First, the charges at the channel bottom flow to the ground, and then the charges from successively higher sections of the channel flow toward the ground. The effect is similar to that of sand flowing in an hourglass in the sense that while the sand flows downward the effect of this flow is felt at higher and higher sections of the hourglass. In the lightning channel the electrons flow toward the ground, while the effect of this action, the luminosity, is seen to move up the channel at speeds approximately one-third that of light. This is the time when damage occurs to objects that are struck by lightning. Peak currents typically reach 20,000 amperes and occasionally exceed 200,000 amperes. The channel expands at supersonic speed to a luminous diameter of perhaps 5 or 6 cm, and the acoustic wave resulting from this expansion is eventually heard as thunder. Pressures in the channel may reach several tens of atmospheres, and temperatures exceed 25,000° K. These peak currents, temperatures, and pressures in the channel are generally attained in a few microseconds; then they diminish as the current decreases to one-half its maximum value in 40 or 50 microseconds. The luminous return stroke takes only 70 microseconds to reach the cloud base from the ground. Current continues to flow in the channel for a few hundreds of microseconds. In less than a millisecond, the return stroke phase is over.

It might be expected that the completion of this return stroke phase would end the lightning event. Often it does, but usually another phase begins in 40 milliseconds with the emergence of a dart of light from the cloud base. It typically follows the old channel and moves at a speed 1/100 that of light. There is no stepping, just a downward progression of a luminous dart 50 m in length carrying a few hundred to a thousand amperes and depositing a few coulombs of charge on the channel. No branching occurs, and the dart usually contacts the same point as the previous leader process. Consequently, the cloud is again short-circuited to the ground and another return stroke occurs. Typically, three or four leader-return phases will occur to produce a lightning flash with an average duration of 0.2 seconds. However, 26 leader-return stroke events have been reported in a lightning flash that lasted 2 seconds.

About 20% of the cloud-to-ground discharges are characterized by more than one ground contact point in the same lightning flash. This occurs because the dart leader encounters a section of the channel "older" than 100 milliseconds and consequently forges a new stepped leader to the ground, thereby increasing the potential damage that could be caused by the flash.

The cloud-to-ground discharge is not the only lightning flash that involves the ground. Studies at New York City's Empire State Building in the late 1930s revealed that lightning sometimes begins at the ground and propagates to the clouds. A detailed study of these triggered-lightning discharges was recently made by Karl Berger at the Mt. San Salvatore Lightning Observatory, near Lugano, Switzerland. There, two instrumented towers, each 70 m high, on Mt. San Salvatore above Lake Lugano, are involved in more than 100 lightning flashes each year. More than 80% of these are triggered lightning, and the remainder are "normal" cloud-to-ground discharges of the type just described. A detailed study of the photographic and electric current characteristics reveals that the triggered lightning begins with a stepped leader propagating from the tower top. It may carry either positive or negative charge. The upward-propagating leaders exhibit many of the same characteristics seen in downward leaders. A difference becomes apparent, however, when the upward leader contacts the cloud. No return stroke occurs. The channel luminosity increases along the entire channel, and there is an increase in the current. If a subsequent stroke occurs, it is initiated by a dart leader that begins in the cloud and propagates toward the tower. A normal dart leader return-stroke sequence then occurs in all subsequent strokes.

Not all triggered lightning involves objects on or near the ground. H. T. Harrison of United Air Lines reported in a one-year survey of airline flights that thunderstorms were conspicuously absent in many of the incidences of lightning discharges involving airplanes. Pilots characteristically separate their reports into lightning strikes and static discharges, the latter occurring primarily in non-thunderstorm situations. Visible damage was reported in one-third of the incidents. There was no apparent difference between the damage patterns that occurred during thunderstorm conditions and those that occurred in non-thunderstorm circumstances. It would appear, therefore, that "static discharges" are, in fact, lightning that is initiated or triggered by the presence of the airplane.

By far the most common type of lightning is that which occurs within a cloud, called an intracloud lightning flash. The ratio of this type to cloud-to-ground lightning is estimated to be 10:1 in tropical latitudes, decreasing to lower values at higher latitudes. In Norway, at latitude 60° N, the intracloud and cloud-to-ground frequencies are approximately the same.

Our knowledge of the physical processes occurring in intracloud discharges is meager. A leader process occurs and propagates within a cloud at average velocities of 2×10^4 m per sec. During this time the cloud has a low luminosity that is punctuated by random periods of brightness associated with the leader reaching pockets of net charge within the cloud. No return stroke is observed to occur. The electrical charge transferred is estimated to be similar to that in a cloud-to-ground flash—20 coulombs—but measurements indicate that it can range from 0.3 to 100 coulombs. The vertical extent of the intracloud

Photos, courtesy, Richard E. Orville, SUNYA

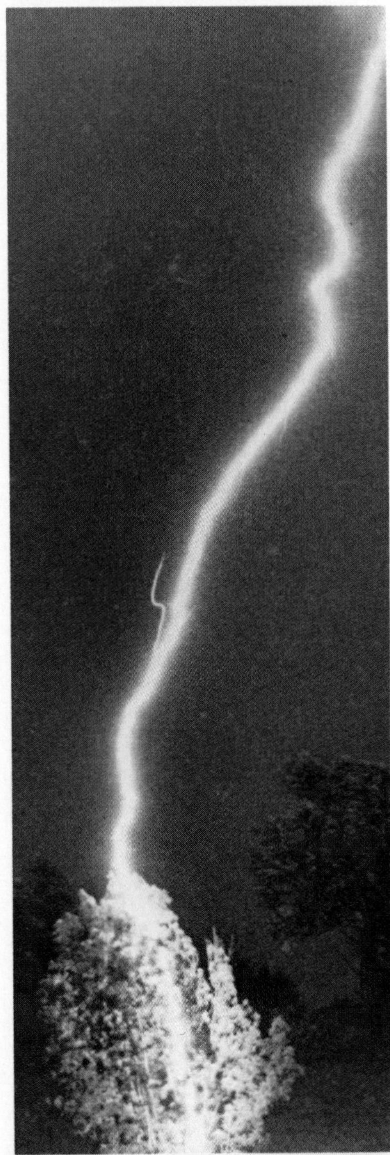

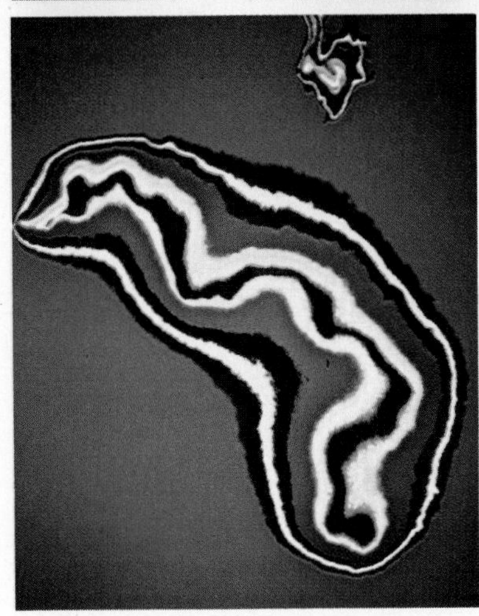

discharge averages 0.6 km, whereas the horizontal extent is 1–10 km with a mean of 3 km (1 km = 0.62 mi). Thus, the intracloud discharge appears to be predominantly a horizontal flash within a cloud.

Aural and visual effects

By far the most dramatic evidence of a lightning flash is the thunder that is heard and the light flash that is seen. It should not be surprising that scientists have developed extensive instrumentation and techniques to record these emissions. From these recordings the physical characteristics of the flash have been deduced.

Arthur Few of Rice University, Houston, Texas, made a detailed study of the thunder spectrum, or acoustical signature, of the lightning flash and mapped its three-dimensional path in space. He found in several cases that the normal cloud-to-ground flash has a long horizontal section that may be ten times longer in the cloud than the short vertical section that can be observed below the cloud base. Furthermore, this horizontal section is largely within a horizontal slice of the atmosphere that is defined by the isotherms, or levels of constant temperature, of 0° C and −10° C. Thus, it would appear that the lightning originated in a region of the cloud known to contain both ice crystals and water drops.

A rough estimate of the distance to a flash and its channel length can be obtained in the following way. First, the time interval between the light emissions and the thunder is recorded in seconds, and then the duration of the thunder is measured. Dividing the time interval by three yields the distance to the channel in kilometers; dividing the duration by three yields the minimum length of the channel. The basis of this simple rule is the great difference between the speed of propagation of light and that of sound. Furthermore, the lightning channel is an instantaneous but geometrically distributed source of sound such that the thunder from the various parts of the channel will arrive with a time delay determined by the distance to the channel segment. Consequently, the duration of the thunder in seconds, when divided by three, provides the minimum length of the lightning channel in kilometers.

It is difficult to specify the exact distance over which one could expect to hear thunder, as the propagation characteristics of the atmosphere are a complex function of wind, temperature, and the height of the lightning source. The distance that thunder can be heard is typically no more than 25 km and may be considerably less.

The visual effects of the lightning flash are always impressive and have led to reports of many different types of lightning. These, however, can mostly be explained by recalling the previously discussed fundamental properties of a flash. For example, "forked lightning" refers to the cloud-to-ground flash with its many branches, and "bead lightning" to residual luminous sections of the lightning channel whose light emissions come from thicker sections of the channel and therefore remain luminous for a longer time, giving the appearance of luminous beads. "Heat lightning" refers to the distant flashes of a

Ira Richolson

Between- and within-cloud lightning. The latter is the most common variety of lightning, occurring about ten times more frequently than cloud-to-ground flashes in tropical latitudes. In far northern and southern regions the frequencies of occurrence are approximately equal.

thunderstorm that appear red and from which no thunder is heard. The channel is red for the same reason that the setting Sun is red: the blue light from the flash is scattered more than the red. No thunder is heard because the sound waves are refracted up into the atmosphere and pass overhead. "Sheet lightning" refers to the luminous appearance of a cloud when an intracloud flash occurs. The channel is not visible and the cloud, illuminated from within, resembles a white sheet.

Occurrence and distribution

Thunderstorms can occur wherever moist air is present and a mechanism exists for moving this air through large vertical distances in the atmosphere to produce cooling and condensation. The greatest frequency of lightning is reported to be on the island of Java, where thunderstorms occur on the average of 223 days of the year. Within the United States, lightning occurs most frequently in Florida. There is an ample supply of moisture from the surrounding water, and the solar heating of the land provides a heat source to lift the moist air. The resulting cumulonimbus clouds extending to many kilometers in height produce regions in Florida that report approximately 90 thunderstorm days per year.

An area of low thunderstorm frequency is found along the West Coast of the United States. Although the Pacific Ocean provides the region with a sufficient source of moisture for thunderstorms, the area is dominated by an anticyclone, or high-pressure area, which suppresses convection. Consequently, the West Coast is favored with fair weather. H. T. Harrison, mentioned earlier in the discussion on triggered lightning, noted a disproportionately high frequency of lightning strikes involving airplanes in the vicinity of airports on the West

Coast. He suggested the interesting possibility that the few thunderstorm days that are reported from airport meteorological offices in that region may in part be the result of aircraft above the airport triggering the discharge. It would be interesting if further research substantiates the occurrence of these "artificial thunderstorms."

Destructiveness

By far the greatest damage is done when lightning strikes the earth's surface and causes loss of life. More than 100 persons each year are killed in the U.S., down from an average of over 400 a year in the early decades of this century. Recent research indicates that the loss of life can be further reduced.

When lightning strikes the human body, it causes burns and tissue destruction. These may not necessarily cause death. Far more serious is the loss of respiration and of the rhythmic beat of the heart (ventricular fibrillation). If either of these conditions are present, the body suffers irreversible damage unless first aid is given within five minutes. In the absence of breathing, artificial respiration should be given. In the absence of a heartbeat, heart action can be stimulated by placing the victim on his back and pressing firmly on his chest with the heel of the hand once every second or slightly faster. In this way, prompt first aid may "reverse" death by lightning.

Precautionary measures, however, remain the surest way to avoid becoming a lightning death statistic. Stay inside during a thunderstorm and keep away from electrical appliances. The safest place is inside a metal structure. Therefore, if traveling by car in a thunderstorm, one should remain inside it. If caught in the open, do not seek shelter under isolated trees but crouch so as to cover only a small area of the ground. Keep your feet together. If they are apart, the lightning current from a nearby strike can flow up one leg and down the other because this path offers less resistance than the ground between your feet. Remember that getting wet is far preferable to participating in a lightning flash!

Lightning strikes to property cause damage that exceeds $100 million per year. A home in an area with 20 to 30 thunderstorm days per year can expect a strike on the average of once every 100 years. You cannot prevent lightning from striking your home, but you can provide a safe path to the ground by installing a lightning rod with a conductor to the earth, where it is safely grounded. It is surprising that Benjamin Franklin's invention of the rod more than 200 years ago has in principle remained unchanged; it is still the most reliable way to protect a home against lightning.

Cloud-to-ground discharges in the U.S. start, on the average, 10,000 forest fires, of which 95% are extinguished while small. It is the remaining 3–5% which get out of control that cause 95% of the damage. These relatively few fires can be extremely destructive. In 1970 alone, large fires burned parts of southern California to the extent that the region was declared a disaster area. Fourteen lives were lost, 800 homes and buildings were destroyed, and 600,000 acres of timber and watershed cover burned.

Marvin Dodge of the California Department of Conservation warned that additional disastrous fires can be expected because of the accumulation of dead fuels taking place in the wildland areas of the western United States. It is in the areas protected by the government that the accumulation is greatest and the hazard of a conflagration increases. Dodge argued that prescribed burning may be the solution and that critics fail to recognize the difference between high-intensity wildfires that destroy everything and low-intensity fires that may cause little or no damage. A similar but more subtle view was expressed by Alan Taylor of the U.S. Forest Service. Taylor recognized that a few lightning fires become holocausts and that these certainly require research toward their control. He emphasized, however, that the greatest challenge is to gain biological insight concerning the roles of lightning and fire in plant and animal communities. This knowledge should then be coupled with man's technological capabilities to prevent disasters associated with lightning and at the same time allow this agent of change to pursue its natural course. Thus, it would appear that in the 1970s the principles of Smokey the Bear were yielding to evidence that lightning and its associated fires may play a natural role in the ecological balance of forests.

The future

Efforts to predict and control lightning are still within the realm of intensive basic research. Experiments to modify the thunderstorm and to reduce the harmful effects of lightning are few in number and inconclusive to date. Studies in the past have been performed by individual research institutes working alone to determine the physical properties of the lightning discharge. Yet, prediction and control can only come after understanding is gained of the complicated meteorological patterns of air flow and precipitation that are associated with the buildup of charged regions in the thundercloud. This will require a large coordinated scientific effort using the latest technological advances to study all aspects of the thunderstorm simultaneously. Such an effort will bring the goal of prediction, and perhaps limited control, of lightning within the realm of applied technology.

The World of Superhardness

by Herbert M. Strong and Francis P. Bundy

Performing tasks that nothing else can, the superhard materials—diamond and cubic boron nitride—have been of great value to modern industrial civilization.

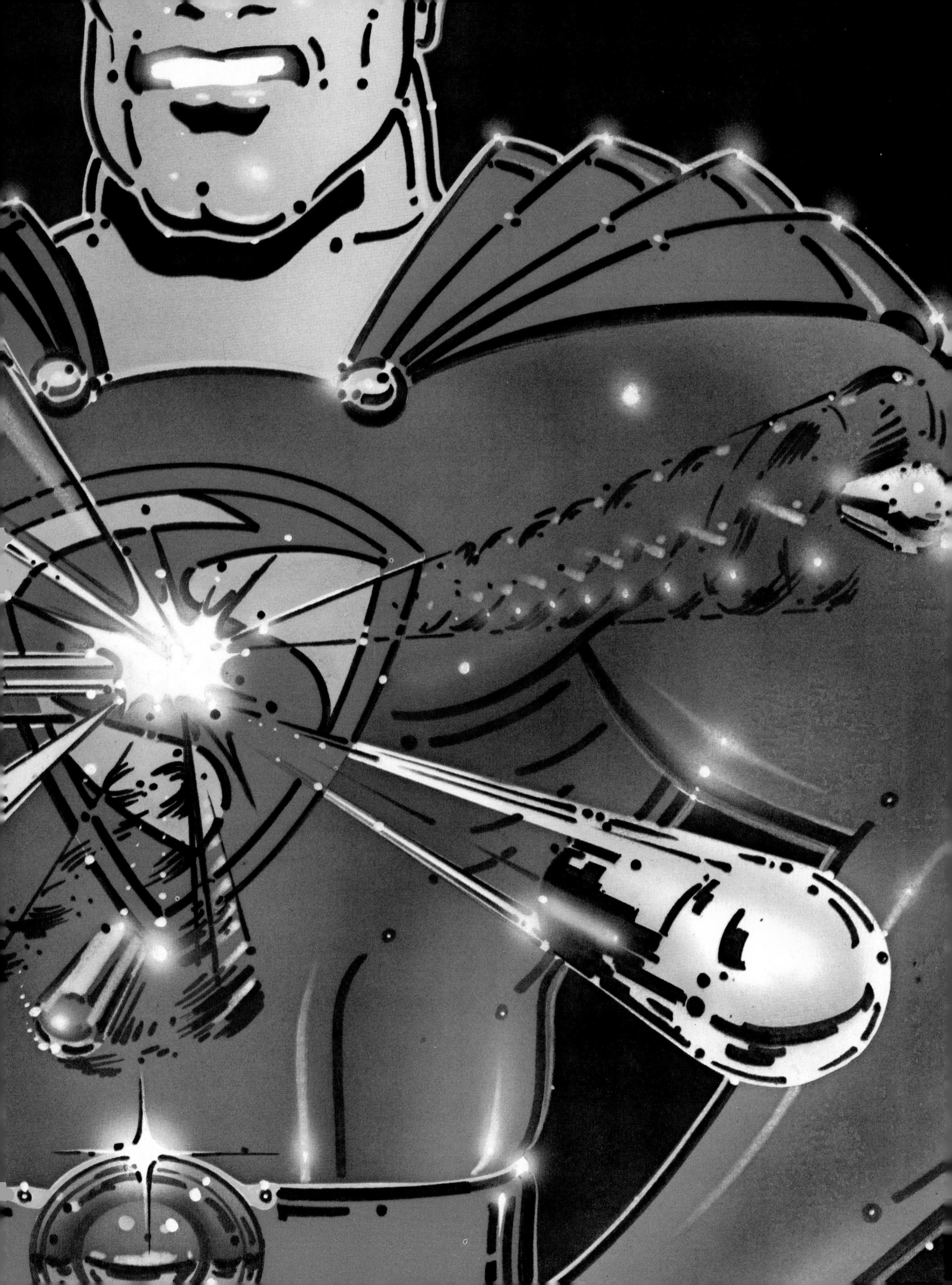

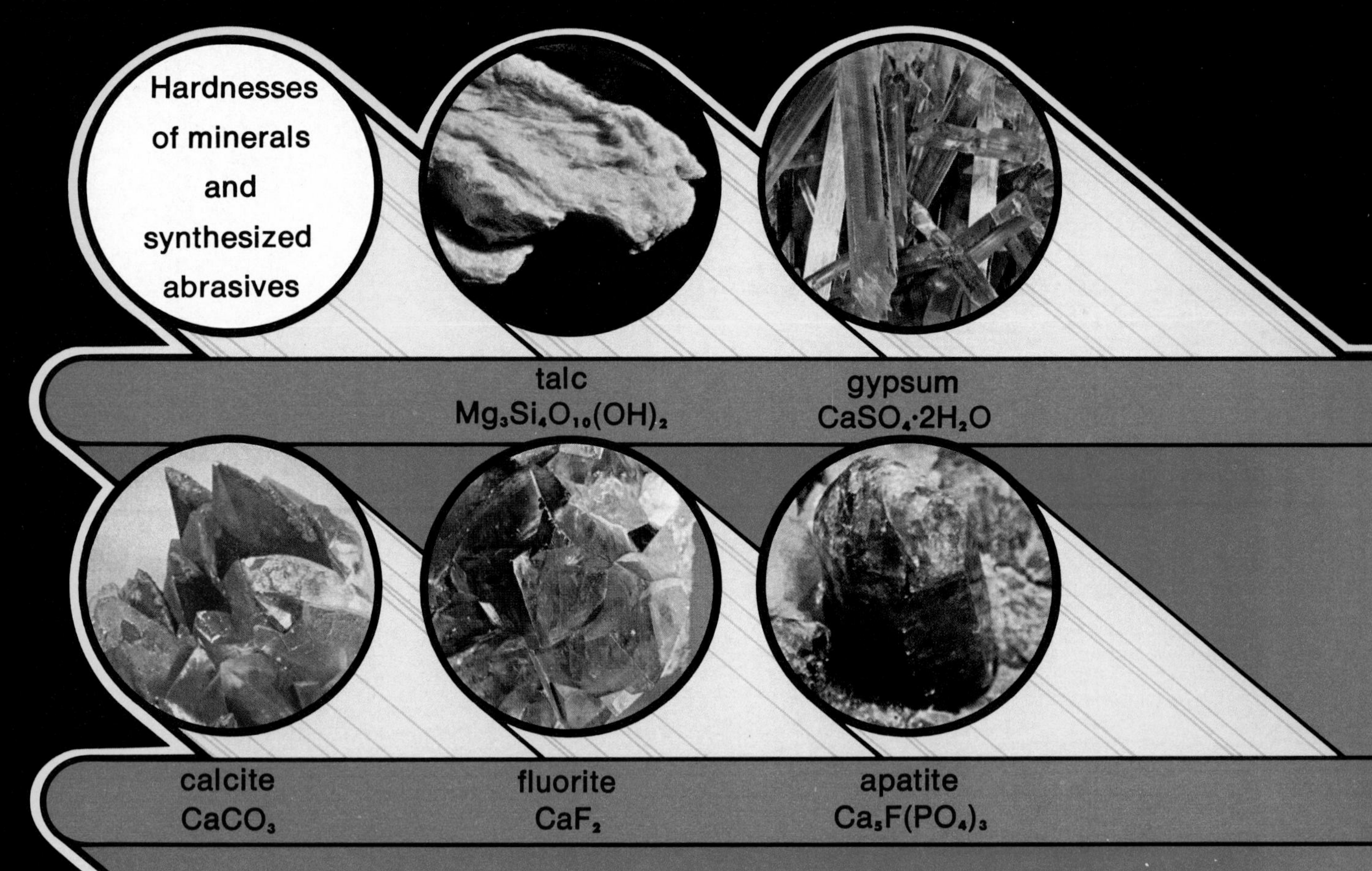

To the desert camel rider, superhard materials have little significance. To the riders of jet aircraft and automobiles, superhard materials have great, if indirect, significance. If such materials did not exist, the manufacture of the parts for these and many other modern conveniences would be more expensive, more difficult, and, in some cases, nearly impossible. The use of superhard materials and their adaptation to a continuing flow of new fabrication problems are integral parts of a developing technology.

Diamond and cubic boron nitride are the "superhard" materials. They are set apart from the commonly used cutting and abrasive materials because they are many times harder than their nearest competitors, such as corundum (aluminum oxide), carborundum (silicon carbide), and cemented tungsten carbide. These latter crystalline substances differ little in hardness among themselves, whereas diamond and cubic boron nitride constitute a unique class.

Hard materials are hard because of the quality and geometric arrangement of their chemical bonds. All very hard materials are also brittle, like glass, but have much greater strength than glass. Though brittle, diamond is nature's hardest and strongest material.

HERBERT M. STRONG and FRANCIS P. BUNDY *are, respectively, retired and active physicists at the General Electric Research and Development Center, Schenectady, New York.*

(Overleaf) Illustration by Peter Lloyd

Measurement of hardness

Qualitatively, hardness is conceived as the capability for one material to dent or scratch another softer material. This concept suggested in 1812 to the German mineralogist Friedrich Mohs the possibility of a

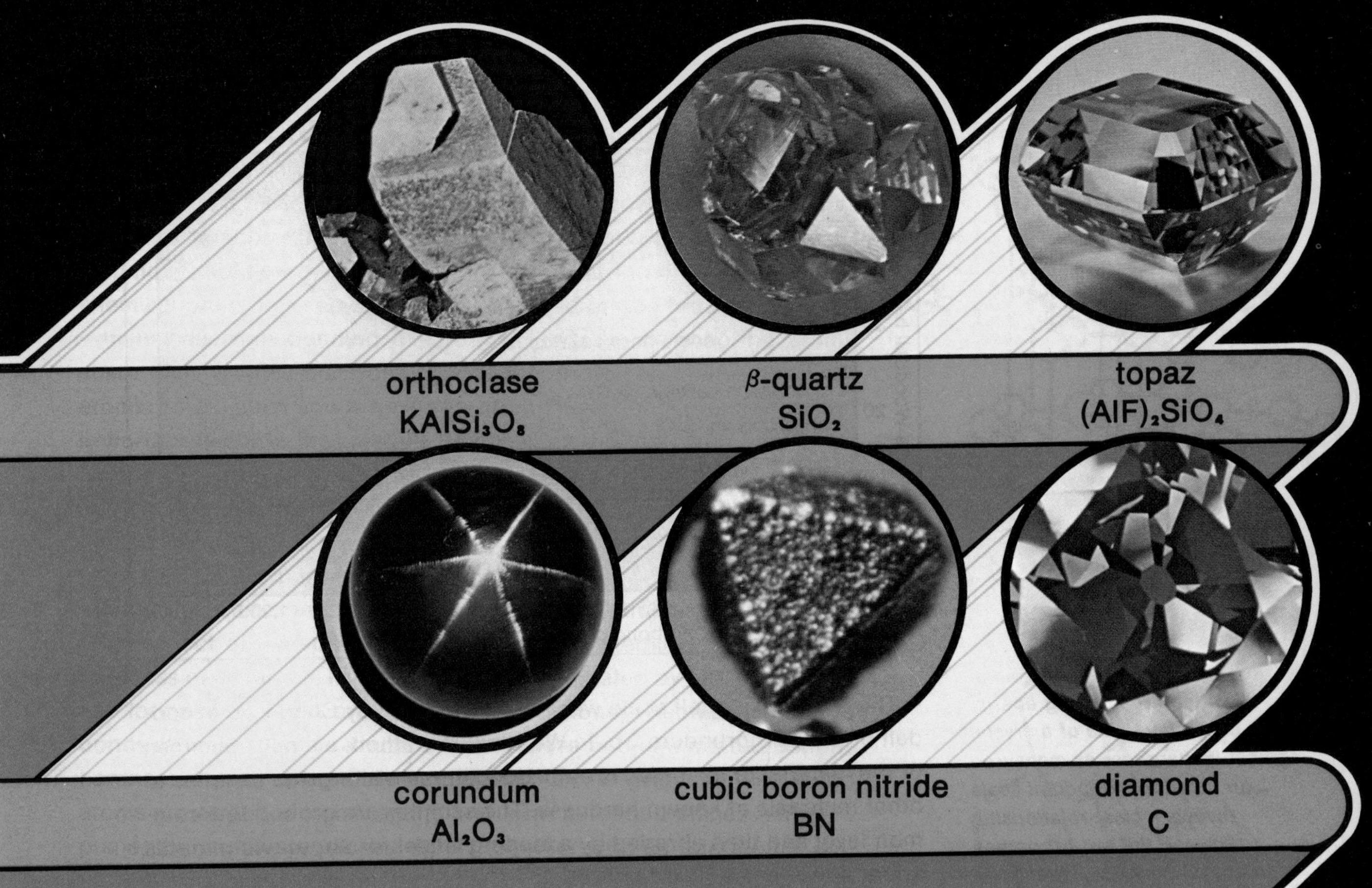

hardness scale for minerals based on "scratch order." Mohs' ten standard minerals are arranged in ascending order of scratch hardness from talc, with a value of one, to diamond, which is ten. Other substances may have their hardnesses classified by comparing their scratch resistance with those of Mohs' standard minerals.

Hardness of ten minerals plus synthetic cubic boron nitride can be compared on three scales, the Mohs, Knoop, and Mohs-Wooddell. The values, respectively, are: talc 1, no assigned value, 1; gypsum 2, 32, 2; calcite 3, 135, 3; fluorite 4, 163, 4; apatite 5, 430, 5; orthoclase 6, 560, 6; β-quartz 7, 820, 7; topaz 8, 1,340, 8; corundum 9, 2,100, 9; cubic boron nitride, no assigned value, 4,500, 25; and diamond 10, 9,000, 42.5.

The Mohs scale compares fairly well with modern hardness tests in the range from one to nine, where 99% of all substances lie. Each step, one to nine, corresponds approximately to a 60% increase in indentation hardness. The scale, however, gives no hint that diamond at ten is five times harder than corundum at nine. In order to make more meaningful quantitative studies in the technologically important range above nine, a quantitative type of hardness test must be used instead of the simple Mohs scratch test. For this purpose, standardized indentation or abrasion tests have been devised.

The Knoop scale, adopted by the U.S. Bureau of Standards in 1939, measures the hardness of a material by making an indentation in its surface. Sharp, elongated, pyramid-shaped diamond "indenters" are pressed against the surface of the material to be measured at a set force. Hardness is calculated in terms of the stress, in kilograms per square millimeter, based on the force applied and the area of the indentation produced. On this scale, the hardnesses of corundum, cubic boron nitride, and diamond are 2,100, 4,500, and approximately 9,000 kg per sq mm, respectively. The hardest steels have hardnesses of about 1,000 kg per sq mm.

Courtesy, General Electric Specialty Materials Department

Tool with a cutting edge made from synthesized polycrystalline diamond processes an aluminum alloy.

conductivity and the velocity of sound through it are high; the coefficient of friction is low; and it provides a good "window" for (is transparent to) the infrared region of the spectrum. Except as a semiconductor, diamond excels all other elements in these properties.

Diamond doped with boron becomes a p-type semiconductor (one in which the hole conduction—absence of electrons—exceeds the electron conduction). However, the technology for controlled doping of pure diamond during growth of the crystal has not been entirely successful. Nor has it been possible to dope diamond for n-type conduction (the reverse of p-type). Consequently, the field of semiconducting devices is dominated by other members of the diamond cubic series: silicon, germanium, gallium arsenide.

The thermal conductivity of pure Type II diamond is the greatest of any known material in a wide temperature range above 90° K (−183° C). At room temperature its thermal conductivity is five times greater than that of copper. For this reason tiny diamond plates are being used as mounts for semiconducting diodes made of gallium arsenide, in order to spread and carry away the intense heat such diodes generate during use. If diamond could be successfully doped, it could conceivably serve as an ideal self-cooled diode.

The calculated thermal conductivity of a perfect cubic boron nitride crystal at 25° C (77° F) is about two-thirds that of diamond and about three times that of copper. High heat conductivity is often thought to be an exclusive property of metals, especially copper, silver, and gold. In metals, heat is transmitted by drift of the free electron gas that characterizes the metallic state. Because diamond is an insulator, there are no free electrons to carry heat. Instead, heat energy is carried from atom to atom by the vibrational energy of carbon atoms. The high rigidity of the lattice, together with the light weight and rapid response of the carbon atoms, is responsible for diamond's high thermal conduction. The same properties are responsible for the high velocity of sound in diamond.

The high thermal conductivity of Type II diamond together with its infrared transparency has made it attractive as an exit window for high-power laser beam generators. In this case diamond is apparently the only material that can be used.

The velocity of sound in diamond is 18,000 m per sec. By comparison, the velocities of sound in air, steel, and crystalline boron are, respectively, 344, 5,000, and 14,000 m per sec. The estimated velocity of sound in cubic boron nitride is 15,000 m per sec. There are possible uses for this property in electronic circuit acoustic delay lines. An electronic signal would be passed through the diamond as an acoustic signal and then transformed back to an electronic signal, thereby briefly delaying its passage.

Gem-size diamonds will always have high aesthetic value. As their unique properties are explored and more technological applications are made, they will also be of increasing value to industry. Many now-unrecognized uses will probably be found.

Courtesy, General Electric Specialty Materials Department

Grinding wheel with abrasive made of cubic boron nitride works on hardened steel.

Future research and prospects

Can there be a harder material than diamond? There is no fundamental principle that excludes this possibility. The requirement seems clear. It is that such a crystal must have symmetrical covalent bonding with bonding energy density greater than that of diamond. The periodic table of elements has been well explored with this in mind. The likelihood of finding a combination of elements or a crystal form of a particular element that meets this requirement seems remote.

It has been suggested that a form of carbon denser than diamond might be harder than diamond. It is known that silicon and germanium, the sister elements of carbon, do go into more dense forms at very high pressure; these forms, however, are tinlike metals that are not very hard. It is probable that diamond carbon would also transform to a denser tinlike metallic structure at several hundred thousand atmospheres of pressure, but being a metal it could not be very hard. Furthermore, as already known for silicon and germanium metallic forms, metallic carbon would probably not survive decompression to ordinary pressures.

The prospects for finding a material that is harder than the presently known superhard materials appear poor. The likelihood of further progress in adapting diamond and cubic boron nitride to new machining problems is good. The polycrystalline compacts of these materials have introduced a new degree of flexibility for solving cutting problems that bodes well for the future of machining technology.

The Earth's Changing Climate

by Hubert H. Lamb

Throughout its history the Earth has experienced climatic fluctuations of varying degree and duration. Current observations indicate that another shift may now be in its early stages.

Jay Simon

Climate is always changing. The fluctuations of weather and climate take place on all time scales, from the gusts and lulls of the wind, which occur within a fraction of a minute, to the shifts of regime over hundreds of millions of years associated with continental drift and wandering of the poles. Fortunately, however, for the development of life on the Earth and in particular for the evolution of contemporary life forms, the range of the temperature changes has been limited. Since the first appearance of life, there have presumably always been regions of the Earth where the air and water temperature remained generally between 20° and 30° C (70° and 85° F) and extensive regions where the limits of 0° and about 35° C (32° and about 95° F) were never far exceeded nor for more than a few hours at a time. This limited range bears witness that the Sun must have been a reasonably constant star and the Earth a planet hospitable to the survival and spread of life. Nevertheless, the changes in the Earth's climate that have occurred have brought innumerable local and regional disasters and have repeatedly challenged the tolerance and adaptability of man and of all other living things.

The greatest and quickest changes of climate, and of the environment dependent on climate, have been connected with the onset and ending of ice ages and with fluctuations near the ice margin in times when the Earth is partly glaciated, as it is at present. The mean temperature in Iceland, for example, rose by about 2° C during the global warming between A.D. 1850 and 1950, and the corresponding change averaged over the whole Earth was a rise of around 0.5° C. These shifts not only moved the limits of the vegetation belts equatorward or poleward, and up and down the mountains, but also affected the rainfall in continental interiors and the position and development of the desert belts. As the ice masses built up on land or melted away, sea level changed by as much as 100 m between the extremes of ice age and interglacial climate.

The greatest losses of life directly attributable to weather and climate have resulted from coastal floods, specifically those produced by storm winds and tidal surges at times when the world sea level had been rising during preceding decades or centuries of warm climate. The histories of China, the Bay of Bengal, and the North Sea coasts of Europe provide many instances in which hundreds of thousands of people have died in such sea floods.

Whether the climatic stress for mankind, and for animals and plants, presented itself in the form of a change of the prevailing temperatures, or aridity, or in a greater incidence of floods and storms, survival often depended on migration. In his more recent history, man has enabled himself and his crops and animals to survive and flourish beyond the limits previously imposed by nature with the aid of artificial indoor climates and of irrigation; this ability to transcend nature's limitations now depends, however, on the use of a great deal of energy (especially oil) and often on the use of "fossil" water in underground strata, which may also be limited.

HUBERT H. LAMB *is Director of the Climatic Research Unit, University of East Anglia. An original version of this article was written at the request of the Royal Swedish Academy of Sciences and was published in Swedish in the science periodical* Forskning och Framsteg, *No. 1, 1975.*

(Overleaf) Photograph by Jay Simon

Causes of climatic variation

The causes of climatic change can be classified in four general categories. The first chiefly concerns changes of the Earth's geography. It includes drift of the continents, which change their positions relative to each other and to the poles; uplift and erosion, which change the magnitude and disposition of mountain barriers and thereby affect the flow of the winds; and changes during the Earth's history in the total mass and chemical composition of the atmosphere and oceans. This category also includes variations in the energy output of the Sun and changes in the heat flow from the Earth's interior (probably almost always of minor importance). The first three items in this group deal with changes that generally become significant only over tens or hundreds of millions of years.

The second category of causes of climatic change comprises cyclical variations in the Earth's orbital arrangements. These include the angle of tilt of the Earth's rotation axis to the plane of the orbit, which varies by a few degrees over a cycle of 40,000 years and changes the latitudes of the tropics and polar circles and the angle of elevation of the Sun. Also in this category is the precession of the equinoxes, the progressive change in the position of the Earth in its elliptical orbit for any given time of the year. This 21,000-year cycle alters the distance of the Earth from the Sun at any given season. There is also a 100,000-year cycle in which changes in the ellipticity of the Earth's orbit affect the yearly variance of the Earth's distance from the Sun.

The third category consists of changes in the transparency of the Earth's atmosphere to incoming (mostly short-wave) and to outgoing (mostly long-wave) radiation. The best-demonstrated effects under this heading are those that have followed volcanic explosions which throw great quantities of fine dust into the stratosphere, creating a veil there which characteristically spreads over the Earth and lasts for two to three (occasionally as long as seven) years. While the veil lasts, temperatures rise in the stratosphere, due to direct absorption of solar radiation there, and are lowered at the surface of the Earth, due to loss of incoming short-wave radiation. Changes of cloudiness and in the atmosphere's content of water vapor, carbon dioxide, and other substances that are not transparent to radiation on some wavelengths also affect the radiation balance. Some of these vary as a result of the weather itself. Others are increasingly contributed by human activities (though perhaps not yet in sufficient quantities to affect climate).

The fourth category comprises the changes in the amounts of heat absorbed and given off at the surface of the Earth. These are due to variations in the extent of ice and snow, in the distribution of vegetation and of waterlogged or parched ground, and in the amount of anomalously warm or cold water on the ocean surfaces as a result of variations in the amount of sunshine or upwelling, respectively. These changes, like some mentioned in the third category, are produced by the weather itself and may in some cases increase the likelihood of persistence of the weather pattern that produced them.

Jay Simon

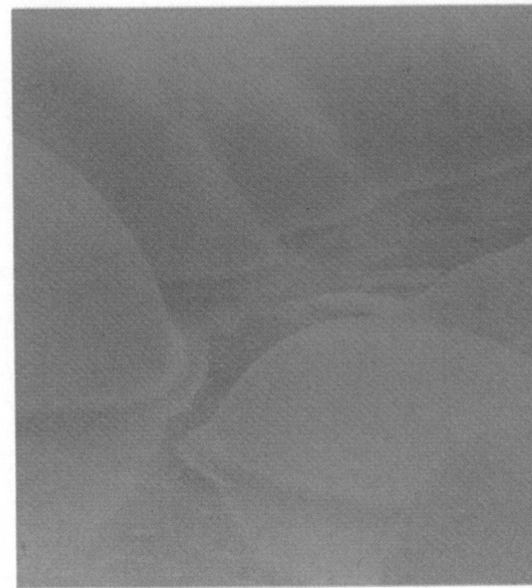

Colorado landscapes reflect varying climatic conditions. Above, snow over rocks forms mounds on an iced-over river in Boulder Canyon. Wind-piled dunes on the opposite page are in the Great Sand Dunes National Monument in the southern part of the state.

Courtesy, Naval Research Laboratory

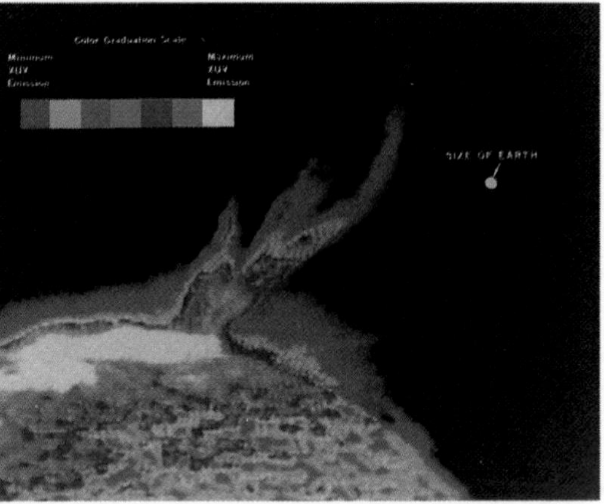

Color picture, electronically processed from a black-and-white photograph taken in the light of ionized helium by the Skylab spectroheliograph, shows an eruptive prominence as it reached a distance of 350,000 mi from the Sun's surface, on Aug. 21, 1973. Such pictures are obtainable only outside the Earth's atmosphere, which absorbs the ultraviolet light of helium. Though Earth is totally dependent on the Sun, it receives only one part in 2.2 billion of that star's radiant energy.

Climate during the last billion years

Through the longest stretches of geological time the Earth had only warm climates, with no great polar ice sheets, though it is now thought that there may have been many periods during which any landmasses that were near the poles bore "permanent" ice. If one accepts this view, the development of the greater ice ages chiefly depended on continental drift to place a continent at one or the other of the Earth's poles or, at least, to put landmasses of continental extent in high latitudes. However, the matter seems to be also affected by astronomical and solar variations. Evidence of the occurrences of greatest extent of ice in the past points to a fairly regular interval of nearly 300 million years, which may be explainable by gravitational effects of the rotating galaxy upon the Sun's activity and output. And in each of the times when extensive ice occurred there seem to have been alternations between ice-age conditions and interglacial periods, which as in the Quaternary (the last one million years approximately) may be attributed to the effects of the Earth's varying orbital arrangements (and tilt of the polar axis) upon the gain and loss of radiation from the Sun in summer and winter.

For reasons probably among those listed but which cannot yet be determined, there were important fluctuations that brought cooler climates at times during the Mesozoic Era (between 225 million and 65 million years ago). One of these, about the end of the Cretaceous Period and beginning of the Cenozoic Era (about 65 million years ago), may have been associated with the extinction of the dinosaurs, which were presumably not adapted to withstand the cold as well as their warmblooded contemporaries.

During the Tertiary Period of the Cenozoic (65 million to 2.5 million years ago), a greater and more persistent cooling set in. In the late Tertiary (Pliocene Epoch) superposed fluctuations on a time scale of about 40,000 years can be traced. These fluctuations were presumably associated with the cyclical variation of the Earth's axial tilt (obliquity), and some of them are now thought to have produced conditions that

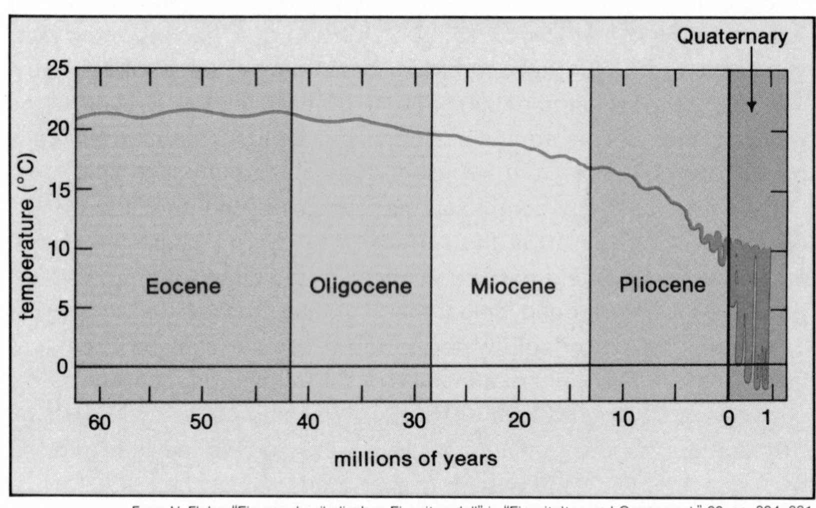

Average temperature in the middle latitudes of the Northern Hemisphere over the last 60 million years declined gradually until the Pliocene when it began to drop more sharply. Climate historians have increased the estimated number of ice-age/interglacial fluctuations in the Quaternary to about ten since this diagram was drawn. The time scale is enlarged for the last one million years.

From H. Flohn, "Ein geophysikalisches Eiszeitmodell" in "Eiszeitalter und Gegenwart," 20, pp. 204–231, Öhringen (Württemberg), Oct. 1969 began to approach the severity of the Quaternary (or Pleistocene) ice ages. The latest evidence suggests that it was during the Pliocene that the first races of man emerged as distinct species differing from the apes. Perhaps it was a superior ability to adapt to the changes of climate and of the terrestrial environment, even in the warmer regions of the Earth, that enabled the new species to thrive. There is little doubt that it was the emergence during the Tertiary of a geography similar to that of the present, with a south polar continent and an almost complete land ring about the North Pole, that made possible the development of increasingly extensive ice sheets, thus cooling the oceans and therewith the whole Earth.

Recent evidence from oxygen-isotope measurements on carbonate-bearing sediments of the Pacific Ocean near the Equator suggests that the end of the Pliocene and beginning of Pleistocene time is marked by a sudden change in the pattern of oscillations: from that time onward the temperature history is dominated by oscillations of close to 100,000 years in extent. Whether this means that some astronomical event occurred that made the variations in the ellipticity of the Earth's orbit more important than before is not yet clear, but the periodicity seems to coincide with those variations. The 40,000-year periodicity is still present but seems to play a somewhat subsidiary role. This latest evidence suggests that there have been about ten major developments of glaciation during the last one million years of the Quaternary Period, though of varying severity and, as always, with every kind of shorter-term fluctuation superposed.

Courtesy, Naval Research Laboratory

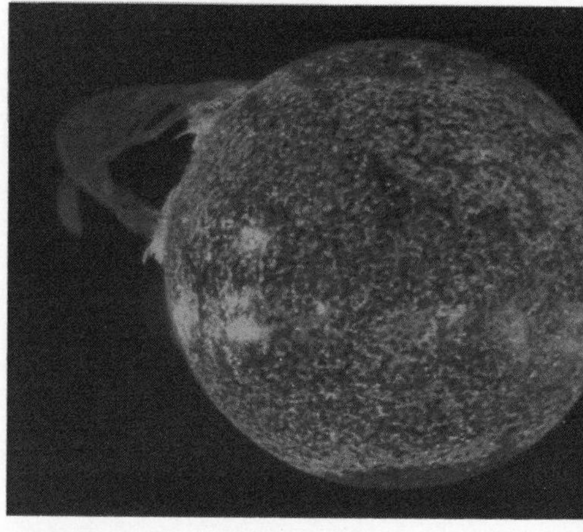

Like a gigantic, twisting ribbon, an eruptive prominence rises from the surface of the Sun. Taken Dec. 19, 1973, by the extreme ultraviolet spectroheliograph aboard the Skylab space laboratory, the photograph dramatically illustrates the constant activity taking place in Earth's nearest star, the ultimate source of all its energy, of climate, and hence of life itself.

The last ice age and early man

The history of the last glaciation is indicated in the diagram on p. 186 by a curve that represents the changes of prevailing temperature of the warmest month of the year (July) in The Netherlands. This pattern was derived from the changing composition of the flora shown by the counts of different species among the fossil pollen. Oxygen-isotope analysis of the ice taken from a boring in the Greenland ice sheet gives a record over the last 100,000 years which seems to parallel this in so many details that the results tend to corroborate each other, though the magnitude of the temperature changes indicated for north Greenland is larger than in The Netherlands. From the ocean bed in the tropical Atlantic comes similar corroboration, but there the temperatures (derived from oxygen-isotope measurements and from counts of the different species of marine microfauna) show a smaller range.

During the last glaciation primitive men made what was probably an easy living by hunting the large grazing animals—reindeer, bison, mammoth—on the open steppe-tundra lands in what is now France and elsewhere on the European plains. They have left a record of their life and of this fauna in cave-wall paintings at Lascaux, France, and Altamira, Spain. During the last glaciation the first men probably entered the Americas, traveling from Asia about 35,000–15,000 years ago over the broad grassy lowland laid bare in what is now the Bering

SCALA

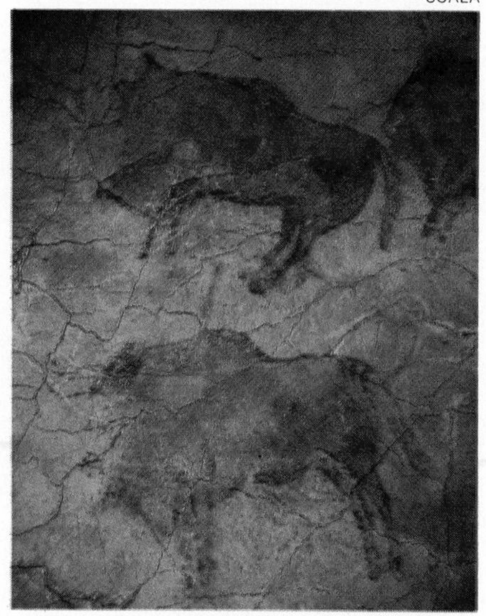

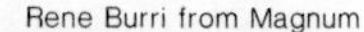

Rene Burri from Magnum

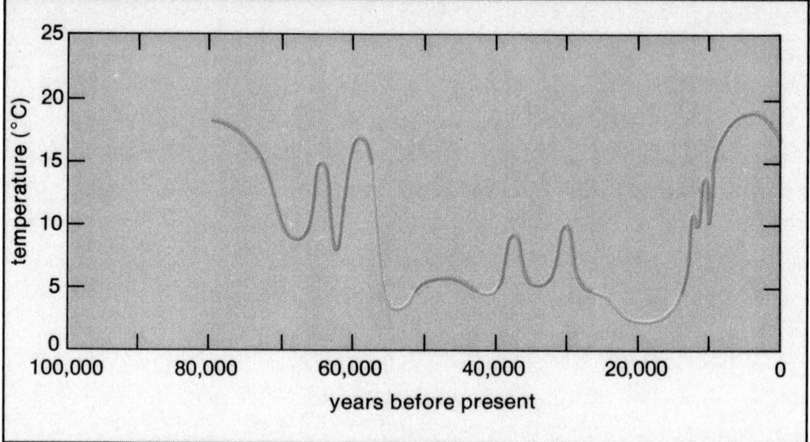

Average temperature of July in The Netherlands over the last 80,000 years, as derived from observations of fossil pollen, shows the history of the last period of glaciation (right). During this period primitive man hunted the large grazing animals common to the time and left a record of his activities in paintings on cave walls in Spain (above) and France (above right).

Strait and Bering Sea, thanks to the drop of world sea level. The same circumstance probably allowed the first aboriginal men to pass from Asia to Australia approximately 25,000 years ago.

After the ice age

The postglacial climatic regime developed rapidly, particularly in Europe, though there was at least one drastic setback when for a period of 500 or 600 years, in the 9th millennium before Christ, glacial conditions returned or readvanced. With the postglacial warming there came a time when the rivers were swollen enormously, particularly in summer, by the melting ice. Gravels and sand were rapidly deposited and quickly produced thick deposits in some places; lakes formed and sometimes quickly silted up and completely disappeared. The landscape was changing rapidly, but the greatest change for the human population and for the animals they hunted was the disappearance of the open plains, as the forest advanced northward in Europe and North America.

Species of trees whose pollen and seed are light and are transported far by the winds, or else whose seed is spread by birds, soon replaced the open grassland and tundra. The first immigrant species in middle and northern Europe were generally birch and pine. These arrived very early—about 10,000 years ago—as far north as Denmark. It took thousands of years for the oaks and elms and beeches to spread from their ice-age refuges, beyond the mountains in southern Europe, and first gain an entry into, and then replace, the earlier established types of woodland in central and northern Europe. Forest fires, and later (in the last 6,000 years) the clearings produced by man, provided opportunities for change in the composition of the forests. In North America the ice-age stands of the more warmth-demanding tree types were not remote as in Europe but were merely a little farther south on the great plains. Therefore, the arrival of the forest farther north was correspondingly more rapid: probably only a few decades for the first trees and a later changeover in the course of 100–300 years to the types that eventually dominated the deciduous woodlands.

Man seems to have adapted to these changes more successfully than did the animals. The ranges of both moved northward, but the extinction of various species (perhaps including the mammoth) and the disappearance of others from Europe, northern Asia, and North America was probably due to the reduction of their numbers by man.

Other great changes in the landscape were brought about by the rapid rise of sea level, which over some thousands of years proceeded at a rate averaging one meter per century. This ultimately separated Britain and Ireland from continental Europe, created the Baltic Sea and The Sound (Øresund), greatly enlarged the Mediterranean, and caused the loss of vast areas of coastal flatlands in many parts of the world that had previously been easily (and perhaps, therefore, densely) inhabited. It seems likely that there was great loss of life and of the primitive industrial sites for making salt by evaporating seawater. Indeed, it has even been suggested that in this way the end of the ice age probably brought about one of those occasions, rare in history, when the total population of mankind was significantly reduced. This is almost certainly the origin of some of the widespread legends of a flood disaster in the early history of mankind.

The postglacial warming took place so quickly and so extensively that by about 6000 B.C., and thereafter for perhaps 5,000 years, most of the world was warmer than it is now. Although it took until well after 4000 B.C. for the last of the great North American ice sheet to disappear, ultimately the forest spread about 100–200 km (60–125 mi) north of its present limit in northern Canada.

As the warm temperatures moved north, all the climatic belts seem to have moved north too, so that the summer monsoon rains reached farther north into the Sahara than they do now. The levels of Lake Chad and other lakes in tropical Africa rose, and some of them became much larger than they are today: this, and the accompanying rains, also may be the origin of some of the flood legends.

From J. C. Schofield and H. R. Thompson, "New Zealand Journal of Geology and Geophysics," vol. 7, pp. 359–370 (1964)

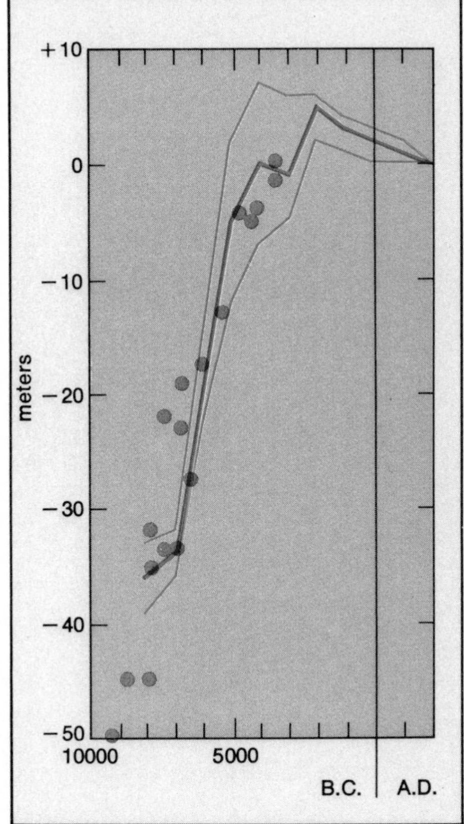

Rapid rise of the world sea level during the last 10,000 years resulted from the postglacial warming trend that began about 10000 B.C. The dots represent individual dated estimates from various parts of the world, and the curves indicate the most probable sequences and the outer limits of all values for the different dates.

Maps from H. L. Crutcher and O. M. Davis, "Navy Marine Climatic Atlas of the World," vol. 8, NAVAIR 50-1C-54; U.S. Naval Weather Service Command

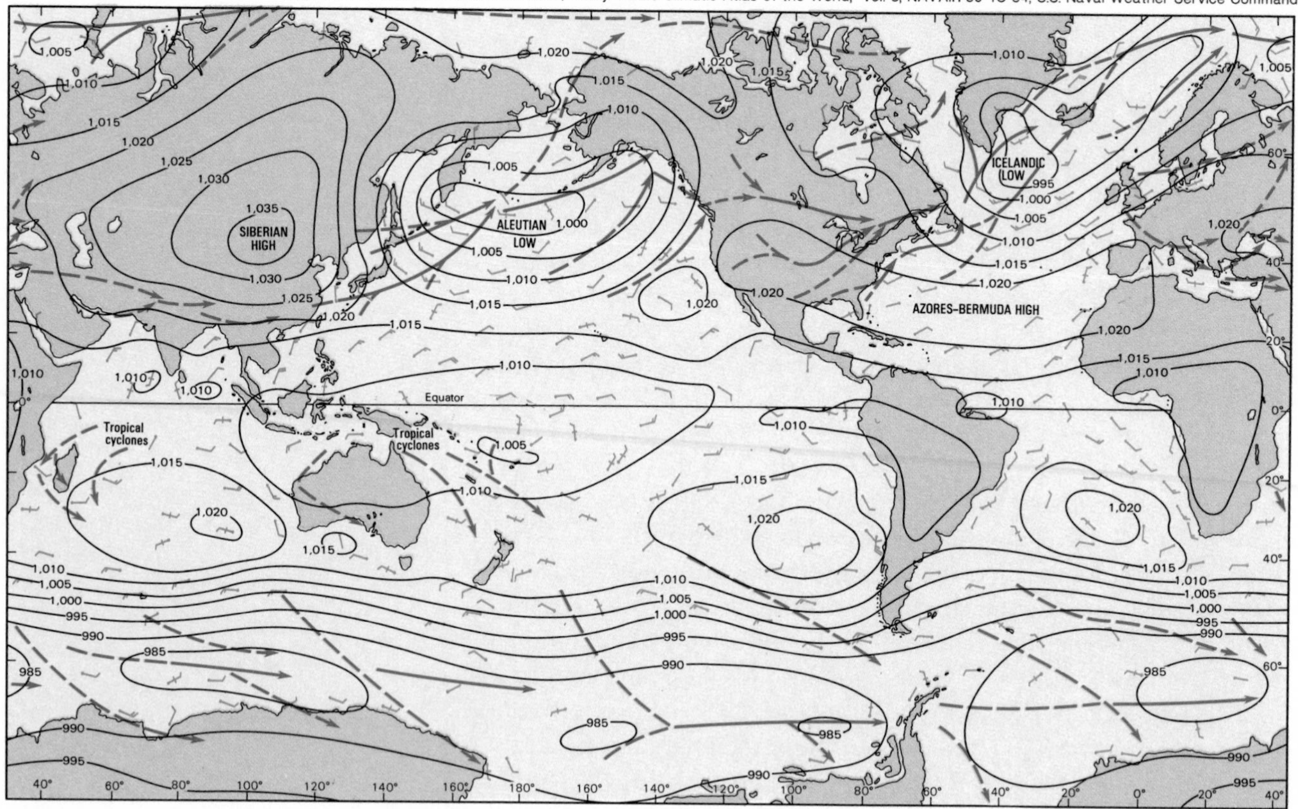

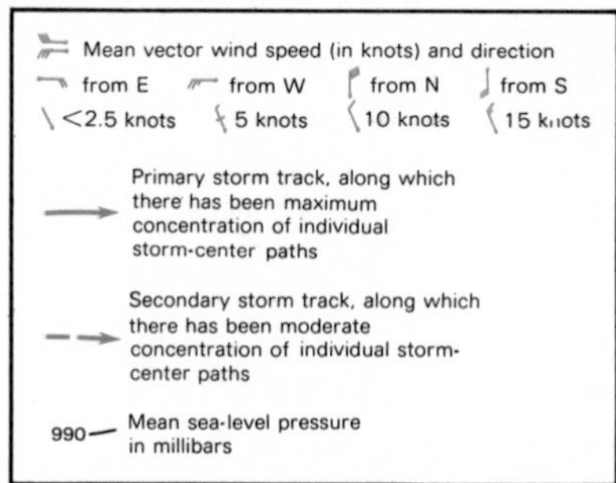

World distribution of mean sea-level pressure (in millibars) for January (above) and for July (opposite page), and primary and secondary storm tracks. Also shown are the general speeds and directions of the global winds.

Many thousands of years earlier, during the glaciation, Lake Chad had been an enormous inland sea (as big as the Caspian Sea today) extending from latitude 10° to 18° N and between longitudes 14° and 20° E. This indicates that under ice-age conditions the climatic zones were displaced toward the Equator, especially in the Atlantic, American, and European sectors, where the ice sheets were most extensive. Therefore, with rainfall from the fronts of the cyclonic depressions traveling east from the Atlantic—the depression centers probably mostly passed through the Mediterranean and from there northeast into Russia and Siberia—the Sahara and the other desert and nearly desert areas of the Near and Middle East received some regular rainfall. Evaporation would have been much less than now because of the lower temperatures and cloudier skies. For these reasons the ice-age climatic regime left a legacy for thousands of years afterward in the form of a higher level of underground water in the subsoil and rock strata of the present deserts than now exists. Radiocarbon tests have indicated that much of the water in the oases and in the water-bearing strata under those deserts today is about 20,000 to 25,000 years old.

There is evidence of a much drier phase in the Sahara and in the Arabian desert, and a lower level of Lake Chad, in the earliest postglacial stages. But in the warmest postglacial times there was some renewal of the moisture by means of a climatic pattern that allowed the summer monsoon rains to penetrate farther north.

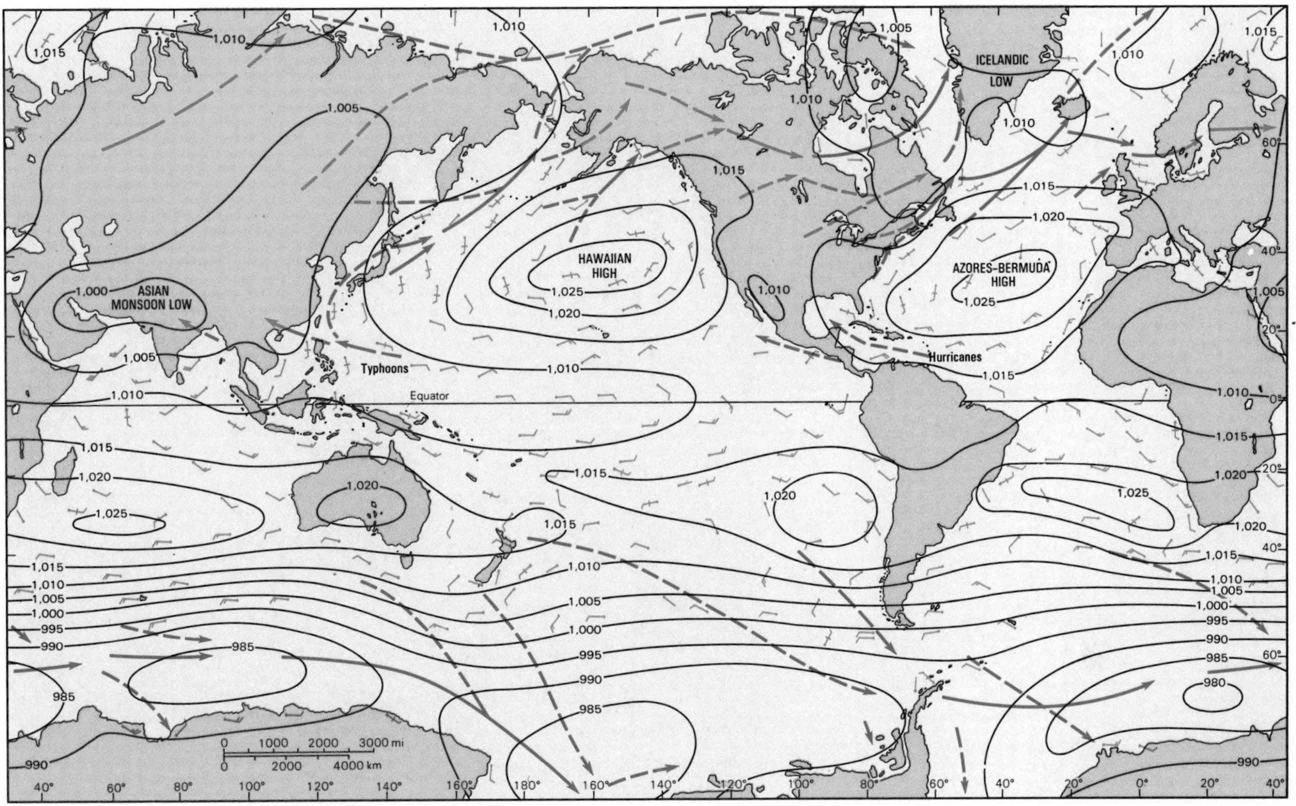

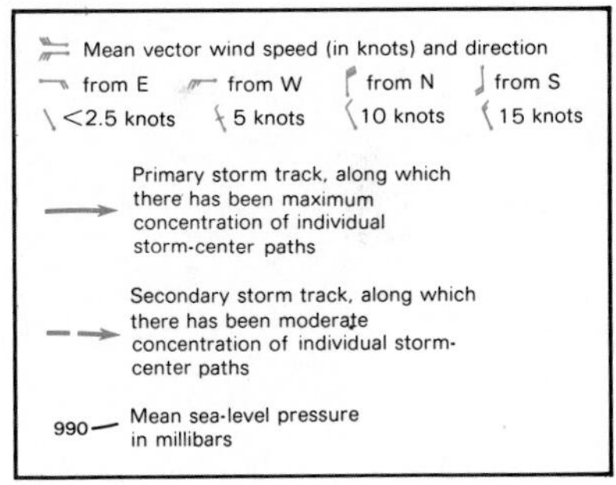

The development of civilization

The early civilizations in Palestine, in the Nile Valley, in Mesopotamia, and in the Indus Valley area and the vegetation that they knew certainly owed something to the great abundance of water stored from an earlier age, as described above. The oases were more extensive and, to judge by the rock drawings from about 5000 to 3000 B.C. found in places in the heart of the Sahara, many species of animals were still able to migrate across what is now a desert region. Climatic fluctuations during those times, however, probably caused variations in the frequency of floodwaters in the wadis (desert streambeds that are normally dry) and in the sizes of the great rivers (as is known to have happened in the case of the lakes). This seems to have been the cause of expansion of population and settlement into areas that are now desert on the fringes of Palestine during two moist phases, around 6000 and 3000 B.C. In both cases the maximum expansion was short-lived. Soon after 3000 B.C. actual records of the time show that the level of the annual floods of the Nile River dropped, and after 2200 B.C. there came at roughly 200-year intervals some sequences of years when the level was so low as to cause starvation in Upper Egypt. The worst of all these times may have been about 1200 B.C., when it is suspected that a widespread increase of aridity provoked migrations of peoples throughout the Near East and an invasion of Egypt by the ancient Libyans and their allies.

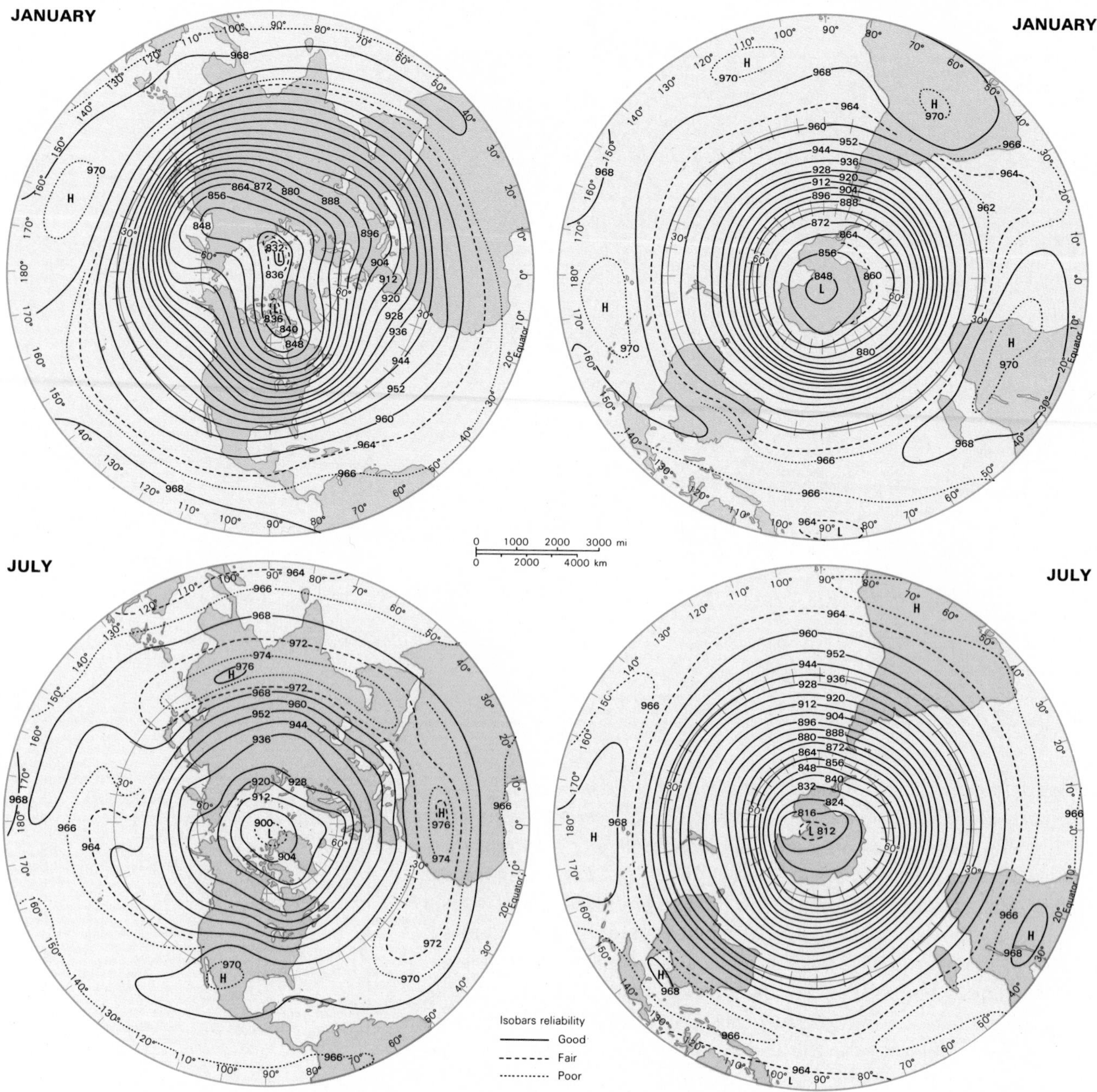

Contour lines are drawn of the height of the isobaric surface where the pressure is 300 millibars. The winds of the upper troposphere (an altitude of about nine kilometers) parallel these contours. The marked seasonal change in pattern in the Northern Hemisphere can be noted.

From H. L. Crutcher, R. L. Jenne, H. van Loon, and J. J. Taljaard, "Climate of the Upper Air," part 1, "Southern Hemisphere," NAVAIR 50-1C-55; U.S. Naval Weather Service Command

The conditions of the postglacial warmest millennia led to an enormous increase in the area of bamboo (Bambuseae) growth beyond its present natural limit in China, where it then abounded over most of the great lowland plain of the Yangtze and the Huang Ho. One deduces that the prevailing temperatures in that part of the world were then as much as 2° C warmer than now over the whole year and 5° C warmer than now in winter. (A 2° C excess over modern values seems to have prevailed in many parts of the world, including Sweden.) Since the bamboo plant provides both convenient building and writing material as well as the use of its shoots for food, it may be that this was important in the early development of a high civilization in China.

Kim Hart from Black Star

Rock drawings in Norway date from about 2000 B.C., near the end of the warmest era. They indicate the use of boats similar in shape to the Viking ships of a later age but without sails.

The warmest millennia also saw the spread of peoples and highly skilled cultures into northern Europe. About 2000 B.C., toward the end of the warmest era, rock drawings in Norway testify to the use of boats (similar in shape to the Viking ships of a later age, but without sail) and skis, and it is known that human settlement reached as far north as latitude 69° N. Agriculture, with wheat and barley, had spread to Denmark and to southern parts of Sweden and Norway about 1,000 years earlier. The numerous stone circles, or temples, built about 2000 B.C., some of which seem also to have served as astronomical observatories (probably for calendar fixing, to determine the normal round of the seasons), were widely distributed from Brittany in northwest France and in western parts of the British Isles to the Outer Hebrides and Orkney Islands. They indicate a main line of communications by sea, over the fringes of the Atlantic, up to near 60° N and probably on into western Norway.

About 2000 B.C. the recession of the forests from the coasts of northwest Scotland and Orkney suggests that the climate was beginning to become windier—which probably points to the beginnings of a renewed cooling of the Arctic—and the seas were becoming rougher. Archaeological evidence from the time suggests that it was then that the main population concentrations in Britain and Scandinavia for the first time shifted to the east side. The glaciers in the Alps had receded before 2000 B.C., and perhaps only then, after so many prevailingly warm millennia, did they reach their postglacial minimum extent. It seems likely that transcontinental routes of travel became more widely developed than ever before, including routes over the Alpine passes. Between 2000 and 1000 B.C. salt was mined in the mountains at the Iron Age site near Hallstatt in Austria, and gold was mined high in the Hohe Tauern range in mines that were later abandoned when the glaciers readvanced.

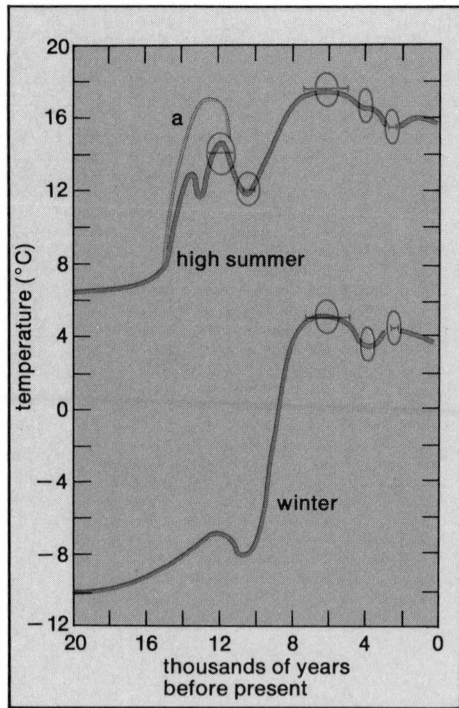

Trend of prevailing temperatures in central England during high summer (July and August) and winter (December, January, and February) is shown for the last 20,000 years. The oval figures indicate the range of uncertainty of the dates and temperatures derived at various times, and the horizontal bars within them show the probable duration of the indicated temperature levels. Values for the earliest millennia rely largely on fossil plant evidence; the alternate line (a) of the summer curve indicates a somewhat different temperature history suggested by fossil beetles.

Neoglacial cooling and subsequent fluctuations

About, or soon after, 1000 B.C. a sharp cooling of world climates was in progress. By 500 B.C. the prevailing temperatures in Europe had been lowered by 2° to 3° C from their previous warmest level. Winter snows and frosts and storminess must have become much more severe in Scandinavia. It has been maintained that it was this that gave rise to the northern legends of the *fimbulvetr* (an exceptionally long winter heralding the end of the world), and the twilight of the ancient gods and the heroic life of long ago, as well as to similar legends of an icy hell in central Asia (Altai region). Certainly there is historical evidence, in the writings of ancient Greek and Roman authors, that there were southward movements of peoples from central and northern Europe in the last millennium before Christ, as well as successive waves of Celtic and Teutonic peoples moving westward over the European plain to France and the British Isles.

During the time of the Roman Empire, and indeed from about 100 B.C. until about A.D. 400, the world's climate seems to have again been becoming gradually warmer, drier, and more stable. The sea level was rising, and grapes and olives were cultivated farther north than before (including districts where even in Italy such agriculture had not succeeded in the 2nd millennium B.C.). Trajan's stone-piered bridge over the Danube, built in A.D. 104–105, survived for almost 170 years without disturbance by ice. It is reasonable to suppose, as Ellsworth Huntington suggested, that it was increasing dryness and failure of the pastures used by nomads on the eastern steppes that set in motion from that region the *Völkerwanderungen*, long known as the barbarian invasions of Europe, which brought about the downfall of the Roman Empire in the West. Indeed, direct evidence of such dryness is the particularly low level of the Caspian Sea at about that time.

A colder period followed, with storms and wetness and more severe winters, particularly between about A.D. 550 and 800. This was followed in its turn by a remarkable warming, which for several hundred years seems to have restored the temperatures in northern Europe and Greenland to near their warmest postglacial level. Agriculture and human settlement spread rapidly farther north and up the valleys and mountainsides in Scandinavia and northern Britain to levels never before occupied, including some places that became marshy or so exposed after A.D. 1300 that they have never been cultivated again. (In England these were the places where the Black Death and subsequent ravages of the plague struck hardest; they were also the places where the population had declined most in the years of bad harvests and famine about A.D. 1315.) It seems likely that the great Viking voyages of discovery and colonization in Iceland, Greenland, and beyond, between A.D. 800 and 1000–1200, were favored by the absence of sea ice and by calmer seas than prevailed in later centuries.

Evidence of climatic worsening in the late Middle Ages can be traced in the records of an increase of the Arctic sea ice, ice on the Baltic, increased storminess in the North Sea, lowering of the tree line on the mountains in central Europe and in California and of the limits of cultivation in many places, and advances of the glaciers in the Alps and elsewhere. First symptoms of the deterioration in Europe seem to have occurred erratically between 1210 and 1320; severe phases followed, particularly around 1430–70 and between 1550 and 1700. A general decline of average temperature level by almost 1.5° C between the 13th and 17th centuries can be deduced. Though there were always occasional warm years (sometimes accompanied by severe droughts) and some warmer decades, the general decline and the variability must have borne hard on the relatively primitive economy and the health of the peoples of Europe and elsewhere.

The cooling was worldwide. (Associations with changes in the amount of radioactive carbon in the Earth's atmosphere suggest a solar origin.) Some of its effects included the dying out of the Norse colony in Greenland, unrest in the Highlands of Scotland, retreat of the agricultural peoples who had been spreading up the Mississippi River Valley in North America, and the abandoning of the cultivation of oranges in the Kwangsi Province of China. There was a similar southward shift of the limit of vineyard cultivation in Europe. The Southern Hemisphere was similarly affected.

Climate in modern times and the activity of man

The cooling trend was reversed about 1700, and, apart from the setback that followed the remarkable warmth of the 1730s and some smaller fluctuations in the 19th century, the history of the climate from the time when thermometers came into general use until the 1940s was a one-way trend toward ever increasing warmth. The warming became much stronger about 1900. In the Arctic, where it was strongest of all and caused a reduction of about 10% in the extent of sea ice, the greatest warming took place after 1920. The climatic change in the first half of the 20th century could well be described as a general improvement of world climates because, in addition to the increasing warmth, significantly more rainfall was reaching the continental interiors (except the United States Middle West), the summer rains extended farther north into the southern fringe of the Sahara, and the Indian monsoon rarely failed. These aspects seem to have been linked with general intensification of the global wind circulation. They certainly eased the problem of producing enough food for the world's increasing population.

The coincidence of the latest period of general warming of world climates with the industrial revolution and of the more intense warming with the industrial growth earlier in the present century led to the suggestion that the warming was due to man's production of carbon dioxide. Since the 1890s there has been a 10–15% increase in the proportion of this gas, which acts as a trap for the Earth's outgoing radiation, in the atmosphere. But since 1945–50 the global average temperature has been falling again, despite an even greater production of carbon dioxide by human activity. This suggests a natural climatic fluctuation that is strong enough to outweigh the effect of the increasing amount of carbon dioxide. With better knowledge of the past record of climate, it may become possible to identify the nature of this most recent fluctuation or the entire series of cyclic changes to which it belongs.

Analysis of the longest available series of weather data, and of data related to weather such as the yearly layers in the ice of the Greenland ice sheet and the yearly growth rings of trees, indicates the presence of quasi-periodic, or cyclic, oscillations, each of which probably corresponds to the normal time scale of some process in the atmosphere, the oceans, or the cosmic environment. The period lengths that are most commonly found are about 2.2, 5–6, 9, 11, 19, 22–23, 50, 90, 170–200, and 400 years; and periods of about 700, 1,300 and 2,000–2,600 years are increasingly suggested by workers who have studied series of exceptional length. Not many of these cyclic processes, except perhaps the 2.2-year cycle and those of 200 years and longer, are often strong enough to appear dominant. They all interfere with each other. The 9-year and 19-year cycles, and perhaps a cycle of about 2,000 years, are probably caused by changes in the range of strength of the combined tidal force exerted by Sun and Moon. Many of the others are often supposed to be due to cyclic variations of solar output, but some may be no more than the frequencies of "beats" (as in musical sounds) between cycles of shorter length.

Latest computations of the increase of man's output of carbon dioxide, including its redistribution between atmosphere and oceans and its effects upon radiation exchanges in the atmosphere, indicate that doubling it would be expected to raise the overall average temperatures prevailing at the surface of the Earth by about 1.9° C.

The future

The increasing scale of man's activities and the variety of ways by which he pollutes the atmosphere are causing anxiety, not least as regards the possible effects on local and, ultimately, on global climate. To chemical pollutants injected near the Earth's surface from industry, traffic, and domestic heating must be added pollution of the stratosphere by high-flying aircraft and rockets (and, it could be, by dust and ashes from the use of nuclear bombs in war). Some of the substances introduced into the atmosphere may one day significantly affect the radiation balance. A straightforward projection of continuity of the growth of man's population, industrial activity, and demands indicates that within about 100–200 years at the longest, the output of heat will come to have a dominating effect on world climate. It has even been suggested by a leading Soviet scientist that, at that time, it may become necessary to spread artificial dust veils in the stratosphere in order to reduce the incoming solar radiation and thereby control the heat and prevent melting of the great ice sheets on land, which would lead to a disastrous rise of world sea level. Despite these anxieties, however, it appears that at the present time natural fluctuations are still the dominant element in the Earth's climatic changes. It is important, therefore, that research should also be devoted to improving knowledge of the past record of climate and gaining better understanding of these natural changes.

The seriousness of these fluctuations in a world that is now so heavily populated that the food reserves (since 1971) have been declining sharply, even in years with good harvests in the main food-producing regions, can hardly be overemphasized. The current climatic trend, which has produced a colder Arctic and increased ice since about 1950, seems to have caused (on average) a slight equatorward shift of the main zones of the global wind circulation and the climatic belts that accompany them. This, in turn, has produced the droughts, by now continued over many years, just south of the Sahara and an increased frequency of failures of the Indian monsoon. In middle and higher latitudes it has altered the character of the prevailing wind circulation patterns, with more frequent blocking anticyclones that persist for weeks or months in varying locations; this change too has given rise to more frequent droughts.

Because of the variability of position of the blocking anticyclones there are correspondingly great differences in the kinds of long spells of weather experienced in one location in middle latitudes during the same season in different years. This variability of the seasons affects both rainfall and temperature. It has caused bad harvests in the central Asian grainlands of the Soviet Union (1972) and in China. The incidence of such failures in some of the world's main grain-producing areas (about one year in four since 1960) must place limits on the total production of food. Because there is no margin of grain-growing capacity remaining, years of great climatic fluctuation or change—whether caused by nature or by man—can no longer occur without bringing dire shortages.

Dendrochronology: History from Tree Rings

by Harold C. Fritts

The growth rings of gnarled and weatherworn pines in the American Southwest have opened vast new possibilities in the study of the Earth's environment.

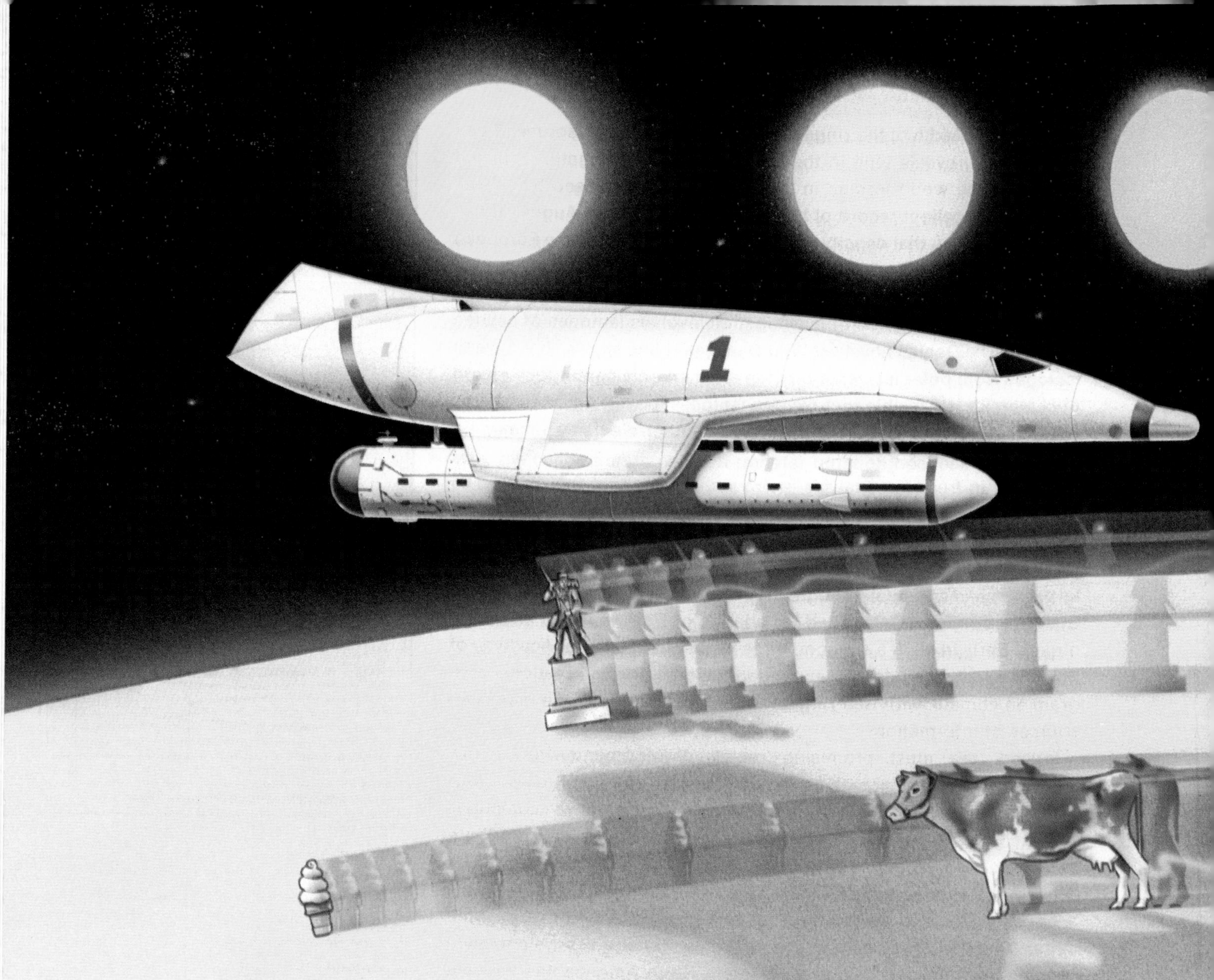

The Speed of Light: A Physical Boundary?

by Fred Hoyle

A basic tenet of physical science is that no velocity can exceed the speed of light. While generally affirming the validity of this principle, physicists have found that it is subject to certain limitations.

Youssi

At the International Congress of Arts and Science, held at St. Louis, Missouri, in September 1904, the French scientist Henri Poincaré commented on his research as follows: "From all the results there must arise an entirely new dynamics, which will be characterized above all by the rule, that no velocity can exceed the speed of light." The speed of light as a physical boundary has remained a basic principle of physical scientists. In recent years, however, research has indicated that there are limitations to this rule.

The present article examines first what is meant by the speed of light, *c*, and how *c* is to be measured. It may be thought that the value of *c* is so well known, about 300,000 km (186,000 mi) per sec, that such a discussion is hardly necessary. But the statement $c = 300,000$ km per sec prompts an important question: Does the same numerical value hold good throughout the universe? Unless this question can be answered unequivocally, discussion is necessary, for only then will it be possible to go on to the second part of the article, namely, to the nature of the physical boundary imposed by the speed of light.

Measuring the speed of light

The classical 19th-century method of measuring the speed of light is illustrated in figure 1. The method depends on the relative position of the points E and E_1. The position of E is measured with the mirror R not rotating. Thus, with R at right angles to the light rays from S, the rays are returned toward S and some of them are reflected to E by the half-silvered mirror G. On the other hand, E_1 is measured with the mirror R rotating. The rays travel from S to position 1 of mirror R. They are reflected to the mirror M, which returns the light to R. Because of the time required for the light to propagate from R to M and back again, R moves slightly away from position 1, causing the rays to be returned to G along a slightly different path; this, in turn, causes E_1 to be slightly displaced from E. By measuring the amount of this displacement and by relating it to the distances from R to M and from R to G, and to the rotation rate of R, the travel time, and thus the speed of light, can be calculated.

On first consideration, it would seem that this method gives an operational procedure for measuring *c*, not very accurately it is true but at least to a valid approximation. Yet this impression is illusory. How are the distances R to M and R to G in figure 1 to be measured, and how is the rotation rate of R to be determined? The classical answer is that distances are measured with a ruler, usually graduated in "meters," and that intervals of time are measured with a clock graduated in "seconds." Before the mid-20th century, the "second" was determined by dividing the rotation period of the Earth into 24 hours, each hour containing 3,600 seconds. Using these measurements, and at the same time neglecting experimental inaccuracies, one would obtain the result $c = 299,792,500$ m per sec.

However, a critic can argue as follows. The graduation of a ruler into "meters" is an arbitrary prescription, and so is the division of the day

__FRED HOYLE__ is an eminent mathematician and astronomer, noted as well for his popular science writing. He currently is Sherman Fairchild Distinguished Scholar at the California Institute of Technology.

(Overleaf) Illustration by John Youssi

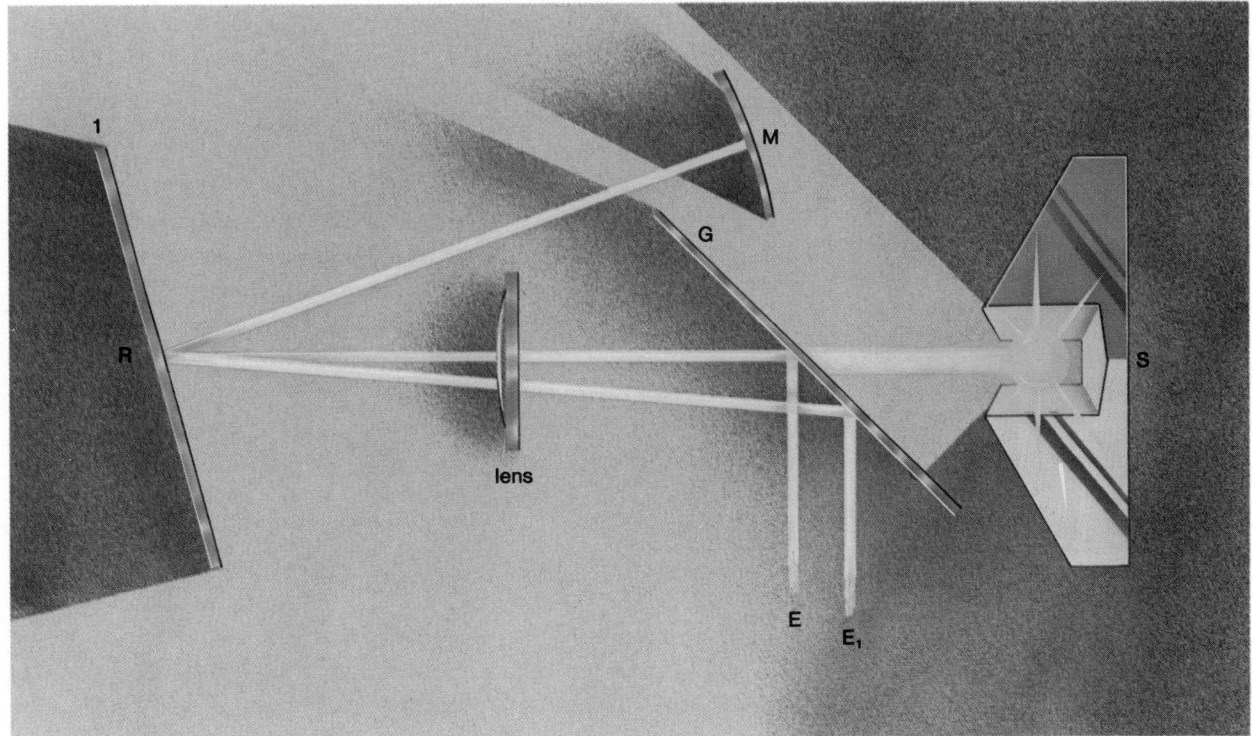

Figure 1. Measuring the speed of light by the 19th-century method of classical physics requires a fixed mirror M, a rotating mirror R, and a half-silvered mirror G. The method depends on the relative positions of the points E and E_1.

into 86,400 "seconds." Therefore, the experiment of figure 1 only has used the properties of light to relate these two arbitrary choices, through the expression

$$c \times 1 \text{ sec} = 299{,}792{,}500 \text{ m}$$

The choice of a ruler or a clock with different gradations would certainly change the number on the right-hand side of this equation. Thus, there is evidently nothing absolute about 299,792,500. If the speed of light is to have an absolute quality, the quality cannot lie in this particular number.

What then is the absolute quality of the speed of light? Suppose we seek such a quality in the following manner: Let the procedure of figure 1 be carried out in different places and at different moments in time. With the meter and the second always defined in the same way, the speed of light is always 299,792,500 m per sec. This proposition contains the germ of a useful idea, but the details remain open to criticism, for how can we be certain that the meter and the second really do remain the same? By imagining the same ruler and the same rotating Earth to be transported all over the universe? Such transport is impossible, not only in practice but also in principle, for terrestrial pieces of material cannot be transported everywhere, not even in the abstract. Near a neutron star the gravitational field is so strong that both the ruler and the Earth would collapse into a tiny smear of material. Of course, we could imagine regions of strong gravitation being avoided in the application of the above proposition, but such a position would be too weak to be worth pursuing. If we are to consider

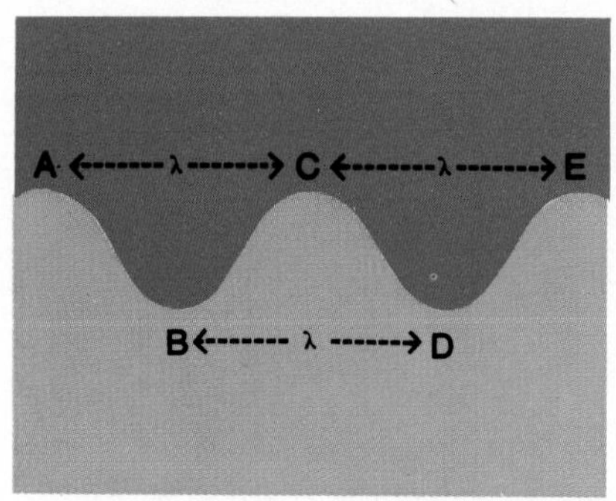

Figure 2. An oscillation occurs at every point of a wave; λ denotes wavelength, the distance between two adjacent wave crests or troughs.

Figure 3. Movement of a float shows that the oscillation at any one point is completed in the same time that the whole wave takes to move through a distance equal to λ.

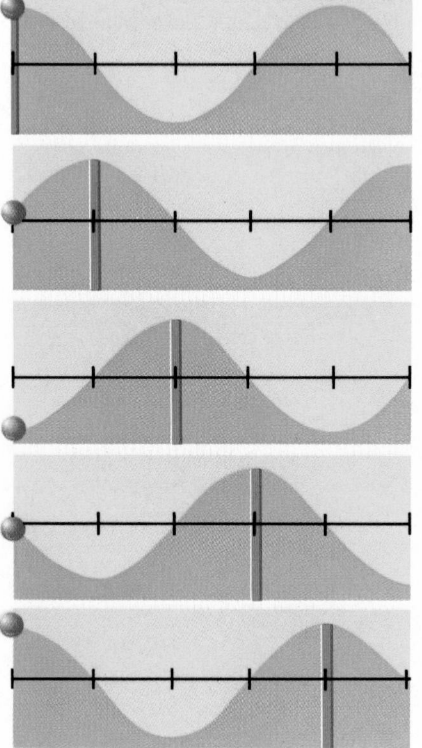

the speed of light as a physical boundary, it is precisely in such cases as regions of strong gravitation that we shall be interested.

At present, physicists do not use rulers for measuring spatial distances. Nor do they base time measurements on the spinning Earth. Radiation emitted in a specific transition of a specific kind of atom—one of cesium—is used for time. The radiation has a wave structure with certain properties that it shares with an ordinary water wave. These similar properties are the ones used for obtaining measurement units, with respect to both spatial distances and time. It is sufficient, therefore, to discuss the measurement method using a simple water wave rather than the more complex radiation wave.

A wave has two basic properties. First, at each point there is an oscillation; in the case of a water wave the water moves up and down. This is easily shown by putting a float on top of the water. Second, there is a spatial correlation between the up-and-down motions at different points. This is illustrated in figure 2. A peak at A is followed by a trough at B, and that trough is then followed by another peak at C, and so on. Not only is there an oscillation at each point taken by itself, but different points have an orderly relation with respect to each other. This spatial ordering is measured by the wavelength (λ), the distance between two adjacent wave crests or two adjacent troughs. As time proceeds, the whole spatial pattern moves along as shown in figure 3. The effect of this motion is to produce the oscillation at each separate point. At one moment, at a given place, the wave is up, and at a later moment it is down. The time required to complete the oscillation at each point is simply the time required for the wave to travel through a distance equal to the wavelength. If the speed of the wave is c, then the time required for the wave to move through the distance λ is simply λ / c. This is the time that a float placed on the water takes to move from its highest to its lowest position and back again.

The number of oscillations to occur per unit time is known as the frequency (ν); therefore, $1/\nu$ is the time for a single oscillation. From the considerations of the previous paragraph $1/\nu$ must, therefore, be the same as λ / c, so that $c = \lambda\nu$. This relation holds just as well for light as for water waves.

Suppose now that we elect to use such a wave for establishing units both of spatial distance and of time. The wave itself is defined as that which results from atoms of a specific kind undergoing a specific transition. The wavelength λ can be taken as giving a unit of spatial distance, while the oscillation time $1/\nu$ gives the unit of time. In such a system of measurement we have

$$\lambda = 1,\ 1/\nu = 1;\ \text{therefore},\ \nu = 1$$

With both λ and ν thus unity, the relation $c = \lambda\nu$ gives $c = 1$. Inevitably, in such a system of measurement the speed of the wave is unity. This is true whatever kind of atom and whatever transition of it was used to produce the wave.

The position reached thus far in this attempt to give an absolute quality to the speed of light might seem frustrating. In using radiation

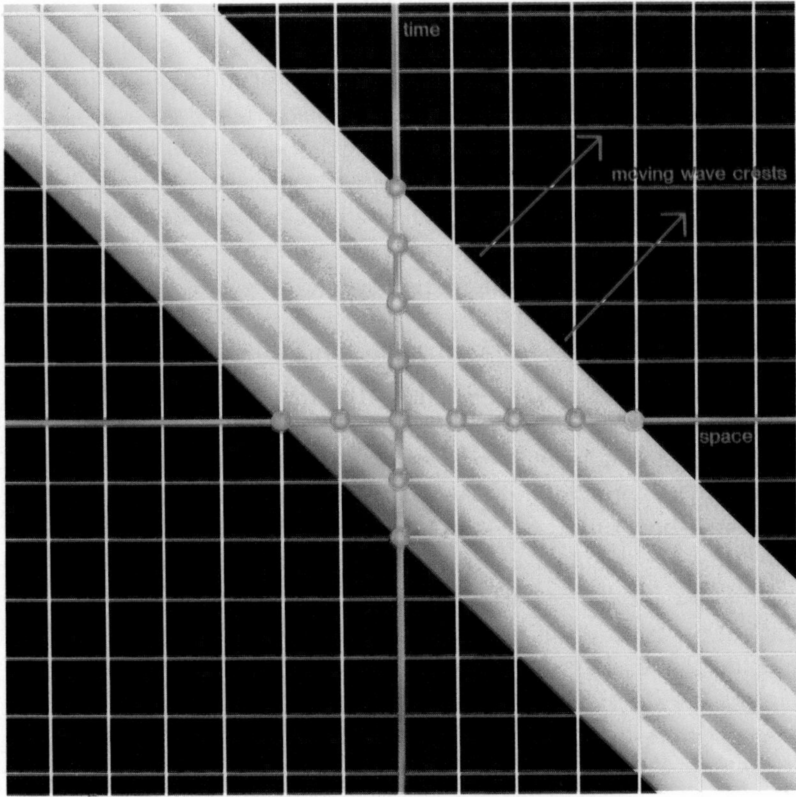

Figure 4. Moving wave crests serve to determine units of spatial distance and time.

of a definite wavelength—whether a radio wave, infrared, light, ultraviolet, X-rays, or gamma rays—to determine units both of time and of spatial distance, the speed of light would always be determined as unity. The situation is that shown in figure 4, with the radiation always propagating at an angle of 45° in a local space-time diagram. Of course, different waves could be used, one to define a unit of time and the other to define a unit of distance. The speed of light would not then be unity, but its value would tell us only about the way the two waves happened to be chosen; the value would tell us nothing about the light itself.

In spite of this apparent frustration, however, several notable advantages have been gained. The speed of light certainly does have an absolute quality; it is precisely equal to 1 everywhere. There is no requirement for the transport of rulers or of the spinning Earth. Every observer can establish the result $c = 1$ for himself, by generating radiation from atoms of a specific kind taken from his own neighborhood. Different observers are not even required to choose the same kind of atom. And with $c = 1$ for every observer, a standard of comparison is established everywhere throughout the universe. Speeds of particles can be determined relative to the speed of light, and it can be asked in an unambiguous way whether in such a comparison the speed of light is ever exceeded.

Local problems

The speed of light is believed to represent a physical boundary in a sense that is precisely defined. The first requirement of the discussion is that the region of space-time under consideration not be too large. The implication of this restriction will appear when nonlocal problems involving large regions of space-time are discussed.

The belief in the speed of light as a physical boundary dates from the first few years of the 20th century. Hendrik Lorentz was at that time concerned with relating the frequency and the direction of propagation of radiation emitted by moving atoms to the frequency and direction of radiation from stationary atoms—the problems of the Doppler shift and of aberration, respectively. These problems were completely solved by Einstein in a famous paper published in 1905. Yet even before Einstein's paper, Poincaré had enunciated his "Principle of Relativity" according to which the speed of light was to be regarded as a physical boundary.

Poincaré's statement can be expressed in the following way. Light emitted from a source S always travels at 45° to the time direction in a diagram of the form of figure 4. Remembering that space-time really has four dimensions rather than the two dimensions of figure 4, the light rays from S must determine a three-dimensional cone known as the light cone. Lines from the vertex of the cone at S make an angle of 45° with the time axis. Such lines can be extended backward so as to form a double cone, as is indicated schematically in figure 5. If S is considered to lie on the path of a particle, then the whole trajectory of the particle lies within the double cone from S. Moreover, this property holds for every point on the trajectory of the particle. This is the meaning of the statement that the velocity of the particle is less than the speed of light.

Poincaré's statement, quoted at the beginning of this article, called for a new form of dynamics in which the situation of figure 5 would always hold good. This form of dynamics was given by Max Planck in 1906. According to Planck's new dynamics the mass of a particle depended on the form of its trajectory. The mass changed with the direction of motion, becoming infinitely large as the trajectory turned to a 45° angle with the time axis. Since an infinite mass would imply infinite energy given to the particle, it was concluded that no particle could ever be turned literally to the 45° angle. Thus, Poincaré's statement was seen to be correct.

These ideas have not been changed over the past 70 years. Indeed, they have been amply confirmed by experiment, sometimes in remarkable ways that could scarcely have been expected in the early years of the century. Perhaps the most remarkable confirmation has come from cosmic rays, which contain protons having masses some 10^{11} times larger than the mass of a proton at rest. Such protons in the cosmic rays move in trajectories very close to the limiting angle of 45°, but they are still not quite at that limit.

The main difference between the present-day point of view and that

Figure 5 (above left). Trajectory of a particle passing through S lies wholly within the double cone defined by the propagation of light from S when the light is traveling in a vacuum.

Figure 6 (above). When light is propagated in a medium (glass) other than a vacuum, a fast-moving particle can follow a trajectory outside the double cone. The units of time and spatial distance are considered as being established by light in the medium.

of Poincaré, Einstein, and Planck is that, whereas the early workers believed their statements to be true generally, today scientists apply certain limitations. The most important of these limitations is the one already noted, that the considerations apply locally. This limitation will be considered later but other limitations of a less basic nature will be discussed in this section.

Suppose a light wave is propagated through glass. Provided the glass is of uniform texture and does not contain air bubbles, the wave has a well-defined frequency and a well-defined wavelength. It can be used to determine units of spatial distance and of time for the region of space-time occupied by the glass. The double cone of figure 5 can again be drawn for a light source S within the glass, and once again the lines from S that define the cone all make an angle of 45° with the time direction. But in this case our precept that the trajectory of any particle through S must lie wholly within the cone is wrong. A highly energetic cosmic-ray particle that happened to pass through the glass, and through S, could have a trajectory outside the double cone, *i.e.,* a velocity greater than that of light through glass, as in figure 6. This situation actually happens, and it has been widely studied experimentally. Electrically charged particles with the property of figure 6 emit radiation of an unusual kind, known as Cherenkov radiation, a circumstance that is frequently made use of in modern high-energy physics.

To restore the situation in which the trajectory of any particle through S lies within the cone, we must add the condition that the light cone of figure 5 is to be considered in a vacuum. The speed of light then reasserts itself as a physical boundary. In practice, the space and time units are determined with respect to radiation in a vacuum rather than with respect to radiation propagating through a medium, as in figure 6. With the units determined in a vacuum figure 6 would need to

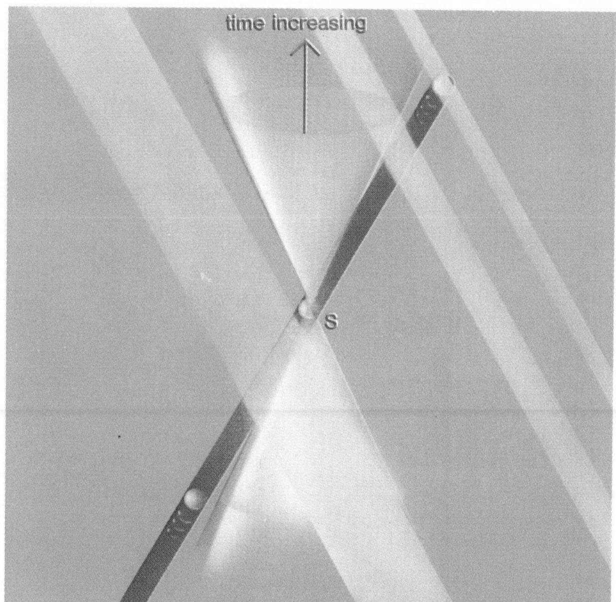

Figure 7 (above). When the units of time and spatial distance are established in a vacuum, the double cone given by light in a medium is narrowed. A fast-moving particle then follows a trajectory that makes an angle of less than 45° with the time direction. In figure 8 (above right) light passes from a distant observer at K to an observer at S. Particles at both K and S have trajectories lying within their respective cones.

be redrawn, as in figure 7, with the cone narrowed and with the cosmic-ray particle trajectory now making an angle of less than 45° with the time axis.

It is possible to consider what the properties of a particle would be if the situation of figure 6 occurred even in a vacuum; that is, if the speed of light was not a physical boundary. This problem has been considered quite deeply in recent years. However, the particles, called tachyons by theoretical physicists, are hypothetical and have not been found experimentally. If they were, Poincaré's statement of 1904 would no longer be valid.

The discussion has so far been entirely concerned with classical particles. For completeness, a somewhat curious situation in quantum mechanics must be mentioned. Instead of going from the point S of figure 5 by one particular defined trajectory, as does a particle in classical physics, a particle in quantum mechanics has many possible trajectories through S, for which probabilities can be calculated. The situation concerning the speed of light as a physical boundary would be the same in quantum mechanics as in classical physics if the probabilities for all trajectories through S going outside the light cone of figure 5 turned out to be zero. This is essentially true when figure 5 is taken on a coarse enough scale, but on a fine scale (small time intervals of less than 10^{-20} sec) the probabilities for such paths are not zero. Thus on a fine enough scale, and according to the way quantum mechanics is usually formulated, the speed of light is not a physical boundary. However, one can argue that the usual calculation of probabilities includes trajectories of a nonlocal nature and that, if these nonlocal trajectories were excluded, the speed of light as a physical boundary would be reasserted. This would be the writer's interpretation of the situation.

Nonlocal problems

Returning to classical physics, let us see to what extent the situation of figure 5 can be extended to nonlocal problems. Suppose there exists a distant point K with S on the light cone of K, so that an observer at S can receive a message from an observer at K. The situation is illustrated in figure 8, with the same limitation in the trajectory of a particle at K as on a particle at S. Both must lie within their respective cones, with the cones both defined by lines that make a 45° angle with the time direction.

Now suppose figure 8 to be drawn on a sheet of rubber, and imagine the rubber to be stretched in the region between K and S so as to change figure 8 to figure 9. Such a situation would be analogous to one in which (1) an observer at S makes space-time measurements in such a way as to give the space-time diagram of figure 5 for his own locality; and (2) gravitation produces the distortion in the region between K and S. What the observer at S cannot do when strong gravitational fields exist in the region between K and S is to maintain figure 5 outside his own locality.

Likewise, an observer at K could make his space-time measurements in such a way as to give the space diagram of figure 5 for his locality, but then the distortion between K and S would affect the situation near S, as in figure 10. In the presence of strong gravitation, neither observer would be able to maintain the simple form of figure 8 on a nonlocal scale.

The concept that particle trajectories can never make angles greater than 45° with the time direction must now be abandoned. Although the 45° rule continues to apply in figure 9 for the locality of S, it no longer applies to particle trajectories in the neighborhood of K. In the system of measurement set up by S, trajectories near K can be so distorted

Figure 9 (above left). Distortion in region between K and S produces a situation at K which causes a particle to have a trajectory at an angle greater than 45° with the time measurement system established by S. In contrast, when the time measurement system is established by K, the distortion affects the situation at S, as in figure 10 (above). Such distortion and the distortion at K in figure 9 are not necessarily symmetrical; thus, the particle trajectory at S in figure 10 may not make an angle greater than 45° with the measurement system at K.

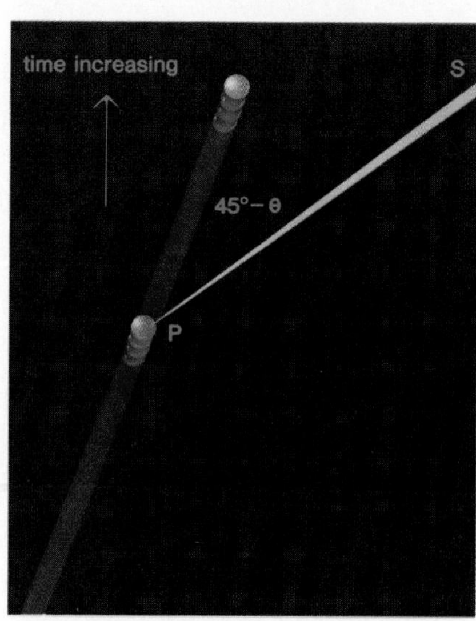

Figure 11. Light from moving atoms at P is received by the observer at S.

that they make angles greater than 45° with the time direction. Notice, however, that the basic requirement continues to hold for a particle at K: the trajectory still lies within the light cone at K. In this critical sense the concept that a particle can never move faster than light continues to hold good, no matter how great the distortion effect.

In addition to the gravitational distortion between K and S, other severe distortion could be introduced near K such as would occur if a massive dense body were located near K. There would still be no change of the basic physical property: Particle trajectories at K would continue to be contained within the light cone at K. So far as is known, this rule is never violated.

The Doppler shift

Let us return to the local situation of figures 4 and 5 and ask how the physicist verifies experimentally that particle trajectories never turn to angles of more than 45° with the time direction. In figure 11, an observer at S is considered to receive radiation from a collection of atoms of a particular kind that are undergoing a particular transition. The frequency of the radiation is then well defined for the observer at S, who measures it to be, say, ν_p. The observer then compares ν_p with the frequency ν_{lab} of radiation produced by stationary atoms of the same kind undergoing the same transition in a laboratory at S. When the atoms P are in motion toward S, it is found that ν_p is greater than ν_{lab}; when P is in recession from S, the opposite situation occurs. In the special situation in which the atoms P move directly toward or directly away from S, there is a relation between ν_p, ν_{lab}, and the angle θ that the trajectory of the atoms at P makes with the time direction established by S. The relation in question is the one found by Einstein in his paper of 1905, namely

$$\frac{\nu_p}{\nu_{lab}} = \begin{cases} \left(\dfrac{1+\tan\theta}{1-\tan\theta}\right)^{1/2}, & \text{approach} \\ \left(\dfrac{1-\tan\theta}{1+\tan\theta}\right)^{1/2}, & \text{recession} \end{cases}$$

This result has the interesting property that as θ goes to 45° (and $\tan\theta$ therefore goes to unity) ν_p goes to infinity for approach and to zero for recession. In practice, ν_p never quite goes to zero nor does it become unlimitedly large. This shows that θ never quite goes to 45°—which was the experimental problem to be solved.

In regard to the nonlocal case, once gravitation produces the distortion encountered in passing from figure 8 to figure 9, Einstein's 1905 relation between ν_p, ν_{lab}, and θ cannot be used. The above equations are not just wrong, they are meaningless. Much confusion has arisen in astronomy and cosmology through attempts to use them, and this mistake must be guarded against.

Nevertheless, the observer S in figure 9 can measure ν_p from atoms P situated near the observer K, and he can seek to obtain information by comparing ν_p with ν_{lab}. The procedures he must follow are different from the Doppler shift described above: They form the science of cosmology. While a discussion of cosmology in any depth is beyond the scope of this article, some general comments by way of conclusion may be of interest.

Observations in astronomy often give ν_p less than ν_{lab}. Such observations are known as red shifts. Distant galaxies show red-shift effects, and the more distant the galaxy the greater the red shift. Still more distant galaxies than those actually observed would presumably give ν_p still smaller compared to ν_{lab}. Can there be galaxies for which ν_p declines to nothing at all? The answer to this question is affirmative in all well-known systems of cosmology. Can there be still more distant galaxies? The answer to this further question is affirmative in some systems of cosmology but not in others.

A situation somewhat similar to this can be found in the development of a black hole. Let atoms P at K in figure 9 lie at the surface of an imploding object. As the resulting black hole develops, ν_p decreases more and more in comparison to ν_{lab}. Eventually ν_p declines to zero, and communication between K and S then ceases. No further message can pass between observers at K and S. Must an observer at S then assume that K has ceased to exist? According to the usual ideas of Einstein's theory of gravitation, the answer is no. Communication to the outside world has ceased, but K continues to exist at least until it plunges into the space-time singularity associated with the black hole.

Erroneously, what is sometimes done in this black hole situation is to use the above relation between ν_p, ν_{lab}, and θ. With ν_p having plunged to zero, it is incorrectly argued that θ has gone to 45° and that, in the subsequent evolution of the black hole toward its eventual singularity, θ has increased above 45° and, therefore, the speed of light has been exceeded. This application of the purely local relation among ν_p, ν_{lab}, and θ is, however, both physically wrong and conceptually misleading. The speed of light as a physical boundary is not vitiated in this way and it remains as real today as in Poincaré's day.

The Natural Satellites

by Isaac Asimov

Long ignored by astronomers, the natural satellites of the solar system are now being examined by space probes and with large telescopes, revealing new and sometimes unexpected findings.

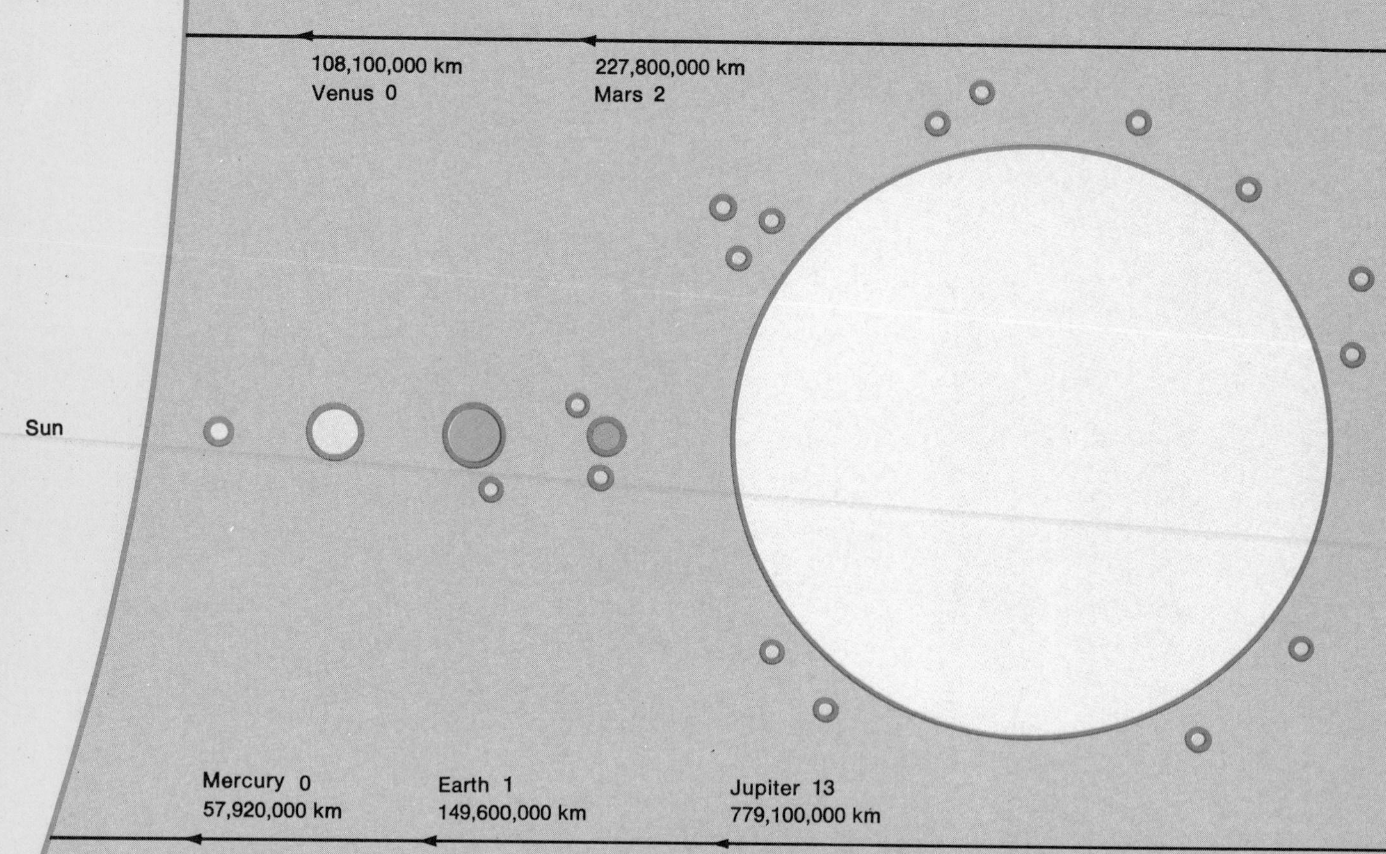

The natural satellites of the solar system suffer by their association with the far more glamorous planets that they circle. The satellites are attendants, hangers-on, commonly dismissed in a final paragraph after a long discussion of the planet. In recent years, however, this has begun to change. The new age of space exploration has led to interest in a number of the satellites in their own right. This is true even for the one natural satellite that has always seemed a distinct world to us by reason of its closeness, our own Moon.

The Moon

In many ways the Moon is the most remarkable of the natural satellites, quite apart from its relationship to the Earth. Among the 33 known satellites of the solar system, it is the closest to the Sun, because those planets closer to the Sun than the Earth, namely Venus and Mercury, have no satellites. In relation to the world it circles, moreover, the Moon is by far the largest of all the satellites. Its diameter is 0.27 and its mass is 0.0123 those of the Earth. No other satellite even approaches such values. In comparison to the planet it circles, the Moon is about ten times as massive as its closest rival. Thus, the Earth and its Moon can reasonably be spoken of as a "double planet," something unique in the solar system.

This closeness in size has given rise to some speculation that the Moon is too large to be a true satellite of the Earth, that it must have been a planet in its own right at one time and been captured by the

ISAAC ASIMOV *is Associate Professor of Biochemistry at the Boston University School of Medicine and a prolific author of scientific texts and science fiction.*

(Overleaf) Courtesy, NASA; illustrations by Dave Beckes

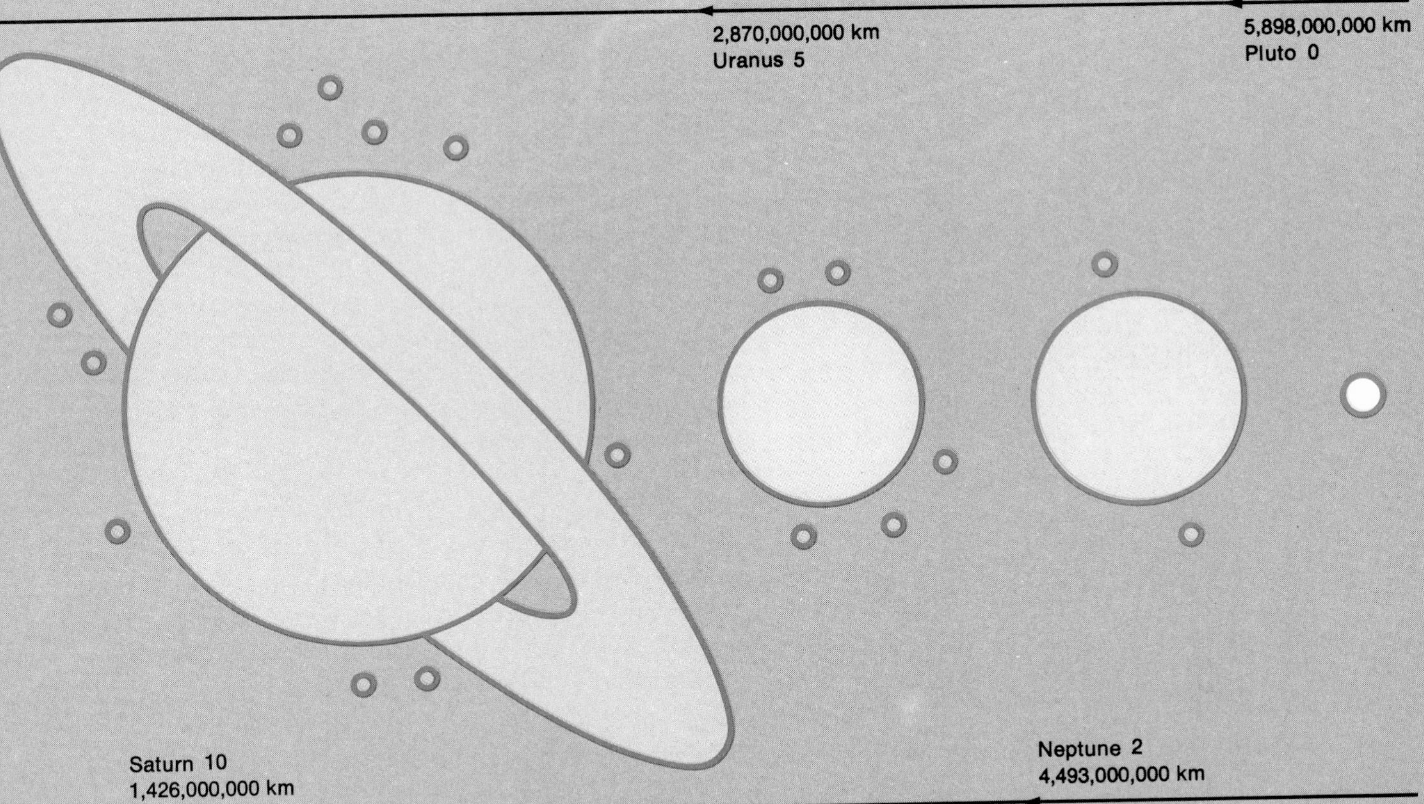

Earth. True satellites, which were formed in the same process that shaped the planet and which coalesced out of the outer edges of the dust cloud that condensed to form the planet, would be expected to revolve in the plane of their planet's equator and in the same direction as the planet rotates. They would also be expected to have nearly circular orbits. These properties hold true for 20 of the 33 satellites.

Captured satellites would be expected to have eccentric orbits that are inclined, sometimes greatly, to the plane of their planet's equator. They would not necessarily be moving in the same direction as their planet rotates but might be retrograde; that is, moving in the opposite direction. They would generally be considerably farther from the planet than are the true satellites, and would also be smaller. Ten of the 33 satellites are generally accepted as having been captured. If the 20 undoubted true satellites are added to these, one can see that there are three remaining satellites that fall into a doubtful category. Among these is our Moon.

The eccentricity of the Moon's orbit is 0.055, a value greater than that for the undoubted true satellites, and its inclination to the Earth's equatorial plane varies from 18° to 28° and is only 5° from the plane of the Earth's orbit around the Sun—a characteristic more appropriate to a planet than to a satellite. The Moon is also unusually far from the Earth. Its mean distance of 384,400 km (238,900 mi) is 60 times the Earth's radius; none of the true satellites approaches that distance in terms of the radius of its primary. (This may not be as important as it

The planets of the solar system and their natural satellites are shown in order of their distance from the Sun. The sizes of the planets relative to one another are depicted but not their distances from one another and from the Sun. Only the number of satellites for each planet is shown; their relative sizes and distances from their planet and each other are not portrayed.

Bettmann Archive

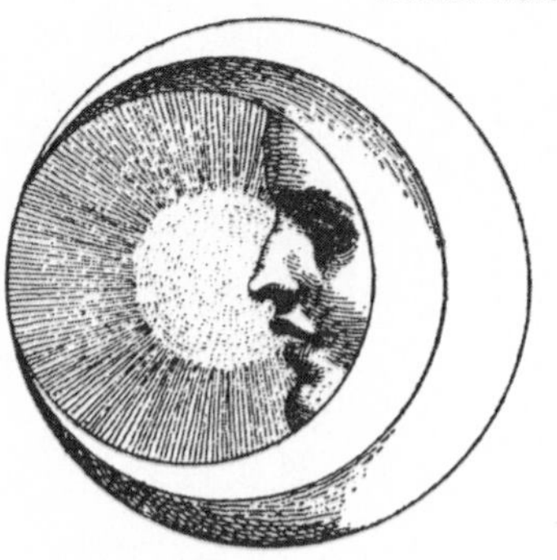

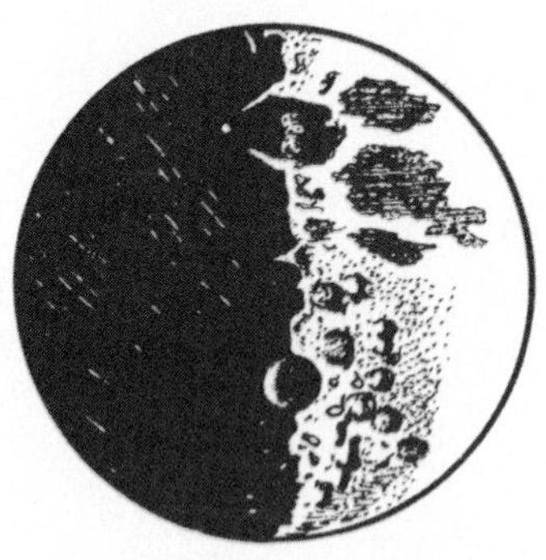

Fanciful early rendition of the Man in the Moon bears a resemblance to the drawing below published by Galileo in 1610. Galileo theorized that the dark spots might be bodies of water. (Bottom) Picture published in the New York Sun *in 1835 was described as "A View of the Inhabitants of the Moon" as seen through the telescope of Sir John Herschel. The picture and accompanying story were soon found to be a hoax perpetrated by British-born essayist Richard Adams Locke.*

sounds, however, for the Moon is slowly retreating from the Earth and must have been considerably closer at some former time.)

Lunar rock samples obtained by successful flights to the Moon show that its crust is at least 4.6 billion years old, indicating that it reached its present state in the early period of the formation of the solar system. The crust seems to lack water almost completely, however, and to differ from the Earth's crust in being considerably poorer in those elements whose common compounds are volatile (form vapors at relatively low temperatures). Combined with the common occurrence of glassy grains on the lunar surface, this seems to indicate that the Moon has experienced periods of surface heating of a kind which did not occur in the Earth's crust.

The lunar crust may have been heated by the meteoric bombardments it suffered during the process of its formation, however, and this was an effect from which the Earth's crust was protected by the atmosphere and ocean. In that case, the differences are consistent with formation of the Earth and Moon at the same time in approximately the same place, and remaining together in the same system thereafter. For some time after its formation (perhaps in the Earth's neighborhood), the Moon may have traveled for some reason in an elliptical planetary orbit that carried it considerably closer to the Sun at perihelion and nearer the Earth's orbit at aphelion. The chief argument against this point of view is that the mechanics of capturing a body the size of the Moon by a body as small as the Earth, and thus making a satellite out of a planet, involves conditions too proscribed to be credible.

Astronomers hope that it may one day be possible to analyze the crust of the planet Mercury, which also lacks a significant atmosphere and which is much closer to the Sun than are the Earth and Moon. A comparison of the lunar crust with that of Mercury might prove to have an important bearing on the question of whether or not the Earth captured the Moon.

In one respect, the Moon is unique in the Earth's sky. It is the only heavenly body that presents to us only one face. The far side of this nearest of objects was an impenetrable mystery until October 1959, when a Soviet lunar probe passed beyond it and sent back the first crude photographs of that far side. Since then, more sophisticated probes have succeeded in mapping the entire surface of the Moon in great and accurate detail.

This mapping has revealed that the Moon has a surface asymmetry. The far side is heavily cratered but lacks the "maria," or "seas," that cover a large fraction of the near side. These seas, roughly circular,

(Opposite, top and center right) courtesy, NASA; (center left) courtesy, Brookhaven National Laboratory; (bottom) NASA from Black Star

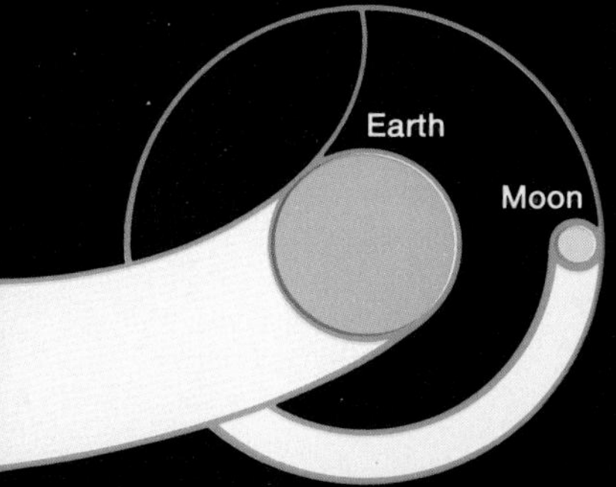

Rocky, cratered nature of the Moon's surface is revealed in photographs taken during the U.S. manned Apollo spacecraft program. Far side of the Moon (top left) was photographed from the command module of Apollo 15. Astronauts from Apollo 11 brought back to Earth a brecciated basalt, whose crystalline texture is shown in polarized light (center left), while the boulder with multiple cracks (center right) was photographed by the men of Apollo 17. Apollo lunar module is shown at bottom.

Courtesy, NASA

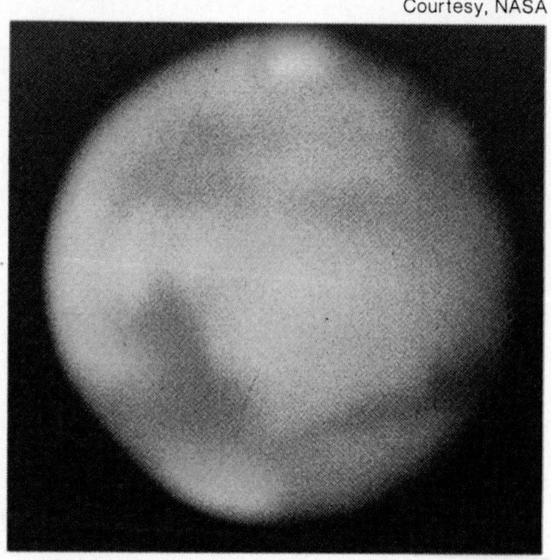

(Above) Mars as seen through the Catalina Observatory 61-in. telescope in 1967. The white spot at the top is the north polar cap, and other white markings are clouds or haze. (Below) Drawing by Italian astronomer Giovanni Schiaparelli in 1888 shows the canali *that he observed on Mars. His observations led to speculation that these channels might have been built by intelligent beings, but space probes in the 1960s revealed them to be alignments of large craters.*

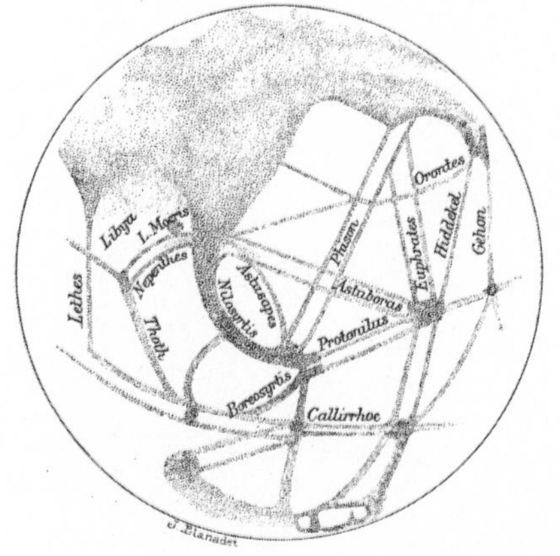

large, and relatively uncratered, were formed by catastrophic processes about 500 million years after the rest of the Moon was formed. The seas may have been formed at a time when the Earth's presence exerted an influence on the Moon, perhaps for the first time—when the capture was effected.

Phobos and Deimos

Mars, the next planet beyond the Earth from the Sun, has two satellites. These are insignificant bodies, so small that they were not discovered until 1877. Then, at a time when Mars was making a particularly close approach to the Earth, the U.S. astronomer Asaph Hall systematically scoured the neighborhood of the planet in search of possible satellites. He finally gave up, but his wife urged him to try one more night. He did, and that was the night he sighted Deimos, the first of the discoveries.

The two satellites are named Phobos and Deimos ("fear" and "terror") after the sons of the war god Ares (Mars) in Greek mythology. Both move around Mars in nearly circular orbits in the equatorial plane and may be considered true satellites. They are remarkable for their closeness to their primary. Phobos, the inner satellite, is only 9,350 km from the center of Mars, or 2.75 Martian radii (1 km = 0.62 mi). It is only 6,000 km above the Martian surface, and its period of revolution is 7 hours and 39 minutes, the shortest for any of the 33 satellites. From the dimness of the two satellites, it was clear that they were very small bodies, and prior to 1971 nothing but their orbital characteristics was known.

In December 1971 the U.S. Mars probe Mariner 9 sent back photographs of the satellites and revealed them to be heavily cratered. This was the first time any satellite, other than our Moon, had been seen from a close distance. In bodies as small as the Martian satellites the feeble gravitation is insufficient to force matter into a spherical shape, and the impacts that formed the craters clearly broke off bits of the satellites and increased their irregularity. Based on the cratering, astronomers deduce that the satellites are solid rock, were formed early in the history of the solar system, and were, perhaps, considerably larger bodies at first.

Phobos and Deimos are irregularly ovoid, with shapes remarkably like potatoes. In radius Phobos has a long axis of 13.5 km, an intermediate one of 10.7 km, and a short one of 9.6 km. The corresponding figures for Deimos are 7.5, 6.0, and 5.5 km. (For comparison, the island of Manhattan is 22.6 km long and 3.7 km wide.) Phobos has a volume of 5,810 cu km and Deimos one of 1,040 cu km. The volume of our single Moon is over a million times as great as that of Phobos and Deimos combined.

Both Martian satellites have a low albedo (proportion of light reflected by the surface) and seem to be composed of a dark rock such as basalt. Each keeps the same face toward Mars at all times, so that their periods of rotation are equal to their periods of revolution, as in

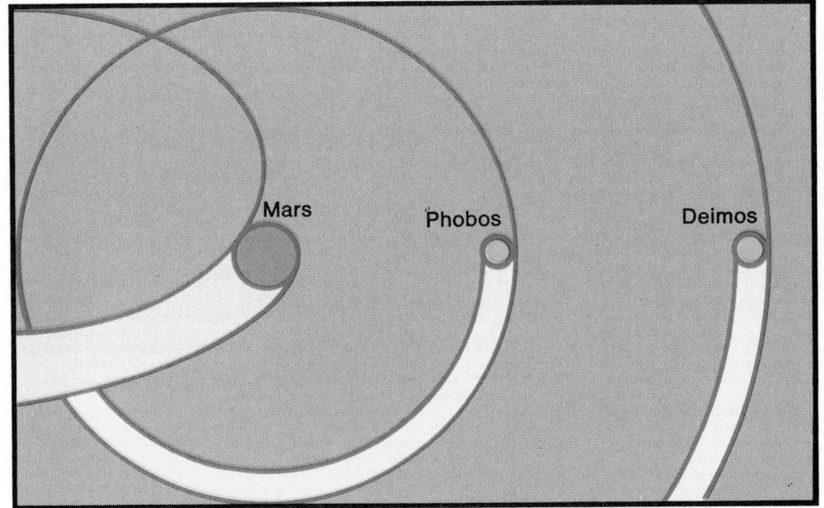

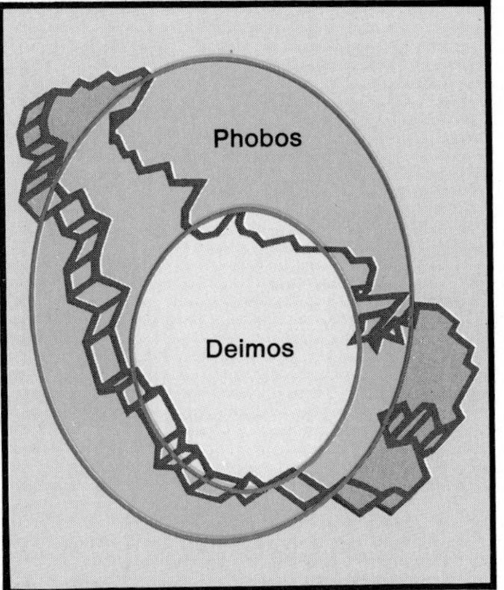

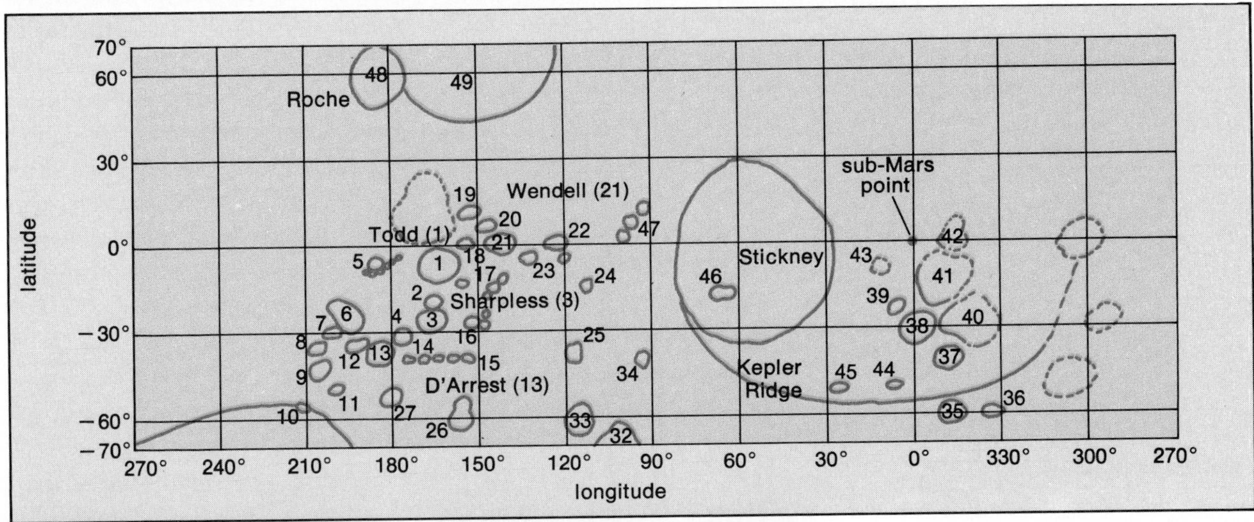

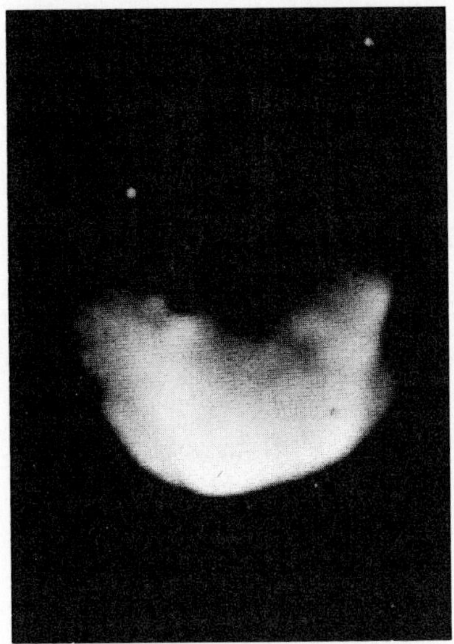

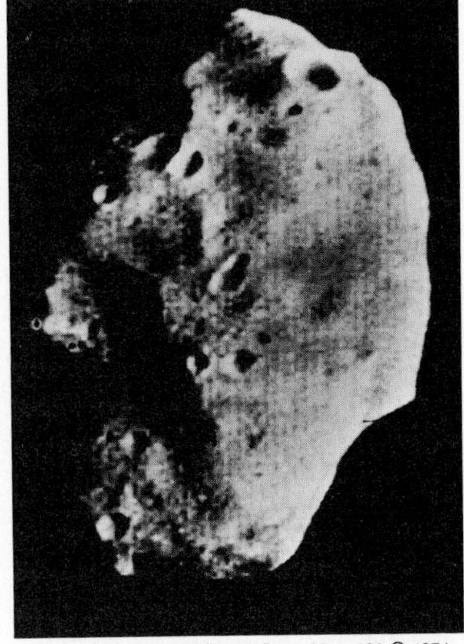

Phobos and Deimos revolve around Mars in almost circular orbits (top left). The sizes of the two moons can be seen by superimposing them on the island of Malta, which has an area of 95 sq mi (top right). Map of Phobos (above), based on Mariner 9 photographs, shows the range of crater sizes; the crater Stickney measures over 8 km across. Photographs of Deimos (far left) and Phobos (left) taken by Mariner 9 reveal the wear of billions of years of bombardment by meteorites.

Courtesy, NASA

(Above) Jupiter's Great Red Spot looks like a Cyclopean eye in the photograph of the planet taken by Pioneer 11 from a distance of 1.1 million km (660,000 mi). (Below) The positions of Jupiter's moons were drawn by Galileo on the basis of a week of observation in January 1610.

the case of the Moon. For both Phobos and Deimos, the long axis points toward Mars, the intermediate axis is perpendicular to it and in the orbital plane, and the short axis is perpendicular to both the others. This is in accord with the prediction of gravitational theory.

Based on the photographs of Phobos, a preliminary map of the satellite was prepared. It showed approximately 50 craters, some of which were named. One of the larger ones, near the satellite's south pole, was, inevitably, named Hall. The largest, on the side facing Mars, was named Stickney, the maiden name of Hall's wife. The blow that formed Stickney seems to have produced a split in the rock that extends several kilometers eastward. It has been named Kepler Ridge, after the German astronomer Johannes Kepler, who first determined the true orbit of Mars.

Satellites of Jupiter

The most far-flung of all the satellite systems is that of Jupiter, the next planet beyond Mars. This is not surprising, because Jupiter is by far the largest and most massive of the planets.

Four of the Jovian satellites are large ones; extending outward from Jupiter, these are Io, Europa, Ganymede, and Callisto. They are collectively called the Galilean satellites because they were discovered by the Italian astronomer Galileo in 1610. Other than the Moon, they were the first satellites to be detected. Their diameters, in kilometers, are 3,650 (Io), 2,980 (Europa), 5,250 (Ganymede), and 4,900 (Callisto). Europa is somewhat smaller than the Moon, Io is roughly Moon-sized, while the other two are larger. Ganymede, the largest of the Galileans, has 1¼ times the volume of Mercury. Mercury, however, is a denser body, and its mass is more than twice that of Ganymede.

From the Earth, the Galilean satellites can be seen through large telescopes as small disks, and faint markings have been used to show that the satellites keep one face to Jupiter as they revolve around it. On Dec. 4, 1973, the U.S. Jupiter probe Pioneer 10 took photographs of Ganymede from a distance of about 750,000 km. These seemed to show a mare more than 750 km wide in the north polar region and another some 1,300 km across in the equatorial region. There also appeared to be a few large craters.

Ganymede's effect on the path of a subsequent probe, Pioneer 11, allowed the mass of the satellite to be calculated with unprecedented precision. Based on this calculation and on the volume of Ganymede, the satellite was shown to have a density of about 1.9 g per cc. This would indicate that Ganymede cannot be a ball of rock, because that would require a density of more than 3 g per cc (as is the case for Io, Europa, and our Moon). Ganymede, therefore, must contain sizable quantities of low-density frozen volatiles, such as water, ammonia, and possibly methane. Callisto, which has a density of only 1.6 g per cc, must be even richer in such ice forms.

The Pioneer 10 photographs seem to show a bright region near Ganymede's south pole. Astronomers have speculated that it would be

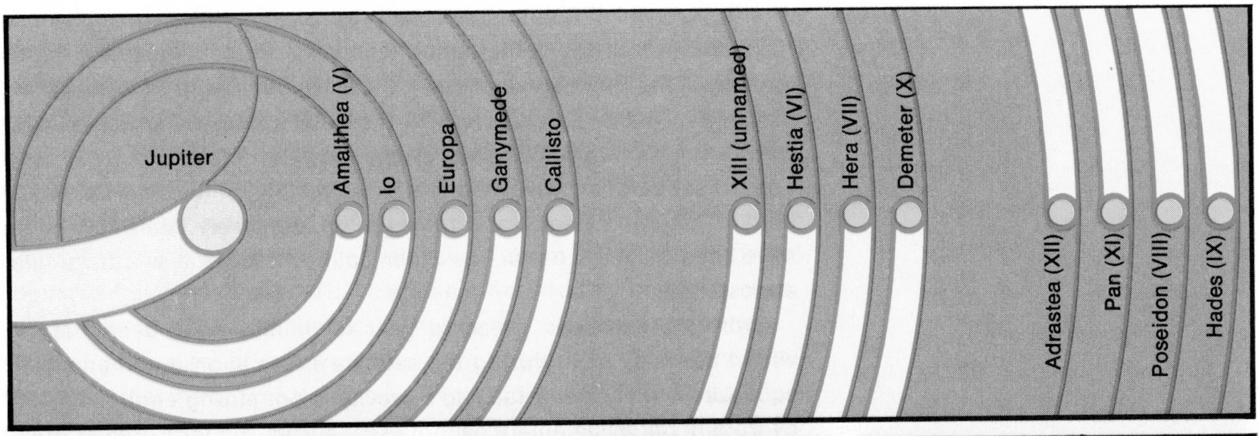

Jupiter
Amalthea (V)
Io
Europa
Ganymede
Callisto
XIII (unnamed)
Hestia (VI)
Hera (VII)
Demeter (X)
Adrastea (XII)
Pan (XI)
Poseidon (VIII)
Hades (IX)

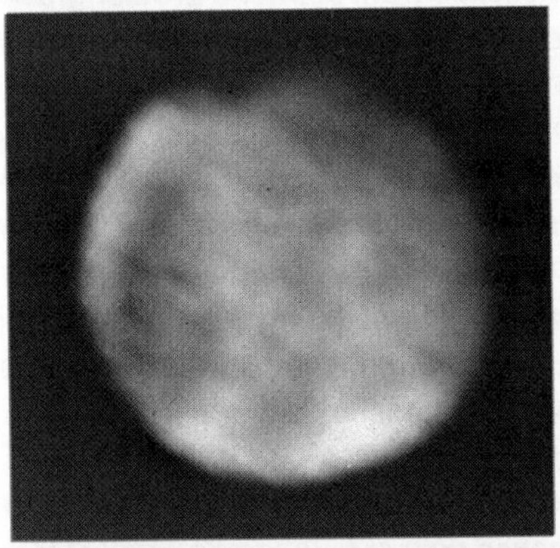

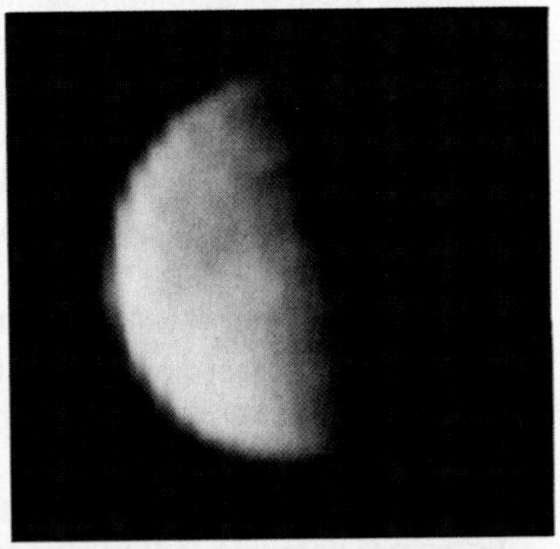

Photos, courtesy, NASA

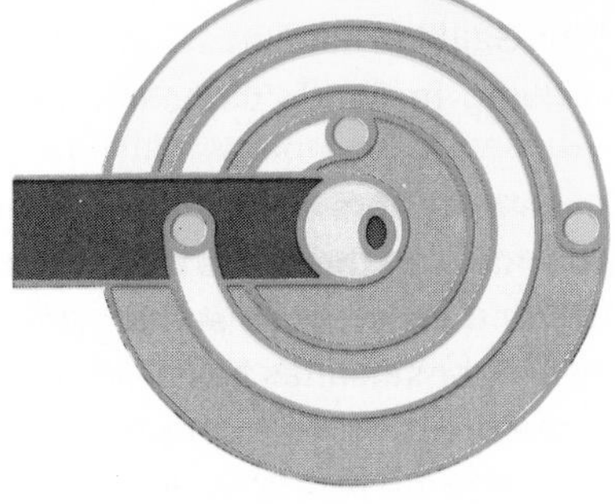

Jupiter's 13 satellites (top) are grouped in three clusters extending outward from the planet. The relative sizes and distances of the moons are not shown. Ganymede, the largest satellite, was photographed by Pioneer 10 from a distance of 750,000 km (above left). Pioneer 10 photo of Europa (above center) clearly shows the satellite's bright and dark regions. Io can be seen above and to the right of Jupiter in the Pioneer 11 photograph above, taken 2.3 million km from the planet. Shadow of Io appears as a sharply defined spot on the disk of Jupiter in the Pioneer 10 photograph at left; Ganymede can be seen at the upper right in the photograph. The diagram below indicates how the shadow is cast.

Courtesy, Lowell Observatory

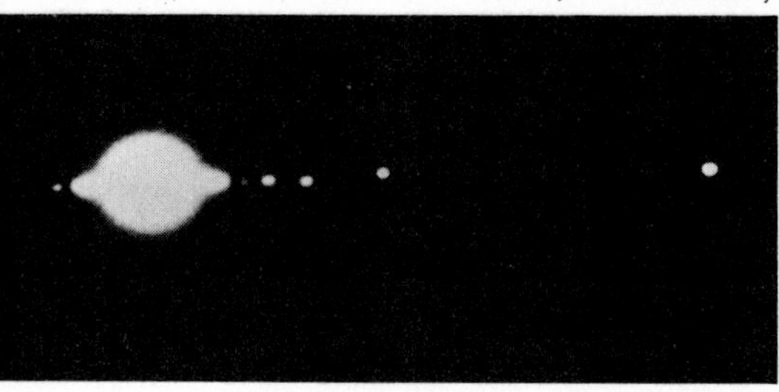

(Above) Saturn and its six brightest moons, Mimas, Enceladus, Tethys, Dione, Rhea, and Titan, were photographed in 1921 at the Lowell Observatory, Flagstaff, Arizona. Below are four drawings of Saturn as it appeared to early astronomers before the nature of the rings was determined; they are by (top to bottom) Galileo in 1610, Christoph Scheiner in 1614, Pierre Gassendi in 1645, and Johannes Hevelius in 1649.

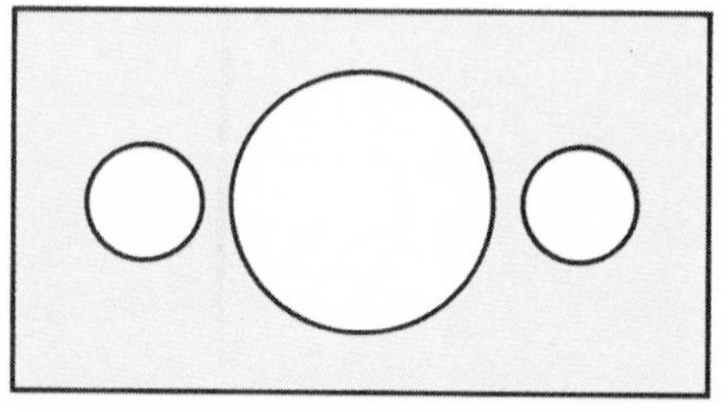

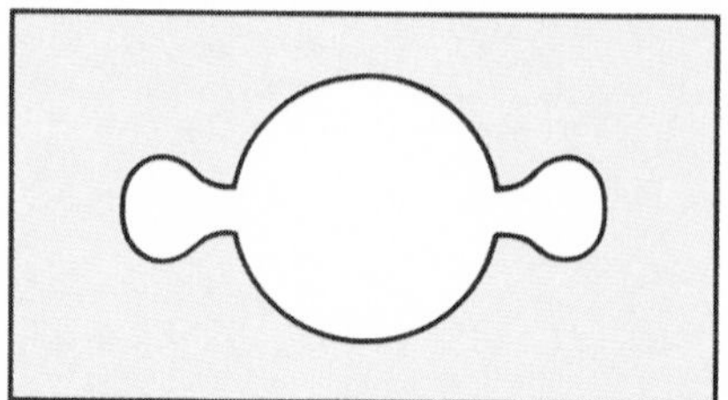

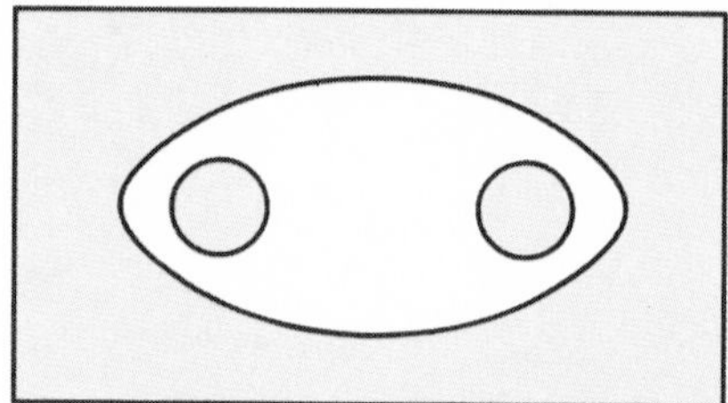

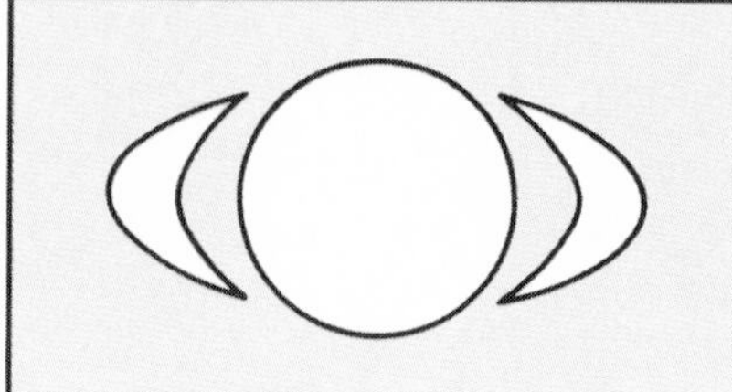

sphere in more than a trace amount. Its atmosphere is almost surely thicker than that of Mars and may even approach that of the Earth. Titan also serves as the source for hydrogen gas, which forms a torus around Saturn in Titan's orbit just as Io's hydrogen forms a torus around Jupiter in Io's orbit. Like Io, Titan is orange in color.

Saturn's nine other satellites are of moderate size. The smallest is the outermost one, namely Phoebe, which has a diameter of about 200 km. Its average distance from Saturn is 13 million km, which is nearly four times as great as that of the next outermost satellite, Iapetus. Its great distance, the eccentricity of its orbit, and the fact that it revolves around Saturn in retrograde fashion make it seem certain that Phoebe is a captured asteroid.

Iapetus, which may be as much as 1,750 km in diameter, is the second largest of Saturn's satellites. It has an orbit inclined to the plane of Saturn's equator by 14.7°, causing astronomers to be uncertain (as in the case of our Moon) as to whether it is a true satellite or a captured body. Iapetus is six times as bright when it is west of Saturn and exposes one hemisphere to us as when it is east of Saturn and exposes the other (assuming that it always turns the same face to Saturn as it revolves.) It is not known why this is so, but perhaps one hemisphere is predominantly icy and the other predominantly rocky.

The inner eight Saturnian satellites rotate in the plane of Saturn's equator and would seem to be true satellites, although the orbits of Titan and Hyperion are rather eccentric. The most recently discovered of the Saturnian satellites is the closest, Janus. It is only 157,000 km from Saturn's center (2.6 planetary radii) and completes its revolution in 18 hours. Because it is the first of the satellites in order, it was named for the Roman god of beginnings.

Within Janus' orbit are Saturn's rings: flat and wide, they encircle Saturn in its equatorial plane. The extreme width of the rings, measuring across Saturn, is 270,000 km, but they are not more than 15 km thick. The rings are seen at various angles during the course of Saturn's 29.5-year period of revolution about the Sun. Twice during each Saturnian revolution they can be seen edge-on from the Earth; at those times, they are so thin that they disappear from view. It is when they disappear and their brightness is eliminated that the inner Saturnian satellites are most readily visible; when the rings were edge-on in 1966, Janus was discovered. The orbit of Janus is only 21,000 km beyond the outermost edge of the rings.

The rings are not solid but are a collection of innumerable particles, each of which may be considered a separate satellite of Saturn. There are no sharp boundaries to the rings, but those regions where the swarm is thick enough to see from the Earth extend inward to about 75,000 km from the center of Saturn (1.25 radii) or only about 14,500 km above the planet's cloud layer. The innermost particles of Saturn's rings revolve around the planet in eight hours. None of the 33 satellites of the solar system is as close to its primary in terms of the planetary radius as are the innermost particles of Saturn's rings.

Courtesy, Lunar and Planetary Laboratory, University of Arizona

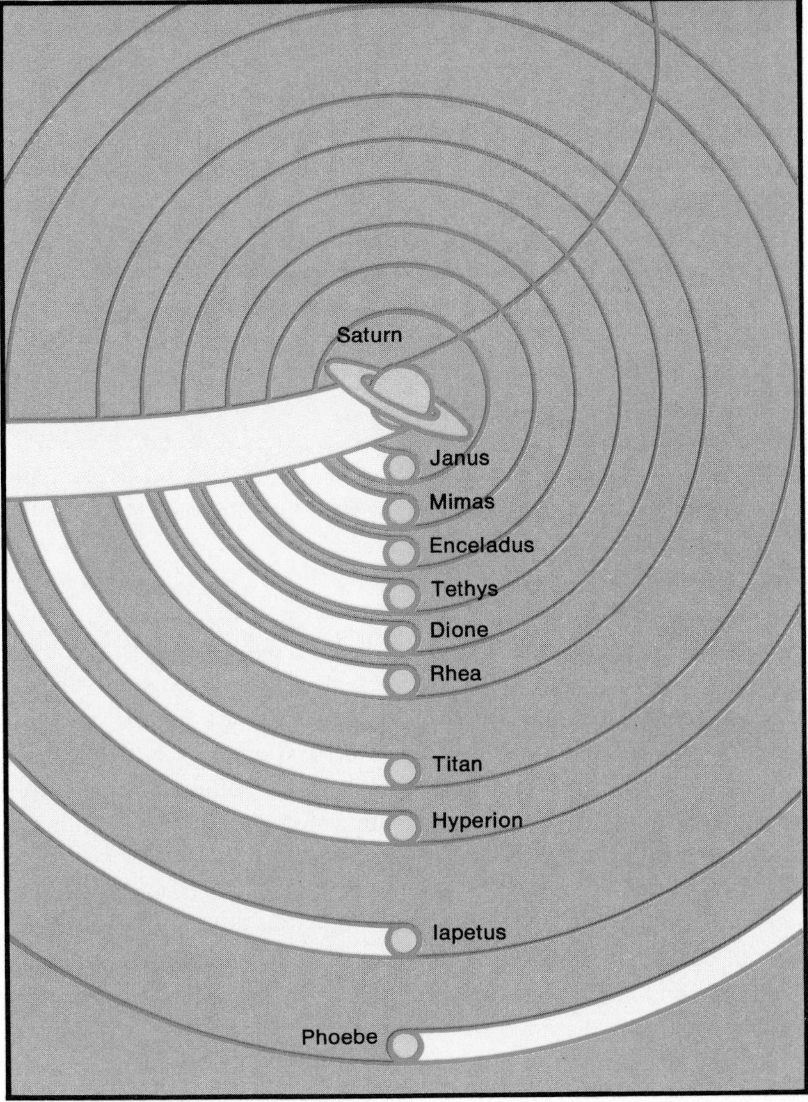

(Left) Saturn as photographed by the 155-cm reflector of the Catalina Observatory in March 1974. The ring plane is tilted 26.9° to the line of sight. The ten satellites of Saturn are shown in their order of distance from the planet; relative sizes of these moons are not indicated.

Courtesy, Lick Observatory

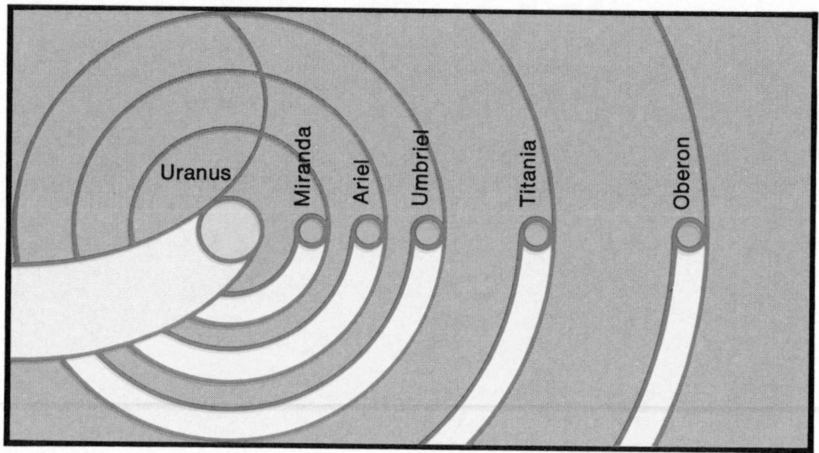

(Above) Uranus and three of its moons, Titania, Oberon, and Umbriel, as seen through the 120-in. reflecting telescope at the Lick Observatory near San Jose, California. (Above right) The five satellites of Uranus are shown in order of their distance from the planet; relative sizes are not depicted.

Uranus and Neptune

The planet Uranus has a compact and, on the whole, unremarkable family of five satellites. Their names, reading outward from the planet, are Miranda, Ariel, Umbriel, Titania, and Oberon. These are not drawn from Greek mythology, but are, rather, the names of characters in the works of Shakespeare and Alexander Pope.

Of the next four planets beyond Mars, sometimes collectively termed the Jovian planets because of their mutual resemblance, Uranus is the only one that does not have a large satellite. The biggest is Titania, which has a diameter of about 1,100 km. Uranus is also the only Jovian planet without at least one captured satellite. All five of the Uranian satellites revolve in the equatorial plane of the planet in orbits of virtually no eccentricity. To be sure, the five satellites all revolve in a plane that is nearly at right angles to the plane of Uranus' revolution about the Sun, and no other planet has satellites of which this can be said. However, this is because Uranus' equatorial plane is at nearly right angles to the plane of its revolution. Thus, the peculiarity is that of the planet rather than of its satellites.

Neptune has two satellites, and each is, in its own way, remarkable. The outer satellite, Nereid, named for the ocean nymphs who, like Poseidon (Neptune) in the Greek myths, lived beneath the sea, has the earmarks of a captured satellite. It is small, perhaps 300 km in diameter, and has an eccentric orbit inclined almost 28° to the plane of Neptune's equator. Its eccentricity is 0.76, greater than that of any object in the solar system other than comets and a few asteroids. This means that, although the average distance of Nereid from Neptune is 5,550,000 km, the orbit is so elongated that at one point it comes to within 1.4 million km of the planet and at the other extreme recedes to a distance of 9.5 million km. The inner satellite, Triton, named for the son of Poseidon, is large, 3,700 km in diameter. It is a little bigger than the Moon, and larger in comparison with its primary than any satellite but the Moon. Its distance from Neptune is 355,000 km, just slightly less than the distance of the Moon from the Earth. Unlike the Moon, Triton has an orbit that is almost perfectly circular.

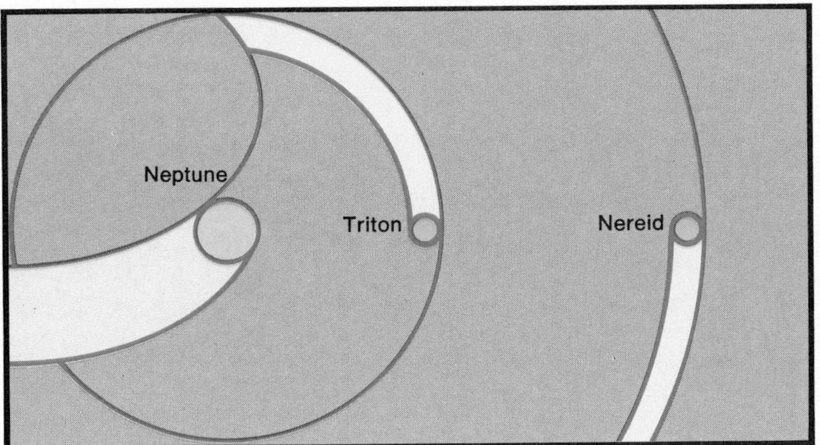

Courtesy, Lick Observatory

(Above) Neptune and its largest satellite, Triton, as seen through the 120-in. reflecting telescope at the Lick Observatory near San Jose, California. (Above left) The two satellites of Neptune are shown in order of their distance from the planet; relative sizes are not indicated.

Triton appears in most respects to be a true satellite, but the plane of its orbit is at an angle of about 20° to Neptune's equatorial plane, about equal to the tipping of the Moon's orbit. Also, Triton's motion is retrograde, in the direction opposed to Neptune's spin. If Triton, like a true satellite, was formed on the fringes of the Neptune dust cloud, it should revolve in the direction of Neptune's spin. If, on the other hand, it is a captured satellite, then what impossibly lucky stroke put it into an almost exactly circular orbit? The problem remains unsolved.

The question of Pluto

The outermost known planet of the solar system, Pluto, has no moon. Its peculiarities, however, have led some scientists to theorize that, perhaps like our Moon, Pluto may have pursued two careers in its lifetime—as a natural satellite and as a planet. Its diameter is small (about 6,000 km) compared with those of the other outer planets, and its tiny mass and lack of appreciable atmosphere contribute to its moonlike character. In addition, Pluto's orbit is tilted at an unusually large angle to the mean plane of the solar system and is so eccentric that at perihelion it actually passes within the orbit of Neptune.

These characteristics suggest that Pluto did not coalesce directly into a planetary orbit from the lens-shaped dust cloud that formed the solar system, but may have evolved as a moon of Neptune. If this is true, then other questions need answering: Did Pluto gradually retreat from Neptune, as the Moon is retreating from the Earth, until it eventually escaped the gravitational pull of its planet? Or did some catastrophic event wrench Pluto from its orbit about Neptune and send it spinning around the Sun? This is another mystery of the natural satellites that astronomers hope one day to resolve.

E OF ELEMENTS
LITHIUM
TUNGSTEN
IRON
CURIUM
CALCIUM
Neon
CESIUM
CHLORINE
IN GOD WE TRUST
1973
RADON
TIN
youssi

Periodicity of the Chemical Elements

by Glenn T. Seaborg

Among the most important discoveries in science is that a recurring pattern is exhibited by the properties of the chemical elements when they are arranged in a table in order of increasing atomic number.

The recurrence of properties of the chemical elements when they are arranged in a periodic table is one of the most useful observations in the history of human thought. The culmination of the chemical knowledge, insight, and intuition of many investigators, this periodic law has made possible correlations and predictions that have been crucial to the advance of science. Like so many scientific discoveries, it became possible only after a background of understanding had been achieved; then, within a few years, many innovative thinkers were ready to contribute to a conceptual breakthrough. As more information has accumulated, improvements to the periodic table have been made and during the last quarter-century some exciting extensions and future prospects have emerged.

Discovery of the elements

The history of discovery of the chemical elements is a fascinating story that began during prehistoric times, is continuing today, and promises future developments as well. A number of the elements were known during biblical times, although of course the present concept of the chemical elements was not recognized. The metals gold, silver, copper, lead, iron, tin, mercury, and possibly zinc, as well as the nonmetals sulfur and carbon, were all well known and described some 2,000 years ago. In fact, it is a matter of record that a number of these were known as long as 5,000 years ago, and some perhaps even earlier. Nothing is known about the discoverers, reference to these substances appearing only generally in the early writings of mankind.

The next elements to appear within the scope of man's knowledge are mentioned by the alchemists—those mystical searchers for methods to transform common material into gold—who in their wandering study of substances were the forerunners of the modern chemists. These investigators succeeded in identifying arsenic, antimony, and bismuth between the 12th and 16th centuries, but again it is not possible to identify with certainty the names of the discoverers. Glimmerings of recognition of a few other substances, now recognized as elements, also were evident during or prior to this period; early mention of "white gold," now identified as platinum, is an example.

The first man to be credited with the discovery of a chemical element is a German alchemist named Hennig Brand, who in 1669 identified what is now understood to be phosphorus. This marks the beginning of the period when detailed records were made of the circumstances surrounding the discovery of elements.

During the 18th century 15 additional elements were discovered, and scientists began to develop a modern concept of the chemical element, although their understanding was somewhat vague. The Greek philosopher Democritus had speculated that an "atom" was the smallest unit of an elemental entity, but the thinking of other Greek philosophers dominated over the years, particularly their classification of all matter in terms of air, fire, earth, and water. In 1789 the French chemist Antoine-Laurent Lavoisier published his famous *Traité élémentaire de chimie*, which included what is considered to be the first true table of chemical elements. In the following decades the work of the British scientist John Dalton, the Swedish chemist Jöns Jacob Berzelius, and many others led to an understanding of the atomic nature of matter and to the determination of a number of elemental atomic weights, information that became crucial to the subsequent conception of the periodic table.

Mendeleyev's periodic table

The discovery of most of the remaining elements, about 50 in all, took place during the 19th century. As information on more elements became available, it seemed natural to seek correlations of chemical properties with their atomic weights, because this was the most obvious variable for distinguishing chemical elements. The name that is preeminent as the originator of the periodic table is Dmitry Ivanovich Mendeleyev, a Russian chemist. On March 6, 1869, Mendeleyev, through his associate Nikolay Menshutkin, presented a paper to the Russian Chemical Society in St. Petersburg (now Leningrad) which showed that the elements display a periodicity of chemical properties when they are arranged in the order of their atomic weights in a rectangular matrix array. Although Lothar Meyer of Germany, on the basis of independent work, published a similar system in 1870, Mendeleyev went further in deducing errors in measured atomic weights and especially in predicting the chemical properties of undiscovered chemical elements on the basis of his correlation.

***GLENN T. SEABORG** is University Professor of Chemistry, University of California, and Associate Director of the Lawrence Berkeley Laboratory. A Nobel laureate in chemistry, he is also former chairman of the U.S. Atomic Energy Commission.*

(Overleaf) Illustration by John Youssi

row	group I — R_2O	group II — RO	group III — R_2O_3	group IV RH_4 RO_2	group V RH_3 R_2O_5	group VI RH_2 RO_3	group VII RH R_2O_7	group VIII — RO_4
1	H=1							
2	Li=7	Be=9.4	B=11	C=12	N=14	O=16	F=19	
3	Na=23	Mg=24	Al=27.3	Si=28	P=31	S=32	Cl=35.5	
4	K=39	Ca=40	—=44	Ti=48	V=51	Cr=52	Mn=55	Fe=56 Co=59 Ni=59 Cu=63
5	(Cu=63)	Zn=65	—=68	—=72	As=75	Se=78	Br=80	
6	Rb=85	Sr=87	?Yt=88	Zr=90	Nb=94	Mo=96	—=100	Ru=104 Rh=104 Pd=106 Ag=108
7	(Ag=108)	Cd=112	In=113	Sn=118	Sb=122	Te=125	I=127	
8	Cs=133	Ba=137	?Di=138	?Ce=140	—	—	—	— — — —
9	(–)	—	—	—	—	—	—	
10	—	—	?Er=178	?La=180	Ta=182	W=184	—	Os=195 Ir=197 Pt=198 Au=199
11	(Au=199)	Hg=200	Ti=204	Pb=207	Bi=208		—	
12	—	—	—	Th=231	—	U=240	—	— — — —

Figure 1. Periodic table of Dmitry I. Mendeleyev, published in 1871, demonstrated that chemical elements display recurring properties when arranged in order of their atomic weights in a rectangular matrix.

Figure 1 reproduces Mendeleyev's periodic table as published in 1871, which incorporates improvements in the original version of 1869. Each entry shows the individual element's symbol and atomic weight, starting with the lightest element (hydrogen) at H = 1 at the upper left. The atomic weight increases as the rectangular matrix builds up row by row and in repetitive periods of eight, until the heaviest element (uranium) is reached at U = 240. Elements with similar chemical properties fall in each family group (or vertical column). It should be noted in particular that the table predicts the existence and chemical properties of the elements with atomic weights of 44, 68, and 72. These correspond to the elements scandium, gallium, and germanium, which were actually discovered later, during the period from 1875 to 1886, and were found to have the chemical properties that Mendeleyev had predicted.

First revisions to the table

By the end of the first third of the 20th century the total number of elements had increased to 88, including a family of so-called noble gases which was fitted into the scheme by the addition of another family group (vertical column), and a series of elements—the rare earths, or lanthanides—located in the place of a single element (lanthanum). By this time the work of Sir Joseph Thomson, Ernest Rutherford, Niels Bohr, H. G. J. Moseley, Werner Heisenberg, Erwin Schrödinger, and others on the structure of the atom and its nucleus provided the final argument to support the framework of the periodic table. They developed the concepts of atomic number and electronic structure to such a degree that these, rather than atomic weight, could be correlated with each chemical element's position in the table.

These scientists discovered that the atom consists of a positively charged nucleus comprised of protons (fundamental particles carrying one unit of positive charge with a mass about 1,850 times that of the electron) and neutrons (fundamental particles with roughly the same mass as protons but electrically neutral). This nucleus is surrounded by orbiting electrons, each of which has a unit of negative charge and which are loosely referred to as occurring as groups in shells and subshells. The atomic number (*Z*) of an element is determined by the number of protons in the nucleus of the atom, which is equal to the number of surrounding electrons. Isotopes of an element are essentially chemically identical forms of nuclides (atomic species), each containing the same number of protons but differing numbers of neutrons (*N*). Isotopes therefore differ in their mass number (*A*), the integral sum of the nucleons (the collective name for protons and neutrons) in the nucleus.

By the 1930s, with 88 known elements, only four were missing in the sequence from the lightest, hydrogen ($Z = 1$), to the heaviest, uranium ($Z = 92$). Those four had the atomic numbers (*Z*) equal to 43, 61, 85, and 87. Understood to be missing from nature because of their disappearance by radioactive decay, they were synthesized by transmutation (conversion of one element or nuclide into another) and identified within another decade, thus filling in all the gaps in the periodic table.

Many attempts were made to discover naturally occurring elements heavier than uranium, all of which were unsuccessful. Although uranium gradually disappears because of radioactive decay, its rate of decay is so slow that much of the Earth's original inheritance of this element remains. However, the increasing charge on the nucleus for the elements beyond uranium leads to increasing rates of radioactive

Figure 2. Periodic table as it was known just before World War II differs from that of Mendeleyev by using atomic numbers rather than atomic weights, showing more family groups (vertical columns), and incorporating the lanthanide series as a horizontal row at the bottom. It also includes the predicted positions (later proved erroneous) of the undiscovered transuranium elements. The atomic numbers of predicted elements are given in parentheses.

1 H																	2 He
3 Li	4 Be											5 B	6 C	7 N	8 O	9 F	10 Ne
11 Na	12 Mg											13 Al	14 Si	15 P	16 S	17 Cl	18 Ar
19 K	20 Ca	21 Sc	22 Ti	23 V	24 Cr	25 Mn	26 Fe	27 Co	28 Ni	29 Cu	30 Zn	31 Ga	32 Ge	33 As	34 Se	35 Br	36 Kr
37 Rb	38 Sr	39 Y	40 Zr	41 Nb	42 Mo	(43)	44 Ru	45 Rh	46 Pd	47 Ag	48 Cd	49 In	50 Sn	51 Sb	52 Te	53 I	54 Xe
55 Cs	56 Ba	57–71 La–Lu	72 Hf	73 Ta	74 W	75 Re	76 Os	77 Ir	78 Pt	79 Au	80 Hg	81 Tl	82 Pb	83 Bi	84 Po	(85)	86 Rn
(87)	88 Ra	89 Ac	90 Th	91 Pa	92 U	(93)	(94)	(95)	(96)	(97)	(98)	(99)	(100)				
		57 La	58 Ce	59 Pr	60 Nd	(61)	62 Sm	63 Eu	64 Gd	65 Tb	66 Dy	67 Ho	68 Er	69 Tm	70 Yb	71 Lu	

decay. Therefore, the original inheritance of such "transuranium elements" has long since disappeared (except for extremely small, essentially negligible, amounts of one or two of them).

Figure 2 shows a periodic table as it was known just before World War II. Incorporating all the additional elements known by that time, it differs from the original table of Mendeleyev in five important respects: (1) the rectangular matrix is formed by the use of atomic numbers rather than atomic weights; (2) it is an "extended form," showing more family groups (or vertical columns); (3) it incorporates the noble gases as a family group (vertical column) at the right-hand edge; (4) it incorporates the series of rare earth (or lanthanide) elements in a horizontal row at the bottom; and (5) it includes the predicted positions of the undiscovered transuranium elements.

Placement of the transuranium elements in the periodic table as in figure 2 makes it possible to predict the chemical properties of these elements, much as Mendeleyev did when he predicted the chemical properties of scandium, gallium, and germanium years before their discovery. Because the first transuranium element, with the atomic number 93 (element 93), falls in the family group (vertical column) under rhenium (Re), it was expected to have chemical properties similar or homologous to those of rhenium. Similarly, element 94 was expected to be a homologue of osmium (Os); element 95 a homologue of iridium (Ir); element 96 a homologue of platinum (Pt); and so forth. Further investigation made it apparent, however, that this placement of the transuranium elements by a simple extension of the periodic table was wrong.

Synthesis of transuranium elements

Because the heaviest naturally occurring element is uranium, opportunity for an actual investigation of the missing transuranium elements awaited the discovery and development of nuclear transmutation processes that could be applied to their synthesis. Although the alchemist dreamed of ways to transmute one element into another by means of chemical reactions and made claims of success in effecting such transformations, actual success was not achieved until well into the 20th century and then not by the use of chemical reactions. This success depended on the methods of nuclear physics, which utilized much larger energies than those employed in the unsuccessful chemical experiments of the alchemists. Beginning in the 1930s such transmutations were achieved by bombarding target nuclei with positively charged projectile nuclei furnished by particle accelerators, or with neutrons furnished by nuclear reactions produced by accelerators or by the radiations from naturally radioactive elements.

The first attempts to produce elements beyond uranium were made in Italy by Enrico Fermi, Emilio Segrè, and their co-workers. They bombarded uranium with neutrons in 1934 and found a number of radioactive products that seemed to behave as expected for transuranium elements according to predictions based on the then current

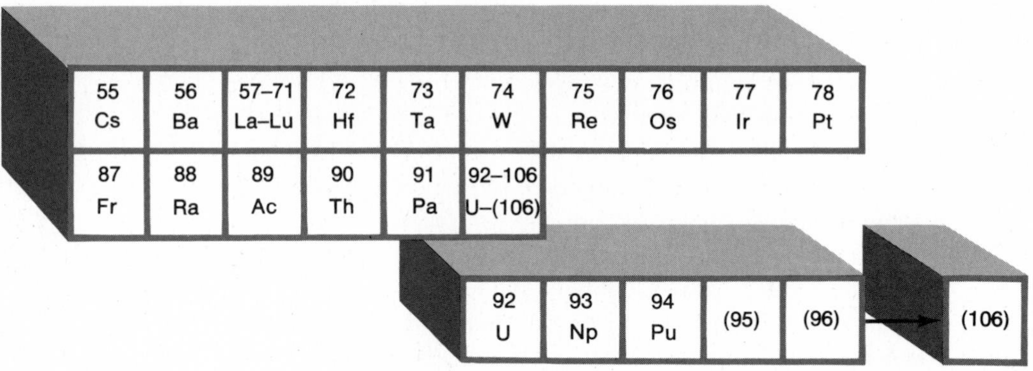

Figure 3. Similar chemical properties of uranium, neptunium, and plutonium, 92–94 on the periodic table, led chemists to believe that there existed a "uranide" series of elements, 92–106, similar to the lanthanide series, 57–71. This belief was proved wrong by the properties of elements 95 and 96, and in 1944 Glenn Seaborg resolved the difficulty by proposing the actinide series of elements, 89–103.

periodic table (figure 2). A group of German investigators, Otto Hahn, Lise Meitner, and Fritz Strassmann—experts in the required methods of radiochemistry—studied these radioactive phenomena in great detail over a period of several years. They became convinced that some of them displayed the chemical properties expected for the transuranium elements with the atomic numbers 93, 94, 95, and 96. Finally, by means of a classical and elegant application of radiochemistry, Hahn and Strassmann late in 1938 showed that these radioactive species were isotopes of medium-weight elements that had been formed by the splitting, or nuclear fission, of uranium. The discovery of the extraordinarily important nuclear fission reaction, which led to the "atomic age," was thus a by-product of man's quest for the transuranium elements in his attempt to extend the periodic table.

With poetic justice, the actual discovery of the first transuranium element in turn resulted from experiments aimed at understanding the nuclear fission process. Experiments at Berkeley, California, to investigate the products of the neutron bombardment of uranium culminated in the chemical identification of a radioactive isotope of the transuranium element with the atomic number 93 by Edwin McMillan and Philip Abelson in 1940. Of great interest from the standpoint of the periodic table was their observation that element 93, for which they suggested the name neptunium (Np), resembles uranium and does not resemble rhenium in its chemical properties. This meant that its placement had to be changed.

"Actinide" elements

The chemical relationship of the neighboring elements uranium and neptunium was immediately reminiscent of the horizontal relationship of the 14 rare earth (lanthanide) elements, placed by themselves along with lanthanum in a row below the main body of the periodic table. In the rare earth elements the electrons successively enter an inner electron subshell that has room for exactly 14 electrons of a special type. These 14 elements, with atomic numbers 58 to 71, all have chemical properties similar to those of lanthanum, and all should fit in the periodic table in the same single space as lanthanum (atomic number 57). They are given the alternate group name "lanthanide" elements.

In view of the similarity of the chemical properties of uranium and neptunium, chemists compared them with the lanthanide elements and conjectured that the expected next inner electron shell might start to fill at neptunium. This shell would continue to be filled until there was a 14-member family group of "uranide" elements with chemical properties similar to those of uranium; the placement in a revised periodic table might be as shown in figure 3. The preliminary study of the chemical properties of element 94, given the name plutonium (Pu), discovered a few months later, seemed to confirm this point of view.

This concept implied that the chemical properties of the next transuranium elements, those with the atomic numbers 95 and 96, should be nearly identical with those of neptunium and plutonium. These assumptions proved wrong, and the initial experiments directed toward the nuclear synthesis and identification of elements 95 and 96 based on this hypothesis failed.

Then, in 1944, Glenn T. Seaborg conceived the idea that perhaps all the known elements heavier than actinium (atomic number 89) were misplaced in the periodic table. He advanced the theory that those elements heavier than actinium might constitute a second series completely analogous to the "lanthanide" elements; it would be called, by analogy, the "actinide" series and would include the 14 elements following actinium, thus ending at element 103. The elements thorium (Th), protactinium (Pa), and uranium (U)—with atomic numbers 90, 91, and 92—would be removed from the positions they occupied in the periodic table before World War II and placed in this second "rare earth" family. This actinide series would be placed as a row of elements below the main body of the table in a position analogous to that of the lanthanide series. Then elements 104, 105, and 106—when discovered—would take over the positions previously held by thorium, protactinium, and uranium.

This concept had great predictive value, and its use to devise chemical identification procedures for many of the transuranium elements, following their synthesis by transmutation reactions, was the key to their discovery. Its success led to its general acceptance by the scientific community and resulted in the form of the periodic table shown in figure 4. This includes all of the presently known transuranium elements (represented by their atomic numbers and chemical symbols) with the actinide series ending at element 103 (lawrencium, Lr).

Beyond the actinides

Of special significance to the actinide concept is the consequent prediction that, beginning with element 104, the transuranium elements should find places in the main body of the periodic table. This crucial prediction, a critical test, was confirmed when element 104 was discovered; the study of its chemical properties confirmed that it is indeed homologous to hafnium, as demanded by its position, and is not an actinide element. Included in figure 4 are predictions for placement of the undiscovered transuranium elements with atomic numbers 107 to 118, put in parentheses to distinguish them from the known elements. (The heaviest known element at the time of this writing, number 106, had not yet been given a name and chemical symbol; groups of U.S. and Soviet scientists each claimed its discovery. This was also the case for elements 104 and 105, represented by the chemical symbols corresponding to the U.S.-suggested names rutherfordium and hahnium.)

Efforts to proceed higher in the periodic table are beset with two types of difficulties: ever increasing rates of radioactive decay of the nucleus of the atom as the atomic number increases, and ever decreasing yields of the transmutation reactions that lead to the synthesis of the new elements. By the time one reaches element 106 the longest radioactive lifetimes are in the region of one second or less, and the yields are as small as one atom per hour. This is the situation in which scientists found themselves as they made plans to synthesize the next transuranium elements, numbers 107, 108, etc., in nuclear reactions by using accelerators that furnish heavy ion projectiles to bombard heavy ion targets.

It is interesting to note, from figure 4, that element 107 is predicted to have chemical properties like those of rhenium (Re). Thus, in a sense, chemists are in the same position as they were 40 years ago, when it was predicted by using the periodic table of that day that the first transuranium element, with atomic number 93, would be similar

Figure 4. Modern form of the periodic table shows the 106 elements discovered through 1975. The elements making up the lanthanide and actinide series are placed in rows below the main body of the table. The predicted locations of still undiscovered transuranium elements are indicated with their atomic numbers in parentheses.

	1 H																	2 He
	3 Li	4 Be											5 B	6 C	7 N	8 O	9 F	10 Ne
	11 Na	12 Mg											13 Al	14 Si	15 P	16 S	17 Cl	18 Ar
	19 K	20 Ca	21 Sc	22 Ti	23 V	24 Cr	25 Mn	26 Fe	27 Co	28 Ni	29 Cu	30 Zn	31 Ga	32 Ge	33 As	34 Se	35 Br	36 Kr
	37 Rb	38 Sr	39 Y	40 Zr	41 Nb	42 Mo	43 Tc	44 Ru	45 Rh	46 Pd	47 Ag	48 Cd	49 In	50 Sn	51 Sb	52 Te	53 I	54 Xe
	55 Cs	56 Ba	57 La	72 Hf	73 Ta	74 W	75 Re	76 Os	77 Ir	78 Pt	79 Au	80 Hg	81 Tl	82 Pb	83 Bi	84 Po	85 At	86 Rn
	87 Fr	88 Ra	89 Ac	104 Rf	105 Ha	106	(107)	(108)	(109)	(110)	(111)	(112)	(113)	(114)	(115)	(116)	(117)	(118)
lanthanides				58 Ce	59 Pr	60 Nd	61 Pm	62 Sm	63 Eu	64 Gd	65 Tb	66 Dy	67 Ho	68 Er	69 Tm	70 Yb	71 Lu	
actinides				90 Th	91 Pa	92 U	93 Np	94 Pu	95 Am	96 Cm	97 Bk	98 Cf	99 Es	100 Fm	101 Md	102 No	103 Lr	

to rhenium. During the intervening years scientists synthesized and identified 14 transuranium elements, accounting for the filling of 14 places in an inner electron shell in order to get on to the real homologue of rhenium.

"Island of stability"

A prime requirement for the existence of a chemical element is the stability of its nucleus. It is the increasing nuclear instability of the heavy transuranium elements that seems likely to set an upper limit to the periodic table. However, just as some chemical elements—the noble gases—have extra chemical stability because of their configurations of closed shells of electrons, so do some nuclei of atoms have extra stability because of closed shells of protons or neutrons. Such closed shells, also referred to as "magic numbers," have long been known to exist for nuclei that contain certain numbers of neutrons (N) or protons (Z)—namely 2, 8, 20, 28, 50, and 82 for the region below uranium. There is apparently no reason why such magic numbers should be confined to nuclei lighter than uranium, and many theoretical calculations have suggested the existence of closed shells at $Z = 114$ and $N = 184$. These calculations suggest increased stability, as reflected by longer lifetimes for radioactive decay, for a range of nuclei in the neighborhood of these magic numbers. This should result in an "island of stability" as shown in the representation of figure 5, in which regions of known or predicted nuclear stability are depicted as land masses in a sea of instability representing forms of radioactive decay. The predicted inhabitants of this "island of stability" are referred to as "superheavy" elements.

A perplexing remaining question is whether there exist nuclear reactions capable of synthesizing such superheavy elements. Such reactions might be achieved by bombarding target nuclei with sufficiently energetic projectiles consisting of heavy ions, possibly very heavy ions. By the mid-1970s such projectiles were becoming available as the result of the operation of several particle accelerators. Unfortunately, the yield of the desired product nuclei is predicted to be

Figure 5. Regions of known or predicted nuclear stability appear as land masses in a "sea" of instability. Such stability is found in nuclei containing closed shells of protons or neutrons. These shells, also called "magic numbers," occur in nuclei containing certain quantities of protons (Z) or neutrons (N): 2, 8, 20, 28, 50, and 82; many chemists have predicted that they also occur at Z=114 and N=184.

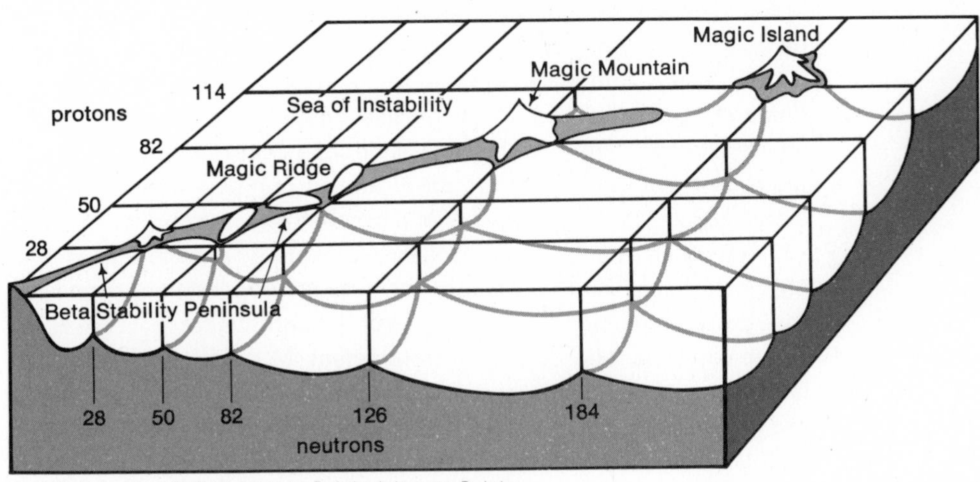

Courtesy, University of California, Lawrence Berkeley Laboratory, Berkeley

1 H																	2 He
3 Li	4 Be											5 B	6 C	7 N	8 O	9 F	10 Ne
11 Na	12 Mg											13 Al	14 Si	15 P	16 S	17 Cl	18 Ar
19 K	20 Ca	21 Sc	22 Ti	23 V	24 Cr	25 Mn	26 Fe	27 Co	28 Ni	29 Cu	30 Zn	31 Ga	32 Ge	33 As	34 Se	35 Br	36 Kr
37 Rb	38 Sr	39 Y	40 Zr	41 Nb	42 Mo	43 Tc	44 Ru	45 Rh	46 Pd	47 Ag	48 Cd	49 In	50 Sn	51 Sb	52 Te	53 I	54 Xe
55 Cs	56 Ba	57 La	72 Hf	73 Ta	74 W	75 Re	76 Os	77 Ir	78 Pt	79 Au	80 Hg	81 Tl	82 Pb	83 Bi	84 Po	85 At	86 Rn
87 Fr	88 Ra	89 Ac	104 Rf	105 Ha	106	(107)	(108)	(109)	(110)	(111)	(112)	(113)	(114)	(115)	(116)	(117)	(118)
(119)	(120)	(121)	(154)	(155)	(156)	(157)	(158)	(159)	(160)	(161)	(162)	(163)	(164)	(165)	(166)	(167)	(168)

lanthanides	58 Ce	59 Pr	60 Nd	61 Pm	62 Sm	63 Eu	64 Gd	65 Tb	66 Dy	67 Ho	68 Er	69 Tm	70 Yb	71 Lu
actinides	90 Th	91 Pa	92 U	93 Np	94 Pu	95 Am	96 Cm	97 Bk	98 Cf	99 Es	100 Fm	101 Md	102 No	103 Lr
superactinides	(122)	(123)	(124)											(153)

Figure 6. Futuristic periodic table shows predicted locations, in parentheses, for transuranium elements as heavy as atomic number 168. A third "rare earth" series, the superactinides, is predicted for elements 122–153 and is shown as a horizontal row at the bottom of the table.

very small, possibly too small to detect. An inherent difficulty in the synthesis of superheavy nuclei in the island of stability is the probable requirement, which it may not be possible to satisfy, that there be a sufficient number of neutrons as well as protons in the product nucleus. Even with the availability of a projectile nucleus sufficiently large to contain the required number of neutrons, the repulsion between the positive charge on the projectile nucleus and the positive charge on the target nucleus may prevent the nuclear fusion required in the synthesis reaction. The feasibility of this route of synthesis can be investigated only by future experiments.

Despite the anticipated difficulties, attempts are being made in a number of laboratories to synthesize and identify superheavy elements. Success might lead to a considerable extension of the periodic table, perhaps covering a region of atomic numbers extending from about 110 to 115, or perhaps even a broader region from about 108 to 120. Scientists hope to identify such elements by chemical as well as physical means, and to compare the measured chemical properties with those predicted through the use of the periodic table. From figure 4, for example, one can see that element 110 is predicted to have chemical properties like those of platinum (Pt), element 111 should be like gold (Au), element 112 like mercury (Hg), element 113 like thallium (Tl), and element 114 like lead (Pb).

If the radioactive decay lifetime of a superheavy element should turn out to be extremely long, it might have survived from the Earth's original inheritance of elements and be found in natural sources. Although so long a lifetime is unlikely, adventurous scientists have pursued this remote possibility. For example, they have searched platinum ores for element 110 and lead ores for element 114. The results of such investigations have established that the concentration of these elements, if they are present, is extremely small—much less than one part in a trillion parts of ore. Most likely, they are not present at all. Cosmic rays, meteorites, and Moon rocks also have been examined with generally negative results.

Future prospects

Encouraged by the prospects for an island of stability, however dim, scientists have projected a grandiose periodic table, as shown in figure 6. The predicted positions of elements as heavy as atomic number 168 are indicated (with the atomic numbers of undiscovered elements again enclosed in parentheses). This, of course, extends far beyond any region where scientists might reasonably expect to synthesize and identify new elements. Of interest in this table is the predicted third "rare earth" series, the "superactinides," included as a row of 32 elements at the bottom of the periodic table along with the lanthanide and actinide series.

The periodic table has served as a guiding principle for a tremendous amount of chemical research and understanding since its inception a little more than one hundred years ago. With the advent of new elements and new knowledge it has been expanded and changed, but always within its basic original framework. Exciting additions may well be made to it in the future.

FOR ADDITIONAL READING

Leicester, Henry M., *The Historical Background of Chemistry* (John Wiley & Sons, Inc., 1956).

Seaborg, Glenn T., *Man-Made Transuranium Elements*, Foundations of Modern General Chemistry Series (Prentice-Hall, Inc., 1963).

Seaborg, Glenn T., and Bloom, Justin L., "The Synthetic Elements: IV," *Scientific American* (April 1969, pp. 56–67).

van Spronsen, J. W., *The Periodic System of Chemical Elements* (American Elsevier Publishing Co., Inc., 1969).

Science Year in Review Contents

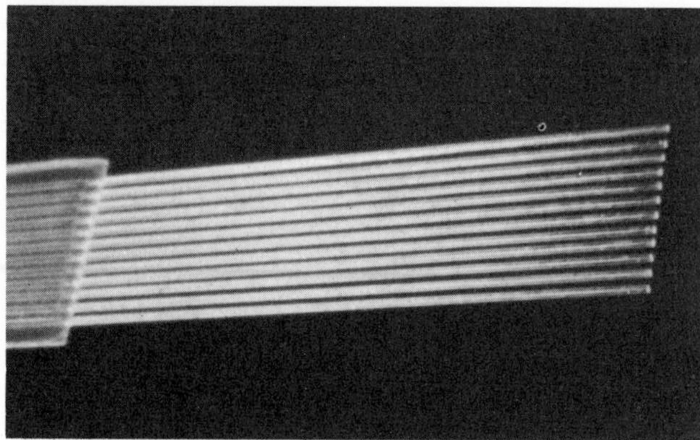

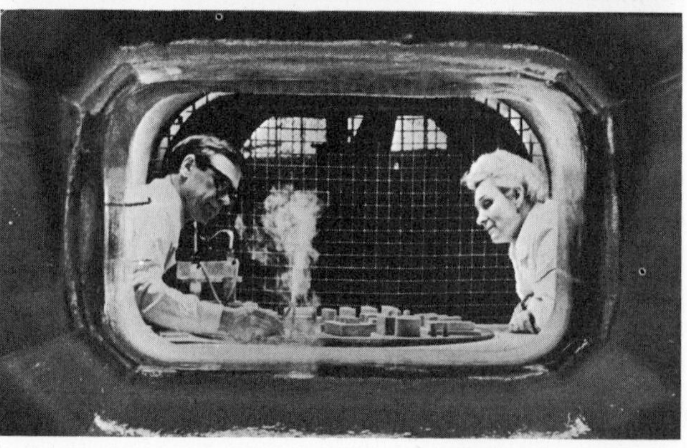

Contributors to the Science Year in Review

Joseph Ashbrook *Astronomy*. Editor, *Sky and Telescope,* Cambridge, Mass.

Fred Basolo *Chemistry: Inorganic chemistry.* Professor of Chemistry, Northwestern University, Evanston, Ill.

Louis J. Battan *Earth sciences: Atmospheric sciences.* Director, Institute of Atmospheric Physics, University of Arizona, Tucson.

Nathaniel I. Berlin *Medical sciences: Cancer research.* Director, Cancer Center, Northwestern University, Chicago, Ill.

Richard B. Bernstein *Chemistry: Physical chemistry.* W. T. Doherty Professor of Chemistry and Professor of Physics, University of Texas, Austin.

Harold Borko *Information sciences: Information systems.* Professor, Graduate School of Library and Information Science, University of California, Los Angeles.

D. Allan Bromley *Physics: Nuclear physics.* Henry Ford II Professor and Chairman, Department of Physics, and Director, A. W. Wright Nuclear Structure Laboratory, Yale University, New Haven, Conn.

F. C. Durant III *Information sciences: Satellite systems.* Assistant Director, National Air and Space Museum, Smithsonian Institution, Washington, D.C.

Robert G. Eagon *Life sciences: Microbiology.* Professor of Microbiology, University of Georgia, Athens.

Gerald Feinberg *Physics: High-energy physics.* Professor of Physics, Columbia University, New York, N.Y.

David R. Gaskell *Materials sciences: Metallurgy.* Associate Professor of Metallurgy, University of Pennsylvania, Philadelphia.

Robert Geddes *Architecture and civil engineering.* W. R. Kenan Professor of Architecture and Dean of the School of Architecture and Urban Planning, Princeton University, Princeton, N.J.

L. A. Heindl *Earth sciences: Hydrology.* Executive Secretary, U.S. National Committee for the International Hydrological Decade, National Research Council, Washington, D.C.

David P. Hill *Earth sciences: Geophysics.* Geophysicist, National Center for Earthquake Research, U.S. Geological Survey, Menlo Park, Calif.

Richard S. Johnston *Space exploration: Manned flight.* Director of Life Sciences, NASA Johnson Space Center, Houston, Texas.

John Patrick Jordan *Food and agriculture: The world food crisis.* Director, Colorado State University Experiment Station, Fort Collins.

Lou Joseph *Medical sciences: Dentistry.* Manager of Media Relations, Bureau of Public Information, American Dental Association, Chicago, Ill.

Ernest R. Kretzmer *Information sciences: Communications systems.* Director, Data Communications Technology and Applications Laboratory, Bell Telephone Laboratories Inc., Holmdel, N.J.

Edward R. Lady *Mechanical engineering.* Professor of Mechanical Engineering, University of Michigan, Ann Arbor.

Mina W. Lamb *Food and agriculture: Nutrition.* Margaret W. Weeks Professor, Department of Food and Nutrition, Texas Tech University, Lubbock.

John G. Lepp *Life sciences: Zoology.* Dean, West Virginia Institute of Technology, Montgomery.

Howard J. Lewis *U.S. science policy.* Director, Office of Information, National Academy of Sciences, Washington, D.C.

Melvin H. Marx *Psychology.* Professor of Psychology, University of Missouri, Columbia.

Walter Modell *Medical sciences: Pharmacology for the future.* Emeritus Professor of Pharmacology, Cornell University Medical College, New York, N.Y.

Raymond Lee Owens *Anthropology.* Assistant Professor of Anthropology, University of Texas, Austin.

Dan Pedoe *Mathematics.* Professor of Mathematics, University of Minnesota, Minneapolis.

Willard J. Pierson *Earth sciences: Oceanography.* Director, CUNY Institute of Marine and Atmospheric Sciences, City University of New York.

Froelich Rainey *Archaeology.* Professor of Anthropology and Director of the University Museum, University of Pennsylvania, Philadelphia.

George Rapp, Jr. *Earth sciences: Geology and geochemistry.* Associate Professor of Geology, University of Minnesota, Minneapolis.

Arnold W. Ravin *Life sciences: Molecular biology.* Professor of Biology and Microbiology, University of Chicago.

John R. Rice *Information sciences: Computers.* Professor of Mathematics and Computer Science, Purdue University, West Lafayette, Ind.

Byron T. Scott *Medical sciences: General medicine.* Editor, *Medical Communications* and Assistant Professor of Magazine Journalism, Ohio University, Athens.

Mitchell R. Sharpe *Space exploration: Space probes.* Historian, Alabama Space and Rocket Center, Huntsville.

Albert J. Smith *Life sciences: Botany.* Associate Professor of Biology, Wheaton College, Wheaton, Ill.

Dorothy P. Smith *Chemistry: Applied chemistry.* Manager, News Service, American Chemical Society, Washington, D.C.

Frank A. Smith *Transportation.* Senior Vice-President —Research, Transportation Association of America, Washington, D.C.

J. Frederick Smithcors *Medical sciences: Veterinary medicine.* Editor, American Veterinary Publications, Inc., Santa Barbara, Calif.

William E. Spicer *Physics: Solid-state physics.* Professor of Electrical Engineering and Materials Science, Stanford University, Stanford, Calif.

Leon M. Stock *Chemistry: Organic chemistry.* Professor of Chemistry, University of Chicago.

Norman M. Tallan *Materials sciences: Ceramics.* Director, Metallurgy and Ceramics Research, Aerospace Research Laboratories, Wright-Patterson Air Force Base, Ohio.

Kenneth E. F. Watt *Environment.* Professor of Zoology and Environmental Studies, University of California, Davis.

James A. West *Energy.* Associate Assistant Administrator, Federal Energy Administration, Washington, D.C.

Frederick Wooten *Optical engineering.* Professor of Applied Science, University of California, Davis.

Anthropology

Fred Eggan's opening essay in the *Annual Review of Anthropology, Vol. 3* (1974), provides a valuable perspective on the current state of anthropology by means of a review of his own career, which began about 50 years ago at the University of Chicago. Eggan experienced many of the currents that shaped anthropology in the United States. Following the completion of a dissertation on the Hopi, Eggan received one of the dozen Ph.D.s in anthropology given in the U.S. in 1933. There were no jobs available, and so with a borrowed car and a small grant he carried out additional research on Choctaw, Cheyenne, and Arapaho kinship terminology and behavior. In 1934 U.S. anthropologist Robert Redfield was successful in obtaining funding for Eggan to carry out a study of the Tinguian in the Philippines, which at that time was a U.S. colony.

Eggan returned from the Philippines to accept a position as an instructor at the University of Chicago. When World War II broke out, he was engaged in a pilot study of food and nutrition for the U.S. Department of the Interior, and he soon went to Washington, D.C., to work for the Board of Economic Warfare. He later conducted research for Manuel Quezon, president of the Philippine government-in-exile.

After a period in the Army, spent at the School for Military Government at Charlottesville, Va., Eggan developed and ran a Civil Affairs Training School for the Far East. Near the end of World War II he briefly served as a U.S. cultural affairs officer for the Philippines before returning to the "hard work" of teaching at the University of Chicago. Eggan's reflections on the current job crunch, the "information explosion," and changing theoretical currents in anthropology provide some convenient touchstones for this review.

The Ph.D. boom. The very rapid expansion of post–World War II academic departments of anthropology, including those granting the Ph.D., continued despite the increasing scarcity of jobs. The *Guide to Departments of Anthropology 1974–75* indicated that 385 Ph.D.s in anthropology were granted to 119 women and 266 men in the past academic year, more than 30 times the number of those who took degrees with Eggan in 1933. Even so, the job market for most of those graduates was more favorable than it was for Eggan and his fellow anthropologists in the depths of the Great Depression. Moreover, 95% of all U.S. anthropologists occupied academic positions in 1975, in contrast to the high percentage of those who were in "applied" positions before the end of World War II. This is not likely to remain the case, as Ernestine Friedl, president of the American Anthropological Association, pointed out upon assuming office in the fall of 1974. She cited manpower studies sponsored by the association that suggested, on the basis of demographic trends and extrapolations from past behavior of students seeking training, that only about one-quarter of the Ph.D.s in anthropology would be employed in academic institutions by 1990.

While jobs for anthropologists were shrinking, more Ph.D.s were being produced. In the five years 1970–74 the number of Ph.D.s more than doubled as compared with the previous five years; and of the 3,041 Ph.D.s granted since 1947 nearly half were granted in the last five years. In March 1975 the Executive Office of the American Anthropological Association made available a publication, *Anthropology and Jobs: A Guide for Undergraduates*, with information about fields of work for which B.A.s in anthropology are in an excellent competitive position. It also provided information on fields that require a B.A. in anthropology plus additional special training.

One new employment opportunity resulted from the announcement by the Peace Corps in December 1974 of its desire to hire anthropologists for staff positions in many of the 60 overseas countries in which it serves. There were also efforts to apportion more equitably the jobs available among qualified candidates. For example, the updated *Roster of Women Anthropologists 1974–75* was sent out to all those who ordered copies of the *Guide to Departments of Anthropology* for the purpose of "expediting the location and job recruitment of qualified female anthropologists."

The "information explosion." Eggan observed that the information in most academic disciplines has been doubling in volume every decade but that for anthropology the rate of increase seems to have been even more rapid. A glance at the standard bibliography for serial publications suggests that the rate of growth has been about 150% every ten years. The periodicals listed in UNESCO's *International Bibliography of the Social Sciences—Anthropology,* which attempts to include all serious anthropological publications in all languages, grew from 550 in 1955 to well over 1,500 by 1965. In subsequent issues a more selective policy was employed by the UNESCO listing, but the continued pace of growth in information in the field was apparent from the most recent effort to compile a comprehensive list of periodicals: *Serial Publications in Anthropology,* edited by Sol Tax and Francis X. Grollig (1973). This publication includes close to 3,000 entries but is regarded as only a beginning by its editors.

The number of books written by anthropologists

Courtesy, Dr. E. Richard Sorenson

Mother and children of the Fore people in the New Guinea highlands. The Fore permit infants to have almost continuous physical contact with their mothers or other caretakers, even while sleeping. After the birth of a second child, a toddler wishing such contact is not put off.

also increased considerably. From April 1974 through April 1975 the preeminent anthropology journal, *American Anthropologist*, received more than 780 books for review. In December 1974 *Anthropologist* editor Robert Manners announced that economic considerations would restrict pages available for contributed materials to under 900 per year for at least the next few years, delaying the publication of many reviews already received. He indicated that in the future more stringent standards would have to be adopted for the publication of reviews and also revealed that a new format was being considered. Among its innovations would be the inclusion of some short reviews or notes of approximately 50 words each.

For most anthropologists the reviews, which in 1975 averaged about 800 words each, were already too short to be very meaningful. In addition, most books had already been in circulation for two or three years before they were reviewed by the *Anthropologist*. In this context, *Reviews in Anthropology,* a quarterly that began publication in February 1974, was especially welcome. Editors Gretel H. Pelto and Pertti J. Pelto chose a review board of 18 anthropologists from all fields of specialization to help in selecting and reviewing more than 100 works each year. The books selected were to be taken from all fields of anthropology as well as from other disciplines. The *Reviews* also aimed to include a significant number of non-North American books and monographs. Each book review would average between 3,000 and 4,000 words, discussing the book in considerable detail, comparing it to other works, and elaborating its strengths, weaknesses, and implications. In addition, most of the books were to be reviewed within a year of publication.

Anthropologists continued to wrestle with the difficulties of making anthropological data available in data banks. Custodians of the Human Relations Area Files (HRAF), which since the late 1940s has been the best and most readily available anthropological retrieval system in existence, announced in 1974 that they were in the process of coding for computer use all material in the file. In the future a user will be able simply to indicate to HRAF the program or hypothesis that he wishes to test; HRAF will submit it to a computer and send the user the printouts. Such a development will be welcome, despite the many problems of reliability, sampling, and coding yet to be resolved.

New theoretical currents. In reviewing a number of influential recent works in anthropology, Eggan characterized Clifford Geertz's *The Interpretation of Cultures* (1973; reviewed in this space in the *1975 Yearbook of Science and the Future*) as "closer to the center of anthropology" than the others. It appeared to be a very much appreciated center, for in 1974 the American Academy of Arts and Sciences awarded Geertz its first prize for "a major contribution to the social sciences in the last decade." At the 1974 annual meeting of the American Sociological Association, Geertz was also awarded (jointly with sociologist Christopher Jencks) the Sorokin Award as "author of a publication which contributed in an outstanding degree to the progress of sociology during the two preceding years."

The best sampling of the array of perspectives (some contradictory) operating within social and cultural anthropology is found within the *Handbook of Social and Cultural Anthropology*, edited by John J. Honigmann (1973). Honigmann's volume represents an effort to make an inventory of current anthropological theory and research similar to that undertaken 20 years before by A. L. Kroeber in *Anthropology Today* (1953) but with some interesting differences. The 1953 volume attempted to cover "all of anthropology" in 50 chapters, while the *Handbook* took 28 to cover just social and cultural anthropology. The *Handbook* is actually somewhat greater in total bulk than its predecessor, 1,295 pages compared with 966, a difference well justified by the expansion of the fields covered since 1953. The 1953 volume appeared with a companion book entitled *An Appraisal of Anthropology Today*, which contained the edited symposium discussion of the inventory papers. Unfortunately, no such volume accompanies the *Handbook* though long review articles by Marvin Harris and Richard R. Randolph in *Reviews in Anthropology*, May 1974, help to provide the beginnings of such an appraisal.

The aim of the *Handbook* is to treat the specialties within social and cultural anthropology broadly enough so that "a graduate student or professional unfamiliar with the subject can become acquainted through his own efforts." Though not exhaustive, the subjects treated in the *Handbook* indicate most of the present major subdisciplinary specializations in social and cultural anthropology. In addition, an opening chapter by Fred W. Voget is devoted to the "History of Cultural Anthropology" and seeks to provide a framework for the other papers.

Voget sums up the history of cultural anthropology in a final sentence: "The humanistic-descriptive and scientific-processual strains in anthropology have resulted in a schizoid personality of sorts, but the major thrust has been toward the humanistic end of the spectrum." From the contents of the *Handbook*, however, it would seem that the opposite conclusion is justified, despite the appreciative reception recently given to the work of Geertz. Because the papers, often with very different theoretical focuses, stand in isolation rather than interaction with each other, generalizations about them are difficult. Nonetheless, some trends may be noted. Since 1953 contributions from linguistics have helped to define the areas of "ethnoscience," "cognitive anthropology," and "new ethnography" represented in the *Handbook*. In addition, the impact of computers and formal models, linguistic and mathematical, is much more in evidence than was the case 20 years ago.

One of the gaps in the *Handbook* acknowledged by Honigmann is the lack of a chapter on "biological determinants of human behavior." Two excellent recent publications on one aspect of this subject can be noted, however: *Race Differences in Intelligence,* by John Loehlin, James Spuhler, and Gardner Lindzey (1975), and *Race and IQ,* edited by Ashley Montagu (1975). The latter book is a collection of articles by biologists, geneticists, psychologists, and educators underlining two principal objections to the arguments for a direct link between race and IQ: (1) the concept of "race" used in such arguments was discredited two or three decades ago; and (2) IQ scores do not actually measure general intelligence but simply predict success in middle-class (predominantly white) public schools.

Though not a direct replacement for Honigmann's missing chapter, Robin Fox's *Encounter with Anthropology* (1973) certainly stands out as one of the most insightful and readable of recent attempts to restore a unity of direction to anthropology, which he sees now as a "rag bag of odds and ends held together by sentiment and dynastic interest." Fox reviewed the 14 rather frantic and productive years he spent trying to become an anthropologist while perhaps succeeding more as a "renegade philosopher" and a "radical" in the sense that he calls for a radical restructuring of anthropology. After many observations on the "profession of anthropology" Fox proposes that "there should be an anthropology that is simply a division of evolutionary biology, itself a branch of ethics. To do this will be to have a vision of a science of man that would truly be a science of all mankind, past and present, civilized and savage, hominid and protohominid; a vision, that is of course Darwin's, of a science of life that will pull man back into nature without robbing him of his dignity."

—Raymond Lee Owens

Archaeology

Archaeology has not escaped the effect of the drastic economic, social, and political changes presently taking place everywhere. Growing concern with national cultural heritage—something quite apart from intense nationalism—and growing wealth in what was known as the underdeveloped world are shifting archaeological activity into new fields, new interests, and somewhat different management. Available funds increasingly go toward restoration and conservation of objects and monuments, as in Italy, rather than into new excavations. The Arabs now have the means to concentrate both excavations and restora-

Wide World

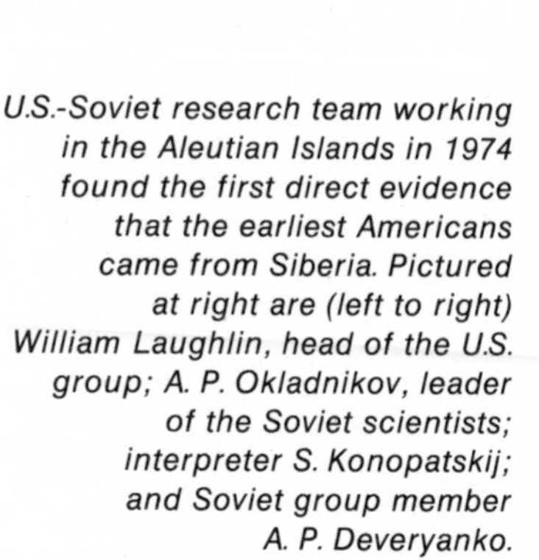

U.S.-Soviet research team working in the Aleutian Islands in 1974 found the first direct evidence that the earliest Americans came from Siberia. Pictured at right are (left to right) William Laughlin, head of the U.S. group; A. P. Okladnikov, leader of the Soviet scientists; interpreter S. Konopatskij; and Soviet group member A. P. Deveryanko.

tion upon Islamic remains, which are of primary concern to them. Indonesia spends large sums upon the restoration of Borobudur and also, currently, upon the infrastructure of a national archaeological center to concentrate archaeological research under Indonesian management rather than that of foreign expeditions.

America. Radiocarbon dating continues to demonstrate an ever greater age for the remains of man in various regions of the continent. A University of Pittsburgh team has announced carbon-14 dates for a deposit of stone tools near Avella in Pennsylvania as early as 13000 B.C., 1,000 years older than any other known site in the eastern U.S. Certainly such dates in the East imply a much earlier arrival of man in America from Asia via the Bering Straits region. These early people in Pennsylvania must have lived close to the edge of the last great glacier and thus achieved the technology to cope with a glacial age. In this connection, a study published in *Science* in February 1975 by J. P. Kennett and R. C. Thunell advances the theory that the Ice Age was a period of exceptional volcanic activity, probably the result of changing pressures on the Earth's surface and seabeds as ice sheets formed and melted. Volcanic ash falls can be helpful in dating and preserving human remains, and an unusually active volcanic period could well effect the movement and dispersal of men and animals.

The method that Ice Age inhabitants of America must have used to kill mammoths, animals standing 9–14 ft high at the shoulder, has long puzzled archaeologists but now may be explained by the discovery of bone lance shafts in a Clovis-period burial at the Anzick site in southwestern Montana. These are foreshafts something like those used with Eskimo harpoons but they are beveled at one end, which suggests that stone heads were lashed to the foreshafts which were in turn lashed to a spear shaft so that spearhead and foreshaft would become detached and remain in the body of the animal in the manner of a harpoon. The 10,000–12,000-year-old grave contained 100 bone and stone artifacts and two skeletons covered with red ochre—the only Clovis-period burial that is known thus far.

In Alaska a Soviet-U.S. expedition headed by William Laughlin and A. P. Okladnikov discovered 9,000-year-old stone artifacts on Anangula Island in the Aleutians. They are related to a type of stone blades found in the Gobi Desert and are believed to be the first direct link between the Aleutian Islands and central Asia via the presumed land bridge which joined these regions.

The now famous Koster site near Kampsville, Ill., being excavated by a team from Northwestern University under the direction of Stuart Struever, continued to produce a most remarkable record of ancient life in the Midwest. Probes made in 1974 show the debris of human occupation extending to a depth of 44 ft. Excavations have now reached down to the 11th horizon, dating to 6500 B.C., and the probes to lower horizons indicate an occupation that may be as early as 10000–8000 B.C. Study of the collections shows that the early inhabitants lived in wooden huts (the earliest known houses in America) in a settlement of about 50 persons extending over a half acre. They were hunters and collectors depending upon deer, elk, bear, waterfowl, fish, and nuts. There is no evidence of fighting until the last period of occupation about A.D. 1000. There is evidence for the domestication of dogs by 6500 B.C.

Asia. Excavations at Ban Chiang in northeastern Thailand, directed by Chester Gorman of the University of Pennsylvania's University Museum and Pisit Charoenwongsa of Thailand's Department of Fine Arts, begun in 1974, have confirmed the extraordinary age of bronze metallurgy in Southeast Asia and also the predicted complex archaeological record in that region. The low mound at Ban Chiang, about one kilometer in diameter, is essentially a cemetery with many thousands of graves lying in at least three major horizons. The two lower levels produce many cast bronze objects and the upper level extensive iron objects. Radiocarbon dates, corresponding with thermoluminescence (TL) dates from the now famous Ban Chiang pottery, range from 3600 to 200 B.C., but different cuts in the mound show somewhat different strata and different material, and it is reasonable to assume that the oldest levels of the mound are not yet located. The oldest TL dates for pottery of uncertain horizon are in the range of 4600 B.C.

The long search for a Greek city site in Bactria, northern Afghanistan, or for concrete evidence of Greek settlement there following Alexander's conquest in 329 B.C., was rewarded by the French discovery and excavation of a site called Ai Khanoum ("Moon lady" in the Uzbek tongue). French archaeologists in Afghanistan had been looking for such a site since the early 1930s. While hunting in 1963, Mohammad Zahir Shah, king of Afghanistan, was presented with some stones bearing Greek inscriptions. He informed the Délégation Archéologique Française, which sent archaeologists to the site. But it was near the then-sensitive Soviet border on the Amu Darya (Oxus River) and actual excavation was delayed for some years. Now Paul Bernard of the Délégation Française observes that Ai Khanoum has at long last unveiled Greco-Bactrian civilization. Found there are Corinthian capitals, roof tiles, sarcophagi, sculpture, and Greek inscriptions that identify these Bactrian people with the Hellenistic world. The city wall is about ten kilometers in length. A vast irrigation system and a dense population of the valley indicate a concentrated urban area.

Pottery and grave at Ban Chiang in Thailand. Excavations at the site have revealed a complex archaeological record dating back to at least 3600 B.C. and have also confirmed the extraordinary age of bronze metallurgy in Southeast Asia.

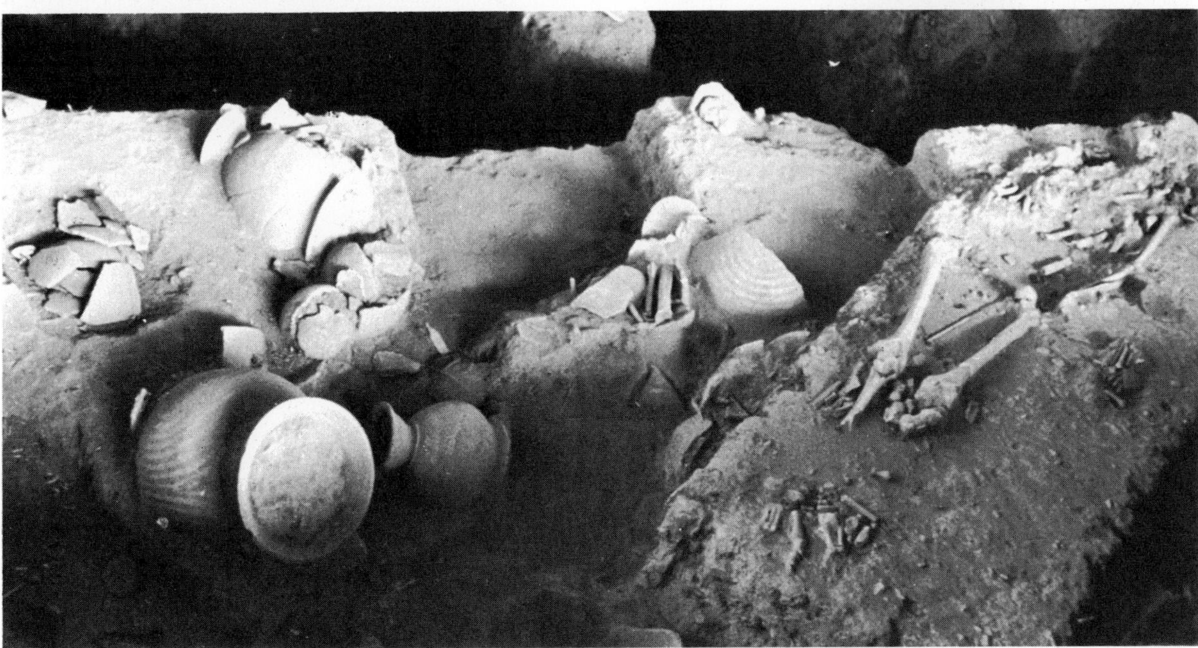

Photos, courtesy, The University Museum, University of Pennsylvania

Near and Middle East. The systematic excavation of a Phoenician city in the homeland, rather than at the sites of Phoenician colonies in the Western Mediterranean, during the past season has again produced most significant results. Excavation of a shrine of the 7th and 6th centuries B.C. (the first Iron Age sanctuary to be found on the Lebanese coast—the Phoenician homeland) disclosed many cultic objects such as figurines, carved ivories, masks, amulets, cosmetic equipment, and lamps. Among them was a fragment of ivory containing a four-line inscription that reads, "This statue made Shillem, son of Mabaal, son of Izai for Tanit-Ashtart." That fragment solves an old problem about the Tanit cult—famous in Carthage because it demanded the sacrifice of children—and for the first time proves the goddess was worshiped in the home country and presumably exported to the colonies, such as Carthage.

Fighting in Cyprus has disrupted extensive archaeological research not only in Cyprus but also in Greece and Turkey. Turkey, in particular, had become one of the most active fields for archaeological research anywhere in the world. War, plus a strong national reaction to the illicit export of antiquities, has drastically reduced all foreign excavations. The accidental death of Spyridon Marinatos on Thera (Santorini) has also interrupted the very important excavation of the Minoan site on that island.

In Egypt the discovery of Horemheb's tomb (late 18th dynasty) at Saqqarah concludes a search carried on by the archaeologist Walter Emery for many years before his death in March 1971. In January 1975 Geoffrey Martin (Egypt Exploration Society of London), A. Klasens (Leiden Museum, The Netherlands), and Ali el-Khouly (Egyptian Antiquities Service) announced the discovery of the tomb a few miles south of the sacred animal necropolis excavated by Emery. Horemheb was not actually buried at this site but in the Valley of the Kings at Thebes. Unusually fine relief sculptures and inscriptions from this Saqqarah tomb were looted in the 19th century and are now in museums in Leiden, London, and Bologna, Italy.

Also in Egypt an international team of scholars led by James Robinson of the Claremont (Calif.) Institute for Antiquity and Christianity is about to complete the translation of early Christian writings discovered in a cave near Nag Hammadi on the Nile below Luxor. This is a collection of leather-bound books found in a huge storage jar in the cave. Robbers who found the books tore them up and distributed the pages for sale. Many years were required to run down fragments of the collection spread all over the world. All but one of the texts have been recovered. Apparently most of the volumes were completed just after A.D. 350 by monks under the influence of Gnostic Christian ideas.

Sarepta, an Iron Age city of Phoenicia located on the coast of present-day Lebanon, is being excavated by the University Museum of the University of Pennsylvania. Soundings of lower levels at the site show the plan of an earlier Bronze Age city of the 16th century B.C.

Courtesy, The University Museum, University of Pennsylvania

Courtesy, The Cleveland Museum of Natural History

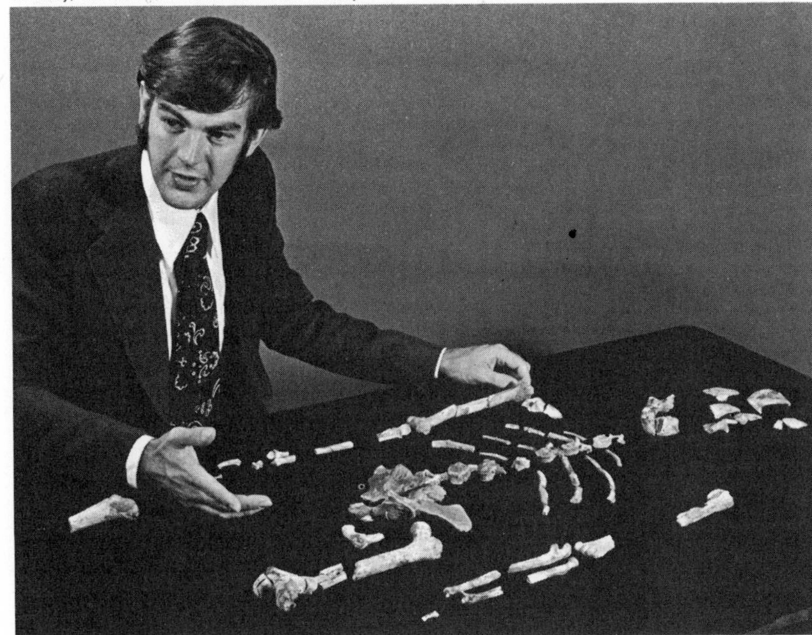

Donald Johanson displays bones of a three-million-year-old woman he found in Ethiopia in 1974. The woman is not believed to be in the direct line of descent to modern man but instead is more closely related to the australopithecines—"near men."

Europe. Excavations this year by the Belli Arti Society of Siena at Greppe di Sant'Angelo near Cerveteri (ancient Caere) in Italy have uncovered two impressive Etruscan tombs of unusual significance. Found there was a statue of Tuchulca—one of the two gods of the underworld—and a number of inscribed tablets. Etruscan inscriptions are rare (which accounts for the difficulty in translating the language) and no other statue of Tuchulca is known.

Regrettably, these new excavations will be closed for lack of funds. The two new tombs are apparently on the edge of a large funerary complex (near the famous necropolis of Cerveteri), and it is almost inevitable that they will be looted by the *clandestini*. A recent police raid of grave robbers in the Cerveteri area netted gold and ceramics valued at over 700 million lire. Estimates are that 500 tombs a year are being looted in this one region. One hopeful and likely possibility is the claim of the local people that the great numbers of Etruscan art objects now being sold abroad are all forgeries.

Also in Italy publications have recently appeared that describe the very extensive excavations of the site of Sybaris in Calabria. The excavation, under the direction of the superintendent of antiquities of Italy, is quite certainly the most expensive, and in some ways the most elaborate, ever carried out in Italy. The site, located in 1968 by a U.S.-Italian team using a cesium magnetometer and auger drills, lies up to 30 ft deep and several feet under the water table. Excavation can be done only after very extensive pumping with a well-point system, and with large earth-moving machines. Hence, excavations so far have cost something over the equivalent of $2 million.

In England the discovery of the wreck of the British warship "Colossus" has been reported by Roland Morris, a marine archaeologist. For undersea archaeology this wreck is peculiarly interesting because it was carrying an extensive collection of 5th and 4th century B.C. Greek objects collected by Sir William Hamilton when he was British envoy to Naples in the late 18th century. Ten cases of Greek vases, sculptures, terra-cottas, etc., were saved when the vessel was wrecked off the Isles of Scilly in 1798, but another eight crates were lost. The wreck is embedded in silt at depths of 40 to 75 ft; hence there is a good chance that the Greek material will be preserved.

Africa. The astonishing number of fossil hominid remains discovered in East Africa in recent years is totally revising our ideas of the age and evolution of man. In 1974–75 additional, and upsetting, discoveries were made in the rich areas of Lake Rudolf in Kenya, Olduvai Gorge in Tanzania, and particularly in Ethiopia.

Two discoveries in the Afar Depression in northeastern Ethiopia are a good example of what is happening. Donald C. Johanson of the Cleveland Museum of Natural History and Case Western Reserve University, Maurice Taieb of France, and Alemeyu Asfew of Ethiopia headed an expedition that found bones of one individual not in the direct line of human descent (but related) and those of another individual believed to be in that direct line and thus included in the genus *Homo*.

Most unsettling is the one accepted as genus *Homo* because, judging from the potassium-argon method of dating, the bones must be three million to four million years old. Richard Leakey had shaken many anthropologists in 1972 by announcing such a human fossil from 2.6 million years ago, and other examples of such a creature are now being found; thus, the evidence for the very great age of man's direct ancestors is good, although an age of three million to four million years is still very much of a surprise to many scientists.

The other individual, a female 18 to 20 years of age at death, is thought to be not in the direct line of descent to modern man but in some way more related to the australopithecines—"near men." One of the remarkable aspects of this find is the number of bone fragments that apparently are from one individual. There are 40 of these, including parts of the skull, jaw, teeth, vertebrae, arm and leg bones, pelvis, and ankle and foot bones. With these it is possible to describe a creature standing only about three feet high, walking erect, and with teeth resembling *Australopithecus*. Potassium-argon dating gives an age of about three million years.

There can be little doubt at this point that several manlike and other *Homo* species were contemporary in very ancient times. Some were to become extinct, and others were to evolve into modern man. Moreover, recent excavations at Lake Rudolf are confirming that these truly ancient genus *Homo* types were tool makers.

Techniques. There are now eight or nine possible atomic nuclear methods of dating ancient materials that have been explored since the original radiocarbon method proved effective. But in 1974 a dating technique in a wholly different field was demonstrated. It has been given the rather formidable name of aspartic acid racemization reaction, but it is in effect a method based upon amino acids in fossil bones. Fossil materials have been found to contain D-amino acids, and the proportion of D-amino acids to L-amino acids increases with the age of the fossil. Unfortunately, the racemization reaction is temperature dependent and therefore involves some estimate of the temperature history. Some promising checks with carbon-14 dates were made, and the method appears to have promise.

The Applied Science Center for Archaeology (MASCA, at the University of Pennsylvania's University Museum) is currently engaged with the Stanford Research Institute (Menlo Park, Calif.) in testing a new soil-penetrating radar system. Initial tests in Chaco Canyon, N.M., show a pattern of radar echoes that indicate it will be possible to detect buried walls and other archaeological phenomena. Soil profiles up to ten meters deep can be recorded. Tests continued in the White Mountains of California in search for bristlecone pine logs buried in alluvial fans.

Since MASCA is also engaged in perfecting the correction factors for radiocarbon dating, one reason for the radar tests in the White Mountains is to discover "fossil" logs which can extend the tree-ring record beyond the 8,200 years now worked out. A systematic correction system for carbon-14 dates depends upon their correlation with tree-ring records, so that an extension of corrected dates beyond 8,200 years ago depends upon the discovery of such ancient bristlecone pine logs. Measurements made in 1974 indicate that the dielectric constants between the pines and the surrounding alluvium are sufficient to provide a contrast for radar.

A. Simopoulas of the Nuclear Research Center in Athens has recently suggested the possibility of using Mössbauer spectroscopy for dating ancient ceramics. This is based upon the shift of iron (III) to iron (II) and the degree of disintegration of iron oxides in the clay, which is a function of time. It may be possible by a comparison with pottery of known age to work out a measurable disintegration rate in terms of time.

—Froelich Rainey

Architecture and civil engineering

The year 1974 marked the centenary of the birth of Auguste Perret, a French builder architect of the first structures (an apartment house in 1903 and a garage in 1905 in Paris) in which the reinforced-concrete framework became an essential element of architectural form. The centenary was marked by a major exhibition in France and Great Britain, featuring Perret's famous aphorisms:

> Architecture is the art of organizing space. It is in and through building that it finds expression.
>
> He who conceals a pillar makes a blunder. He who builds a sham pillar commits a crime.
>
> The great buildings of today have a skeleton, a framework of steel or of reinforced concrete. The framework is to the building what the skeleton is to an animal.

The evolution of modern architecture continued to be a subject of lively debate. Is modern architecture alive or dead? Is it in its revivalist phase, recycling itself; or is it in a post-modern phase? Is it local and regional, or is it cosmopolitan and international?

Courtesy, Reynolds Metals Company, photo by Nathaniel Lieberman

88 Pine St. office building (right) in New York City, seen here from the East River, won a major architectural design award in 1974.

The phrase "International Style" was first coined in 1932 and referred specifically to the work of the previous ten years by architects in 15 European countries and the United States. This was what might be termed the heroic period of modern architecture; the heroes included Alvar Aalto (Finland), Le Corbusier (France), and Walter Gropius and Ludwig Mies van der Rohe (Germany). The aesthetic of their buildings was influenced by the visual structures of Cubism. In the case of Le Corbusier, working with the painter Amédée Ozenfant, the prewar work was described as Purism; the revival or continuation of this Purism was one of the seminal directions of thought and work in 1975.

It is perhaps characteristic of architecture that some of its most powerful new directions are given unfortunate names (like Gothic or Baroque, which were originally terms of disrespect for the new architecture of their times). So it is that Le Corbusier's remarks at the inauguration of his building the Unité d'habitation in Marseilles, France, referring to the use of the crude aesthetic of raw concrete, the *béton brut*, led to the coining of the phrase "the new brutalism" for postwar architecture. But whatever its name, the new architecture is truly international, with a variety of cultural expressions in such different places as Japan, Italy, Scandinavia, Great Britain, and the Americas. Some recently completed examples are evidence of the vitality of modern architecture: in West Germany, the Free University of Berlin, designed by the architects Georges Candilis, Alexis Josic, and Shadrach Woods; in Britain, the Olivetti training center at Haslemere, by James Stirling; in Japan, the Koba Kindergarten, by Fumihiko Maki; and in the United States, the Science Center at Harvard University, designed by José Luis Sert.

The 1974 R. S. Reynolds Memorial Award, one of the most prestigious in architecture, was presented to James Ingo Freed, a member of I. M. Pei & Partners, for his design of an office building at 88 Pine St. in New York City. The jury, from the American Institute of Architects, stated that 88 Pine was "not only a handsomely designed urban building; it is also a type of structure that rarely receives the care and attention that this one was given by a master designer." The exterior of the building was clad with aluminum organically coated in white.

Courtesy, Harvard University News Office

Science Center at Harvard University, designed by José Luis Sert, exemplifies the vigor of recent international architecture.

Industrialization of buildings. The tension between the arts, the crafts, and the machine has been one of the sources of the modern movement in design, recalling John Ruskin, William Morris, Frank Lloyd Wright, and Walter Gropius. In 1975 the questions remained as lively as ever: What is the proper relationship between design and industry? How can the industrial production process be used to serve man's social, economic, and aesthetic goals for the built environment? Some recent developments are important: the use of a "systems building" method in the construction process, and the growth of pre-engineered metal building systems.

The systems building method uses the performance requirements of a structure as the basis for the selection of its particular subsystems (such as the structure, the heating and ventilating, the interior partitioning, the exterior skin). An early example of systems building was the Crystal Palace of London, designed by Joseph Paxton and fabricated in 1851 of modular, standardized, coordinated subsystems of metal and glass. Outstanding recent examples include the systems used for post-World War II schools in England; for public schools in California and in Toronto, Ont.; for the Social Security Administration buildings; for the New Jersey state colleges; and for the Marburg University Building in West Germany. For the coherence of their designs these buildings rely characteristically on the coordination of the subsystem elements into an "open" system that allows for interchangeability, change, and growth.

In the U.S. there has been great growth in the manufacturing of pre-engineered interrelated metal building systems. Originally, these steel structures were fabricated for agricultural purposes, such as grain bins and farm buildings; later, they were expanded to include ready-made metal buildings for other industries, and in the post-World War II years they have been used for all nonresidential construction. Since 1965 the metal building industry has outpaced the entire nonresidential construction industry by more than 3 to 1; in 1974, metal building manufactures accounted for $770 million in sales of structural, roof, and wall subsystems.

Renewal of cities. The relationship between the new architecture and the new planning of cities is complicated. On the one hand, whole districts of cities are being planned, designed, and built as new, coordinated urban bodies. And, on the other hand, whole districts of cities are being preserved, reconstructed, adapted to new lives in a continuous incremental process.

The recent Reynolds Awards for Community Architecture (the previous awards having been for the new towns of Cumbernauld, Scotland, in 1967,

Head House and 2nd Street Market in the Society Hill district of Philadelphia bears witness to the effectiveness of the urban restoration programs taking place in many cities. The building is shown before restoration at left and afterward at right.

Photos, courtesy, Redevelopment Authority, City of Philadelphia

and Beersheba, Israel, in 1970) went to the old city (Altstadt) district of Munich, West Germany. The new architecture in the district included a 900-meter pedestrian street distinguished by varying widths and arcades and plazas. It was woven into the fabric of the old city, extending on an east-west axis from the Karlstor (old city gate), past several medieval churches, to the Rathaus and Marienplatz.

In Paris the old district of Le Marais was in the process of restoration in 1974 and 1975. Most of the district was being rebuilt or preserved in its historic appearance; in some cases, new architecture was being constructed as the in-fill of the urban fabric.

In Jerusalem plans were developed for the reconstruction of the ancient Jewish Quarter, based on the original Roman axes that had been obscured since medieval times. In order to restore the original form of open and enclosed spaces at certain points, prefabricated concrete arches were to be used to provide a new vaulting system for the street arcades.

There were many examples of old urban districts in the U.S. that were being renewed. An outstanding example of the mixture of restoration, preservation, and new architecture is the Society Hill district in Philadelphia. A large collection of 18th- and 19th-century brick townhouses was restored; the streets and walkways were brick-paved and landscaped; and new architecture was introduced to harmonize with the old. Although the new buildings were rarely distinguished architecturally, the district as a whole became a lively and humane townscape.

Entirely new urban districts were being built in the U.S., often taking advantage of the sites of the now-obsolete urban transport systems of the past century. In many major cities these transport areas (including railroad yards and ports) were being rebuilt for new uses. In New York City, Battery Park City and Manhattan Landing were being developed on sites created by landfill along the old shore lines of lower Manhattan. Illinois Center in Chicago was being built over the tracks of the Illinois Central Railroad near the downtown Loop area. In Philadelphia the historic Delaware River port area was being rebuilt as the multi-use Penn's Landing district. These and other developments comprised mixtures of commercial and residential buildings.

Research. Two areas of research are particularly interesting to review briefly because they illustrate the ways in which research methods can be imported from other disciplines to improve the understanding of buildings. The application to architecture of the social and behavioral sciences provided the focus of the International Architectural Psychology Conference, held in 1974 at Lund, Sweden, and the Environmental Design Research Association (EDRA) Conference, held each year in the U.S. The research topics reviewed at Lund included environmental aesthetics, building assessment, outdoor behavior, and environmental simulation. At EDRA in 1974 the topics included man-environment relations, liveability and satisfactions of housing, childhood and elderly environments, and behavior settings. A major issue confronting these research conferences was the integration of social behavioral research into the process of designing and planning buildings.

Courtesy, Robert L. Geddes

Photoelastic model study of Gothic cathedral structure was devised by Robert Mark of Princeton University as a method of applying sophisticated structural analysis to the study of historical buildings.

The application of sophisticated photoelastic structural analysis models to the study of historical buildings was the work of Robert Mark at Princeton University. His structural analysis of Gothic cathedrals, especially the comparison of those at Bourges and Chartres, was widely discussed by architects, engineers, and historians. His most recent work sought to explain the reasons for the famous collapse of the vaults of the Beauvais, France, cathedral in 1284. The research was based on experiments on scale-models of the Beauvais structure, and on an examination of the existing material on the site. The results differed significantly from previous explanations and added much to the understanding of the relationship between structural design and architectural form.

—Robert Geddes

Astronomy

During the past year, astronomy achieved notable observational advances in solar-system studies, thanks in part to spacecraft. These successes tended to overshadow major theoretical problems, such as the interpretations of red shifts and high-energy phenomena, where progress was slower.

Large telescopes. A dramatic increase in the effectiveness of ground-based astronomical observation was under way, with a number of large-telescope projects coming to fruition almost simultaneously. The largest optical telescope in the world is the 6-m (236-in.) reflector near Zelenchukskaya, U.S.S.R., in the Caucasus Mountains. It was expected to be operational soon, after long delays arising in part from its novel computer-controlled altazimuth mounting.

Most of the new very large telescopes were being erected in the Southern Hemisphere because most of the existing great observatories were in northern latitudes, leaving the southern sky relatively unattended. Thus, the new 3.9-m (154-in.) Anglo-Australian telescope is located at Siding Spring, New South Wales, Australia. It became operational in 1974. Also at that site is a 48-in. Schmidt telescope, nearly a twin of the Palomar Schmidt in California.

In Chile the 4-m (158-in.) reflector of the Cerro Tololo Inter-American Observatory took highly successful trial photographs in October 1974. Testing and final adjustments of this instrument were expected to require a year. It is very similar to the 4-m telescope completed in 1973 at Kitt Peak National Observatory in Arizona.

Also in the Chilean Andes and only 83 mi N of Cerro Tololo is the European Southern Observatory, at La Silla. Its 3.6-m (142-in.) reflector was scheduled to begin operating in 1976. Nearby, at Las Campanas, Chile, is the 2.56-m (101-in.) du Pont telescope at the Carnegie Institution's Southern Observatory. This sophisticated and versatile reflector, erected early in 1974, can be used with a Gascoigne-type corrector lens to permit stellar photography over a field 2.1° square, a remarkably wide field for a telescope so large.

Not all of the growth in observing capabilities was due to new telescope construction. At least as important was the increased efficiency of existing telescopes because of more sensitive light detectors, new auxiliary equipment, and design improvements. A noteworthy example of the last-mentioned factor is the 1,000-ft radar-radio telescope operated by Cornell University at Arecibo, Puerto Rico. As originally built in 1963, this fixed spherical antenna with an area of 20 ac had a concave collecting surface of wire mesh, for observing at a wavelength of 70 cm (27.6 in.). In late 1974, however, an extensive upgrading of the telescope was completed. Its sensitivity was increased 2,000-fold and it could operate at much shorter wavelengths.

New 4-m (158-in.) reflector telescope (left) at the Cerro Tololo Inter-American Observatory in Chile is the largest in the Southern Hemisphere. Observatory of the Soviet Academy of Sciences (right) in the Caucasus Mountains houses the largest optical telescope in the world, a 6-m (236-in.) reflector.

Courtesy, National Science Foundation

Tass from Sovfoto

Courtesy, Harvard College Observatory

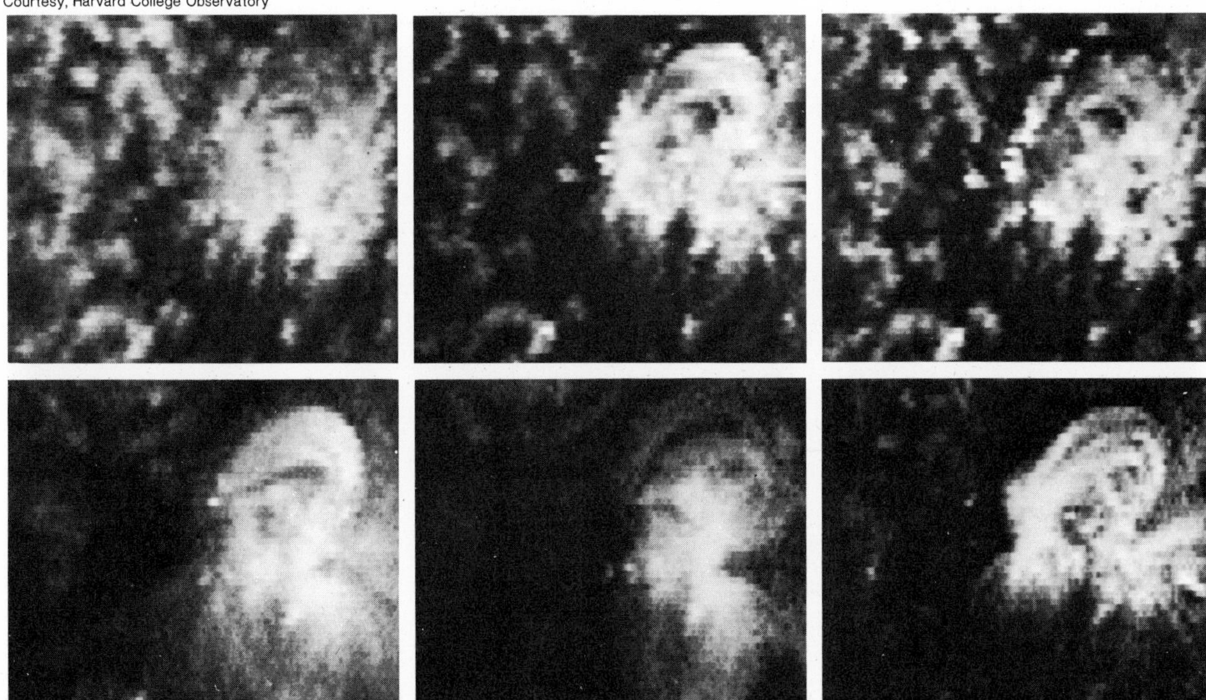

Six spectroheliograms of an active region on the solar disk containing loops of magnetic fields were taken simultaneously by a spectroheliometer on the U.S. Skylab spacecraft. Each was taken in a different wavelength representing a different height in the solar atmosphere (6,500–62,000 mi).

Part of this upgrading consisted of replacing the wire mesh with 38,778 aluminum panels that could be accurately adjusted individually. This allowed the deviations of the surface from a sphere to be small enough to permit observing at wavelengths as short as 4.2 cm. When serving as a radar telescope, the Arecibo instrument used a new high-power 12-cm transmitter, radiating an average of 450 kw, and also employed improved, more sensitive receivers. This system, planned especially for radar surface mapping of Venus during the summer of 1975, was expected to distinguish altitude differences of as small as 300 ft on that planet's surface. Radar mapping of the major satellites of Jupiter and Saturn was expected to be possible, and the properties of Saturn's rings might be examined in detail. More important, perhaps, when the upgraded Arecibo antenna is used passively, as a radio telescope, it should be extremely effective for detecting the weak microwave spectral lines of interstellar molecules.

New solar features. Important results continued to emerge from the wealth of solar observations obtained in 1973–74 by the manned Skylab satellite. Circling the Earth once each 93 minutes, this spacecraft carried, among other experiments, an X-ray spectrograph that formed images of the Sun in radiation of 3.5 to 60 angstroms wavelength (1 angstrom, Å, equals 10^{-10} m).

Conspicuous on such X-ray photographs are numerous temporary bright points, which were studied by L. Golub and his associates. An estimated 1,500 of these bright points appear on the Sun each day, with an average lifetime of eight hours. In a typical case, a bright cloud appears, inside of which a luminous core forms. Cloud and core brighten rapidly, growing to the size of the Earth before starting to fade gradually.

These transient features, which are fairly evenly scattered over the Sun's surface, are believed to be convection cells of hot gases rising from the solar interior. The bright points carry fairly strong magnetic fields (averaging about ten gauss) from the interior to the surface. This flux may be an important, if not the main, contributor to the observed general magnetic field of the Sun.

The first observation of the X-ray bright points on the Sun was made as early as 1969 by G. S. Vaiana with an X-ray telescope carried by a rocket above most of the Earth's atmosphere. However, such a rocket flight gives a "look" of only a few minutes, allowing a much less detailed study of these changing features than did Skylab, which could follow them nearly continuously.

Courtesy, Jet Propulsion Laboratory

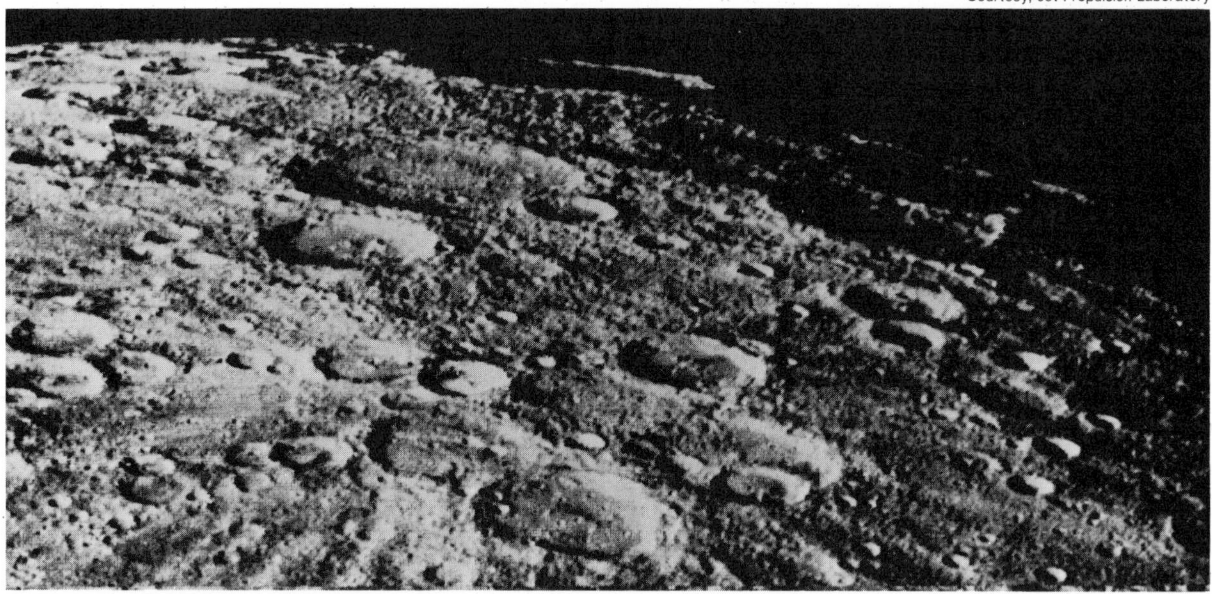

South polar region of Mercury was photographed by Mariner 10 from an altitude of 53,200 mi. Scarps and cliffs are believed to have been formed by shrinkage of the planet in its infancy.

Skylab observations also led to the recognition of solar macrospicules. These are large, bright spikes, observable in far-ultraviolet light, that extend up to a minute of arc or higher above the edge of the Sun's disk. They have lifetimes of up to 40 minutes. Macrospicules were reported early in 1975 by two groups of investigators, from the U.S. Naval Research Laboratory and from Harvard Observatory. Both groups had instruments aboard Skylab for obtaining images of the Sun in the light of individual spectral lines, for example the 304 Å bright line of ionized helium.

The Harvard investigators obtained detailed life histories of two macrospicules that lasted for 15 minutes. Each one was a luminous jet that rose to about 20,000 mi above the Sun's rim and then dropped back at a rate that grew to 85 mi per second by the time the jet faded out.

Macrospicules had been previously noted on small-scale ultraviolet solar pictures taken in 1969 by rocket-borne equipment but were mistaken for bunchings of ordinary spicules. Also, in 1952 the Swiss astronomer Max Waldmeier photographed a giant projection at the Sun's limb that may have been a macrospicule.

Mercury. The historic reputation of Mercury as one of the least understood planets was strikingly reversed by the Mariner 10 spacecraft's three flybys in 1974–75. Years of optical observations of the surface of Mercury even with large telescopes had yielded little of value. In fact, radar techniques were needed for the first determination, as recently as 1965, of the planet's true rotation period (59 days). In 1974, Irwin Shapiro and his colleagues reported radar detection of 5-km vertical relief on the Mercurian surface.

Launched on Nov. 3, 1973, Mariner 10 flew past Venus on Feb. 5, 1974, and reached the vicinity of Mercury for the first time on March 29, 1974. After coming within about 460 mi of the planet's surface, the spacecraft was placed in an elliptical orbit around the Sun with a period of 176 days—just twice Mercury's orbital period. This ensured a second encounter, on Sept. 21, 1974, when the least distance was 29,814 mi. On Mariner's next revolution around the Sun, it flew past Mercury for a third time on March 16, 1975, missing its surface by a scant 198 mi. This third encounter was the last on which scientific observations could be made, for five days later the exhaustion of Mariner's attitude-control fuel left the craft tumbling helplessly, unable to generate electricity because its solar panels no longer faced the Sun.

Mariner 10's twin television cameras secured nearly 5,000 pictures of Mercury, some showing features as small as 200 to 300 ft across. On the first flyby 25% of the planet's surface was mapped; the total was raised to 37% by the second mission and to 57% by the third.

The surface of Mercury was found to be heavily cratered, closely resembling that of the Moon. The largest feature is a circular enclosure 800 mi across, called Caloris Basin, whose floor is laced with fractures and sinuous ridges like those on lunar maria. Astronomers believe that Caloris Basin is the scar left by the ancient impact of a huge meteoroid. A large area of damaged terrain was observed in the region antipodal (diametrically opposite) to Caloris Basin, and is thought to be the result of converging seismic waves produced by the same impact. Seventeen basins larger than 120 mi in diameter were noted on the quarter of Mercury's surface mapped during the first encounter with the planet.

Long scarps or cliffs up to one mile high, some of them extending for hundreds of miles, are prominent features on Mercury. They are believed to be caused by compressional forces on a planetary scale, possibly resulting from cooling. Apparently, Mercury is everywhere covered by a fine-grained porous soil, which, like the Moon's soil, is a very poor conductor of heat. This conclusion was based on measurements of Mercury's thermal radiation, obtained with Mariner 10's infrared radiometer.

Mercury was found to have a magnetic field about one hundred times weaker than that of the Earth. This discovery was made during the first flyby, when the magnetometers aboard Mariner detected a magnetosphere surrounding Mercury analogous to the tear-shaped tail of the Earth that results from the interaction of the solar wind of charged particles with the Earth's magnetic field.

Mercury's magnetic field is generated internally and is not a result of direct impingement of the solar wind upon the planet's surface. This was established by the third encounter with the spacecraft, when the Mariner trajectory passed the night side of Mercury at a high northerly latitude for the first time. The magnetometers again detected the bow shock wave of Mercury and other features of the magnetosphere, at exactly the times predicted on the basis of an internal magnetic field. It remained unknown whether Mercury's magnetic field is generated by permanently magnetized rocks in the planet's interior or by action in a liquid planetary core.

Other results under study included the detection by Mariner's ultraviolet spectrometer of a very tenuous Mercurian atmosphere, consisting largely of neutral helium. The surface pressure on the planet was found to be less than about 2×10^{-9} millibar.

A preliminary value of the mass of Mercury was derived from its gravitational effects on the motion of Mariner 10, as evaluated from radio tracking data. More precise than any previous Earth-based value, the result is 6,023,600 ($\pm$600) : 1 for the ratio of the Sun's mass to Mercury's.

Eros. On Jan. 23, 1975, the minor planet (asteroid) 433, Eros, made its closest approach to the Earth since it was discovered in 1898 by Gustav Witt at Berlin. The minimum distance was only 14 million mi. This was just one day before Eros reached the point of its orbit closest to the Sun.

Astronomers in the U.S., Europe, and Australia seized this favorable opportunity to make extensive coordinated observations of the asteroid, especially of its light variations, infrared radiation, polarimetric properties, and radar echoes. Eros is an irregular spindle-shaped body approximately 22 mi by 10 mi by 4 mi according to one estimate. It makes one rotation around its shortest diameter every 5 hours 16 minutes. Measuring the resulting changes in brightness is a way of studying the shape of Eros. The infrared spectrum gives indications of material composition (apparently silicate rock), while the partial polarization of the sunlight reflected from Eros provides information about its surface texture, which is apparently crumbly or dusty. Unlike the previous favorable apparitions in 1900–01 and 1930–31, when Eros was observed primarily to determine the solar parallax, the emphasis in 1975 was on ascertaining the asteroid's physical properties.

Eros passed directly in front of Kappa Geminorum on the evening of Jan. 23, 1975, the first observed occultation of a naked-eye star by an

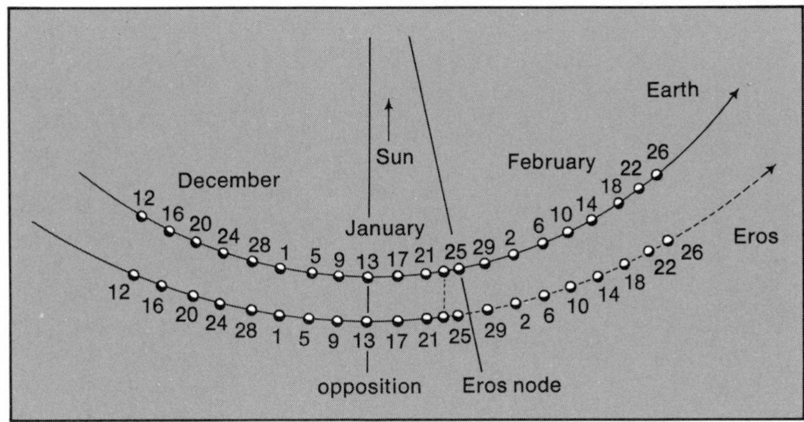

Orbits of the Earth and Eros were 14 million mi apart on Jan. 23, 1975, closer to one another than at any time in more than 80 years. Numbers and dates indicate the positions of the two planets at midnight. The plane of the Earth's orbit is in the plane of the page, whereas that of Eros descends from the left and is shown with a dashed line after passing beyond the plane of Earth's orbit.

Adapted from Keith Hindley, "New Scientist," p. 218, Jan. 23, 1974 asteroid in astronomical history. This occultation was observed along a 10-mi-wide strip extending southeastward through New York and including Great Barrington, Mass., and New Britain, Conn. To watchers with small telescopes within that path, Kappa Geminorum abruptly dimmed to near invisibility and then after about three seconds recovered its light with equal suddenness. The geographical location of the occultation strip was well fixed by ten such observations. A preliminary analysis indicated that Eros was nearly end-on as seen from Earth at the time of occultation, presenting a roughly circular cross section perhaps 14 mi in diameter.

J-XIII. A new satellite of Jupiter, J-XIII, was discovered in September 1974 by Charles T. Kowal of the Hale Observatories during a systematic search with the 48-in. Schmidt telescope on Palomar Mountain in California. On comparing three long-exposure photographs of the region around Jupiter, taken on the nights of September 10–12, he detected an exceedingly faint object that was moving at very nearly the same rate as the planet. At first, however, the nature of Kowal's object was quite uncertain; it might well have been an asteroid having unusual motion or perhaps a very distant comet.

Additional observations, very difficult because the new object was as faint as magnitude 21, were obtained on September 23 and November 9 with the 90-in. reflector at Steward Observatory in Arizona. Other photographs were made at Palomar on October 17, 18, and 21, and on December 12.

J-XIII, the 13th moon of Jupiter, was discovered in September 1974 by Charles T. Kowal of the Hale Observatories. It was the first new natural satellite to be discovered in the solar system since 1966.

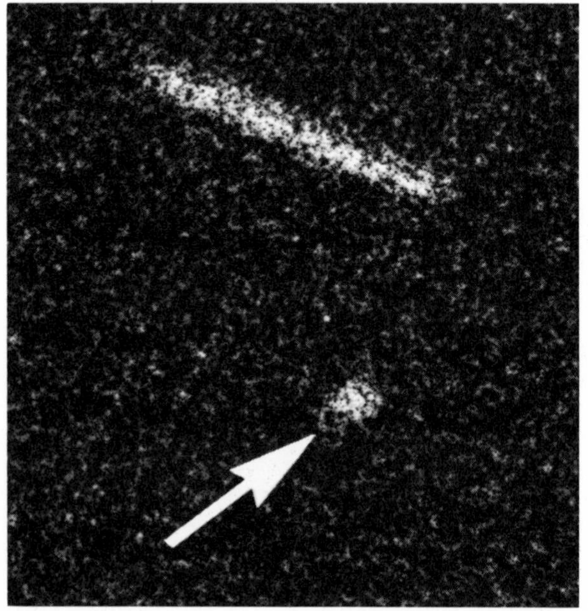

Courtesy, Hale Observatories

Orbit calculations based on all the above observations proved that J-XIII is indeed an outer satellite of Jupiter, with a period of 239.24 days at an average distance from the planet of 6.9 million mi. Its motion in revolving around Jupiter is direct, from west to east. The orbit closely resembles those of satellites VI, VII, and X.

J-XIII became the 33rd known natural satellite in the solar system. It is also the faintest and smallest, being probably only about three miles in diameter. Many such dwarf Jovian moons may await discovery. (*See* Feature Article: VISIT TO A LARGE PLANET: THE PIONEER MISSIONS TO JUPITER.)

Disk of Betelgeuse. The first photographs ever to show the true disk of a star and some indication of its surface features were taken during the year by Roger Lynds and his associates at Kitt Peak National Observatory. Betelgeuse, the star in question, is a first-magnitude red supergiant about 800 times larger than the Sun, yet its disk, only about 0.06 sec of arc in diameter, appears no bigger from the Earth than would a 25-cent piece from a distance of 50 mi.

The pictures were obtained with the 4-m (158-in.) Kitt Peak reflector by a technique called speckle interferometry. Theoretically, the disk of Betelgeuse should be just resolvable with this aperture, but in ordinary practice it is smeared out by atmospheric turbulence. Therefore, very-large-scale images of the star are photographed in nearly monochromatic light with exposures so short (1/125 sec) that the ever-changing disturbances are "frozen." On such a photograph, Betelgeuse appears as a blob about two seconds in diameter made up of a multitude of tiny bright speckles. Each speckle is an image of the star formed by a portion of the telescope mirror. By scanning the star photograph two-dimensionally with a densitometer, and then using a computer to superimpose and combine the speckles, the investigators obtained a single, diffraction-limited image of the star, freed from turbulence-blurring and telescope aberrations.

Betelgeuse photographs made in this way in two different wavelengths showed a slight difference in diameter, in accordance with theoretical prediction. The surface features are barely recognizable bright and dark mottlings, which were thought to be caused by temperature differences associated with rising and falling convection currents of gas in the supergiant's enormously extended, tenuous atmosphere.

Binary pulsar. Pulsars, which reveal themselves by quick, periodic "clicks" of radio emission, are

Adapted from Donald Johnston and Caroline Rand Herron, the "New York Times," © 1974. Reprinted by permission

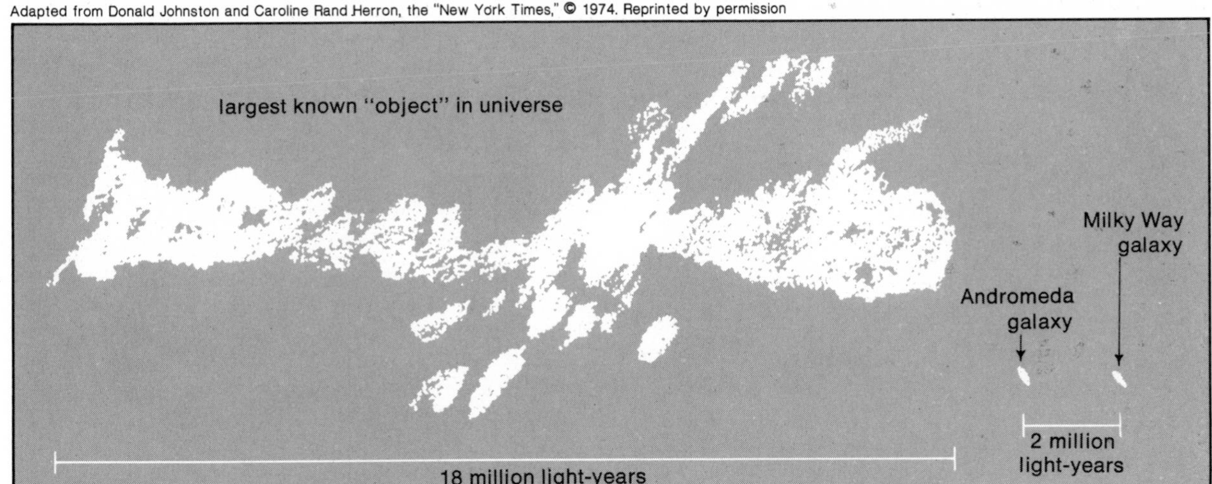

The largest known object in the universe, classified as a radio galaxy, was discovered in 1974. Light rays, traveling at 186,000 miles per second, require 18 million years to cross it.

now known to be rapidly rotating neutron stars—extraordinarily dense collapsed stars only about ten miles in diameter with very strong magnetic fields. Of the more than 140 pulsars discovered since 1967, one that was detected in late 1974 in the constellation Sagitta is of unique interest.

Two University of Massachusetts radio astronomers found the pulsar during a search with the 1,000-ft radio telescope at Arecibo Observatory in Puerto Rico. It emits about 17 pulses per sec, but the pulsation period changes smoothly and continuously between 0.058967 and 0.059405 sec, this cycle requiring 7.8 hours. Clearly, the radio source is one member of a binary system with an orbital period of 7.8 hours, the pulses appearing more widely spaced while the source is receding from the Earth than when it is approaching. The orbit is strongly elliptical (eccentricity 0.61), and the average separation of the two components is about one solar mass (or more, depending on the spatial orientation of the orbit plane). If some way could be found to estimate its orbital inclination to the line of sight and to estimate the mass ratio, this binary may provide the first direct determination of the mass of a neutron star.

The unobserved companion is probably also a tiny, collapsed star, for it is not big enough to block the rhythmic signals of the pulsar when the latter is on the far side of the orbit. Early attempts to detect X-ray pulses by using X-ray telescopes aboard the Copernicus satellite failed.

The pulsation period of 0.059 sec is the shortest known except for the 0.033 sec of the pulsar NP 0532 in the Crab nebula. The latter is known to have originated in a supernova explosion in the year 1054. Analogy suggests that the Sagitta pulsar also is the remnant of a supernova.

Giant radio sources. Vaster even than clusters of galaxies, the largest known objects in the universe are two extended radio sources, known as 3C-236 and DA240. Their nature was established in 1974 by Dutch radio astronomers from Leiden Observatory, working with the big aperture-synthesis radio telescope at Westerbork, The Netherlands. This sophisticated instrument, which consists of 12 dish antennas, each 82 ft in diameter and spread out along an east-west line, feeds data to a computer that produces a high-resolution map of radio-intensity contour lines of the sky area under study.

Observations made at a wavelength of 49 cm showed that 3C-236 is a giant radio galaxy nearly 18 million light-years long. It consists of a strong, compact central source, from which diverge in opposite directions two detached, elongated streamers of radio emission. (This symmetrical triple pattern is common in smaller-scale radio galaxies.) The central source had been identified in 1966 with a 16th-magnitude galaxy whose observed red shift indicates the immense distance from the Earth of 1.7 billion light-years. At that distance, the entire angular length of the radio source mapped at Westerbork (39 minutes of arc) corresponds to an actual length of 18 million light-years. This system is in the constellation Leo Minor. DA240 in the constellation Lynx is somewhat smaller and closer to the Earth. Showing a similar symmetrical structure, it is about 6.5 million light-years in length and is 670 million light-years distant. Like the Leo Minor system, DA240 was probably produced by cataclysmic explosions inside the central body.

These extended systems had escaped early recognition because they lie in "radio confused" regions of the sky that could not be mapped unambiguously with radio telescopes of lesser resolving power. Another factor that delayed recognition was the low surface brightness of some parts of the structures. Nevertheless, A. H. Bridle in 1972 had pointed out the possible huge size of 3C-236.

—Joseph Ashbrook

Chemistry

Much of the research in chemistry during the past year centered on efforts to ease the worldwide food and fuel shortages. Inorganic chemists made progress in developing a new method of nitrogen fixation, and physical chemists worked on methods to store solar energy more efficiently. Other discoveries included a laser fluorescence method for monitoring atmospheric pollutants and the development of antibacterial cotton.

Inorganic chemistry

During the past year research in inorganic chemistry continued to result in important discoveries and contributions toward the further development of science and technology. That inorganic chemistry, once a poor second to organic chemistry, had finally made its mark in the sciences was demonstrated by the awarding of the 1973 Nobel Prize for Chemistry to two inorganic chemists, Ernst Otto Fischer of the Technical University of Munich, West Germany, and Geoffrey Wilkinson of the Imperial College of Science and Technology, London. This was the first time that the prize had been awarded to inorganic chemists since 1913.

Bioinorganic chemistry. Inorganic chemists have recently become interested in biological systems involving metal ions, and this has produced an area of research known as bioinorganic chemistry. This area perhaps generated the most exciting developments in inorganic chemistry during the past year. Some of the more important of these resulted from research concerning oxygen carriers, nitrogen fixation, ferredoxin proteins, and anticancer drugs.

Oxygen carriers. Hemoproteins are porphyrins combined with iron and protein; they include the hemoglobins, myoglobins, and cytochromes. The interaction of oxygen with these complexes is of vital importance to the animal world because it is responsible for the respiratory and metabolic processes. The reversible binding of oxygen with hemoglobin (Hb) and myoglobin (Mb) has long been studied, and much is known about this pro-

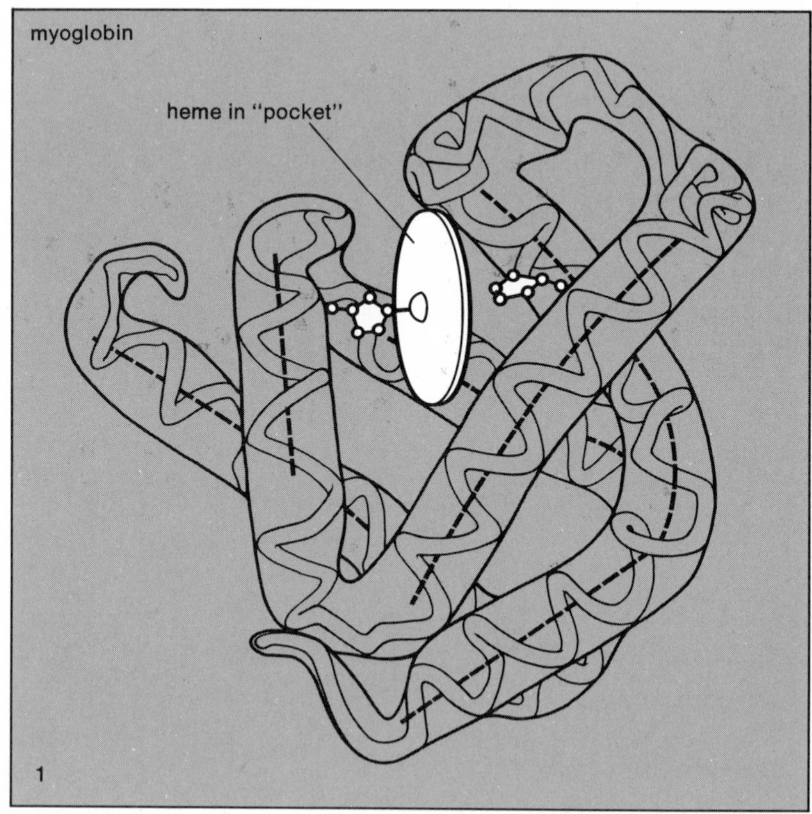

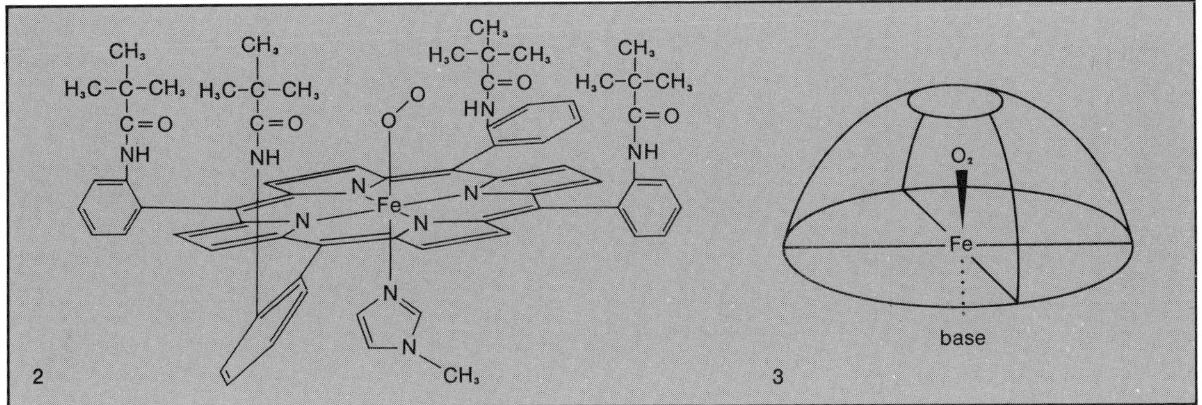

cess. The oxygen binds to iron in the iron-porphyrin complex heme attached to the protein globin; the globin surrounds and forms a "pocket" for the heme to protect it from irreversible oxidation by oxygen (1).

In spite of all that is known about the reversible reaction $Hb + O_2 \leftrightharpoons HbO_2$, some important details are not known because the molecule is very large and complex. Inorganic chemists tried unsuccessfully for many years to prepare an iron(II) (an iron atom that has lost two electrons, Fe^{2+}) compound capable of reversibly binding oxygen, in order to use it as a model for investigations of Hb and Mb. They never succeeded because the irreversible reaction $2Fe(II) + \frac{1}{2}O_2 \rightarrow Fe(III)-O-Fe(III)$ always resulted in the formation of a stable bridged product. What was needed was a molecule designed to afford a protective "pocket" for the iron-porphyrin complex so that the stable product could not form. This was achieved in 1974 by J. P. Collman and his co-workers at Stanford University. They prepared a "picket fence" iron(II)-porphyrin complex (2) that reversibly binds oxygen at room temperature in a manner similar to Hb and Mb. In recent months J. E. Baldwin and his students at Massachusetts Institute of Technology prepared a "capped" iron(II)-porphyrin complex (3), which is also a good model for the biological systems.

Detailed studies of these model compounds, in progress at the year's end, seemed certain to provide valuable information regarding some of the unanswered questions about the complicated biological systems. Research at Northwestern University, Evanston, Ill., demonstrated that the mechanism of oxygen and carbon monoxide interactions with a simple iron(II)-porphyrin complex is analogous to that for the more complicated biological systems. More experimentation was required, however, before the nature of the interaction of oxygen with hemoglobin and myoglobin could adequately be explained.

Nitrogen fixation. Of importance to the life cycles of plants and animals is nitrogen fixation, the process of converting nitrogen in air to compounds of nitrogen. In nature, the metalloenzyme nitrogenase contained in certain biological systems catalyzes this conversion of nitrogen into its compounds. For industry, inorganic chemists just before the start of World War I devised the Haber-Bosch process, the large-scale production of ammonia from nitrogen and hydrogen. This process is used to produce most of the world's fertilizer, but it requires conditions of high temperature and pressure. Unfortunately, the construction of a Haber-Bosch plant is often too expensive for less developed countries.

Inorganic chemists renewed their interest in this problem in the mid-1960s when the first metal-nitrogen compound was discovered. This breakthrough was believed to afford the missing link that would provide a better understanding of how nitrogenase functions, and perhaps point the way toward a new method of nitrogen fixation. As of 1975 some distance from both of those goals remained, but progress had been made.

During the past year A. E. Shilov of the Institute of Chemical Physics in Moscow reported a vanadium system that permits the conversion of nitrogen into ammonia and/or hydrazine under at-

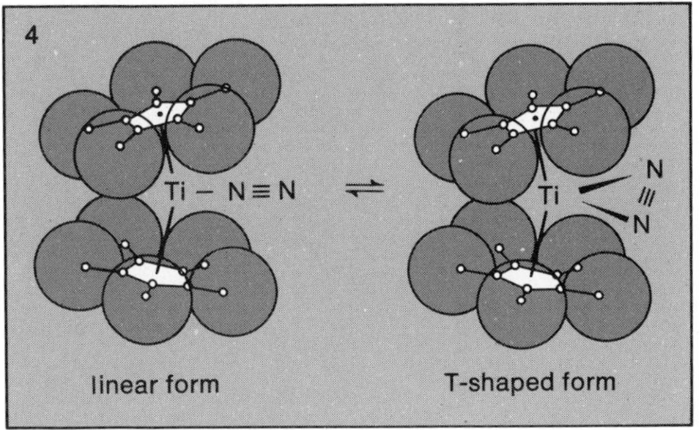

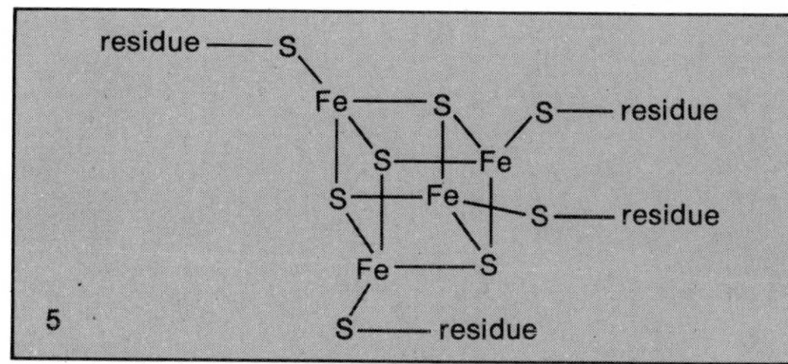

mospheric conditions. Also Joseph Chatt of the University of Sussex, England, reported the first well-defined case of the reaction of nitrogen bound to a metal to give ammonia. His example is particularly significant because the metal is molybdenum, one of two metals contained in nitrogenase.

Also of importance during the year was the research of J. E. Bercaw at the California Institute of Technology. He was able to obtain spectroscopic evidence for the existence of two forms, one linear and one T-shaped, of the bis(η^5-cyclopentadienyl) dinitrogentitanium(II) compound (4), and it appears that only one of the two reacts with acid to yield hydrazine. This accorded with the view that nitrogen is not activated toward reduction when in the linear form but is activated when in the T-shaped form. More experimental work must be done to test this hypothesis.

Ferredoxin proteins. Iron-sulfur proteins constitute one of the classes of metalloproteins and metalloenzymes that participate in many biologically important reactions and present many intriguing problems. A detailed interpretation of the properties of these proteins has been rendered difficult by the absence of well-defined synthetic complexes whose structures and properties resemble those of the active sites of the proteins. This situation was remedied by the research of R. H. Holm and his co-workers at the Massachusetts Institute of Technology. In 1972 they reported the synthesis of an Fe_4S_4 cluster (5) with properties similar to some of the ferredoxins, one group of Fe-S proteins. Thus, inorganic chemistry again supplied a suitable model for the investigation of a much more complicated biological molecule.

Equally significant was the report by Holm and his group that they had been successful in the extrusion of Fe_2S_2 and Fe_4S_4 cores from the active sites of ferredoxin proteins. This achievement makes it possible to compare the natural with the synthetic cores and thus to arrive at a better understanding of the fundamental biological behavior of ferredoxins.

Anticancer drugs. A final important development in bioinorganic chemistry was produced by the laboratory of Barnett Rosenberg at Michigan State University. In 1969 Rosenberg and his co-workers reported that cis-$[Pt(NH_3)_2Cl_2]$, which was first prepared in 1845, is a potent antitumor agent. At the September 1974 national American Chemical Society meeting Rosenberg reported that the compound was being tested against human cancers in many hospitals throughout the world. These early tests over a broad spectrum of tumors in terminally ill patients showed encouraging results.

One-dimensional compounds. Inorganic chemistry has contributed for many years to the science and technology of solid materials such as alloys and ceramics. In recent years considerable re-

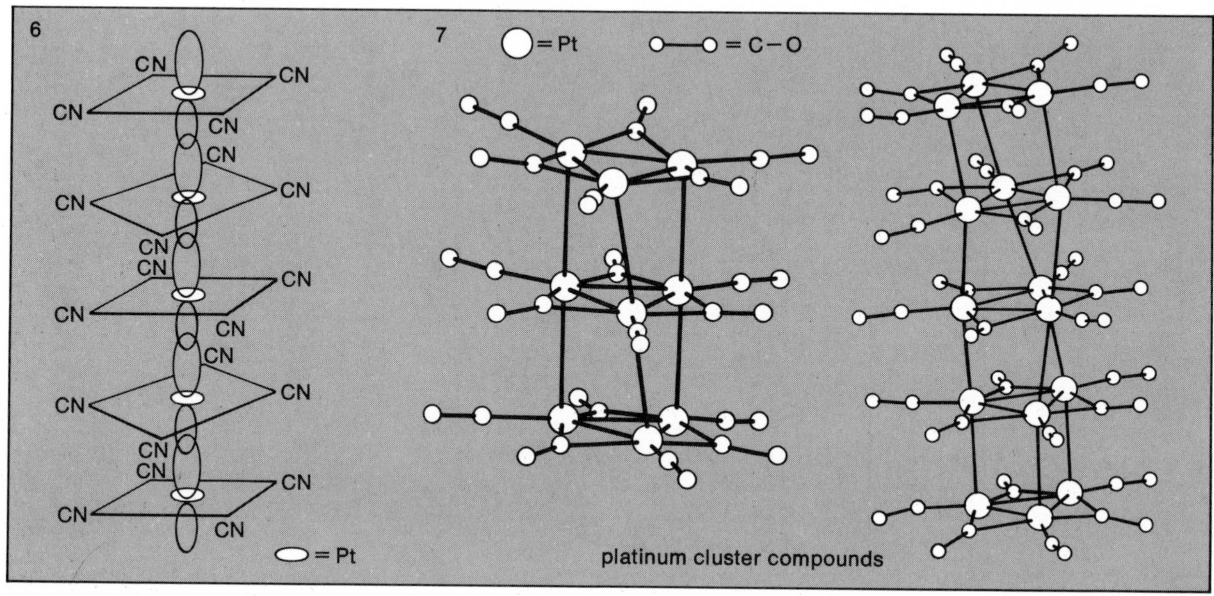

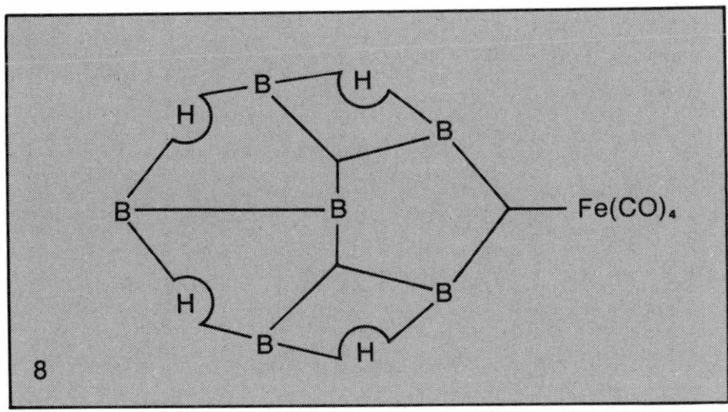

search activity has been aimed at the design and synthesis of organic and inorganic compounds that behave like one-dimensional metals. The platinum compound $K_2Pt(CN)_4Br_{0.3} \cdot 3H_2O$ (6) was reported in 1968 by Klaus Krogmann of the Fridericiana University of Karlsruhe, West Germany, to have desirable one-dimensional electrical and optical properties. This observation sparked research activity to find more such materials in many laboratories throughout the world. Several such solid materials were reported at the 167th national meeting of the American Chemical Society in April 1974, and it was shown that the compounds stack in the solid to allow extensive metal-metal interaction along the metal-metal axis. This means that good "metal-like" electrical conductivity occurs in that one dimension but is not found perpendicular to it. Such behavior is unique because ordinary electrical conductivity by metals such as copper is the same in all directions.

Alan MacDiarmid and his research associates at the University of Pennsylvania in recent months obtained a very pure sulfur nitride fibrous polymer with the formula $(SN)_x$. Although this is a nonmetallic inorganic polymer, it has electrical and optical properties similar to those characteristic of a one-dimensional metal. This is believed to be due in part to the presence of vacant electronic orbitals in the *d* subshells; these vacancies permit the ready flow of electrons in the direction of the linear $S-N-(S-N)_x-S-N$ chain in a manner similar to the electron flow along the chain of platinum atoms in $K_2Pt(CN)_4Br_{0.3} \cdot 3H_2O$.

Metal atom chemistry. The rapid emergence of metal atom chemistry as a powerful synthetic technique culminated in a conference on this subject in 1974 in West Germany. Prior to the development of this technique the method of chemical synthesis had always been to start with bulk elements and/or compounds. Elements and compounds contain strong chemical bonds holding atoms together, and these bonds must be broken to form new bonds in the synthesis of new compounds. Because the breaking of chemical bonds often requires considerable energy, it is sometimes difficult or impossible to prepare the desired compound by a classical method of synthesis.

In metal atom chemistry the metal is heated sufficiently (often above 2,000° F) to break the metal-metal bonds and thereby yield very reactive individual gaseous metal atoms inside a high-vacuum system. The reaction vessel also contains a gaseous compound that readily reacts with the metal vapor to give the desired organometallic product, which is collected in a colder portion of the vessel. One of the pioneers of this technique was P. L. Timms of the University of Bristol, England. He and others used the method to prepare zerovalent compounds that are either difficult or impossible to prepare by other means, such as bis(cyclooctadiene)iron, bis(benzene)titanium, and tris(butadiene)molybdenum. (Zerovalent compounds are those which contain one or more atoms that maintain essentially the same share of orbiting electrons they would have in their uncombined, or zerovalent, state.) Apart from the importance of the preparation of new compounds from the standpoint of fundamental chemistry, these materials held promise for industrial applications. For example, some of the compounds have the potential for strong catalytic activity and might prove useful in olefin polymerization, the process that is used in the manufacture of polyethylene for bags and containers.

Metal cluster compounds are also of interest because of their possible use as catalysts and because they may help chemists understand metal catalysts in commercial use. The latter are compounds that contain three or more metals chemically bonded together in a cluster. A significant advance was made in 1974 by P. Chini at the University of Milan, Italy, who did very difficult chemistry required to prepare extended nickel and platinum carbonyl clusters (7). The catalytic properties of these unusual compounds were under investigation at the year's end.

The boron hydrides, compounds of boron and hydrogen that were investigated as high-energy fuels by the U.S. National Aeronautics and Space Administration, continued to be an active area of research in inorganic chemistry. A significant fundamental discovery was reported by A. Davison of the Massachusetts Institute of Technology. He and his students reported a variety of transition-metal complexes of B_6H_{10} in which the metal has become inserted into the unique basal boron-boron bond to form a three-center two-electron bond (8). Although this type of reaction and these new compounds are of scientific importance, chemists could not yet assess their possible technological significance.

—Fred Basolo

Organic chemistry

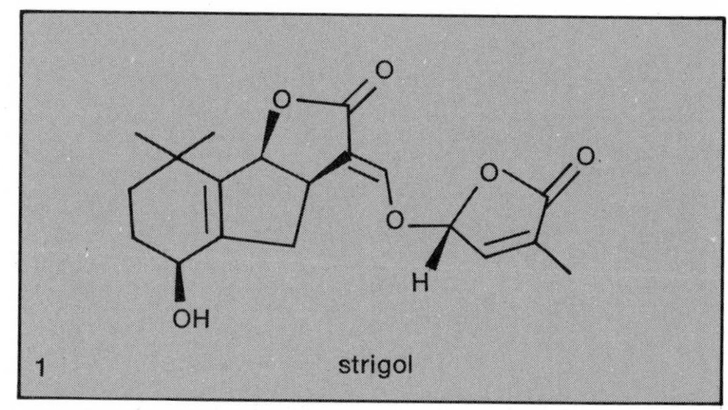

1 strigol

2 vermiculine

Changing patterns of energy utilization suggest that coal gas and synthesis gas, which are rich in hydrogen and carbon monoxide, will become increasingly important raw materials for the chemical industry. This consideration prompted new research on the catalytic conversion of these gases to compounds of commercial importance. While much of this work was in a preliminary stage in 1975, it became evident that organometallic compounds catalyze the formation of many desirable products. Derivatives of rhodium and rhenium, for example, catalyze the insertion of carbon monoxide into organic molecules. Study of this chemistry led recently to the construction of facilities for the manufacture of ethylene glycol from hydrogen and carbon monoxide ($3H_2 + 2CO \rightarrow HOCH_2CH_2OH$) and for the preparation of acetic acid from methyl alcohol and carbon monoxide ($CH_3OH + CO \rightarrow CH_3CO_2H$).

Organic synthesis. Interest in the discovery of new methods for organic synthesis remained high. During the past year research groups around the world described an impressive array of new reactions, reagents, and catalysts for the preparation of both simple and complex molecules. New synthetic methods using organic derivatives of aluminum, copper, chromium, iron, tin, nickel, titanium, rhodium, zirconium, tantalum, and tungsten were reported. The work on the organocopper reagents was particularly noteworthy. These compounds, generally formulated as lithium dialkylcuprates, $LiCuR_2$, were widely used as selective alkylating agents. Techniques for the preparation of organometallic compounds developed apace. Work by R. D. Rieke at the University of North Carolina revealed that zinc and indium as well as magnesium could be produced in a finely divided and highly reactive form by the reduction of a salt with potassium in a dry argon atmosphere. In this form magnesium reacts rapidly with even the most inert organic halides, enabling the preparation of Grignard reagents.

Nonmetallic reagents received equal attention. Reagents based on the chemistry of organic compounds containing boron, phosphorus, selenium, silicon, and sulfur were described in 1974 and 1975. The striking contributions of H. C. Brown of Purdue University, Indiana, on the chemistry of the alkylboranes prompted many further investigations of these reagents. Organosilicon, organosulfur, and organoselenium compounds were proving to be valuable intermediates for organic synthesis. To illustrate, B. M. Trost at the University of Wisconsin found that an olefin could be converted to an ester by the reaction of a π-allylpalladium complex with the sodium salt of methyl phenylsulfinylacetate. The many new nonmetallic reagents and organometallic compounds are now regularly considered in the design of syntheses for complex molecules.

Many naturally occurring organic compounds important in the biological and medical sciences were synthesized in the laboratory during the past year. A Japanese group headed by S. Marumo at Nagoya University described the synthesis of disparlure, the sex attractant emitted by the female gypsy moth (*Porthetria dispar*). This interesting

3

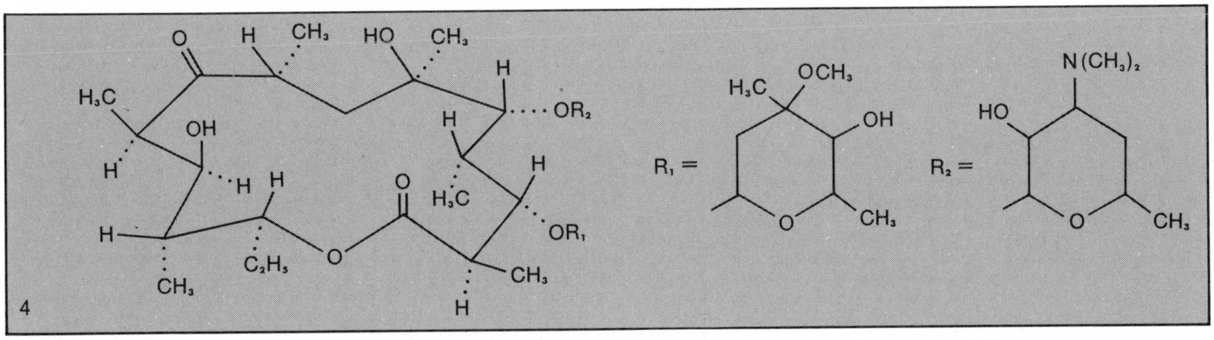

substance was synthesized to establish which of a pair of isomers is the compound secreted by the female moth. It was found that both isomers attract the male, but one of them does so about 10,000 times more effectively than its mirror image.

R. A. Raphael and his associates at the University of Cambridge described the synthesis of an intensely active germination stimulant known as strigol (1). This compound was prepared to verify the structure that had been assigned to it. In general, it is no longer necessary to synthesize natural products to prove their structures; these are now most economically established by spectroscopic and crystallographic methods. Thus, the structures of gnididin and of two other potent antileukemic compounds from the shrub *Gnidia lamprantha* were established by S. M. Kuphan and his associates by chemical and spectroscopic analyses. R. K. Boeckman, Jr., and J. Clardy used X-ray methods to establish the structure of vermiculine (2), a novel macrolide dilactone antibiotic from *Penicillium vermiculatum*.

Antitumor agents. Because the modern methods of mass spectrometry, carbon and proton nuclear magnetic resonance, and X-ray crystallography are so powerful, chemists are now largely relieved of the laborious task of deducing these structures from the products of step-by-step breakdown. As a result they can devote their attention to the problems involved in the development of practical synthetic methods. Many of the antitumor agents discovered in the recent past under the aegis of the National Cancer Institute were obtained from plants in remote areas of the world. Often tons of material had to be harvested to enable the extraction and eventually the isolation of a minute quantity of the effective agent. Mishandling at any stage could waste an entire year's crop. In this situation, a laboratory synthesis can provide an alternative source of supply for these extremely important substances. In addition, a practical synthetic route provides, as a bonus, a variety of closely related molecules that sometimes prove to be as valuable as the natural material.

These considerations were a central feature in the work of E. J. Corey on the synthesis of macrocyclic ring compounds. To accomplish these difficult reactions, Corey and his associates developed a general method (3) for the conversion of ω-hydroxyalkanoic acids to lactones using the 2-pyridinethiol esters to activate simultaneously both the hydroxyl and carboxyl groups.

Corey's use of the novel 2-pyridinethiol esters was suggested by the prior work of T. Mukaiyama at the Tokyo Institute of Technology. Mukaiyama

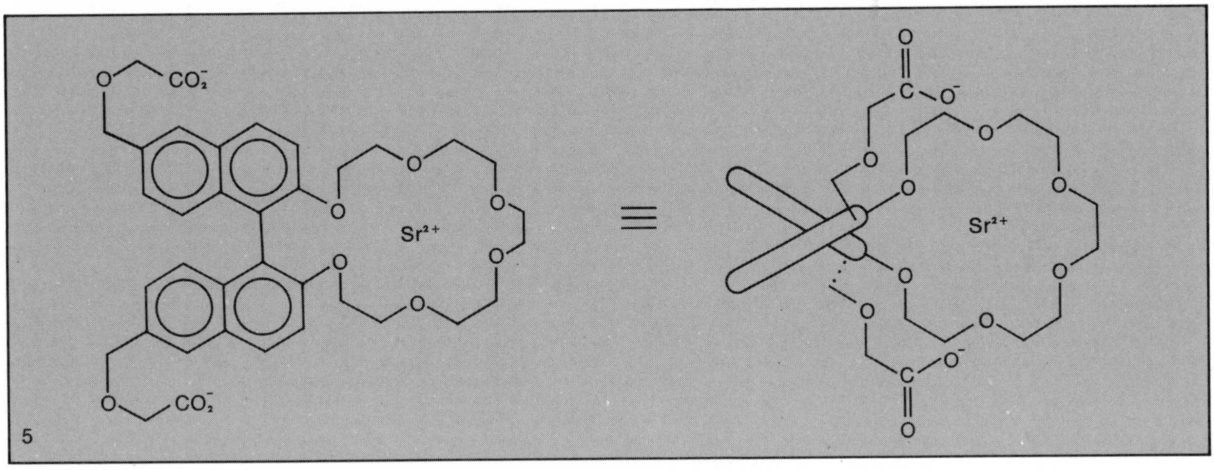

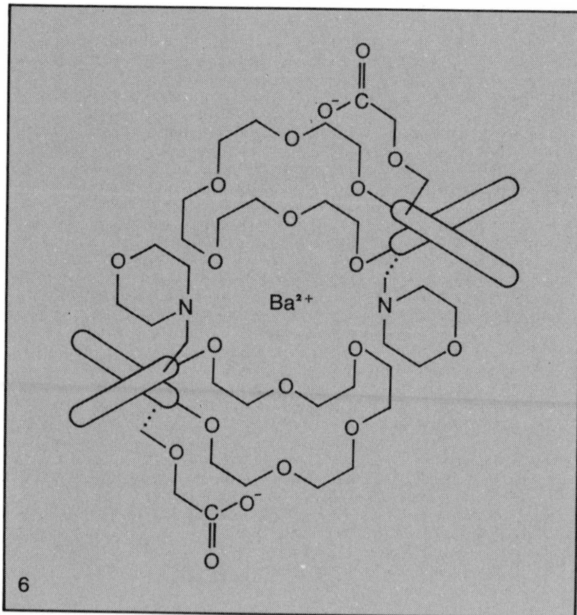

6

$$\text{H}_2\text{C}\!<\!\begin{matrix}\text{CH}_2\\|\\\text{CH}_2\end{matrix} \rightarrow \text{H}_2\text{C}\!<\!\begin{matrix}\text{CH}_2\bullet\\\text{CH}_2\bullet\end{matrix} \rightarrow \text{HC}\!\begin{matrix}=\text{CH}_2\\-\text{CH}_3\end{matrix}$$

7

exploited the chemistry of 2-pyridinethiol esters in the preparation of peptides and ketones such as jasmone. Corey and his group subsequently found that this rather mild reaction could be employed in the synthesis of many macrocyclic molecules, including the lactone of prostaglandin $F_{2\alpha}$ and vermiculine (2). Erythromycin B (4) free of the two sugar residues ($R_1 = R_2 = H$) was also obtained by a ring closure of this kind.

Selective binding of cations. Exciting developments occurred in the use of cyclic polyethers ("crown" ethers) as structural backbones in the construction of extremely effective reagents for the selective binding of cations (ions that carry one or more positive charges). D. J. Cram and his associates at the University of California at Los Angeles reported that cations are often powerfully coordinated in specifically designed host molecules. For example, the barium and strontium cations in the complexes (5) and (6) cannot be readily separated from the hosts.

Much work remained to be done in this area. It was clear, however, that many new reagents highly specific for certain ions would be studied in the future for the selective separation of metal ions from water. Potential applications range from the mining of the ocean to the removal of toxic radioactive compounds from the wastes of nuclear reactors.

Transient intermediate species. Studies of the chemistry of transient intermediate species in chemical reactions continued to yield fundamental information useful in the design of new reactions. Computer technology played an ever-expanding role in this work. Calculations based on a variety of semiempirical quantum mechanical methods were regularly performed to estimate the energy content of the unstable intermediates and to clarify the interpretation of experimental results. Promising results were reported by M. J. S. Dewar of the University of Texas concerning his attempt to develop a quantum mechanical treatment that is sufficiently simple, accurate, and reliable to serve as a practical guide for the discussion of the chemistry of transient intermediates and for a more complete characterization of reaction pathways for organic molecules. Extension of these methods for the analysis of the chemistry of organometallic compounds and for the chemistry of electronically excited molecules may prove even more fruitful.

A clear demonstration that cyclobutadiene is an intermediate in organic reactions was provided by J. Rebek. He found that this molecule could be generated on one solid surface and then transferred through solution to a second solid surface, where the reaction occurred. This novel three-phase system offers a convenient procedure for the detection of unstable intermediates and can readily be extended to other problems.

Free radicals. Important advances occurred in the characterization of free radicals. Groups under the direction of J. K. Kochi at Indiana University, P. J. Krusic at the Du Pont Co., M. C. R. Symons at the University of Leicester, England, D. E. Wood at the University of Connecticut, K. U. Ingold at the National Research Council of Canada, and R. O. C. Norman at Oxford University used clever methods for the generation of unstable free radicals at low temperature. The radicals were then studied by electron paramagnetic resonance spectroscopy to yield important new data concerning their stereochemistry. Those free radicals containing carbon and having electronegative substituents such as oxygen and fluorine atoms were found to possess a pyramidal structure and undergo rapid inversion. (Electronegative atoms are those with a relatively large affinity for electrons in joining other atoms to form compounds.) The situation for alkyl radicals was discovered to be similar except that the deviation from planarity is smaller.

The role of biradicals (free radicals with two unpaired electrons) in the mechanisms of organic reactions has long been debated. Some chemists believe that biradicals may be intermediates in the ring-opening reaction (7) of cyclopropane. Theoretical arguments based on molecular orbital calculations did not resolve the question. Recent work by G. L. Closs at the University of Chicago indicated that the simple alkyl biradicals can be observed at temperatures below 20° K and that these biradicals apparently possess a triplet ground state. Again, much work remained to be done, but this discovery suggests that even the most reactive intermediates can be studied under appropriate conditions.

—Leon M. Stock

Physical chemistry

The subject of physical chemistry is a broad one, covering both experimental and theoretical studies of molecules (and their interactions) in the gaseous, liquid, and solid states. It involves chemical equilibrium and chemical dynamics as well as structural and nuclear chemistry. As in all areas of science, what is classified as fundamental today will often be considered practical application tomorrow. In the following discussion some of each is presented in an attempt to convey the current flavor of research progress in physical chemistry.

Rates of reactions. Since the early 1930s chemists have awaited a fully theoretical prediction of the rate of a chemical reaction. Such a calculation would make use only of Coulomb's law (describing the force of attraction between two point charges), the Schrödinger equation (the equation governing particle waves that forms the basic law of wave mechanics), and, of course, the masses and atomic numbers of the participating atoms. The principles of the procedure have been known for some time: calculate the "potential energy hypersurface" (determined by plotting the electrical forces between the atoms against their interatomic distances); solve the quantum mechanical scattering equations (describing the change in direction of particles owing to their collisions with other particles or systems); and then evaluate the reaction probability and the so-called reaction cross section (the measure of the probability of an encounter between particles such as will lead to a specific effect; this leads directly to the rate of the reaction).

Until 1974 the scattering calculations for any chemical reaction had been too difficult to execute. For the simple atom-exchange reaction of $H + H_2$, the potential energy surface had been available for some time, but only approximate scattering computations could be made. During the past year, however, accurate three-dimensional calculations of reactive scattering for this system were carried out independently by two groups, that of Aron Kuppermann at the California Institute of Technology in Pasadena, and of Robert Wyatt at the University of Texas in Austin. The chemical dynamics of that reaction are now quantitatively understood, starting from first principles.

Van der Waals molecules. In the field of structural chemistry a novel development was the determination of the interatomic configurations of weakly bound "van der Waals" molecules. For a long time it had been known that dimers (molecular species formed by the union of two like atoms or molecules) of heavy noble gases existed—*e.g.*, Ar_2, Kr_2, and Xe_2—and more recently weakly bound adducts (products of addition reactions between molecules) such as $HCl \cdot Ar$, $H_2 \cdot Ar$, $O_2 \cdot Ar$, $N_2 \cdot N_2$, and $Cl_2 \cdot Cl_2$ have been observed by the methods of mass spectrometry and/or mass spectroscopy.

Largely from the work of William Klemperer and his colleagues at Harvard University, the structures of several of these van der Waals molecules were accurately determined by molecular-beam electric resonance. In this method molecules such as $Ar \cdot HCl$ are obtained by the expansion of Ar-HCl mixtures through a nozzle, forming a supersonic molecular beam. Their radio-frequency and microwave spectra are then measured and interpreted. Recently the structure of ArClF was so determined: it was found to be linear, with an argon-chlorine bond length of 3.33 Å (1 Å, an angstrom unit, equals 10^{-10} m). The equilibrium configuration of the ArHF and ArHCl molecules is also essentially linear, possessing the arrangement Ar—H—X.

Lasers. Chemical lasers have been of great practical as well as theoretical interest since their discovery in 1964 by George Pimentel and his co-workers at the University of California at Berkeley. They rely upon the phenomenon known as "population inversion" in fast, exothermic (energy-evolving) chemical reactions. (In population inversion the normal ratio of population of two different energy states is reversed; in this case it means that more atoms are in a higher energy state than in a lower one.) These lasers had been confined to the infrared region of the spectrum, since population inversion is usually vibrational and rotational. Michael J. Berry and his colleagues at the University of Wisconsin at Madison recently discovered a new chemical laser system, based on the photolysis of HCN, in which the population inversion arises from an electronically excited state of the product. Electronic-transition chemical lasers in the visible and/or ultraviolet regions would be of potential practical significance, especially in the area of laser-induced nuclear fusion.

Richard Zare and co-workers at Columbia University, New York, N.Y., developed a tunable-laser fluorescence method to analyze the internal states of molecules and the presence of trace molecules. (A tunable laser is one which can be adjusted to emit over a range of frequencies.) A narrow-band tunable laser is frequency-swept over the wavelength range in which the molecules of interest absorb. When the laser frequency coincides with a molecular absorption line, the molecules undergo a transition to an excited electronic state, from which they subsequently fluoresce. The fluorescence intensity as a function of laser wavelength can be related to the relative populations of the various internal states. The technique was used to identify minute quantities of trace molecules such as aflatoxins (dangerous mold metabolites in food products) and to monitor atmospheric pollutants.

Physical chemistry in the liquid state was advanced by the introduction of the mode-locked laser technique (providing picosecond pulses) developed by Peter Rentzepis at the Bell Telephone Laboratories in New Jersey in 1967. Kenneth Eisenthal and colleagues at the IBM Research Division Laboratory in California carried out a direct time-dependent study of the "cage effect" in the recombination of iodine atoms in liquid solvents (such as CCl_4). Dissolved I_2 was photodissociated (fragmented into neutral atoms or molecules by the action of light) by a five picosecond laser pulse at a wavelength of 530 nanometers, and then the regenerated I_2 concentration was followed by measuring the absorption of probe pulses over a total time span of about a nanosecond following the dissociation. Unlike the situation in the gas phase, in which the probability of original partner recombination is negligible, the recombination of original atoms in the liquid phase is greatly enhanced by interactions with the solvent molecules that preclude "escape" of the fragments. This should be of general importance for many reactions in solution.

Solar energy storage. A contribution to the problem of solar energy storage was made during the year by Norman Sutin and co-workers at the Brookhaven National Laboratory, Upton, N.Y. The photodissociation of water into its elements $H_2O + h\nu \rightarrow H_2 + \frac{1}{2}O_2$ offers an attractive means of solar energy utilization. Although photons from a large part of the solar spectrum possess sufficient energy to dissociate water, photodissociation does not occur directly. However, using an electrochemical cell, K. Honda and his colleagues in Japan succeeded in photodissociating water in 1972 with light of wavelengths of less than 400 nanometers. When an n-type titanium dioxide semiconductor anode dipping into aqueous alkali was illuminated, oxygen was liberated and hydrogen formed at the platinum cathode.

The approach of the Brookhaven scientists was different in that it involved the use of transition-metal complexes, which absorb light over most of the solar spectrum. The complexes are the ML_3^{2+} type, where L is a bipyridine derivative and M is a transition metal such as iron(II) or ruthenium(II). The charge-transfer excited states of these complexes, $^*ML_3^{2+}$, formed upon absorption of visible light, are strong reducing agents, capable of reducing H_2O or H^+ to hydrogen with concomitant production of ML_3^{3+}. The latter, a strong oxidizing agent, can be reduced by OH^- to form oxygen and regenerate the starting complex ML_3^{2+}. This combination of properties (the photoreactivity of the reduced form and the ability of the oxidized form to oxidize OH^-) affords the possibility of photodecomposing water into H_2 and O_2 in an electrochemical cell. The use of the $ML_3^{2+/3+}$ couple is catalytic, only light and water being consumed in the net reaction.

Nuclear chemistry. Highlights in the field of nuclear chemistry in 1974 were two reports of the synthesis of element 106. One was carried out in the Soviet Union by Georgi N. Flerov and his colleagues, who bombarded a lead target with chromium ions. The other synthesis was done by Glenn Seaborg, Albert Ghiorso, and co-workers at the Lawrence Radiation Laboratory of the University of California at Berkeley via bombardment of californium by oxygen ions. The two groups produced different isotopes of the same element: the half-life of the Soviet product was 4–10 milliseconds, and that of the U.S. was 0.9 sec.

Chlorofluoromethanes and the ozone layer. A report of recent developments in physical chemistry would be incomplete without mention of the photochemical-kinetic problem of the influence of chlorofluoromethanes upon the protective ozone layer in the stratosphere. (The cumulative world production of two such chlorofluoromethanes, CCl_3F and CCl_2F_2, used for refrigerants and aerosol propellants, had by 1975 reached an estimated six million tons.) Frank Rowland and Mario Molina of the University of California at Irvine pointed out that these very inert, water-insoluble compounds have a very long life in the troposphere and would diffuse upward slowly into the stratosphere. There they would be photolyzed by the solar ultraviolet radiation to produce chlorine atoms, which could initiate a catalytic chain reaction leading to a net destruction of ozone and regenerating the Cl atoms: $Cl + O_3 \rightarrow ClO + O_2$; $ClO + O \rightarrow Cl + O_2$.

Calculations by Rowland, checked independently by other groups, suggested that significant depletion of the steady-state ozone concentration in the stratosphere could result from this process.

(Chlorofluoromethanes as well as CCl_4 have already been detected in the troposphere by James Lovelock of the University of Reading, England.) Because scientists believe that the effect of thinning the ozone ultraviolet-light "shield" would be quite harmful to health and well-being, the physical chemistry of the chlorofluoromethanes in the Earth's atmosphere is deserving of further careful study.

—Richard B. Bernstein

Applied chemistry

The world fuel and food crises spurred research to alleviate both shortages and to become less dependent upon petroleum. In addition to seeking alternate sources of protein, fuel oil, and gas, chemists began searching for alternate sources of so-called "petrochemicals," the petroleum-derived chemicals used to make most synthetic rubbers and also many fabrics, plastics, adhesives, paints, dyes, and pesticides.

Because the two potential shortages, food and petroleum, concerned carbon-containing materials, it was inevitable that the same organic raw materials or resources would be proposed to alleviate both problems. This was beginning to be apparent during the past year. Many oil companies began developing single-cell protein (SCP) based on methanol, natural gas (methane), and normal (straight-carbon-chain) paraffins from petroleum. Meanwhile, Seattle, Wash., was converting its municipal cellulosic wastes into methanol, which it was using as a gasoline additive.

Separate proposals were made for the use of low-grade wood, agricultural wastes, and municipal cellulosic wastes as sources alternately of energy, materials for plastics production, and feedstocks for single-cell protein. With the proliferating suggestions for use of cellulosic wastes, some support developed during the year for Glenn Seaborg's Recycle Society, in which wastes become primary sources while natural resources assume a backup position.

Wood to replace petroleum. All the essential building blocks to make plastics and synthetic rubber can be derived from wood. The conversion of wood, the traditional source for organic chemicals before the advent of inexpensive oil, into useful chemicals is neither new nor difficult. Although total U.S. fuel needs cannot be met with wood, there is enough renewable low-grade timber in the U.S. to meet the material needs of the entire U.S. petrochemical industry, according to Irving S. Goldstein of North Carolina State University's School of Forest Resources. The amount required would be about 50 million tons a year, or half that consumed each year in wood-pulp production. Moreover, there would be little conflict with the pulping industry because chemical processing can use the low-quality wood that is ignored or discarded by the pulp manufacturers. For instance, the scrubby mesquite shrub in the southwestern U.S. or the great expanse of unused cull trees ("green junk") on cutover land would be ideal for conversion into chemicals.

Sidney Harris

"What I'd like to do is transmute some carbon, sulfur, oil and clay into polystyrene."

To convert the petrochemical industry to wood, new integrated timber-processing plants would have to be built. These plants, in order to be economical, would have to convert the three basic natural polymers in wood—cellulose, lignin, and hemicellulose—into their monomers. Such conversion is technically feasible and was approaching economic viability.

Cellulose and hemicellulose are carbohydrate polymers built from simple sugars. Cellulose, the major component of wood, can be broken down by means of long-established procedures into glucose, which then can be converted to ethanol by fermentation. In turn, ethanol can be converted to ethylene and butadiene, basic chemicals for the production of plastics, synthetic rubbers, and fibers. The step that limits the yield, which chemists believed could be improved by research, is the conversion to glucose, in which only about 50% of the cellulose is effectively utilized.

The shorter chain hemicellulose is more readily broken down into simple sugars (pentoses) than cellulose is into glucose. The pentoses can be converted in 75% yield to furfural, a basic chemical for plastics.

As of 1975 the technology of converting lignin to useful phenols was not well worked out. (Lignin is a three-dimensional polymer formed of phenylpropane units that holds the cellulose fibers together.) No commercial-scale production of phenolic compounds had been attempted, but because the price of phenol had quadrupled in recent months the outlook for producing phenols from lignin profitably was good.

To develop such integrated processing plants economically will probably take decades. Because wood is renewable, it appears to be the only reliable raw material for the petrochemical industry.

Another proposal for wood as an energy source, called the energy plantation, suggests wood as a solid fuel for electricity generation or conversion of wood to synthetic natural gas (SNG). These plantations must be located in areas with at least 20 in. of rain per year. This includes all of the U.S. in the Eastern and Central time zones.

A plantation could be on land unsuitable for dirt farming and would use a mixture of at least two plants grown at high density, 3,000 to 4,000 plants per acre, on short harvest cycles of three or four years. Plants must resprout from stumps, have rapid early growth, and reproduce in such a way that six harvests can be made from one planting. Suitable plants include sugarcane, hybrid poplar, cottonwood, alder, locust, and soft maple. The estimated cost for producing SNG from an established energy plantation is competitive with deriving gas from coal, but in 1975 both of these were higher than the cost of natural gas.

Plant crops as replacement for petroleum. Renewable plant crops also yield high quantities per acre of valuable raw materials for plastics. In fact, the yield and quality of cellulose from such plants can be far superior to that from trees. Their starches and pentosans can be converted into potentially useful new polymers and plastics, according to Rudolph Deanin of the Lowell (Mass.) Technological Institute, or can be converted into conventional "petrochemical" raw materials. The unsaturated vegetable oils are useful in vinyl, alkyl, polyester, polyurethane, and polyamide polymers and in plasticizers. Some companies reportedly were using vegetable oils from soybeans, sunflowers, and sunflower seeds as materials for producing nylon.

An example of such a plant is *Crambe abyssinica*, a rich source of raw materials for special moisture-resistant nylons. A relative of wild mustard and flax and common in Mediterranean countries, the plant can grow well in Indiana and Illinois, according to pilot plant studies of Purdue University, Lafayette, Indiana. About 28% of the crambe seed is extractable oil that is a rich source of the diamine and diacid monomers used in the synthesis of polyamides. Fatty acids obtained from the oil have a high content (55–60%) of erucic acid. This substance inhibits the absorption of water and is responsible for the moisture-resistant properties of the final nylon product. Fracture of the double bond in erucic acid and oleic acid (15% of the oil) results in dibasic carboxylic acids—brassylic and azelaic acids, respectively—which are suitable for use as nylon monomers. The nylons obtained consist of monomers of considerably higher molecular weight than those in most nylons presently marketed.

Single-cell proteins. With analyses showing large potential markets but uneasy consumer acceptance of single-cell proteins, many oil companies and at least one brewer, Anheuser-Busch, proceeded with SCP production primarily for use in animal feeds. Although the feasibility of producing SCP from all types of cellulosic wastes was demonstrated, the major commercial ventures being planned were utilizing homogeneous substrates to minimize problems. The commercial production of SCP from heterogeneous cellulosic wastes appeared to be far in the future.

Supplying protein by the production of SCP requires essentially no land, is unaffected by seasonal variations, and features high growth rates. Bacteria, yeast, or other fungi can be used. Of the three, bacteria supply the most protein. Also, protein from bacteria most closely resembles that from animals.

The most inexpensive feedstock for producing SCP, oil, has been abandoned because of the problems presented by polynuclear aromatic contaminants, such as the cancer-producing benzo(a)pyrene. Many producers stopped using processes based on normal paraffins from oil because of consumer fears of similar taints. Early in 1975 chemists at Gulf Research and Development Co. in Pittsburgh, Pa., reported 1–11 parts per billion of polynuclear aromatic compounds in yeast grown on normal paraffins. When they analyzed the normal paraffin, however, they detected no such compounds. They suggested that the potentially harmful substances may be introduced from the air or from certain additives used in the process; an antifoam food-grade additive was shown to contain these substances.

Recent SCP ventures in Japan and Great Britain were being based on more expensive alcohol and natural gas instead of petroleum. The companies, however, isolated organisms that will grow rapidly on these substrates at 40° C (104° F). This means a substantial saving in cooling costs, because heat is generated during fermentation. Because of the natural-gas shortage in the United States, these substrates probably will not be widely used in the U.S. until the possible advent of large-scale coal gasification in the late 1980s.

In Japan, Mitsubishi Petrochemical Co. selected ethanol as the substrate. The firm's researchers isolated a yeast strain that can grow both rapidly and in high yields (95%) under unusual conditions of temperature (40° C) and in an acidic culture medium. Microbes likely to contaminate the process cannot tolerate the acid conditions. A pilot plant capable of producing 100 metric tons of SCP per year went into operation. Mitsubishi Gas Chemical Co., on the other hand, selected a bacteria-methane process. Production of SCP from these processes in commercial quantities was expected in or after 1978.

In the U.K., Shell Research Ltd. announced an efficient way to utilize methane as the carbon source. No known yeasts grow on methane, but several species of bacteria do. In a new approach to the problem, Shell used a structured, mixed bacterial culture on a continuous basis at fermentation temperatures in excess of 42° C. The white, bland protein that was produced successfully underwent initial toxicological and nutritional testing. Its nutritive value reportedly approached that of best quality white fish meal. A commercial plant was planned for the early 1980s.

Also in Britain, Imperial Chemical Industries Ltd., in choosing a methanol-bacterium system, found a novel way to cause the bacteria to adhere to one another in large agglomerates for easy separation. The selected bacterium will proliferate at 40° C.

A fermentation process useful as a disposal method for pulping wastes, with residual high-protein solids as a valuable by-product, was proposed by Donald L. Crawford of George Mason University in Fairfax, Va. He isolated a filamentous microorganism that can utilize insoluble lignocellulosic wastes as a source of carbon and energy in a reasonable time to yield SCP of good nutritional quality. The heat-loving actinomycete rapidly metabolizes pulping wastes containing up to 10% lignin and significantly degrades wastes containing up to 18% lignin.

Direct use of ordinary photosynthetic green and blue-green algae for animal food and experimental human food was suggested by R. K. Robinson of the University of Reading, England. Algae could be cultured efficiently in giant artificial lagoons of concrete and polyethylene, yielding 20,000 lb of protein per acre, he claimed. The temperature would be controlled and water circulated throughout the lagoon. The large area of shallow water, steady sunlight, and warm temperatures required imply a tropical or subtropical location. After harvesting and drying, the algae, having high protein content, could be used directly in animal feed or in experimental soups and curries, according to Robinson.

Antibacterial cotton. Two chemical processes that render cotton bactericidal were developed at the Southern Regional Research Center of the U.S. Department of Agriculture in New Orleans. Designed originally for hospital operating rooms, where synthetic materials cannot be permitted, the antibacterial cotton can be used in sheets and towels in order to prevent the spread of infection in institutions.

In the first method, cotton is chlorinated until 5% of the cotton by weight is chlorine. In the second method, thiocyanate groups are substituted for the chlorine atoms, and a "thiocyanated cotton" results. Both types of treated cotton showed antibacterial action when tested with *Staphylococcus aureus* and *Escherichia coli* and subjected to home laundry testing.

A new project was started to extend this work to germicidal finishes for apparel. The scientists, led by Tyrone Vigo, were looking for a finish to combat the bacteria responsible for perspiration odor; that is, to develop a treated cotton that will inhibit body odor.

—Dorothy P. Smith

Earth sciences

Research on the huge crustal plates that underlie the Earth's surface dominated the year in the Earth sciences. Project FAMOUS explored the Mid-Atlantic Ridge, considered to be the boundary between two of the plates, and scientists aboard a Soviet ship studied a similar region in the Sea of Okhotsk. In other developments, analysis of lunar rocks continued; geophysicists intensified their search for new deposits of gas and oil and new sources of geothermal energy; and atmospheric scientists completed the observational phase of a major experiment in the tropical Atlantic.

Atmospheric sciences

The story of the year in the atmospheric sciences was the successful completion of the observational phase of GATE (GARP [Global Atmospheric Research Program] Atlantic Tropical Experiment), the intensive experimental study of the tropical atmosphere. The results of this international program are expected to have far-reaching implications for many aspects of the atmospheric sciences. The knowledge should improve man's understanding of the behavior of the Earth's atmosphere and lead to improved weather forecasts. The information gathered in GATE should also be valuable for scientists studying regional and global climate.

It is becoming increasingly clear that adequate supplies of fresh water and food and the consumption of fuel depend on the climate. This makes it essential to develop a much clearer understanding of the factors governing the climate and to learn how to make useful predictions of future weather conditions.

The need for energy conservation because of decreasing fuel supplies appears to be reducing anxieties about air pollution. On the other hand, there continues to be a great deal of concern about possible effects of the exhaust gases of supersonic transport airplanes and of Freon gas on the ozone layer in the upper atmosphere; this layer protects man from harmful quantities of ultraviolet radiation from the Sun.

Progress in the development of a reliable technology for augmenting rain and snow and mitigating the hazards of violent storms was slow, as might be expected from the modest scale of the research effort. There were, however, some important scientific and sociological developments relevant to weather modification. Of particular note was a proposal in the United Nations to adopt a resolution that would prohibit the use of weather modification techniques for other than peaceful purposes.

Global Atmospheric Research Program. One of the most significant and successful scientific programs ever mounted on a truly international scale, GARP continued to make steady progress toward its ultimate goals. GARP started in the mid-1960s and has had the enthusiastic cooperation of many

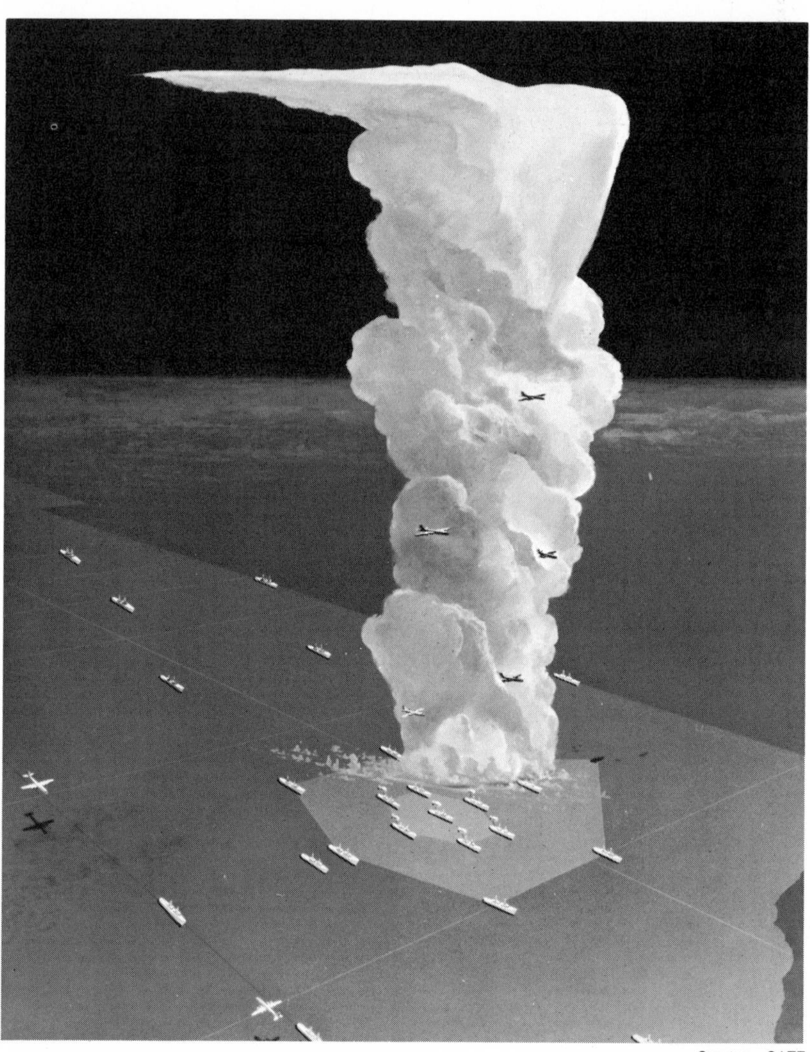

The GATE project, which took place in 1974, was concentrated in the eastern Atlantic off the coast of Africa, where instrument-bearing ships and aircraft were used to study tropical cloud formations, as depicted in the drawing.

Courtesy, GATE

countries in the Northern and Southern hemispheres, particularly the U.S., the Soviet Union, and nations in Western Europe.

A primary objective of GARP is the formulation of an advanced mathematical model of the Earth's atmosphere that could be used to make reliable weather forecasts one to two weeks in advance. The second objective is to develop a better understanding of, and a satisfactory theory of, global climate. In order to achieve these goals, research on many fronts is required. Scientists have devised ever more sophisticated mathematical models of the general circulation of the atmosphere, taking into account such items as: the interaction of atmosphere, ocean, and land; radiation transfer through the atmosphere; the formation of clouds, rain, and snow; and the development, expansion, and contraction of snow- and ice-covered areas.

In order to achieve the goals of GARP it is necessary to obtain measurements of the properties of the atmosphere and oceans over the entire Earth. This has brought about efforts by the participating nations concerning the use of satellites for atmospheric and oceanic observations and for communication of data to collecting stations. It is the analysis of the collected data, of course, which will lead to improved forecasting.

Many aspects of the interactions of the tropical oceans and atmosphere and the role of clouds in transporting water vapor were poorly understood at the start of the Global Atmospheric Research Program. As a result, a series of field experiments of increasing size and complexity were designed and conducted. In 1967 a group of scientists, organized by the National Center for Atmospheric Research in Colorado, carried out an intensive study of convective process analyses over the Line Islands, which are just north of the equator due south of Hawaii.

A much larger observational program, over a region of about 100,000 sq mi and centered just east of Barbados in the western Atlantic, was conducted during 1969. It involved the use of specially instrumented ships, airplanes, buoys, and weather satellites as well as conventional equipment. This project, called BOMEX (Barbados Oceanographic and Meteorological Experiment), in addition to producing a great deal of important information, was extremely useful in planning the substantially larger and more ambitious atmospheric program known as GATE.

The experimental region for GATE extended from latitude 10° S to 20° N and from Mexico eastward across the Atlantic Ocean to the Somali Basin of the Indian Ocean. Most of the observations made from ships and airplanes were concentrated over an area of about 200,000 sq mi in the eastern

Sidney Harris

"The forecast for today is Ice Age, followed in 10,000 years by fair and warmer."

Atlantic about 600 mi off the coast of Africa. The field program extended from June 15 to Sept. 25, 1974, and was directed from a control center at Dakar in Senegal. It involved about 4,000 scientists and technicians from about 72 nations. Observations were made from about 40 ships, more than 60 buoys, 13 aircraft, and 6 weather satellites. This impressive array of people and equipment was coordinated by the International Scientific and Management Group directed by Joachim P. Kuettner of the U.S. Yuri Tarbeev of the Soviet Union served as deputy director.

A massive body of unique atmospheric and oceanic observations was collected during GATE. In 1975 scientists were in the process of analyzing it. The findings are expected to yield a great deal of new information on weather processes in the tropics. More important, the analyses should make it possible to understand better than ever before how conditions and processes in the low-latitude atmosphere and oceans influence overall air motions in the Earth's atmosphere.

In addition, the lessons of GATE are expected to be of great value in planning the First GARP Global Experiment (FGGE). This experiment should, for the first time, obtain extensive measurements of the properties of the atmosphere and ocean over the entire globe. Plans formulated during the year called for FGGE to be carried out during 1978–1979.

Climate and climate change. There is considerable evidence that during this century the Earth has experienced unusually warm conditions. They have allowed long growing seasons and bountiful harvests of food and fiber. The warmth has melted sea ice, allowing ocean shipping at high latitudes. It is now, however, reasonably well established that over the Earth as a whole the average temperature of the lower atmosphere has been in a cooling trend since about 1945, and that most of the cooling occurred in the Northern Hemisphere.

The slowly falling global temperatures have been accompanied by droughts, some of them having dramatic consequences. In 1972 droughts were experienced in the Soviet Union, Asia, Africa, and Australia, and the resulting decrease in grain production caused a serious depletion of the world's food supplies. The tragic consequences of drought in the Sahel zone of Africa have been well publicized.

The need to learn more about the Earth's climate and the factors controlling it is now well recognized. Will the cooling of the last few decades continue and lead ultimately to a new "ice age"? A few scientists appeared to accept such speculations, but it should be recognized that they are only speculations and should not be given the weight of a forecast that is based on a realistic theory of the nature of climate.

Has the cooling been caused by human interference with the geophysical environment, or is it one more random perturbation of global climate of the kind that has occurred many times during the history of the Earth? This question cannot be answered satisfactorily until scientists develop a physical model of global climate much better than the ones existing today.

It is known that the climate depends on many interacting factors, including the quantity of solar energy; the composition of the atmosphere (gases, aerosols, clouds); the characteristics of the ocean; and the overall reflection properties of the Earth. Arguments that the climate is changing because of the observed increases in carbon dioxide produced by the burning of fossil fuels, or the increase of atmospheric particles by human or volcanic activities, are unacceptable unless they are incorporated into a theory taking into account

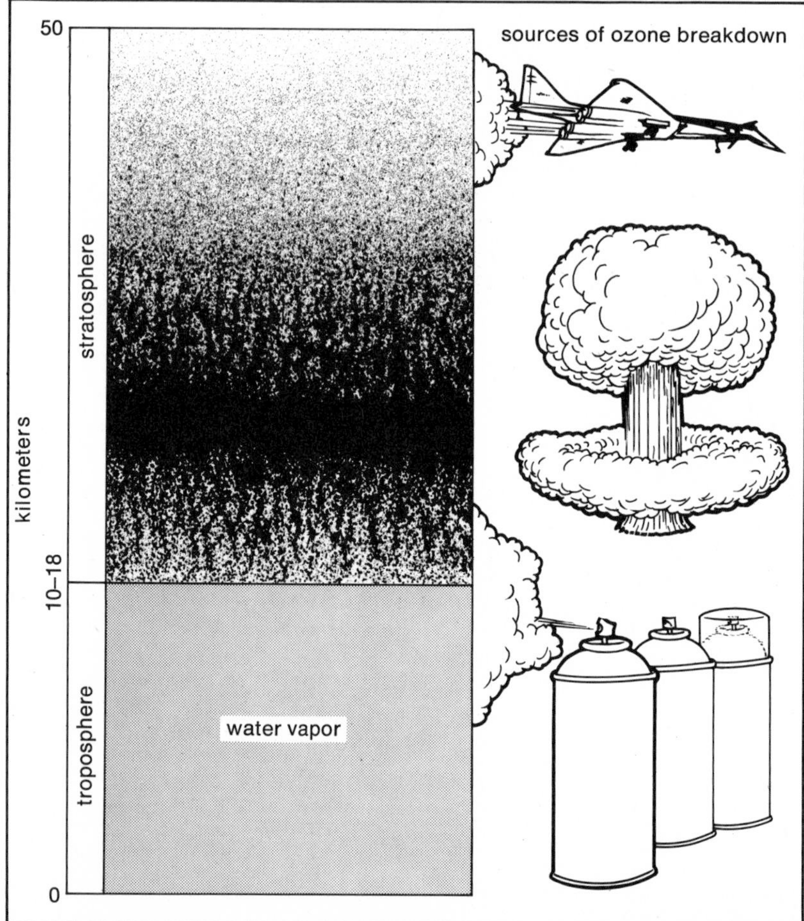

Danger to the protective ozone layer over the Earth's surface is posed by products of 20th-century technology. Supersonic transports and nuclear bomb tests act to destroy ozone by releasing nitrogen oxides. Most harmful is the aerosol spray can. The fluorocarbons that it releases rise to the stratosphere, where ultraviolet radiation causes them, in turn, to release chlorine atoms that may be six times more efficient than nitrogen oxides in destroying ozone.

the complex linkages of the many processes going on in the atmosphere and at the atmospheric interface with the solid and liquid surfaces of the Earth.

In early 1975 the U.S. National Academy of Sciences, recognizing the importance of learning more about this subject, issued a report entitled *Understanding Climatic Change*. The chief recommendation calls for the "immediate adoption and development of a coherent National Climatic Research Program with appropriate international coordination." (*See* Feature Article: THE EARTH'S CHANGING CLIMATE.)

The ozone layer. It has long been known that in the zone above the Earth's surface between about 10 and 50 km (6 and 30 mi) there is a relatively high concentration of ozone, the gas whose molecules are made up of three atoms of oxygen. Although the total quantity of ozone is quite small, with maximum concentrations of less than about 10 parts per million, its effects are profound. The ozone molecules are strong absorbers of ultraviolet radiation from the Sun. As a result there is a warm layer of air in the upper atmosphere. The temperature generally increases with height through the ozone layer, reaching a maximum of about −5° C (−23° F) at about 50 km. This compares with an average air temperature at sea level of about 20° C (68° F) and a minimum of about −55° C at about 10 km.

As a result of the absorption by ozone, the quantity of ultraviolet radiation reaching the ground is quite small. This is a fortunate circumstance because this short-wave radiation over an extended period of time can cause skin cancer.

In 1971 one of the chief ecological arguments made in opposition to the development of a fleet of supersonic transport (SST) airplanes in the United States was that the nitrogen oxides from the engine exhausts would react with ozone molecules and lead to a reduction of ozone, which in turn would allow greater amounts of ultraviolet radiation to reach the surface of the Earth. This matter was given a great deal of attention in an extensive study of the effects of stratospheric pollution by aircraft, conducted under the auspices of the Climatic Impact Assessment Program of the U.S. Department of Transportation. It was confirmed that a fleet of approximately 500 Boeing-type SST airplanes, operating on a regular basis, would constitute a hazard to the ozone layer unless there were sufficient advances in engine technology to reduce substantially the production of nitrogen oxides. At the same time the scientists at the Department of Transportation, in a summary of the complete, multivolumed report, concluded that the smaller number (about 30) of French and Soviet SST airplanes expected to go into operation in the near future could cause climatic effects that would be much smaller than now detectable. A report on this subject issued in 1975 by the National Academy of Sciences again stressed that a fleet of 300–400 large SSTs of the type planned in the U.S. would probably reduce ozone by 10% and increase the incidence of skin cancer by 20%, or about 80,000 cases a year.

Another serious threat to the ozone layer received attention in recent months. Researchers concluded that chlorofluoromethane molecules, most of which come from the Freon used in aerosol spray cans and refrigeration systems, can cause a serious degradation of the ozone layer over periods of 10–20 years. In June 1974 Frank S. Rowland and Mario J. Molina of the University of California at Irvine proposed that the so-called fluorocarbons would, through a series of steps, react with ozone and destroy it. Subsequent investigations, made by various other groups, including a U.S. government task force, confirmed these earlier findings and also estimated the effects on ozone concentration of various quantities of Freon released into the atmosphere. The results, if correct, are ominous.

Fluorocarbon molecules are relatively inert chemically and have lifetimes in the atmosphere that exceed ten years. The molecules slowly diffuse upward through the atmosphere, suffering little change until they reach into the stratosphere. There they become dissociated as they absorb ultraviolet radiation. In the process chlorine atoms are made available, and they catalytically destroy ozone molecules. According to some calculations, even if the emission of Freon into the atmosphere were discontinued immediately, the quantity of fluorocarbon molecules already in the atmosphere would be sufficient to continue ozone destruction long into the future with the maximum rate occurring about 1990.

Recent calculations by Michael McElroy and his associates at Harvard University show that if the worldwide production rate of Freon is not curtailed in the near future, the percentage decreases in ozone can reach disastrous levels after about 1985. It is essential to learn, as quickly as possible, whether or not calculations such as these are valid predictions of reality. Are the physical and chemical aspects of the investigation sound? Are there still unknown means for the removal of chlorine atoms from the stratosphere other than by reacting with ozone? Do the available measurements support the theory? What is the relative magnitude of natural sources of chlorine compounds that might react with ozone? What additional measurements are necessary and how soon can they be collected?

In March 1975 the National Academy of Sciences appointed a special committee under the chairmanship of Herbert S. Gutowsky of the University of Illinois to assess the possible effects of chlorofluoromethanes and other stratospheric pollutants on the ozone layer. During the same month H. Guyford Stever, the chairman of the Federal Council for Science and Technology, assigned to the U.S. National Aeronautics and Space Administration responsibility for developing instrumentation and techniques that could measure changes in ozone concentration and determine the composition of a number of other chemical substances in the ozone layer.

Weather modification. A new major weather modification research program, known as the High Plains Cooperative Program, under the direction of the U.S. Bureau of Reclamation, was begun in 1974. The primary goal of the project is to establish a technology for increasing rainfall from summer clouds over the plains of the central United States. It was decided during 1974 that research would be conducted in three regions: Montana, Kansas, and the area of Big Spring, Texas.

In 1972 Project Stormfury, which had been responsible for testing techniques for weakening hurricanes by means of cloud seeding, was deactivated because of the deteriorating condition of some of the project airplanes. Two new aircraft, Lockheed WP-3D Orions, were scheduled for delivery to the project in 1975 and 1976. After the installation of special scientific instruments, they are to be made available for use in hurricane seeding as Project Stormfury is reactivated in 1977. Present plans call for experiments in the western Pacific, where typhoons occur about three times more frequently than their Atlantic hurricane counterparts.

For several years a number of groups in and out of government have been urging the establishment of international rules on the peaceful use of weather modification techniques. On July 3, 1974, the United States and the Soviet Union issued a joint declaration on measures to overcome the dangers resulting from the use of environmental modification techniques for military purposes. Early in November 1974, representatives from the two countries met in Moscow to begin discussions on how the objectives of the joint declaration could be met.

In the meantime, on Sept. 24, 1974, the representative of the Soviet Union at the United Nations submitted a draft resolution prohibiting the use of environmental and climate modification for other than peaceful purposes. On Nov. 22, 1974, the General Assembly voted to give the matter further consideration at its next session.

In May 1975 the seventh congress of the World Meteorological Organization (WMO) was held in Geneva, Switz. At this meeting it was decided to establish an international program to conduct research on the use of cloud-seeding techniques for augmenting rainfall.

—Louis J. Battan

Geological sciences

During the 1970s Earth scientists have had their attention drawn away from their traditional studies of the geological history of the landmasses and the physical and chemical processes that produced the continental rocks. The availability of lunar rocks from the Apollo program opened new avenues to study the origin of the Earth and planetary bodies. On the Earth new deep-sea drilling technology coupled with the new and powerful integrating hypothesis of plate tectonics, which suggested that many long-sought answers to the development of continental rocks were to be found in events taking place in ocean deeps, led many Earth scientists into a detailed study of the forces acting beneath the oceanic crust.

Geology and geochemistry. Equipped with improved reentry drilling machinery, the Deep Sea Drilling Project set about to investigate the origin and properties of the volcanic layer (designated layer 2) of the oceanic crust. Drilling near the Mid-Atlantic Ridge, the staff of the "Glomar Challenger" discovered the presence of a fossiliferous ooze deep within the volcanic sequence. At one site a record 582.5-m core hole was drilled into young volcanic crust, formed in the last three million years. This hole was successfully reentered nine times.

By mid-1975 the drilling had not been able to obtain the desired specimens from the lower oceanic crust, below the oceanic basalt layer 2. Magnetic studies of the rocks retrieved from the drill holes continued to provide data on the rate of seafloor spreading and the accumulation of new oceanic crust from sources in the upper mantle layer of the Earth's interior.

With plate tectonics as the framework within which research in most of the Earth sciences is being conducted, Project FAMOUS (French-American Mid-Ocean Undersea Study) set out to study the processes occurring where the plates are generated (divergent plate boundaries) and where they are consumed (convergent plate boundaries). These regions are narrow oceanic troughs at great water depths. The project sent three submersibles, the French "Archimède" and "Cyana" and the U.S. "Alvin," to make dives 3,600 m deep in the great rift valley of the Mid-Atlantic Ridge to map and sample the young volcanics. These samples and photographs were studied during the year.

The latest survey in the Sea of Okhotsk by the Institute of Oceanology of the Soviet Academy of Sciences was conducted by the cruise of the "Dmitry Mendeleev" during August to October 1974. This important marginal sea borders the northern and western margins of the Pacific Ocean. It is separated from the open ocean by tectonically active island arcs and by deep oceanic trenches. The Sea of Okhotsk thus lies at the junction of the main tectonic elements of the northeast Asian continental landmass and the Pacific Ocean basin. Soviet, Japanese, and U.S. scientists aboard the "Mendeleev" expected that sediment samples and geophysical data taken by the survey would provide the information to unravel the plate-tectonic history of this crucial area.

Other samples from the ocean bottom brought to light sediments that gave evidence for at least eight major glacial cycles in the last 700,000 years. Studies of ocean bottom sediments pushed the beginning of the Pleistocene "glacial age" back to two to three million years before the present. Such detailed studies called into question the currently accepted view of the Pleistocene as four well-defined cycles of continental glaciation with interglacial stages. Other Earth scientists, after a study of the silt and clay deposits laid down on the coast of Maine by the last great North American continental glaciers about 12,000–13,000 years ago, were able to measure the rate of recession of the ice sheet and to describe a brief marine invasion of the Maine coast before crustal rebound again brought the region above sea level.

In a related matter the question of whether or not late Pleistocene but pre-Wisconsin (before about 75,000 years ago) glacial deposits exist in the Andes of South America seemed to be settled in the affirmative. A buried layer of volcanic ash was found in the Cordillera Central of Colombia overlying the outermost of four nested lateral moraines. Fission-track dating placed the age of the ash at about 100,000 years before the present. The discovery and dating of this pre-Wisconsin till indicates that at least a portion of the northern Andes was at an altitude sufficient to support glaciers about 100,000 years ago.

Lunar geologists continued to work on Moon rock samples, seismic data, and photographs in order to update their hypotheses on the sequential development of the lunar crust and interior structure. Data became available that allowed the lunar interior to be divided into five concentric layers: (1) an outer crust of plagioclase-rich rock extending to a depth of about 60 km (35 mi); (2) an upper mantle of olivine and pyroxene-rich rock extend-

Courtesy, Woods Hole Oceanographic Institution

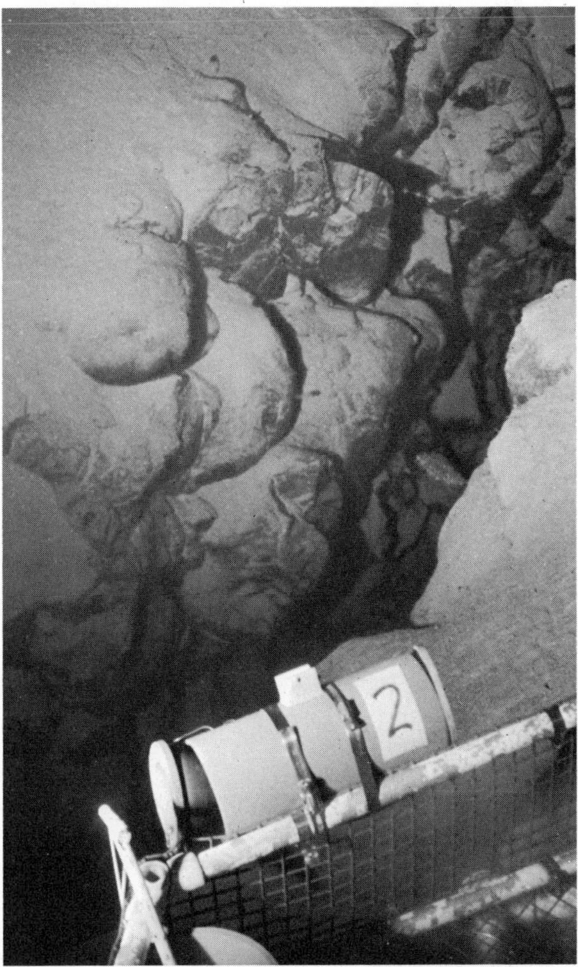

Medium-sized fissure in the Mid-Atlantic Ridge resulting from sea-floor spreading is viewed during Project FAMOUS.

ing from 60 to 300 km; (3) a middle mantle from 300 to 800 km possibly composed of primitive lunar rock from which the outer layers formed during an earlier period of melting; (4) a lower mantle lying at a depth of 800–1,400 km that may be partially molten; and (5) a core, perhaps of molten iron sulfide, below 1,400 km. Other lunar scientists, after five years of organic chemical analyses, reported that they still could find no evidence of amino acids being present on the Moon today or having been present in the lunar past.

Mineralogy. Although much attention was directed skyward and seaward during the first half of the 1970s, the more traditional pursuits of Earth scientists were not abandoned. Equipped with automated single-crystal data-gathering devices, high-speed computers, and powerful mathematical techniques, crystallographers continued to solve complicated mineral structures. Paul Moore unraveled most of the structures of the whole colorful host of secondary pegmatite phosphates and elucidated the crystal chemistry of their formation. And, for coverage of the most abundant rock forming minerals in the Earth's crust, the first two volumes of Joseph V. Smith's *Feldspar Minerals* appeared during 1974.

Mineralogists were also busy describing more than 50 new mineral species, and Akira Kato of the National Science Museum in Tokyo replaced Michael Fleischer as chairman of the International Mineralogical Association's Commission on New Minerals and Mineral Names.

Crystallographers and crystal chemists continued experimental work on the crystal structures of high-pressure minerals and related high-pressure phase transformations (transformations from one crystalline form of a mineral to another) for clues to the materials and processes in the Earth's mantle. Japanese scientists from Osaka University solved the structural relationships among the three polymorphs (different crystalline forms) of Co_2SiO_4, indicating possible models for mantle materials. Other experimenters studied phase transitions in high-pressure silicates.

Mineralogists and petrologists had their share of controversies during the year, including the suggestion by four Cornell University scientists that a piece of natural iron-nickel alloy found in gravel along Josephine Creek in Oregon (and dubbed josephinite) is a piece of the Earth's core, which begins at a depth of 2,900 km (1,800 mi). The composition of the alloy differs from meteorite material and in addition contains garnet crystals (evidence of high pressure). These data led to the suggestion that the rock was transported upward from the core-mantle boundary by a deep mantle plume to a location near the Earth's surface on the Pacific plate, which in turn deposited it on the North American plate when the former was consumed by subduction at the plate margin.

In a related matter, evidence was offered that geothermal and volcanic phenomena at Yellowstone National Park are the result of a crustal hot spot at a junction of major tectonic trends. It was not known, however, whether the hot spot was related to a mantle plume.

Oldest rocks. Questions about the age, location, and genesis of the oldest rocks on Earth continued to occupy geologists. The three-way race for "oldest" among southwestern Minnesota, western Greenland, and southern Africa may become a ten-way or even twenty-way race according to experts who gathered in September 1974 in Minnesota for a conference to trade hypotheses and data.

Their recent work pushed the ages of the oldest Earth rocks back to about 3.8 billion years, ever closer to the 4.1 billion–4.6 billion-year ages of the oldest lunar rocks and meteorites. Most conference participants agreed that rocks 3 billion–3.8 billion years old are more prevalent on the Earth than previously thought and probably will be found eventually in all the Precambrian shield areas (large core areas in the continental masses containing rocks formed in Precambrian time). Field geologists and their laboratory geochronologist colleagues working on these ancient rocks reported that the oldest material most often comes from structurally complex gneiss/granite terrain and that apart from their great age the rock types show no unique properties.

Environmental geochemistry. From their early concerns with toxic concentrations of selenium, lead, and mercury in some environments and a deficiency of iodine and fluorine in others, environmental geochemists have proceeded to a broad understanding of the role of trace elements in nutrition, disease, and health. The geographic distribution of disease incidence has in many cases been correlated with trace-element concentrations in rocks, soils, plants, and natural waters. Trace elements migrate from rocks to soils and waters, then into accumulator plants and animals, and finally into man. Such concerns are basic to the interests of the recently formed Society for Environmental Geochemistry and Health.

Many trace-element concentrations are directly related to industrial activity. During 1974 scientists renewed their efforts to establish reliable baseline data for monitoring the geochemical environment. These efforts were accompanied by increased pessimism about most of the analytical data in the existing scientific literature; many researchers believed that this information is too inaccurate, often because of inadequate techniques and poor standard samples, to be of much value.

Recent research identified natural complexes of selenium that govern the geochemical and biochemical distributions and the environmental mobility of this element. Other findings in recent months showed promise of determining the concentration ranges defining toxicity at one end and, sometimes, deficiency levels at the other end for more than 20 trace elements, including copper, molybdenum, chromium, cadmium, zinc, cobalt, aluminum, and manganese.

Minerals as pollutants continued in the news as the environmental problems surrounding the discharge of taconite tailings into Lake Superior remained unresolved. Asbestoslike amphibole fibers are emitted through the stacks of taconite processing plants and make up a large percentage of the waste sludge dumped by one processor into Lake Superior. The cancer-producing nature of asbestos fibers alarmed residents of the area who drew their water supplies from the lake.

Energy supplies. Throughout the year the argument continued among geologists as to the quantity of oil and gas resources yet to be discovered within U.S. territory. Most oil geologists agreed that perhaps two-thirds of the nation's undiscovered petroleum lies in Alaska and offshore beneath the outer continental shelf. The argument concerned the disparity between the low estimates of M. King Hubbert and other knowledgeable geologists who suggest that U.S. oil resources will not last out the century, and the estimate that had been offered by the U.S. Geological Survey, which is far more optimistic. The latter assumes 35% imports and a 2.5% annual growth in consumption. Hubbert's case was strengthened by his correct prediction in the 1950s that U.S. domestic production would peak in the early 1970s and decline thereafter.

With its obvious impact on U.S. energy policy the accurate assessment of oil and gas reserves occupied an increasing number of geologists in government and industry in 1974. An increase of more than 20% in geological exploration and drilling by U.S. oil companies took place during the year. U.S. Geological Survey personnel undertook an engineering geology study of the route of the trans-Alaska pipeline in order to assess environmental problems involved in getting Alaska crude oil to the U.S. market. Many other Earth scientists turned to the study of oil-bearing shale rock. If the huge U.S. reserves of oil shale (about one-half of the world's total) are to be tapped along with the by-products aluminum and nitrogen, a myriad of geological, hydrological, and soils problems must be solved. Not the least of these is how to dispose of the excess material generated during the processing, which causes the rock that is mined to expand by 20% or more.

Economic geology. Economic geologists continued to use plate-tectonics models to explain the chronology and distribution of metal mineralization (metallogeny, the origin of ores). During the past year investigations of known mineral belts both supported and contradicted current plate-tectonics models. The interest in the relationship between metallogeny and plate tectonics resulted during the year in a NATO Advanced Studies Institute on this topic at St. John's, Newfoundland. Themes discussed at the Institute centered on the evidence for mineralization at the margins of spreading plates and converging plates. Data from ore deposits were presented that supported

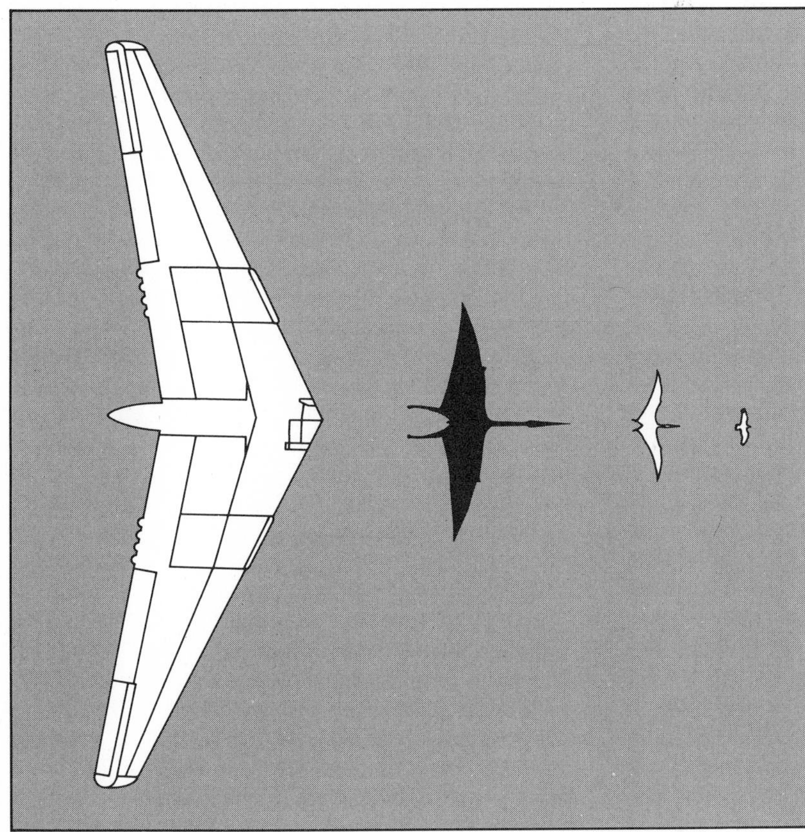

D. A. Lawson, "Pterosaur from the Latest Cretaceous of West Texas: Discovery of the Largest Flying Creature" in "Science," vol. 187, pp. 947–948; cover, March 14, 1975, © AAAS

Pterosaur (second from left), the largest known winged creature, is compared with, left to right, a Northrop YB-49, a pteranodon, and a condor. Remains of three pterosaurs, flying reptiles of the Late Cretaceous Period, were found recently in western Texas.

proposed plate movements to at least the beginning of Paleozoic time (600 million years ago).

Economic geologists found increasing value in applying studies of isotopes to ore deposits. Improved technology resulted in major improvements in the precision of lead-isotope analyses. This led to additional applications of lead isotopes as "fingerprints" for the source of lead in ore deposits and demonstrated that the age of many ore deposits is older than formerly believed.

Isotopic studies also apparently unraveled the old question of whether one or two ages of gold mineralization accounted for the ores in the Homestake mine at Lead, S.D. These deposits, which have produced gold valued at more than $1 billion, are located in Precambrian rocks crosscut by well-documented Tertiary mineralization. Stable isotope studies (sulfur, oxygen, hydrogen, and carbon) followed by a lead-isotope investigation showed that the gold deposits were indigenous to the Precambrian Homestake formation and thus are Precambrian in age.

Paleontology. A major discovery in paleontology was announced during the year. In the Cretaceous (65 million to 136 million years ago) rocks of the Big Bend area of western Texas three partial skeletons were uncovered of a large pterosaur with a conservatively estimated wingspan of 15.5 m. Unlike those of most pterosaurs these remains were found in nonmarine rocks deposited more than 400 km (250 mi) from the nearest ancient sea. These winged reptiles represent the largest creature ever to fly (or soar) above the Earth.

A major area of controversy concerned the way in which these giant pterodactyls took to the air. Some scientists argued that the wings of these creatures were too large to be used for anything but gliding and that the pterodactyls had to launch themselves from some height. Others, possibly noting that the host sediments of the bones showed no evidence of ancient hills nearby, contended that favorable wind conditions and proper flapping could accomplish the aerodynamic lift necessary to get the pterodactyls airborne.

—George Rapp, Jr.

Geophysics. Research stimulated by the theory of plate tectonics, which revolutionized thought on global geologic processes in the late 1960s, and the more recent dilatancy-fluid diffusion model for premonitory earthquake phenomena has led to a period of critical examination and testing in geophysics. At the same time, application of both established and new geophysical methods in the search for energy resources continued to expand in response to increased demands for, and diminishing supplies of, inexpensive, environmentally acceptable energy.

Prospecting for energy. The search for new gas and oil resources remained the dominant effort in geophysical exploration for energy. Worldwide expenditures for geophysical exploration grew markedly during the 1970s, with total annual expenditures passing the $1 billion mark in 1973. More than 90% of this amount was spent on seismic prospecting for petroleum. A large fraction of the recent increase was due to expanded marine seismic exploration of continental shelves and border lands, which are thought to hold large petroleum reserves.

New developments in prospecting technology included the adoption of portable minicomputers for on-the-spot processing of the massive amounts of data generated by the 200–1,000-channel seismic recording systems. These new systems provided much higher resolution of the subsurface structure than had previously been possible, which in turn placed new demands on interpretation methods. Considerable effort in regard to the latter was being devoted to theoretical and computational methods for the analysis of such things as waves scattered from small-scale heterogeneities, the effects of lateral variations in velocity within a formation, and the interpretation of waves reflected from three-dimensional structures. Techniques developed in optical and acoustical holography, for example, were being adapted to use with seismic reflection data as a means of reconstructing complex subsurface velocity structures. Success of the widely publicized bright-spot technique, which relies on the enhanced amplitudes of signals reflected by gas-bearing strata, stimulated research into additional uses of amplitude data for the identification of oil and gas deposits.

The search for new sources of geothermal energy also expanded during the year, although total expenditures remained a small fraction of that spent on the search for petroleum. Many of the techniques developed for petroleum exploration, however, were expected to be effective for geothermal searches as well. The elastic properties of steam-saturated rock are similar to those of gas-saturated rocks in a petroleum reservoir, for example, and the bright-spot technique offers a promising method for detecting vapor-dominated geothermal reservoirs such as exist beneath the Geysers geothermal area in northern California.

Much of the federally sponsored research on methods for geothermal resource exploration and assessment in the U.S. was directed to the study of recognized or suspected geothermal areas. The dual objectives of this research were to evaluate the effectiveness of various geological and geophysical methods for finding geothermal resources and to develop a deeper understanding of the properties of geothermal systems and the conditions under which they occur.

Yellowstone National Park is a prime example of an active geothermal system, and it was one of the several areas in which research on the properties of geothermal systems was being focused. Yellowstone itself, of course, is protected from exploitation by its status as a national park. Preliminary studies of the propagation of seismic waves generated by local earthquakes and recorded on a 27-station seismograph network in Yellowstone showed that shear waves are strongly attenuated along paths that cross the south-central part of the park. (Shear waves, or S-waves, cause shaking motion perpendicular to their direction of propagation; compressional waves, or P-waves, cause shaking motion parallel to their direction of propagation.) These results suggest that a magma chamber or very hot rock may lie within 5–10 km of the surface. H. M. Iyer of the U.S. Geological Survey described evidence that the hot or molten rock may extend to depths of 250 km or more beneath Yellowstone. He based this conclusion upon delays in arrival times of P-waves from distant earthquakes recorded at portable seismograph stations in and around Yellowstone. These results have implications concerning not only the nature of the heat source for Yellowstone's geothermal system but also the broader problem of mantle hot spots and driving mechanisms for plate tectonics.

Heterogeneities in the mantle. Yellowstone, Hawaii, and Iceland are perhaps the best-known examples of some 20 mantle hot spots that have been identified throughout the world. Recent work on the reconstruction of the motion of the major lithospheric (crustal) plates by Bernard Minster and Thomas Jordan while at the California Institute of Technology suggests that these hot spots have maintained relatively stable positions in the mantle with respect to the overriding plates. (The mantle is that part of the Earth between the crust and the core, lying at the depth of 30–3,000 km [19–1,900 mi].) Although it is generally agreed that the hot spots are a fundamental property of the mantle, geophysicists disagreed on their explanation. Of the two widely discussed hypotheses, one suggests that the hot spots are narrow thermal plumes of molten rock rising from near the core-mantle boundary. The other suggests that they are formed by shear melting at depths between 50 and 200 km as the plates move over local heterogeneities in the upper mantle. In the latter case, the positions of the hot spots are thought to be stabilized by a "gravitational anchor" formed as the dense solid residue of the partially molten rock sinks into the mantle. It remains to be seen which of these hypotheses is closer to the truth.

The evidence for low seismic wave velocities extending to great depths under Yellowstone is consistent with the thermal plume idea. Evidence described by Tom McEvilly at the University of California, Berkeley, indicating high average S-wave velocity in the deep mantle beneath Hawaii, however, is more favorable to the gravitational anchor hypothesis. This evidence is based on the travel times of S-waves that were reflected several times between the Earth's surface and the core directly beneath Hawaii. The S-waves used by McEvilly in this measurement were generated by the magnitude 6.2 earthquake that occurred just off the coast of Hawaii on April 26, 1973.

Evidence for large-scale heterogeneities in the mantle was recognized in surface-wave dispersion data in the early 1960s. These data suggested that differences between seismic wave velocities beneath oceans and those beneath continents may extend to depths of 400 km or more into the mantle. With the general acceptance of plate tectonics, however, this evidence faded into the background. Along with plate tectonics, most geophysicists accepted the concept of relatively thin lithospheric plates (50 to 100 km thick) moving over a mantle that, except for local heterogeneities beneath deep ocean trenches and hot spots, was laterally homogeneous below depths of about 200 km. The surface-wave evidence for large-scale heterogeneities in the deep mantle was only recently brought back into focus by Shelton Alexander at Pennsylvania State University.

Observations by Stuart Sipkin and Thomas Jordan at Princeton University that the one-way travel times of S-waves reflected from the core-mantle boundary are about five seconds greater for stations on oceanic islands than for stations on continents support the inferences based on surface-wave data. Taken together, these two lines of evidence strongly point to differences in the S-wave velocity structure beneath oceans and continents, extending to depths at least as great as 400 km (250 mi). This in turn suggests that convection in the mantle associated with plate motions may involve a significant fraction of the total thickness of the mantle.

Earthquake prediction. Earthquake prediction continued to be one of the most active areas of research in solid-earth geophysics. The program expanded internationally during the last year, with working exchanges of U.S. and Soviet scientists and exchanges of visiting delegations of geophysicists between the U.S. and China.

A delegation of ten U.S. geophysicists visited China in October 1974, as part of an exchange program sponsored by the U.S. National Academy of Sciences. They returned impressed by the scale

Courtesy, Jet Propulsion Laboratory, art from the "New York Times" © 1974. Reprinted by permission

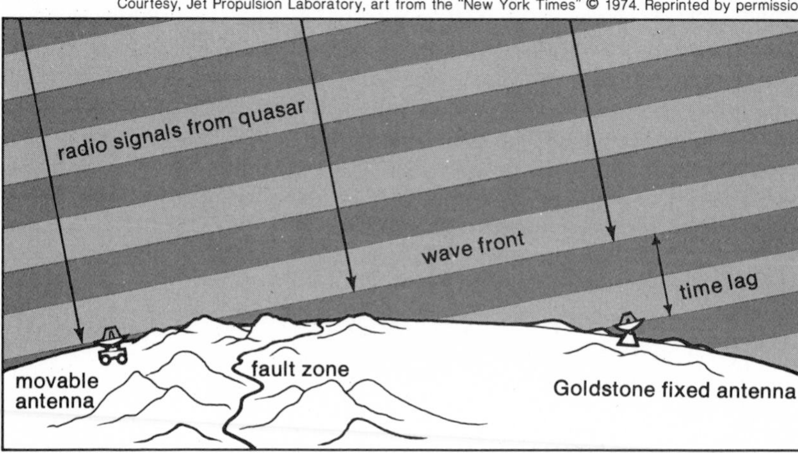

Radio emissions from quasars, billions of light-years distant, are used in California to measure the warping of the Earth—in fractions of an inch—that is believed to precede major earthquakes. Two widely spaced antennas are used for these measurements. The time lag in the arrival of signals at one relative to the other can be compared at various times; changes in the time lag indicate warping.

of the earthquake prediction effort in China, reporting that programs to measure premonitory phenomena associated with earthquakes were being pursued at five provincial centers in the most seismically active parts of China. Each of these centers was carrying out intensive field programs, including measurements of crustal movements, seismic velocities, frequency of occurrence of small earthquakes, radon flux in deep wells, water level and temperature in wells, geomagnetic field variations, and both natural and induced electric fields. Chinese geophysicists claimed to have predicted several earthquakes successfully, but they also emphasized that a number of their predictions had failed. The Chinese strongly directed their effort in earthquake prediction along phenomenological lines; they made little effort to develop theories to explain their observations.

The U.S./U.S.S.R. Working Group exchange program was initiated in 1974 with a joint six-month field session in the Garm region of Soviet Tadzhikistan. Robert Wesson and Ron Kaderabek of the U.S. Geological Survey together with members of the Complex Seismological Expedition of the Institute of Physics of the Earth of the Soviet Academy of Sciences installed eight two-component seismograph stations near the village of Khait. The stations were the same type used routinely by the Geological Survey in California. Initial processing of the data by Wesson and Soviet co-workers revealed a distribution of earthquake locations and fault plane discontinuities that was complex in comparison with the relatively simple situation that occurs along the San Andreas Fault in central California. Further joint field work with these instruments in the Garm region took place in 1975.

Recent earthquake research in the U.S. resulted in somewhat diminished enthusiasm for the dilatancy-fluid diffusion theory, which only a year ago was widely thought to explain most of the reported premonitory phenomena associated with earthquakes. The theory is based on the effect of pore fluids on the strength of rocks under high shear stress, in conjunction with the opening of small cracks (dilatancy) as the rock approaches failure. Some of the new evidence came from a network of tiltmeters along the San Andreas Fault. Malcolm Johnston and his co-workers at the U.S. Geological Survey interpreted the tilt patterns that developed days to weeks before local earthquakes as caused by slow relative displacement (creep) at depth within the fault zone. The dilatancy theory predicts a simple volume expansion in the source region before an earthquake. Evidence for similar creeplike deformation minutes before the great Chilean earthquake of 1960 (magnitude 8.3) was found in long-period surface wave and free oscillation records by Hiroo Kanamori and Don Anderson at the California Institute of Technology. In both cases this relatively smooth, creeplike (aseismic) deformation is thought to be of about the same size as or larger than the sudden seismic slip accompanying the earthquakes.

Laboratory experiments on the failure of rocks under high confining pressure in the presence of fluids were also forcing some reassessment of the dilatancy-fluid diffusion model. But in spite of the accumulating evidence that the model does not represent the dominant process in the Earth before an earthquake, most investigators still agreed that it might play a significant role in what now appears to be a combination of processes.

A magnitude 5.2 earthquake occurred between the San Andreas and Calaveras faults about 10 mi N of Hollister, Calif. on Nov. 28, 1974. Four to six weeks before the earthquake, anomalous patterns developed in the records of two tiltmeters and one magnetometer station in the vicinity of the epicenter. While presenting these data at an informal meeting on November 27, two U.S. Geological Survey scientists pointed out that these anomalies had many properties in common with premonitory anomalies reported before earthquakes elsewhere and indicated that an earthquake in the vicinity of the stations recording the anomalies might be imminent. Analysis of seismic wave travel time data, performed on the day following the earthquake, indicated a weak anomaly in travel times that coincided in time with the magnetic and tilt anomalies before the earthquake. This was the first time that such a variety of precursory phenomena had been found for a single earthquake in the U.S., and the description of the sequence of events generated a wave of optimism in the scientific community regarding progress toward the goal of earthquake prediction. Earthquake prediction remains in an experimental stage, however, and more extensive instrumentation and research must be realized before it will be useful in planning for public safety.

—David P. Hill

Hydrologic sciences

Major achievements in the hydrologic sciences during recent months include the completion of the International Hydrological Decade and the discoveries made during the exploration of the Mid-Atlantic Ridge by Project FAMOUS.

Hydrology. Hydrology reached a certain coming-of-age during the past year. The International Hydrological Decade was completed at the end of 1974, and, almost simultaneously, 88 nations at a UNESCO conference voted that a new open-ended program should be put into effect immediately.

The year 1974 was the 300th anniversary of Pierre Perrault's anonymous publication of evidence that more than enough rain fell on a French drainage basin to account for its runoff. This was the first conclusive evidence for the hydrological cycle as the continuing source of fresh water on the Earth and the beginning of hydrology as a science. The U.S. Army Corps of Engineers marked the 150th anniversary of its responsibility for inland navigation and the beginning of a notable, if sometimes controversial, career of river control and related developments.

The year 1974 was also the tenth anniversary of the Federal Water Resources Research Act and the beginning of the contributions of the Office of Water Resources Research (now the Office of Water Research and Technology—OWRT) in the U.S. Department of the Interior. During those ten years, OWRT broadened the base of hydrological and water-resources research in the country's universities and trained a large body of experienced scientists, professionals, and technicians versed in dealing with the complexity of water-related problems. Research fostered by OWRT during its short history often anticipated the more extensive programs subsequently undertaken by other groups and agencies.

International Hydrological Decade. The IHD was launched on Jan. 1, 1965, under the aegis of UNESCO and directed by a specially constituted Coordinating Council. Its purpose was to advance the study of water resources throughout the world, to promote their rational use, and to advance the training of specialists and technicians in all fields related to hydrology and the use of water.

The program's ultimate objective was to assist all countries to become capable of coping with their own and their mutual or common water problems. The basic scientific problem was the world water balance, but many other scientific and methodological problems were examined as well. It was the largest effort in international cooperation ever attempted in this field.

Before the Decade was over, 107 countries had participated at least to the extent that they had established national committees to coordinate their domestic efforts. In addition, many intergovernmental and nongovernmental agencies participated actively. Special mention should be made of the role taken by the World Meteorological Organization. WMO not only worked jointly with UNESCO in providing leadership to many of the Decade projects but also carried out an extensive publications program.

Overall, the most significant single effect of the IHD was the improvement of understanding among nations of their interdependence in regard to hydrological events, even those that take place outside of their own boundaries. This appreciation extended to the information collected and the research done beyond their national borders. Many large countries, such as the U.S., U.S.S.R., China, and Brazil, believe that large parts of their water problems can be resolved from information collected within their own borders. But even those countries are involved in one way or another with international streams, lakes, and groundwater bodies, to say nothing of their dependence on the transfer of atmospheric moisture that occurs on a global scale.

A second area of major success was the use of participation in the IHD as a reason to expand national hydrological networks and institutional capabilities for processing and applying hydrological data. In the mid-1960s many less developed countries, particularly some of those that had recently achieved independence in Africa, had virtually no national networks for collecting water data. By 1975 almost every country had some sort of a national hydrological network and a hydrological

London Daily Express from Pictorial Parade

Members of SEVER-26, the Soviet High Latitude Expedition, take advantage of the polar spring to explore the Arctic. The expedition planned to record climatic conditions and study ice formation and water channels.

service capable of collecting, processing, and applying water data to its developmental problems.

The central scientific objective of the IHD was an updating of the world water balance. This was accomplished in the U.S.S.R. in 1974. The Soviet National Committee for the IHD, working with data that it collected itself or that was supplied by most IHD member states, prepared a comprehensive volume and atlas, *World Water Balance and Water Resources of the World*. In 1975 this work was only available in Russian, but plans for its translation into English and Spanish were under way.

Although many scientists were disappointed that no outstanding multinational scientific programs were initiated during the Decade and that overall advancement had not been so great as had been hoped, the general satisfaction with IHD as a mechanism for cooperation, liaison, and advancement of hydrological efforts had become so widespread that by 1970 plans were begun for a post-Decade program. In 1972 the 17th General Conference of UNESCO approved the International Hydrological Program (IHP) to begin in 1975. The first Intergovernmental Council of the IHP met in April 1975, in Paris, to establish the first phase of the new effort.

General hydrological events. One of the most extensive droughts in the past century was at least temporarily alleviated in 1974 when normal to above-normal rains fell in many parts of the Sahel region in Africa. In July, for the first time since 1969, rainfall in the area was as much as 7% and, locally, as much as 25% above normal. The rains, however, did little to improve conditions in the extensive areas where lateritic soils exist. There, the runoff tends to collect in flooded lowlands and to evaporate rather than to infiltrate to the soil zone and on down to the groundwater reservoirs.

The construction of wells, under way in 1975 in some parts of the Sahel, can do much to provide water during future dry spells. However, the addition of wells to augment local water resources carries the responsibility of adjusting use patterns to local economics. For example, in many parts of the Sahel, the addition of wells as water sources initially appeared to be entirely beneficial to the cattle economy that supported the local population. With time, however, the cattle drovers became increasingly dependent on the conveniently located wells for water sources and neglected to use the more distant and less accessible water holes. When the drought came, the cattle and their herders, grown used to depending on the new water sources, stayed close to the wells. Consequently, the lands around the wells became overgrazed, while the more distant natural water holes dried up. As a result, the local population lost more cattle than they might have if they had continued their traditional treks from one grazing area to another. Thus, it obviously is not enough to provide new water supplies. They must be integrated into the pattern of land use if they are to achieve the benefits they are planned to provide.

During 1974, while the Sahel was waiting for a break in the drought, rain fell on Australia as it had not done in recorded history. Interior regions that normally total 8 to 10 in. of rainfall per year received four and five times that amount. Streams and rivers flowed that had not done so in the memory of man, and extensive downstream areas were flooded. Lake Eyre, usually an extensive flat salt pan, was filled higher than in a hundred years of record. In many parts of Australia, the surface soil was able to absorb large amounts of the unusual precipitation and many streams continued to flow in 1975, still fed by the groundwater stored during and after the 1974 storms.

Concern for the production of food for the world's increasing population caused scientists to focus considerable attention on the significance of the drought in the Sahel and the excessive rainfall in Australia. Are they within the expected range of variations in the existing climate, or are they indicators of trends in long-term climatic changes? Short-term variations, however catastrophic, can be absorbed to some extent by existing institutional arrangements or accepted through traditional attitudes. A long-term trend, particularly a cooling one, would in time, however, limit the availability of water and reduce the size of food-producing areas. Consequently, an understanding of the significance of these hydrometeorological shifts is required.

Most analyses, based on available meteorological records, indicated that the Sahelian drought and Australian rainfall, however unusual, were within the range of expected variability. Other studies, however, suggested that most large-scale climatic changes have fallen within the range of individual monthly anomalies for the historical record. Thus, the recent unusual events might be precursors of long-term trends. A key problem is that of identifying changes that are simply variations within the expected norm and those that anticipate the future. A second question concerns the amount of time there will be to adjust to any change. At this time, the questions are largely in the realm of scientific discussions based on insufficient information and, perhaps, inadequate concepts. The consequences of not knowing what may occur, and the benefits of knowing, suggest the need for a well-supported long-term investigative effort. (*See* Feature Article: THE EARTH'S CHANGING CLIMATE.)

—L. A. Heindl

Oceanography. The marine sciences reached a stage during 1974–75 where reports summarizing the results of extensive efforts over the previous five or ten years were prepared to document substantial progress in many different areas. This progress was demonstrated by the results of Project FAMOUS and by the many different papers at national and international conferences, such as Oceanology International '75 held in March 1975 at Brighton, Eng., and a meeting on the results of the U.S. Skylab space program, held at Houston, Texas, in June 1975. Also, a series of papers was published on internal waves.

Project FAMOUS. An acronym for French-American Mid-Ocean Undersea Study, Project FAMOUS was a three-year joint program carried out by France, Great Britain, Canada, and the United States. Three deep-sea diving vehicles, the "Archimède," the "Cyana," and the "Alvin," participated along with support ships and the deep-sea drilling vessel, the "Glomar Challenger."

The diving vehicles explored thoroughly certain sections of the seafloor in the rift valley of the Mid-Atlantic Ridge, and approximately 5,250 photographs of the region were obtained. The studies provided additional proof of seafloor spreading

U.S. deep-diving submersible "Alvin" is lowered from the mother ship "Lulu" for one of the 17 dives it made as part of Project FAMOUS.

Theodore L. Sullivan from the "New York Times"

at this rift valley. Samples of newly formed solidified lava as young as 100,000 years were found and then were dated by means of natural radioactivity measurements. This is relatively recent as geological time is measured. Because the seafloor spreads apart by a distance of approximately 1 in. per year in this area, this young sample had traveled by means of the continental-drift mechanism about 50,000 in. or 0.79 mi (half rather than 100,000 in. because each side moves away from the center) since it was formed.

The lava protuberances that were discovered had many strange and peculiar shapes, caused by the solidification of the lava as it entered the sea water. One shape looked like long twisted trunks of elephants. Another resembled pillows, and still others were likened to haystacks. The seafloor appears to spread out from a center line fissure formed at the bottom of the rift valley. Separate fissures widen as they move away from the center, and then, after they travel a sufficient distance from the source, an additional mechanism lifts large portions of the "recently" solidified lava into nearly vertical cliffs on each side of a valley to form the Mid-Atlantic Ridge. According to plate tectonics theory the rift valley is the separation place between the North American plate and the Eurasian plate.

Oceanology International '75. Held at Brighton, Eng., Oceanology International featured national reports by representatives of many different nations, and sections on the developments in the search for and production of offshore oil and gas, on mining of deep-sea minerals, on fisheries technology, and on the science of the environment.

The rich offshore oil and gas deposits in the North Sea, off the coast of Norway, off the coast of Ireland, and in the English Channel were discussed at the conference. This great potential resource had yet to be explored fully. The offshore structures for producing the oil and gas below the seafloor were in various stages of design, construction, and installation, and some had been installed in the North Sea.

One fascinating example of the way that different areas of marine sciences interact was given in a paper by Z. E. Williams. It showed how the continental shelf area around Ireland and Great Britain fits together with the continental shelf off Nova Scotia and Newfoundland and the northern part of Africa, demonstrating that Europe, Greenland, Canada, and Africa were all within just a few miles of each other during a long-past period of geological time. For this reason it seemed likely that oil and gas resources similar to those in the North Sea will be found on the North American shelf in the Grand Banks and Newfoundland shelf areas.

Another important paper was a report from Indonesia by B. Sumantri. The offshore oil recovery platforms near Indonesia were bringing in some natural gas along with the oil. When the oil had to be transported over great distances, the usual procedure was to flare off the gas (that is, to burn it) at the oil platform and to save the oil for shipping to areas where it would be refined. In this case, however, the natural gas from the platforms was sent by means of a pipeline on the seafloor to the land and used as the energy source for the manufacture of fertilizer.

Although not discussed at Oceanology International, it might be noted that the U.S. Supreme Court decided that the federal government rather than the states that border the continental shelf on the east coast of the United States had exclusive rights to exploit offshore oil and gas reserves. This decision opened the way for the possible exploration for oil and the eventual construction of oil drilling platforms on the shelf.

Another important area discussed at Oceanology International was the development of mariculture and of fisheries on a worldwide scale. Mariculture and the intelligent farming of the sea was increasing substantially, and many thousands of dollars worth of food was being produced each year by growing different kinds of seafood (such as shrimp and oysters and certain species of fish) in a controlled environment. This industry was expected to expand considerably by the 1980s.

Another interesting area of study was described in reports by Japanese scientists Takahisa Nemoto and Keji Nasu and Soviet scientist V. P. Bykov. They were studying the harvesting and processing of Antarctic krill and the utilization of it as food for both people and animals. Antarctic krill is the food of the great whales that, though decimated, still inhabit the Antarctic Ocean. It consists of shrimp-like crustaceans and various other small animals that feed on the zooplankton and phytoplankton of the Antarctic Ocean. This krill is rich in fats and protein, and the protein in turn consists of most of the essential amino acids. It has been estimated that somewhere between 10 million and 25 million–50 million metric tons of krill could be harvested on a sustained yield basis annually from the Antarctic Ocean.

Techniques for catching and preserving the krill were described, and Soviet scientists reported their success in processing it into a protein paste called "okean." This paste was described by the Soviets as "a good food product suitable for consumption immediately after it has been [defrosted]. . . . A number of recipes for appetizers, salads, soups, second courses, sausages, and canned food containing protein paste were worked out, and the possibility for further expansion of their production was shown."

The problem of describing the winds and the waves in the North Sea and in the waters surrounding the British Isles was also a subject at the conference. Much of the scientific data used for the design of the various oil drilling platforms in those areas was presented. The oil companies and the nations around this region joined together in order to gather more extensive data on the environmental problems in the North Sea, and studies by many European scientists were to be made.

Skylab conference. The study of the oceans from spacecraft progressed remarkably well during the year. The data obtained from the U.S. Skylab program were analyzed, and the final reports for the U.S. National Aeronautics and Space Administration (NASA) were scheduled to be completed by the end of December 1975. A volume summarizing the accomplishments of the program was to be completed by July 1976. Some of the results of Skylab were already described, as illustrated in the following discussion.

One of the instruments on the Skylab orbiting laboratory was a combination radar altimeter and scanning pencil-beam microwave-radar radiometer that was used over the ocean. The altimeter was able to measure the distance between the spacecraft and the surface of the ocean to an accuracy of almost one meter. The accompanying diagrams show what happened when this altimeter was operated in the vicinity of Puerto Rico and of the Cape Verde Islands.

Figure 1 shows a pass of the altimeter as Skylab moved from north to south toward the island of Puerto Rico. The big spike is the island itself. The wavy line, caused by scatter in the individual measurements, is the distance measured between the spacecraft and the surface of the ocean. The path of the spacecraft was a smooth circular arc with no major changes along it in regard to the distance of the craft from the center of the Earth. The illustration shows that, just north of the island of Puerto Rico, the surface of the ocean moves approximately 12 m closer to the center of the Earth than it is farther north and that the surface of the ocean moves back away from the center of the Earth as the coast of Puerto Rico is approached.

The dotted line in the figure illustrates a fine-scale section of the geoid, the surface around the Earth that is everywhere perpendicular to the direction of gravity and coincides with mean sea level in the oceans. It was obtained by drawing an average line through the altimeter variations. The ocean surface is by definition level along the geoid. If the minor effects of waves and other surface irregularities could be smoothed out, a marble placed anywhere on the surface would not roll.

The altimeter was built specifically to test the idea that the geoid over the ocean could be determined, and the pass over Puerto Rico proved that the instrument could accomplish this. The curve in the figure labeled the Marsh-Vincent geoid is the best estimate from other sources of this surface. The problem of locating the orbit of the spacecraft in coordinates relative to the center of the Earth is quite complicated, and the difference between the curve near −60 m and the curve near −90 m may be an error because of this difficulty. If, however, the curve for the Marsh-Vincent geoid is moved down (to the dotted line) so as to achieve the best possible coincidence with the altimeter trace, it still does not agree with the measurements of the altimeter. This demonstrates that the altimeter will make it possible to improve considerably the measurement of the geoid over the oceans.

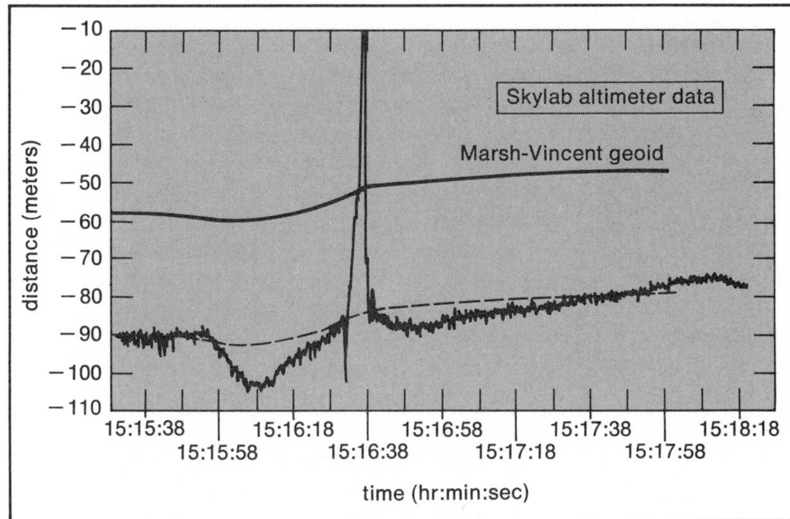

Figure 1. Altimeter data from Skylab during a north–south pass over the Puerto Rican trench. The wavy line is the distance measured between the spacecraft and the ocean's surface, and the tall spike represents the island of Puerto Rico. See text for additional explanation.

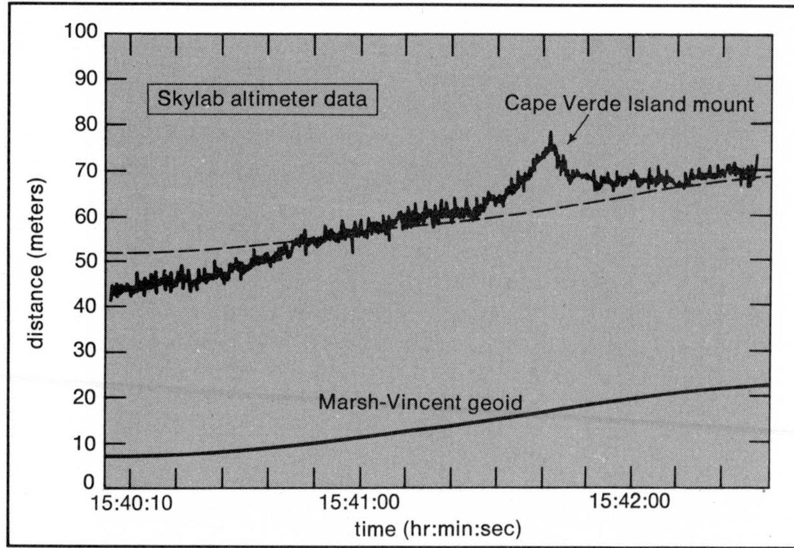

Figure 2. Altimeter data from Skylab during a pass near the Cape Verde Islands. Seamount on the floor of the ocean causes the ocean's surface to move away from the center of the Earth by about 15 m. See text for additional explanation.

Figure 2 shows a pass near the Cape Verde Islands during which the altimeter was always over the ocean. A seamount on the floor of the ocean caused the surface of the ocean to move away from the center of the Earth by a distance of approximately 15 m along this section of the pass. Again the surface of the ocean is level along the smooth curve that will be obtained by drawing an average (dotted) line through the variations of the altimeter measurements.

The other purpose of the altimeter was to measure the speed of the winds over the ocean and to show that such winds could be measured on a systematic basis by a similar, but improved, instrument on future spacecraft. Although the results were preliminary and more refined theories were to be applied to the same data before the final results are published, the experiment demonstrated that it is possible to determine the winds by measuring the amount of radar energy scattered back from the sea surface to a spacecraft. The root mean square difference between the direct measurement of the wind speeds over the ocean and their estimated velocity based on the strength of radar backscatter was less than four knots. This was remarkable because it is difficult to obtain measurements of winds over the ocean to this accuracy, a consequence of the gusty nature of the winds and the problems of using an anemometer on a ship.

In April 1975 a new spacecraft, GEOS-C, was launched with a radar altimeter on it. It took measurements similar to those that have just been described. In a few more years knowledge of the location of the surface of the ocean relative to the center of the Earth should be vastly improved by this program.

Internal waves. As mentioned in the 1975 *Yearbook of Science and the Future*, a session at the American Geophysical Union discussed the problem of internal waves. This resulted in the publication of nine different papers on that subject. Internal waves have been a long-standing and difficult problem for the physical oceanographer. The density of the waters in the ocean varies with depth, and, of course, must increase with depth in order for the ocean to remain stable; otherwise, the ocean turns over and stirs itself. This, in fact, does occur in some places, and it is the reason why the polar regions of the world oceans are so rich in animal and plant life. The turning over that occurs each year brings nutrient material to the surface where sunlight is available so that a chain of biological activity starting with small floating plants can produce the rich fisheries and the abundant krill described previously.

In other parts of the ocean, however, the stable stratification frequently causes the ocean to form layers where the density changes markedly with depth over a relatively short distance. These layers thus form internal interfaces within the ocean. Movement up and down of these layers is similar to the motion of waves on the surface, but in many other ways these internal waves differ markedly from those on the surface. It has been difficult to understand the nature of internal waves and to determine which way they are traveling, how they are generated, and how they are dissipated.

One difficulty with internal waves is that they are observed by a wide variety of different methods, such as towing a string of thermistor chains behind a ship, measuring the change of temperature at a fixed point in the ocean with time, dropping systems down through the water, and others.

Christopher Garrett and Walter Munk sought to reconcile this diverse evidence with a single conceptual model. Their achievement was to interrelate the different kinds of measurements that can be obtained by various techniques and to place them all into one coherent scheme. Although the problem was not completely solved, it was clear that Garrett and Munk had made considerable progress in dealing with it.

—Willard J. Pierson

Energy

Throughout 1974 and early 1975 the United States and many other nations, especially Japan and those in Western Europe, continued to cope with and seek solutions to the world energy crisis and the serious economic problems that it spawned. Even though the five-month Arab oil embargo that precipitated the crisis was ended for the U.S. in March 1974, the major developed industrial nations were left with a 400% unilateral increase in oil prices by the 12-nation group of oil-producing and exporting countries, called the Organization of Petroleum Exporting Countries (OPEC). As of 1975 OPEC controlled more than 60% of the world's most vital and strategic commodity, on which Western Europe, Japan, and, to a much lesser extent, the U.S. depended for their energy supply and economic well-being. By curtailing oil supply, OPEC made its enormous monopoly price increases stick and brought about the most revolutionary transfer of wealth from one group of nations to another group in peacetime economic history.

Until the mid-1960s the United States was the dominant world oil producer and potential emergency supplier to other nations because it had excess oil-producing capacity. This capability and the leading role of U.S. companies in the development of very large foreign oil supplies led to a long period of stable and low-priced energy relative to other sources. Thus, the industrial structure of the world became increasingly dependent on oil, and consumption grew at rapid rates. In the 20-year period from 1954 to 1974, the average annual rate of growth of oil consumption in Western Europe was about 11.5%; in Japan, 14%; in Latin America, Africa, and the Middle East, 7 to 8%; and in the United States, 4.2%.

As world oil consumption increased, the development of new sources in the U.S. lagged, and Middle Eastern and African countries emerged as the largest and most important suppliers of oil to the world. In 1967 the U.S. lost its self-sufficiency in oil supply, and by 1975 the nation's excess producing capacity had dwindled to less than 500,000 barrels per day (BPD). By contrast, the OPEC countries were the source of nearly two-thirds of the non-Communist world's oil supply of nearly 46 million BPD during 1973 and were the only countries with significant (8 million to 10 million barrels per day) excess producing capacity. These developments effectively removed the U.S. as the world price setter and potential source of crude oil to other large consumers during an emergency. Thus, the enormously important power to determine oil prices was moved into OPEC's hands and this organization, in existence since 1960, became a viable economic cartel as a result of its production curtailment during and since the 1973–74 oil embargo.

OPEC exercised this power with a vengeance. By a series of steps, including nationalization, the cartel increased its share of the profits from oil

Depardon from Gamma

Luxury automobiles in a parking area for government officials of Abu Dhabi reflect the prosperity of the oil-rich Middle East.

operations since 1971 from about 60% to approximately 90%. Then, in November and December 1973, the OPEC nations unilaterally increased the price of oil that had sold for about $2.60 per barrel in January 1973 to $11.65 per barrel. By reducing production OPEC maintained this tremendous price increase.

The financial and economic impact of the oil price increase was unprecedented. During 1974 the OPEC countries received an estimated $110 billion in revenue, of which $105 billion was from the sale of oil. This contrasts with total revenues of $25 billion–$30 billion in 1973. Total revenues during 1975 were expected to reach $125 billion. Many of the OPEC countries have small populations (Saudi Arabia, Kuwait, Libya, Iraq, the United Arab Emirates) and despite large expenditures cannot absorb these large revenues. The OPEC countries as a group spent only $28 billion in 1973 and $45 billion–$50 billion in 1974; therefore, they had current account surpluses of approximately $60 billion in 1974. These surpluses for the producers became the deficits the consuming nations had to finance in order to continue their current level of oil consumption. In 1974 the industrial nations incurred a collective payments deficit of $40 billion, while the less developed countries had a deficit of $20 billion.

The growing debt to the producers was being financed through loans and by the sale or mortgage of the productive assets of the oil-consuming countries. Most of the consuming nations reduced their consumption, and this, in turn, reduced their economic activity. The growth rate of the gross national product in most industrial countries either dropped dramatically or actually declined. By 1975 the U.S. and most of the non-Communist world were in the throes of the worst economic recession since World War II.

World energy demand during 1974 leveled off instead of increasing by the usual 6 to 7% per year. In the U.S. total energy demand was 2% less in 1974 than in 1973. Total world oil production rose by only 0.8% during 1974. U.S. crude oil production declined by 4%, and oil product demand declined by 3.3%. Energy conservation programs were adopted and pursued vigorously in many countries.

As a consequence of increased energy prices and in order to lessen their vulnerability to OPEC actions, the principal consuming nations accelerated efforts to develop alternative energy sources. Substantial new non-OPEC oil sources were being developed in the Alaskan North Slope, North Sea, offshore U.S., and other areas. A major shift to nuclear power, coal, tar sands, and other energy sources was being planned.

Project Independence is the U.S. program to achieve these objectives. In November 1974 the Federal Energy Administration, with the assistance of all the major federal energy agencies and with private industry advice, completed the most comprehensive study of alternatives for achieving U.S. energy self-sufficiency ever undertaken. Acting on these findings, U.S. Pres. Gerald Ford in January 1975 proposed a long-range energy program to make the U.S. invulnerable to energy cutoffs of foreign oil by 1985. His proposals called for immediate actions to reduce oil imports by one million barrels per day by the end of 1975 and two million barrels per day by the end of 1977; to reduce energy consumption; and to expand domestic energy production. The program included tough conservation and supply expansion measures; increased taxes and tariffs on crude oil production and imports; a large national security storage program; a massive research program to develop new energy technologies; and a relaxation of environmental protection measures.

For the first time the major oil-consuming nations of the world united in order to deal cooperatively with energy problems. Sixteen nations established the International Energy Agency (IEA) in November 1974 in Paris, and two additional countries joined later. These nations consume four-fifths of the world's petroleum and include the U.S., Canada, Japan, and all the West European Common Market countries except France. They agreed to share available oil supplies in the event of another embargo against the group or any individual member; to take joint action to cut energy consumption; to expand emergency oil stocks; and to cooperate in developing alternate energy sources.

The 24-nation Organization for Economic Cooperation and Development (OECD) completed and published a long-term energy assessment in January 1975. The report concluded that the non-Communist industrialized world has the potential to become essentially energy self-sufficient by 1985 through conservation and increased energy production. This would, however, require drastic reductions in energy consumption and massive energy investments that might conflict with other economic objectives. The OECD estimated that its member nations could become 80% self-sufficient by 1985, a large improvement over the actual energy output of 65% in 1972; however, effective cooperation among the OECD nations would be required to accomplish this goal.

Other significant developments affecting energy during 1974 and early 1975 were: (1) the Energy Research and Development Administration was formed to plan, coordinate, and direct all major energy research activities in a single U.S. agency; (2) construction of the trans-Alaskan pipeline was begun, and up to two million barrels per day of oil from the North Slope were expected to begin to flow to U.S. markets by late 1977; (3) world oil exploration and well drilling operations were greatly expanded, and major new oil finds were reported in Mexico, offshore Brazil, Egypt, Iraq, Ecuador, Peru, Indonesia, China, and in the North Sea; (4) U.S. natural gas production peaked and consumption declined as gas shortages became more widespread and frequent; (5) world coal consumption, despite the energy crisis, increased by only about 2% in 1974; and (6) world electricity generation appeared to have actually declined slightly in 1974.

Petroleum and natural gas

For the first time in recent years world oil demand showed only a slight (less than 1%) annual growth rate in 1974. Pre-embargo world oil demand in September 1973 had reached 58 million barrels per day; however, the average demand was only 55.7 million BPD for 1974. Oil demand in the non-Communist nations, which had risen to nearly 48 million BPD before the embargo, averaged only 45 million in 1974, a decline of 1% from 1973. A nearly 7% increase in petroleum production by the Communist countries offset the decline in non-Communist nations.

U.S. oil demand in 1974 fell by 3.3% to 16.7 million BPD, the first decline since World War II. Crude oil and natural gas liquids production declined by about 4%. Imports averaged nearly 6.1 million barrels per day, less than 2% below those of 1973.

The reduced oil demands resulted from the supply disruptions caused by the Arab oil embargo and production cuts early in 1974, and from the curtailment of consumption due to the tremendous oil price increases. These reduced demands forced voluntary production cuts in OPEC countries amounting to nearly 10 million BPD by March 1975. These restraints were necessary to enable OPEC to sustain its unilateral price increases; the excess revenues of the OPEC nations permitted them to take these actions.

As world oil demand slackened in 1974, it was reflected in a reduction in the output of the OPEC countries, which declined for the first time by about 1% to a level of 30.7 million BPD. The decline was uneven, with the Middle East producers showing a gain of about 3% while output dropped by 8% in the other regions. The largest declines were in Libya, Venezuela, and Kuwait.

Spurred on by the high oil prices and the efforts of consumer nations to diversify their supply sources, oil and gas exploration and development were accelerated throughout the world. The early results were encouraging. Development proceeded at a fast pace in the North Sea, where an estimated 40 billion-barrel reserve was located. The first production from this area was expected to reach the U.K. in the spring of 1975. Production from Norway's Ekofisk field expanded greatly on the completion of the 220-mi, one million-barrels-per-day pipeline to the U.K. Before completion of the pipeline this field was producing about 200,000 BPD, enough to make Norway self-sufficient.

Sverre Borretzen from Camera Press/Photo Trends

Oil rig off the coast of Norway. Production from Norway's Ekofisk field in the North Sea increased considerably during the past year.

An estimated 10 billion-barrel oil reserve was reported to have been discovered in Mexico's Reforma trend in the southern part of that country. As a result, Mexico was able to expand oil production by 16% and resume exports for the first time since World War II. China reported a 30% increase in output and expanded exports to Japan to about 200,000 BPD. Brazil reported its largest oil discovery ever in the offshore Garoupa field, and Iraq announced a large multibillion-barrel oil discovery on the edge of Baghdad. Other significant oil finds were reported in Egypt, Ecuador, Indonesia, and in India's offshore Bombay area.

Capital expenditures by the petroleum industry in the non-Communist world reached an all-time high of $31.7 billion in 1973. Nearly one-half of this spending was for finding and developing new fields, and more than one-third of the total investment was in the U.S. Even this record rate was expected to be exceeded in 1974 and 1975.

In the U.S., oil- and gas-well drilling increased by nearly 20% in 1974, which represented a five-year peak. New wells drilled totaled almost 32,000, and nearly 9,000 of these could be classified as exploratory. More than 35,500 wells were to be completed in 1975, approaching the capacity of the available drilling rigs and equipment. Net growth in crude oil refining capacity amounted to 616,000 barrels per day in 1974.

Estimates of world reserves of natural gas in 1975 ranged from 2,000 trillion to 2,500 trillion cubic feet (Tcf). This is equivalent to about a 50-year supply at the current world rate of marketed natural gas consumption of approximately 46 Tcf. In 1974 the U.S. continued to account for nearly one-half total consumption, while the U.S.S.R. and its eastern European satellites remained in second place with about one-fourth of the total. Consumption increased in Western Europe as important new natural gas supplies were developed in the North Sea. Italy's state oil company discovered large new gas reserves in very deep zones at the Malossa field, located 13 mi E of Milan. First production awaited completion of a gas-treating plant, expected in 1976.

Conventional U.S. natural gas production peaked in 1974, and consumption declined by nearly 4% as shortages and curtailments of supply became frequent. Curtailment of deliveries to firm industrial customers affected about 10% of the 1974–75 heating season supply. The Federal Power Commission (FPC), which regulates interstate gas sales, established priorities that allocated available supplies to residential and commercial customers ahead of industrial users.

In an effort to stimulate and accelerate the development of new natural gas supplies, the FPC issued orders in December 1974 that permitted the national wellhead gas price to increase to 50 cents per thousand cubic feet (Mcf) and eliminated the "old" and "new" distinctions previously used to establish gas prices. The FPC stated that this equal treatment of gas would result eventually in a uniform base price that could be equated to the cost of replacing the gas consumed. The U.S. government proposed complete deregulation of natural gas prices in order to provide the incentives to curtail excessive demand and accelerate the development of supplies.

Escalating gas prices were expected to improve world gas supplies. The U.S.S.R. and Western European nations concluded agreements for increasing deliveries of pipeline gas. Also, imports of liquefied natural gas (LNG) were expected to increase in response to the resumption of LNG deliveries from Libya and new agreements for Algerian LNG supplies. Plans for the development of pipelines to permit delivery of large quantities of gas from the Alaskan North Slope and Canadian Arctic areas to U.S. and Canadian markets were under active consideration.

Smokeless briquettes processed from lignite are being used in Turkey's capital city of Ankara as a countermeasure against air pollution.

Courtesy, United Nations

Coal

The development of coal as an alternative to the shortages and increased price of petroleum received increased attention throughout the year. Even so, 1974 world coal production and consumption increased over 1973 by only an estimated 2% to about 3.3 billion tons. Moreover, this increase occurred mainly in the Communist countries, which together produced an estimated 1.9 billion tons, or 58% of the total.

Despite the energy crisis the European Economic Community (EEC) countries experienced a 10% decline in coal production during 1974, totaling only 243 million tons. The British coal miners' strike early in 1974 contributed to the decline. By the end of 1974, EEC coal output had recovered somewhat. U.S. coal production declined by nearly 1%, to about 590 million tons in 1974. A strike in November and December 1974 during wage negotiations resulted in an estimated 35 million-ton loss in production.

World coal trade continued to expand, exceeding 170 million tons in 1974. The U.S. remained the foremost exporter, accounting for nearly 30% of world exports. Poland, the U.S.S.R., Australia, West Germany, and Canada were the other major exporters. Japan, France, Canada, Italy, West Germany, Belgium, the U.S.S.R., Bulgaria, and Czechoslovakia were the principal importers.

Although coal is the largest energy resource in the U.S., its share of the nation's energy supply remained about the same in 1974 as in 1973. Expansion of coal-mining capacity was inhibited by uncertainties generated by the coal strike, by tightened supplies of oil and gas, and by unresolved environmental issues pertaining to air pollution and surface mining. The Federal Energy Administration proposed rules that would carry out a congressional mandate to prohibit certain power plants and other major fuel-consuming facilities from burning natural gas or petroleum and require them to burn coal. New plants would be required to be designed and constructed to use coal as their primary fuel.

To speed up the development and use of coal as a clean boiler fuel, the Office of Coal Research of the U.S. Energy Research and Development Administration awarded several contracts for the construction of demonstration plants to determine the commercial feasibility of technologies for the conversion of coal to liquid and gaseous fuels.

Electric power

Consistent with the general decline in demand for energy and fuels, world electric power generation and consumption leveled off during 1974 at about 5.5 trillion kw-hr. Although precise data were unavailable, it appeared that the 8% average annual increases recorded for many years by the world's major power consumers might have actually declined slightly in 1974. Electricity generated by U.S. utilities in 1974 declined by less than 1% from the 1,848,000,000,000 kw-hr generated in 1973, according to preliminary estimates.

As in the past, most of the electricity was produced in steam-electric plants using the fossil fuels (coal, oil, and natural gas) as the primary heat source. These fuels continued to account for about 75% of power generation, while hydroelectric sources (about 23%) and nuclear plants (2%)

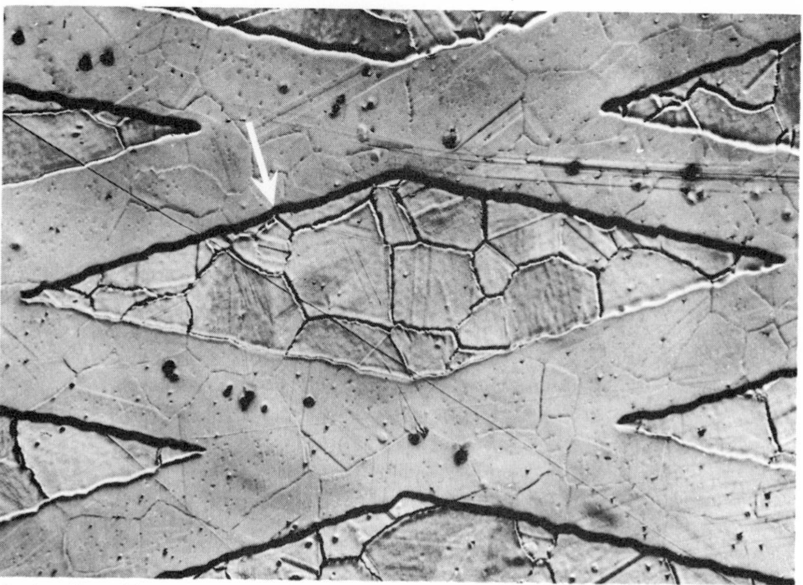

Courtesy, General Electric Research and Development Center

Exposed areas (diamond-shaped regions) of a stainless-steel sample bombarded for one hour by atomic particles traveling at speeds up to 2,500 miles per second swell and rise to form plateaus up to 80 millionths of an inch high. Stainless-steel cores of fast-breeder nuclear reactors would experience the same effect after a year of operation.

Courtesy, "Smithsonian," photo by Ralph Crane

Exhibit of Indian village life, on the steps of the Palace of the Republic in Bucharest, Romania, served as a reminder to the delegates at the World Population Conference of the problems they were considering.

the last several decades. The U.S. situation with respect to domestic supplies of oil was deteriorating. Between 1970 and 1973, U.S. crude oil input to refineries rose from 4,252,000,000 bbl to 4,845,000,000 bbl while, during the same period, U.S. proven crude oil reserves fell from 39 billion to 35.3 billion bbl and the lifetime of proven reserves at the current rate of use declined from 9.17 to 7.29 years. Reserves dropped to 34.3 billion bbl in 1974, despite a 19% increase in drilling. Thus it is genuine resource depletion rather than inadequate exploration that is causing the shortage.

Environment and the world economy. The environmental implications of the increased interdependence of national economies became more apparent during the year. As each nation began to run out of various critical commodities, it had to make up the deficit by expanding its trade with other countries. The political price paid for this was a loss of control over one's own destiny, but there were also marked effects on the domestic environmental system. Thus, the increase in U.S. exports of various food items brought about by the need to pay for imported crude oil forced up not only the domestic price of food but also the price of farmland relative to city land. This seemed likely to result in a lower rate of conversion of farmland to urban uses, higher urban densities, greater volumes of traffic flow on main transportation arteries, and higher relative economic attractiveness of public, as opposed to private, transportation.

To illustrate the magnitude of the change, in 1970 the U.S. exported $42.6 billion worth of goods (all categories) while importing $3.1 billion worth of petroleum. In 1972 these figures rose somewhat, to $49 billion for all exports and $4.8 billion for petroleum imports, but by 1973 they had ballooned to $70.2 billion and $8.1 billion, respectively. Preliminary estimates indicated that oil imports in 1974 cost the U.S. $24 billion. The deeper implication in these figures was that the U.S. economic system was becoming increasingly tightly linked to the global economic system and, in general, the larger the system, the greater will be its instability.

National and global systems instability. Perhaps much of the explanation for the growing concern of environmental scientists with systems instability lay in recent events. Many variables were showing a tendency to instability rarely seen before; national and global unemployment rates, indices of industrial production, sales (*e.g.*, of automobiles), prices, international movements of commodities, and inflation rates were examples. Many environmental scientists were focusing their attention on those properties of large systems that affect their stability. The key research question was: How do you manage a particular system so as to keep fluctuations of the system within desired bounds?

Scientists concerned with environmental issues were beginning to see that the decision-making processes of institutions and the way in which information flows into and out of institutions are key components of environmental systems. In general, the response of national and international institutions to environmental crises during the year was not reassuring. The 140-nation Law of the Sea Conference in Caracas, Venezuela, in the summer of 1974, the World Population Conference in Bucharest, Romania, in August 1974, and U.S. Pres. Gerald Ford's state of the union address all showed the same flaw: emphasis on short-term, political considerations rather than on long-term goals, objective assessment of the facts, and development of rational policy based on the facts. The World Food Conference in Rome in November 1974 was similarly disappointing.

Two documents released in 1975 illustrated the gulf between policymakers and scientists over what constitutes an appropriate strategy for the future: President Ford's state of the union address, released on January 15, and the National Academy of Sciences' 348-page report "Mineral Resources and the Environment," issued on February 11. President Ford urged a sharp reduction in foreign oil imports by 1985 and development of the nation's energy technology and resources so that the U.S. could supply a significant share of the energy needs of the non-Communist world by the year 2000. Also, he "envisions" the construction of 30 major new oil refineries within the next ten years. He emphasized energy conservation but did not mention mass transit or solar heating, although these offer the best hope for effecting a substantial reduction in national energy use in the next decade.

Two main types of relationship exist between gross regional product per capita and the ratio of energy consumption per capita to income per capita. For centrally planned economies (A) the emphasis on heavy industry results in an early peak in energy consumption. As the gross regional product per capita increases for these economies, as in the case of Eastern Europe, the relationship approaches that prevailing in the other regions of the world (B). Energy demand, thus, depends on population, the level of economic activity, and the degree of industrialization.

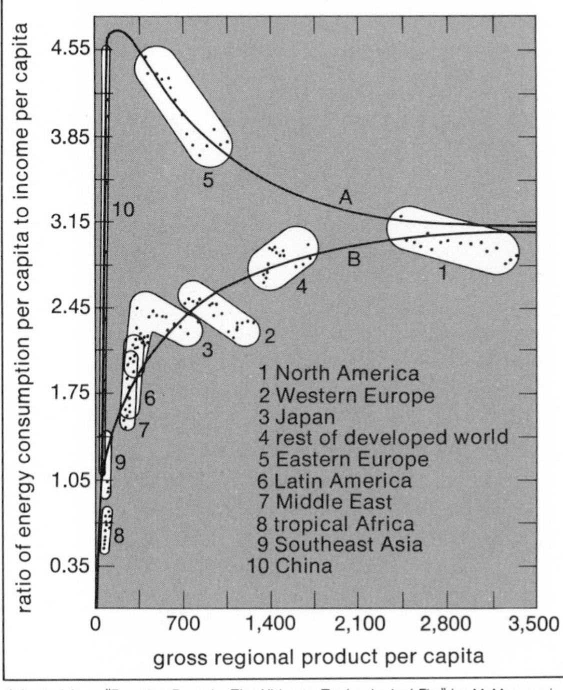

Adapted from "Faustian Bargain: The Ultimate Technological Fix," by M. Mesarovic and E. Pestel in "New Scientist," pp. 724–727, March 20, 1975

The National Academy of Sciences document, on the other hand, indicated that the government's goal of energy self-sufficiency by the 1980s is essentially impossible. Further, the report determined that actual U.S. reserves of oil and gas were, at best, only one-half as large as had been estimated by the U.S. Geological Survey. Even including undiscovered but presumably recoverable deposits of oil and gas, the NAS concluded that the U.S. would run out of these commodities by 2000 or earlier, not 2015–35.

Internationalization of environmental science. A notable trend in environmental science was the internationalization of research and education. For example, UNESCO sponsored a *Curso Internacional de Análisis de Sistemas en Ecología* in Caracas. Classes and research were conducted in two languages by lecturers from five countries, and the students were from nine countries.

The influence of the Club of Rome was growing to an extraordinary degree. The first, highly preliminary report of research sponsored by the club was the controversial *The Limits to Growth*, which by the middle of 1974 had sold three million copies in 27 languages. Subsequently, a number of new research teams began working on global or regional computer models, extending the methodology and adding more detail and realism to the models.

In addition to computer-based models, extremely wide-ranging world economic-environmental-political questions were being considered by international teams at the International Institute for Applied Systems Analysis near Vienna; the International Federation of Institutes of Advanced Study, a group of 20 research institutes; and the East-West Center in Honolulu. Major teams had undertaken large new research projects in Canada, Argentina, and Japan.

However, the most important second-generation output from the club was probably *Mankind at the Turning Point*, by Mihajlo Mesarovic of Case Western Reserve University, Cleveland, Ohio, and Eduard Pestel of Hanover University in West Germany, published in 1974. The new computer model on which it was based was constructed by a team of 50 researchers from nine countries. Unlike the *Limits to Growth* model, which lumped together data for the entire world, the new model simulated developments in ten separate regions, together with the interactions between regions. The output from the model suggests that to avoid various catastrophes, the world must now engage in international cooperation on a grand scale, sharing capital, energy, food, and technology, and that solar energy and hydrogen fuel should be developed as important energy sources. This contrasts strikingly with present national policies.

Some major developments of the last year or so were either stimulated by the Club of Rome or were running parallel courses. Most universities, government, industry, and most conventional large organizations are organized so as to split up the problems of the institution into "manageable" units. Thus, there are subunits within a given organization working on population, agriculture, energy, mineral resources, housing, transportation, land use, the weather, economics, political science, forestry, and so on. A key result of the Club of Rome's activities has been to demonstrate conclusively that this approach to problem solving, in which each component is examined out of context, absolutely guarantees inadequate solutions. But existing organizations, for the most part, are unable to make the organizational changes that would enable them to look at the whole problem.

To fill the need, many new organizations, new types of organizations, or even ad hoc groups were organizing to take a critical look at society from a holistic point of view. For example, the Center for Intercultural Documentation in Cuernavaca, Mexico, with which the well-known social critic Ivan Illich is associated, examined the relationship between the amount of energy use in a society, the speed of vehicles, and the equity of income distribution, and found that high speed and equity are incompatible. The Fundación Bariloche in Argentina, the International Institute for Applied Systems Analysis near Vienna, the East-West Center in Honolulu, and the Center for the Study of Democratic Institutions in Santa Barbara, California, were among the many other groups engaged in similar studies, often using highly quantitative approaches and computer simulation as aids.

Not only were these new types of organizations cutting across disciplinary boundaries, they were also examining problems in a novel way. Their point of view was sufficiently comprehensive to include not only the substantive aspects of all the problems considered, jointly rather than severally, but also the interaction of these problems with the system of human policymaking and leadership-selection procedures. In some studies, interaction between the belief systems of cultures and the resultant effect on the socioeconomic-political-environmental system is explicitly the central target of the research.

Scientific developments. The stability of large systems has become a dominant research issue in environmental science, but the research is of limited practical value unless the equations used are realistic descriptors of the system being studied. The most successful research of this type requires a wide and deep grasp of mathematical stability theory and of the details of the subject matter. To provide this, a number of teams have been formed comprising one or more mathematicians and one or more biologists.

Two such teams were already demonstrating how productive this interdisciplinary union could be. Myron Fiering, a Harvard systems engineer-applied mathematician interested in environmental systems, and C. S. Holling, a University of British Columbia systems ecologist concerned with resource management, began collaboration

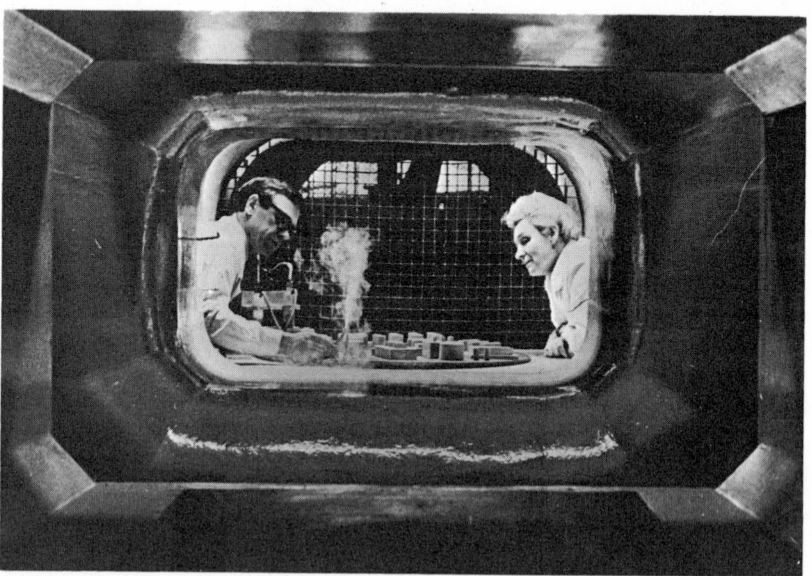

Soviet scientists study air swirls produced by a wind tunnel that is linked to the model of a factory. By using such a model the scientists hope to determine the speed at which air pollution spreads.

London Daily Express from Pictorial Parade

at the International Institute for Applied Systems Analysis. They were interested in developing techniques for the discovery of systems-management policies that are optimal from the point of view, jointly, of biological systems and the goals of society. A particular emphasis was the search for policies that maintain the stability and resilience of systems (the ability to return to equilibrium after a perturbation). Given the current evidence on instability and diminished resilience, this work could not be more timely.

Another international team studying stability included Robert May of Princeton University, an ecologist whose background was primarily in theoretical physics, and G. R. Conway, M. P. Hassell, and T. R. E. Southwood, all systems ecologists at the University of London. Their work was based both on the stability theory of systems of differential equations and on realistic equations describing the behavior of insect parasite-host systems. It was opening up an important new scientific domain, as well as a new era in the control of insect pests and other commercially important biological populations. The main theme of this research was to discover the sets of properties that insect parasites and predators must possess in order to stabilize the system that includes them, the pest insect species, and the plants supporting this animal system that are important to man.

Two recent reports of fieldwork had an important relationship to this study. In reviews of many insect pest-control problems in the tropics, B. J. Wood showed that pesticides may actually cause a pest problem unless they are used very selectively and with an understanding of the ecology of the pest. In a number of cases (*e.g.*, cocoa in Sabah and Ghana; tea in Sri Lanka, Malaysia, and Sumatra; oil palms in West Malaysia; and cotton in Peru), the cause-effect relationship between unwise use of pesticides and pest outbreaks was documented. Removal of the pesticide eliminated the pest; when spraying was resumed in a given area, the pest reappeared there and also spread to surrounding areas. This evidence, disquieting in itself, was even more disturbing when viewed against a background of mounting evidence indicating that many problems supposedly being solved by technology could in fact be solved by removing the existing technology (and, in some cases, by substituting alternate technology that may be much more sophisticated but less highly publicized).

Another important study on the ecology of insect host-parasitoid communities was reported by Don C. Force. For years considerable controversies have raged over two topics in ecology. One concerned the diversity-stability hypothesis: Was it true that an increase in the number of species in a community made the species populations fluctuate less violently? The other concerned the appropriate strategy for the biological control of pests: Was control best accomplished by introducing one species of parasitoid or predator of the pest, or several? Force's experiments and field data provided critical new evidence on both topics. Through a combination of field sampling and laboratory experiments, he showed that where there are several species of parasitoids, the percentage of parasitism is lower than where there is one effective species of parasitoid.

This runs counter to intuition and suggests that biological control might work best where there is only one species of parasitoid. Further, it suggests that an increase in the number of species of parasitoids, by causing an increase in the stability among the parasitoid species, lowers their ability to diminish the amplitude of population fluctuations in the host. Thus, an increase in diversity of species within the whole host-parasitoid system in this case causes a decrease, not an increase, in the stability of the system. This result is particularly noteworthy because it is compatible with some theoretical findings by Robert May, to the effect that increased species diversity does not necessarily produce increased population stability.

Methodological developments. For about two decades, ecologists and environmental scientists have placed increasing reliance on computers to deal with the data resulting from massive field and laboratory studies and to construct complex mathematical models. However, the great cost of computers—for either purchase or rental—has always been a problem for a field not as well endowed as some others. With the appearance of new microcomputers costing a few thousand dollars or less, many of which have the memory and arithmetic capabilities of much larger machines, the difficulty appeared to be solved. It seemed likely that within a few years there would be many computers available, as small as or smaller than a book, into which enormous data files could be fed on cassette tape.

Another noteworthy development was the introduction into ecology of the new curve-fitting methodology developed by J. Matyas. Instead of using analytical-derivative based methods to fit nonlinear equations on computers, this technique utilizes "adaptive random search" methods, which are quicker and simpler to use in most problems of interest to environmentalists. Thus, within a single year, new hardware (microcomputers) was linked with new mathematical routines.

Critical writings. The last few years have been marked by the appearance of a body of writings,

Courtesy, Messerschmitt-Bölkow-Blohm GMBH

"Mufflers" for jet engines, developed by the West German firm Messerschmitt-Bölkow-Blohm, are used during testing after engine repairs or inactivity. They have reduced the noise level by about 23 dB.

strongly critical of science and technology, that comes from inside science, not from without. This literature is relevant to environmental science because it argues that many of the problems occurring in the environment have their root cause in the nature of science and technology itself. One argument is that technology that works well on a small scale (transportation systems, sewerage systems) is doomed to failure when the scale becomes too large because of the increased probability of failure or congestion. Another criticism is that vested interest groups can "capture" science and technology and promote the use of inappropriate methods, such as broad-spectrum insecticides or the freeway-automobile system, that may aggravate the very problems they are supposed to be solving.

A far more penetrating criticism is that the interacting system of science and technology has internal problems related to its own, little understood, inner dynamics. According to this line of reasoning, science and technology gradually become more and more divorced from real-world problems and begin to take on a life of their own. The pretense that they are value free actually conceals some strongly held values; for example, the justification for a particular type of science and technology is to breed the need for more of the same, at a constantly increasing rate.

Such criticisms might not command much attention, except for two things: to an increasing degree, they are coming from technical experts and concern subjects in the writers' fields of expertise, and research in a number of problem areas seems to bear out the criticisms. In transportation, energy generation, and agriculture, to name only three, trends in research appear to grow out of past trends in research rather than in response to social needs.

—Kenneth E. F. Watt

Food and agriculture

National and international attention was focused in 1974–75 on food supplies and agricultural production as a result of a number of crises, including droughts in Africa, depletion of grain reserves, and rising food prices. Food attained a new importance in international trade and relations. The 1974 World Food Conference, held in Rome under the aegis of the United Nations Food and Agriculture Organization (FAO), laid plans to deal on an international cooperative basis with problems of supply, demand, and distribution relative to food production.

The world food crisis

The situation is extremely complex and further complicated by the growing awareness that there exists today a world market for foodstuffs that had not previously been so affected by the vagaries of international politics. In addition, there is confusion on the part of individuals about how they can act effectively in their concern for the hungry. Giving up one meal a week will not supply an African mother with her needed caloric intake, nor will substituting chicken for beef provide food for an undernourished South American farmer. At the same time, however, there is a growing awareness that certain dietary preferences are becoming things of the past as priorities about grain consumption for humans and cattle are realigned. In the U.S. people are coming to terms with problems that are an order of magnitude larger than usual and thus are beginning to realize that the price of tea in China is no longer irrelevant.

Although the world food picture is particularly grim in areas such as Africa, the overall perspective offers some grounds for hope. While one person in six worldwide is undernourished and the daily caloric consumption in less developed countries is still only about two-thirds of the minimum requirement that has been recommended by the FAO, a 2% increase in world grain production, properly distributed according to need, would close the gap. Twice the quantities of grain moved in aid programs since 1965 would eliminate the worst aspects of malnutrition.

Although the world now uses only one-half the land area potentially suitable for crop production, most of the additional land lies outside densely populated countries. In addition, significant inputs of energy would be required to provide market access, water, and the initial capital necessary to bring these areas into production. Therefore a substantial part of future food production gains will have to come from yield-increasing techniques: appropriate fertilizers, improved seed varieties, better cultural practices, and irrigation. In less developed countries, however, increases in population have literally eaten up any food production increases so that there has not yet been a net gain in caloric intake per capita. Distribution, the link between supply and demand, must also be addressed if national and international inequities are to be removed. In the U.S., for example, 62% of each dollar spent by consumers at the retail level for food normally goes to pay for processing and marketing.

U.S. production potential. The U.S. Department of Agriculture (USDA) has studied the potential for domestic food production and finds that U.S. farmers have the capacity to increase their outputs of major agricultural products significantly. Given certain conditions, by 1985 farmers could achieve a 50% increase in feed grain production, a 33% increase in soybeans, a 44% increase in beef cow numbers, a 30% increase in cotton, 400% in peanuts, and 200% in rice. These are potentials and depend on: (1) farm prices favoring increased production; (2) no restrictions on land use; (3) adequate and reasonably priced supplies of gasoline, fertilizers, and other agricultural chemicals; (4) normal growing conditions; and (5) agricultural research commensurate with national and world needs.

Increasing food production. Although agricultural research is conducted around the world, by far the largest investor is the U.S., both internally for domestic production and as a substantial part of its international aid program. In the United States, agricultural research investments from federal, state, and industrial sources approach $1 billion per year, an indication of its high priority in the minds of legislators, businessmen, consumers, and farmers.

Crops. While insecticides, fertilizers, herbicides, and other agricultural chemicals are an integral and continuing part of the U.S. food picture, scientists are working to reduce their use to a minimum. By careful genetic selection, plants can be made to respond to farmers' needs. Nevada Synthetic XX alfalfa, for example, has high host resistance to five major pests and is the first alfalfa with high

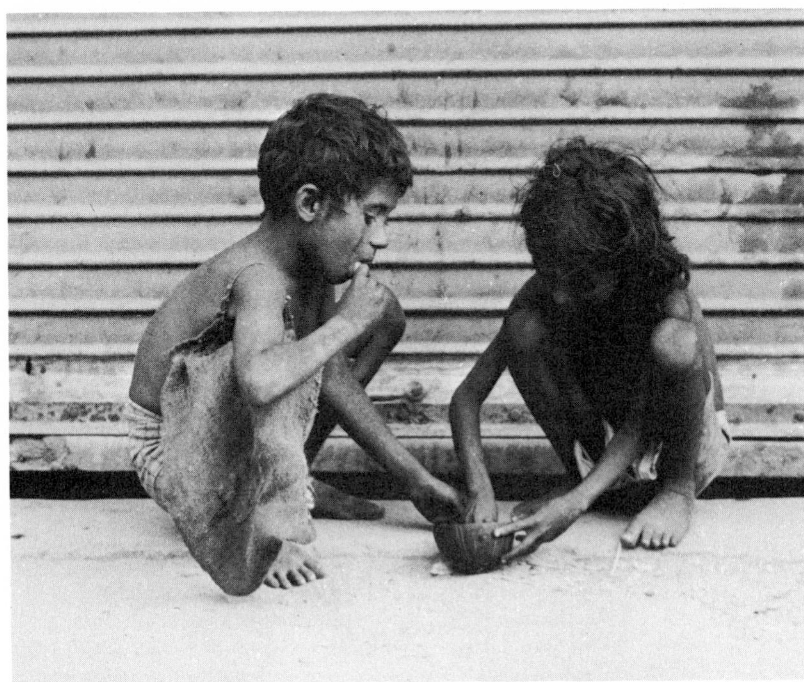

J. P. Laffont from Sygma

Hungry children share food in Dacca, Bangladesh. Severe flooding in Bangladesh in July-August 1974 helped cause a serious food shortage there.

resistance to Northern root-knot nematode. Host resistance is sufficient to resist the most intense infestations of these particular pests and makes the use of pesticides unnecessary for their control. Because of this and related research it has been estimated that insect-resistant varieties return $300 to $600 for each $1 invested in research.

Most persons are aware that insects and diseases continually modify themselves to adapt to changing environmental conditions, including man-made pesticides. The importance of continual vigilance in monitoring these changes can be demonstrated by the discovery at Auburn (Ala.) University of the resistance of *Cercospora* (the leaf-spot organism which attacks peanut plants) to the fungicide Benlate (benomyl). The manufacturer of the fungicide was advised of this and confirmed the observations. The chemical was removed from the approved list, and other companies were able to increase supplies of alternate products sufficiently to meet 1974 needs. Quick action by scientists, businessmen, and farmers saved many millions of dollars for Alabama producers alone, plus additional millions for producers in other states who shared the benefits of the Auburn scientists' alertness. The final beneficiaries, of course, were the consumers, who were spared unnecessary increases in food prices.

While new grain and field crop varieties with superior genetic characteristics are routinely released from agricultural experiment stations and the U.S. Agricultural Research Service, delay between release and widespread adoption of new strains has been a limiting factor in yield improvement. The timely release of several drought-resistant strains prior to the extensive spring drought in the mountain and plains states during 1974, coupled with rapid adoption, was instrumental in providing a near-record wheat crop despite delayed rainfall.

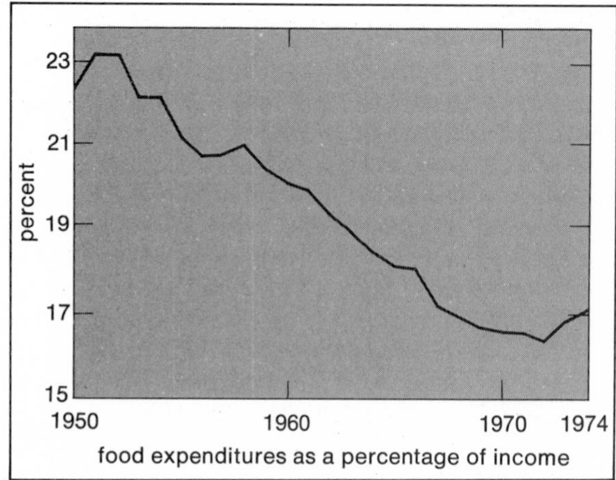

food expenditures as a percentage of income

Courtesy, U.S. Department of Agriculture

The wheat story is illustrative of a basic fact of agricultural research. There are no instant research results. The capacity to respond rapidly to environmental crises depends, in large part, on continuous, day-to-day research which may temporarily show no applicability. The other element in increased yield is the farmer, who must be alert to innovations. Alertness of this kind is responsible for the adaptation of midwestern no-till corn and soybean techniques to potato growing in the Northwest. Scientists at Washington State University have developed a "mini-till" potato planter that ends tillage-related wind erosion, reduces nitrogen leaching, improves water infiltration, and improves the condition of the soil while saving three gallons of gasoline per acre per year.

Crop research includes more efficient and effective use of resources. For instance, discovery that excessive use of some types of nitrogen fertilizer actually decreases sugar content in beets has reduced fertilizer applications for this crop. Since many crops require drying before storage, the substitution of direct solar energy for fossil fuels is under investigation. South Dakota State University and several other universities are pursuing projects in the conservation of farm fuels. Water conservation is another element of resource management essential for increased farm production. Computer-managed irrigation systems, increased use of mulches, and trickle irrigation conserve water and provide more reliable harvests by reducing the effect of changes in the natural water supply.

Processing and marketing research is still another part of the agricultural research picture. Crops in the field must be processed and marketed. For example, the U.S. produces three-fourths of the world's soybeans but approximately 90% of our soybean meal is consumed by livestock. Both the world food situation and technical evidence argue that livestock neither need nor should consume so valuable and complete a plant protein. However, unless methods can be developed that will make soybeans and soybean products acceptable to the American palate, much of the potential for high-quality protein supplementation in the human diet will be lost.

Fishing and mariculture. Until recently it appeared that the world might be able to alleviate the food crisis by using the seemingly inexhaustible resources of the oceans. Although we now recognize that oceans cannot be robbed but must be farmed in much the same manner as land, popular acceptance of marine agriculture has been slow. Recent spectacular decreases in previously reliable catches have focused attention on one of the newer and more exciting aspects of agricultural research. Although the use of water and marine

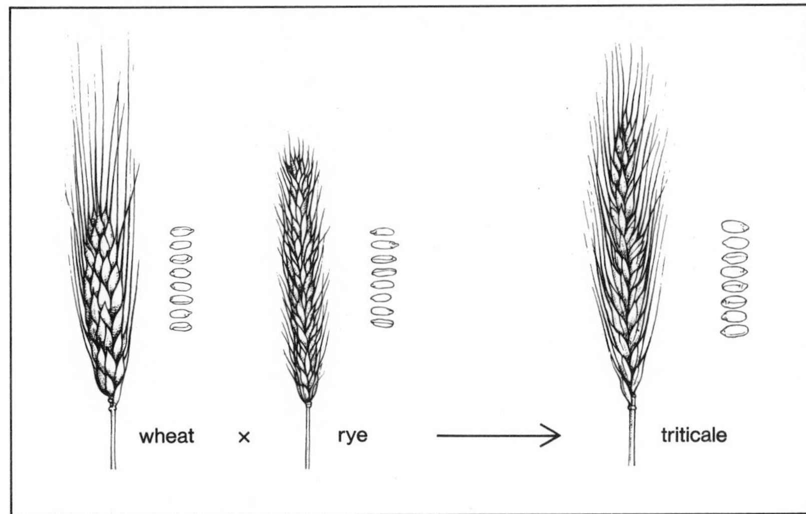

Triticale, a hybrid of wheat and rye, combines the high yield of the former with the ruggedness of the latter. In 1974 it was grown on more than one million acres in 52 countries under environmental conditions ranging from the prairies of Canada to the foothills of the Himalayas.

resources in the food crisis is not a panacea, water-grown foods can become a permanent and significant portion of the diets of all nations.

Marine agriculture is being approached in a variety of ways—from genetics and disease control to processing and crop management—in a variety of places. In New Hampshire, for example, scientists are attempting to hybridize Pacific and Atlantic salmon in order to shorten time to mature Atlantic-adapted salmon. In Oregon two new vaccines have been developed, one injected and the other water-carried, which control vibriosis, a disease that limits aqua farming and hatchery programs. In Louisiana techniques for canning catfish are being developed that prevent undesirable texture and off-flavors. One of the most common and adaptable fish in the U.S., the catfish has not yet become a commerical product because of problems in processing. Recent experience in the Delaware Bay indicates that overfished or polluted marine areas can be reclaimed if carefully managed. The harvest of clams there in 1974 was the greatest since 1956. In addition, farmed marine areas can be protected from floating oil slicks by the application of certain chemicals that increase biodegradation tenfold.

Meat production. Because of growing concerns about grain shortages and world food crises, Americans became conscious of the beef industry, which provides the hamburger and steak in grocers' cases. The beef industry and agricultural scientists responded by stepping up research on more efficient methods of bringing meat to the table. Among the most important findings is the discovery that breeding cattle do better when fed less. Cows maintained at restricted levels of caloric intake survived longer and produced more calves than those maintained at higher caloric intake. As a consequence, cows are allowed to lose weight over the winter, reducing feed and labor costs. In addition, better understanding of the effects of early withdrawal of protein supplements, in some cases as many as 100 days before slaughter, and a general reduction of concentrates have reduced grain use for cattle that are being fattened in feeder lots. The substitution of low-cost roughage and more effective utilization of native pastures have also reduced feed-to-beef ratios without affecting quality.

The use of grass-fed beef will increase in the future, as roughly half the land surface area of the U.S. is rangeland or permanent pasture and is suitable for no other food production use than grazing. The conversion efficiency of range beef compares favorably with wheat, corn, and soybean conversion efficiencies. The number of calories available for food consumption in each of these crops and in range beef is in excess of the number of "cultural calories" (fossil-fuel calories) needed for production. The bulk of the energy used to produce range beef comes directly from the Sun and is captured through photosynthetic conversion in grasses.

Several techniques have been developed to increase palatability of range-finished beef. In a sophisticated version of grandmother pounding the beefsteak, vibrating needles are inserted into beef carcasses, breaking down tough connective tissues. Until recently it was thought that the relationship between fat beef and tenderness was direct, and an effort was made to increase fat marbling to produce a more tender product. It now appears that the connection was indirect and that increased tenderness resulted from the insulatory properties of the fat, which slowed down the cooling of the carcass during aging. As a result a new technique called "cold shortening" can materially increase tenderness in lean carcasses and reduce the need for grains in finishing. A technique developed at Texas A&M University, which requires no additional equipment or energy inputs but only a simple change in handling, hangs carcasses from the aitchbone (pelvis) instead of the hind shank, and a 20% increase in tenderness is achieved.

To free grain for direct human consumption, substitutes for traditional cattle rations are being sought. Grass seed straw and wheat straw can be treated to double their digestibility, alfalfa can be fractionated to improve quality, and new plants such as meadow-foam can be grown where conventional fodders cannot flourish. Feedlot waste can effectively supplement up to 15% of conventional rations. One of the more interesting new fodder supplements is wood pulp, which can supply a major portion of carbohydrates in cattle and sheep diets. With grain prices as high as $120 a ton, a commercial product that markets for $35 a ton is very attractive. In addition, consumption of this fodder by cattle and sheep alleviates pulp and forest slash disposal problems and helps to reduce pollution problems.

Plant pathologist at the International Maize and Wheat Improvement Center in Mexico injects disease inoculum into corn to test its resistance.

Courtesy, Centro Internacional de Mejoramiento de Maíz y Trigo

As a part of the drive for more efficiently produced protein, breeding and reproductive techniques are being intensively studied. At Louisiana State University scientists are working on several techniques for inducing multiple births. In a recent experiment, 62 calves were produced from 48 cows, a 129% calf drop compared to the usual 75–80%. At Colorado State University George Seidel is developing a method that will allow cattle breeders to speed up the dissemination of favorable genetic characteristics. This is accomplished by allowing prize cows to give birth, by implantation of fertilized ova into incubator or "brood" cows, to 30 or more calves each year.

Dairy and poultry. While the role of cholesterol in the human diet is still undetermined, researchers at the University of Georgia have demonstrated lowered levels of cholesterol in eggs through genetic selection. Previous attempts to lower cholesterol in eggs have been made through dietary manipulation.

Studies undertaken in the highlands of Puerto Rico demonstrate that dairy cattle can be carried on well-managed tropical grass pastures. Even when grazed at the rate of one cow per acre, cows produced high milk yields throughout successive lactations. Indications that tropical grass forages may be more digestible than previously thought and that cow nutritive requirements in mild tropical climates are lower than in temperate regions open up many possibilities for new protein sources in the protein-poor tropics.

Horticulture. While the world's attention has been focused on grain supplies and U.S. attention has been focused on the price of beef, other researchers have been working toward increased yield and higher quality in fruits and vegetables. In industrialized and less developed countries alike, 25% of the calories and many essential nutrients come from fruits, vegetables, seeds, and starchy plants.

Traditionally, pears and apples were planted at a density of 30 trees an acre. More recently better results were obtained by a tenfold increase in density to approximately 300 trees an acre. However, a recent breakthrough in the Northwestern U.S. indicates that another tenfold increase in density produces even better results. Trees are planted in rows and pruned to a Christmas-tree or pyramidal shape, which maximizes the sunlight falling on each leaf. In test runs with as many as 3,600 dwarf trees per acre, yields for pears increased from 6 to 8 tons per acre to 20 tons of high-quality fruit, and apples showed a sevenfold increase in production from 10 tons per acre to a phenomenal 74

Oregon Agricultural Experiment Station, Oregon State University

Red Delicious apple trees are planted densely to form a hedge and then are pruned like Christmas trees to afford maximum exposure to the Sun.

tons with no loss of quality. While existing orchards cannot be changed to the high-density pattern, the method of "hedge rowing" fruit trees can substantially increase yields and reduce harvesting costs as the method is adopted.

A research team at the Utah Agricultural Experiment Station has developed an overhead sprinkler system that delays fruit bud development and reduces early spring frost damage, a major cause of crop loss. A computer predicts when winter rest (dormancy) is completed. Then a thermostat automatically turns on sprinklers that cool the buds when rising temperatures threaten to break dormancy. The system is operated until danger of spring frost is past.

Techniques developed to provide disease-free stock for flowers (lilies in Oregon and carnations in Colorado) recently have been transferred directly to vegetable crops. The use of disease-free potato stock promises to increase production substantially, and the technique is transferable to other crops as well.

—John Patrick Jordan

Nutrition

The critical situation regarding food for the people of the world was the concern of the UN World Food Conference in 1974. And in spite of the great current shortage of food in many areas of the world, few countries—among them Canada, Australia, and the U.S.—committed emergency food during the conference. A number of delegations sensed that the special feeding programs were mere "stop-gap" actions, because almost nothing was included about nutrition education for the recipients of the donated foods. Basic reform should bring nutrition education to all levels of society at all ages, should strive for more equal distribution of food, and particularly must provide nutrition education for professionals to improve their insight about food as a contributor to health. Even health professionals tend to overlook at times the role of nutrition in recovery from surgery, burns, extensive radiation or chemotherapy, and other conditions in which a well-nourished body is essential for recovery from therapy or an accident. Major initiatives called for at the food conference included an international fund for agricultural development to channel voluntary contributions toward improving agricultural production in developing countries, which hopefully would receive generous support from oil-rich nations, a global information and early warning system that would alert the world to crop conditions and possible areas of food deficiency, and an annual target of 10 million tons of reserve grain for food aid beginning in 1975.

Increased food demand is primarily due to increased population with the balance attributed to growing affluence that has increased demand for protein-rich meats, milk, and eggs. Harvard University nutritionist Jean Mayer reported that some people feel the necessity of applying triage to the poorest underdeveloped countries, reasoning that some nations are already so poor and overpopulated that it is useless to try to save them and that we should wait to offer help until their population has been reduced by hunger and diseases to a more manageable number. Mayer rejects such an inhumane approach and instead recommends that pressure be put on Western Europe, the U.S.S.R., and the Organization of Petroleum Exporting Countries to do their share in long-range food production and distribution programs, to build processing plants, and to furnish fertilizer for greater food production within needy nations. The initiation of long-range agricultural development along with large-scale measures of population control and reduction of social injustices against the poor, Mayer believes, will contribute to the goal of eliminating hunger and malnutrition for the world at large before the end of the century.

Too often the idea has been that food per se will prevent or solve hunger and malnutrition. Data from studies in the U.S. and other countries do not support this hypothesis. A two-year study reported in the *Journal of Nutrition Education* (January–March 1975) identified a significant quantity of household food waste and its high monetary value. The study, termed Garbage Project, strongly supports the premise that food, affluence, and economic assistance will not solve the problem without education to relieve numerous other aspects of the "ecology of malnutrition." Many studies report data on dietary habits of people of all ages whose suboptimal nutrition relates to rejection of food, to preparation and plate wastes, to flagrant meal skipping (mostly breakfast), to impulsive snacking, and to general behavioral anomalies that interfere with consumption of adequate nutrients from an abundant food supply. In line with this concept is the detrimental effect of monotony of food acceptance, static meal patterns, and limited awareness of the variety of food available which could contribute to eradication of malnutrition, especially of children. Affluent people are highly selective in the food they eat, but starving people can reach a level of lethargy that makes them indifferent to all food and particularly to unfamiliar food. Ercel Eppright, nutritionist formerly at Iowa State University, stressed that "food with man is not just food; it is the crossroads of emotion, religion, tradition and habit." Telling does not constitute education, and food does not nourish if it is not eaten. Even Ralph Nader, the consumer advocate, states that food is the most "intimate consumer product."

In the zeal to feed the underprivileged of the U.S., or even of the world, neither the U.S. nor UN agencies request assurance that food will be eaten by the appropriate people; that they are familiar with the food, like it, and can prepare it; that the children, the ill, or the aged are fed first. Will the persons of high risk be the first to receive needed food? Many cultures have the patriarchal family; the father is ruler of the family or commune and his needs are supplied first. The young children and pregnant women eat whatever is left.

To assist in educating the general public about wise nutrition some concerned advocates have developed "score card" systems for use by consumers to score or rate foods on their nutritional value. For example, one food containing a number of nutrients may score high whereas another may have a single nutrient in high or low concentration and score low; a drink or snack may even have a negative score. Such a scoring device can be highly misleading; a food or drink is valuable only as it meets a need for a certain person. A cereal with vitamin D added may be less desirable than one that has none if the person has no need for an extra amount or uses another adequate source. A high-protein beverage has little value for a thirsty person or one who eats sufficient meat. No evidence or data support the concept that overnutrition is advantageous for a person. Research shows that some megadoses are more risky than others. Overeating continues as a U.S. national habit.

Interest and concern about the outcome of the 1969 White House Conference on Food, Nutrition and Health were expressed during 1974. Progress has been made even if some of the recommendations made during the conference have been very limited in application, especially in relation to educating people about improving their food habits. Supplementary food programs are in operation under control of various administrations in the U.S. government. The most significant current effort is the Special Supplemental Food Program for (pregnant) Women, Infants and Children (under age four), referred to as WIC and directed to meet the needs of high-risk individuals whose families qualify under the Office of Economic Opportunity (OEO) guidelines for poverty. The program is sponsored by the USDA and administered by state health departments to be executed by local groups. The program issues vouchers for the purchase of specified foods: milk, eggs, iron-fortified cereals and formulas for infants, and fruit juices with a specified level of ascorbic acid. The foods supply the needed animal protein, calcium, iron, and ascorbic acid in addition to an assortment of other nutrients.

Nutrition and child development. Since 1969 authoritative information about urgency of nutritional adequacy for normal infant development has had wide distribution by governmental and private agencies through special publications, news media, and other communications systems such as free handouts. Limited research on people usually does not control the variables to a degree that conclusions are unequivocally true. Furthermore, the public and some professionals are reluctant to accept data from laboratory animals as applicable to people. The U.S. has a high infant mortality rate (19 per 1,000 births) among the industrialized nations of the world (12–13 per 1,000 births). In an industrialized urban society infant mortality goes hand in hand with poverty, which has in its complex structure the deficiency of education on infant and child feeding, the lack of accessible and economical health and medical services, insufficient transportation to the major medical facilities, and, as the final "thorn," an

Sidney Harris

unsympathetic, highly organized, sophisticated technical staff with equipment to terrify the people who most need the services.

Contrary to popular belief nutrients and other components are shared by the mother and the fetus during pregnancy; the fetus is not a true parasite, which extracts all of its needs from the host. Fetal development evolves from an interaction of genetic and environmental factors, which include the nutrient supply for the metabolic systems in all its cell structures. If the fetus is malnourished, resulting in low birth weight at full term or when delivered prematurely, brain growth may be affected. The undersized and underdeveloped newborn may grow more slowly and have more illnesses and more limitations in brain development and in behavior.

Nutritional deprivation in the maternal diet prior to and/or during pregnancy can reduce the number of brain cells in the fetus, a situation that cannot be remedied during later growth. Over 90% of brain cells are formed during the last four months of prenatal development and the first four years of childhood; the most rapid postnatal increase in brain cells is during the first two years of age. The mass of the cells may increase but the cell number is established; improved nutrition in later childhood changes neither the cell number nor the mental capacity.

A survey was made of former patients at a well-baby clinic in a midwestern city. From a long list of infants who had been patients at the clinic about ten years previously, 110 were located in 19 public schools. These children were assessed as to growth and developmental progress and school achievement. Of the 110 subjects, 65 had been identified as malnourished when infants and 45 were classed as well-babies. The socioeconomic backgrounds of all the children were similar during infancy and early childhood since all their families had qualified by OEO guidelines for services at the clinic. Most frequent signs of malnutrition judged by pediatricians at the well-baby clinic were anemia, undersized development at birth, and failure to grow and develop satisfactorily. These signs were absent in the well-baby group. Ethnic origin and sex did not correlate with development. Of those who were malnourished when infants, only 25% developed in normal channels of growth, while 47% of the well-babies developed in normal channels. In school performance only 31% of the malnourished infants achieved average or better, whereas 87% of the well-babies achieved average or better.

Trends toward faddism. Harvard nutritionist Mayer asserts that nutrition plays a definite role in delay of death from degenerative diseases but that the role of nutrition in control of death from old age is uncertain. The past century has increased the average life expectancy about 20 years, but since 1950 little gain has been noted. The U.S. population has not been willing to learn to ward off the diseases of old age through better diets. Instead, people of all ages and especially those past age 50 look for miracles and are being exploited by quacks who promise panaceas with megavitamins and natural diet supplements. No evidence indicates that death can be delayed or avoided through any nutritional manipulation. Claims that antioxidants prolong life have not been substantiated in man, and the theory of alteration of genetic substances in cells (DNA) does not imply a nutritional solution.

Cortez F. Enloe, editor of *Nutrition Today*, stressed that exaggerations and distortions for profit are to be expected when the public is eager for knowledge about nutrition, which is the single most important aspect of the health professions in the eyes of the public. After all, people eat and thus they know what is best for them. Emil Mrak, chancellor emeritus at the University of California, Davis, stated, "We are in danger of becoming a nation of food neurotics . . . distrust has fastened itself even on the neighborhood grocer." The increasing amount of misinformation about food and nutrition to reach the public has prompted the American Dietetic Association to publish a position paper in its March 1975 journal, complete with references. Topics included are health foods, natural foods, and organic foods; vitamins, minerals, and dietary supplements; quick-weight-loss diets; vegetarian diets; hypoglycemia; and high-protein diets for athletes. The Nutrition Foundation has published a supplement to *Nutrition Reviews* entitled "Nutrition Misinformation and Faddism" to help combat faddism even among professionals.

Howard Appledorf *et al.* reported an analysis of samples of 24 foods purchased from health food stores and supermarkets. The foods were analyzed for nutritional quality, microbial contamination, and pesticide residues. Only minor differences in proximate composition and microbial content were found, and no pesticides were detected. The health foods were 1.7 times as expensive on the average as supermarket foods.

The U.S. food supply has never been safer, of higher quality, more nutritious, and abundant than at present. Nonetheless, increasing concern for safety and quality of food has gripped the public. R. O. Nesheim, vice-president of research and development at Quaker Oats Co., considers this paradox a major dilemma confronting educators and scientists. Furthermore, only people with satisfied appetites and full stomachs are inclined to be critical. The approach required to alleviate malnutrition of people in the U.S. differs from that for other peoples of the world even though many problems are common to all.

—Mina W. Lamb

Information sciences

Outstanding developments in the information sciences during the past year included improvements in the capacity and speed of computer memory systems, advances in the method of transmitting information through hair-thin glass fibers on very short pulses of light, and the launching of the first United States commercial communications satellites.

Computers

The ever widening applications of computers were demonstrated in 1975 by proposed new laws that would affect almost everyone. These new rules also might place challenging requirements on the hardware and software of computer systems.

Privacy and security. The question of privacy, security, and accuracy in personal files was one of the fastest developing areas involving computers. Privacy in a personal file means that the information will be used only for the intended purpose. It means that a bank cannot provide data to the U.S. Internal Revenue Service or to a credit bureau and that a university cannot provide transcripts to a company unless specific permission is given by the person involved. Security means that the information is safe from fire, flood, and power failures, as well as theft.

A privacy law passed in late 1974 for government files forbade one government agency to use another's files, and required that extensive records be kept of all accesses to personal information. For example, it forbade the Internal Revenue Service, state welfare departments, and state university student aid departments to exchange information. Although this law corrected some abuses, it had some unforeseen consequences. Apparently, for example, it made it illegal to publish a high school honor roll.

Another law pending in 1975 included private as well as government files and required even stronger safeguards. One motivation for these laws was to ensure the reasonable use of private information; another was fear that large computerized files might lead to "big brother" situations like those described in George Orwell's book *Nineteen Eighty-four.* Indeed, the bill introduced in the U.S. Congress was deliberately numbered

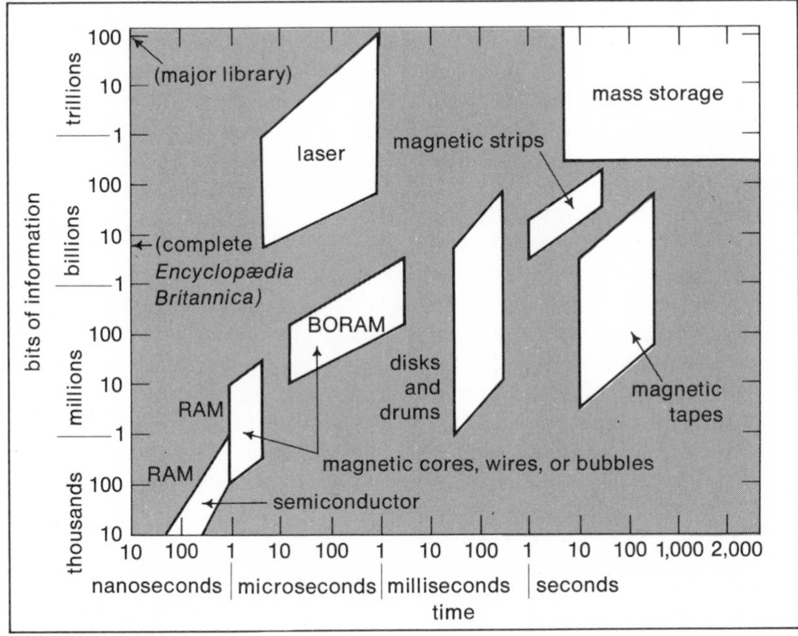

Sizes and access times are indicated for the various types of computer memory systems expected to go into operation from 1975 to 1980.

HR1984. Its consequences for computerized files were not immediately obvious and were an area of intensive study during 1975. Some expressed concern that the proposed law might have unforeseen consequences because it not only requires privacy but also specifies particular methods to protect that privacy.

For an assessment of some potential consequences of HR1984 one can examine its possible impact on the computerized policyholder files of a large insurance company. The law would require that records be kept of all persons who have regular access to each person's information and that the name, time, and reason be recorded for all other persons who use the file. These files are normally available via computer consoles to any one of 20 to 50 operators, and so their names must be included in the "regular" category. An insurance company's management regularly gathers data from its files to study such questions as the risk of fires in southern Idaho and the risk of heart attacks to overweight men of age 44. All the management people involved in these studies must either be considered "regular" (and have their names added to each person's file), or else a record must be kept in every person's file for each such study made. The result would be that the list of accesses to a person's file might become longer than the file itself. If this access record were to be straightforwardly included in each person's file, then the cost of computer memory and processing would be increased by a large amount, approximately 100% or 200%. One hopes that there might be a more clever method to keep these records that would reduce this extra cost to a reasonable amount. This example indicates that an operation which has always observed strict privacy may be required to spend large sums just to continue doing things the way they always have.

Cost is a key issue in this matter, and one must weigh the benefits gained against the costs involved. The 1974 law for government files has been estimated to lead to extra costs of several hundred million dollars a year. Restricting the flow of information also generates another kind of "cost." At present, the U.S. government has the responsibility for many difficult tasks, such as regulating the details of the economy and assuring nondiscrimination in various activities. Intelligent action in these areas requires adequate information about people, thus leading to a possible conflict between the individual's privacy rights and the responsibilities of government.

The main concern with the proposed law is not that it requires privacy but that it exactly specifies the methods to be used to achieve it. It seemed possible, even likely, that for computer files other methods might be both more effective and much cheaper. The real danger in computerized files is not that someone obtains private information about a particular person—that has been possible for about 50 years if one wants to spend approximately $100, for most people more than such information is worth. The danger is that some organization may cheaply obtain improper use of masses of personal information.

HR1984 does take one step to inhibit specifically the massive use of computerized files. It forbids the use of a universal identifier such as the Social Security number. In fact, it forbids any two organizations from using the same identifiers for people. This makes it much more difficult to combine the information in two files and thereby generate new relationships or facts.

The most common method of achieving security from thievery and misuse in computer systems is the use of authorization codes or passwords. This approach is directly applicable to the privacy problem. Each file can be assigned a password, and only persons who know the password can have access to it. For example, all regular users of a file would know the password, and special access can be made by giving passwords that are valid for a short time only. A problem with passwords is that they must be stored in the computer for it to check against, and it has become widely recognized that a clever, industrious, and well placed programmer can obtain the information in a computer system in spite of all precautions.

Three steps can be taken to achieve greater security and to ensure privacy. The first is to make certain that the key people involved with the computer system are trustworthy. Second, one can use levels of authorization. For example, though 50 console operators may need access to the insurance company's files, they do not need access to more than, say, 1,000 records a day. It is not difficult to make the computer system prevent them and others from making massive examinations of the file. Only a very few people in the organization should be authorized to examine or process the entire file.

Finally, there is an interesting technique that makes it pointless to steal the password table. One transforms the user's password before comparing it with the table entry. For example, suppose the password is 425 and the transformation is (along with the results for 425): cube it (76,765,625); discard digits 1, 4, 5, and 7 and then divide by 100 (to obtain 67.65); take the cube root (4.074640213); discard digits 2, 5, 7, and 10 (474,421); square it (225,075,285,241) and take digits 3 through 8 (507,528). It is clear that discovering 425 from the table entry of 507,528 is almost impossible even if one knows the transformation.

Memory systems. The memory has long been a critical component of the computer system, both because of its cost and of the limitations on its performance. There are two distinct types of memory, main and auxiliary, and they are often further subdivided in large complex systems. The main memory contains the programs, data, and results of the current calculations, while the auxiliary memory contains information (programs and data) needed at some future time, perhaps only a hundredth of a second away. The speed of a memory system is measured by the access time it takes to obtain an item of information (bit); the speeds of main memories in 1975 were measured in nanoseconds (billionths of a second). Main memories must not only be very fast to keep up with the central processors (the calculating part of the computer) but they must also be random-access memories; that is, one must be able to obtain any one bit of information as fast as any other.

The information in auxiliary memories is usually organized in blocks or files; such data may consist of the payroll information for a company, the day's transactions in a bank's savings accounts, or a program waiting to be executed. An example of a block-oriented auxiliary system is provided by a magnetic tape. It takes considerable searching along the typical 2,400-ft tape to find the start of a block of information, but once it is located the rest of the block is immediately available. Block-oriented random-access memory (BORAM) is that in which all blocks can be obtained equally quickly; this contrasts with magnetic tapes, where blocks near the start of the tape are obtained much more quickly than blocks at the end.

Memory sizes and speeds. There are two extremes in the type of information in auxiliary memory systems. One is the massive systematic data sets such as those compiled by a large bank, insurance company, or government agency. For example, if the Internal Revenue Service has two pages of information concerning the tax returns of each individual plus 50 pages for each company, then it has about six trillion bits of information. An insurance company with two million policyholders has about 100 billion bits of information. In each case the total information is a massive collection of rather similar sets of data; it can be organized in various systematic ways.

The other extreme is a large collection of completely unrelated items of a wide variety. For example, a large university computing center might have 2,000 students in various courses who will each have 1–20 pages (10,000–200,000 bits) of programs in various languages. In addition, there are normally 200–300 advanced students and faculty members who have much more substantial needs, say 100 pages (or one million bits). Finally, there are one or two dozen people with really large problems and data sets that are 5–100 times as large. Thus, this university requires perhaps two billion bits of memory to save the programs and data of users. This collection of information has no systematic structure and is subject to rapid changes.

Laser memories. An extremely accurate laser beam can be used to etch or burn holes in some material and thereby record information in much the same way as do the holes on a punched card. The information is read with a lower power on the laser beam to see if a hole is located at a particular spot. Laser memories are relatively new, and a trillion-bit memory became commercially available in 1975 for the first time. One example is the Precision Instruments 190, which records on rhodium-coated plastic strips that are 2½ ft long and 5 in. wide. Each strip can contain about 1½ billion bits of information, equivalent to about 25 ordinary-sized books. A trillion-bit memory requires that only 640 data strips be mounted, and an operator can change data strips in about one minute. Thus an ordinary-sized room (16 by 20 ft) could contain ten trillion bits (one trillion of which are available to the computer at any one time), which is equivalent to a library of 150,000 books.

The first laser memories were of the block-oriented variety; thus they were quite slow (requiring many milliseconds) in locating a block but then could read a block of 150,000 bits extremely quickly, about three bits every microsecond. Future designs are expected to reduce the initial delay substantially.

One disadvantage of a laser memory is that once information is recorded on a spot it cannot be changed. Various schemes used to overcome this problem involved making a change by voiding the old information and writing the new information somewhere else. This approach could be implemented in an efficient manner and was acceptable provided the changes did not occur too frequently. Eventually, though, the information had to be copied onto new materials.

Magnetic devices. Memories that utilize magnetic devices are divided into two categories according to whether or not the memory system involves moving parts. The magnetic tape is the oldest of these devices, and it uses basically the same technology as ordinary tape recorders. Magnetic disks, drums, and strips use similar magnetic surfaces but on different physical devices. Because the disks and drums rotate at very high speeds no part of the information is very far from the reading heads, thereby providing quick access times.

Magnetic cores, wires, and bubbles use magnetic materials that are stationary and built in such a way that electronic circuits can have access to or magnetize any particular spot. This property makes these devices suitable for main memories because they can provide both high speed and random access. The magnetic-bubble devices represent a new and promising technology that was just beginning to be put into practical application in 1975. Magnetic bubbles are movable domains of magnetization in certain special materials. One can roughly visualize a bubble as moving in a slot, and its position in the slot indicates whether or not a bit of information is present. Researchers hoped that bubble memories

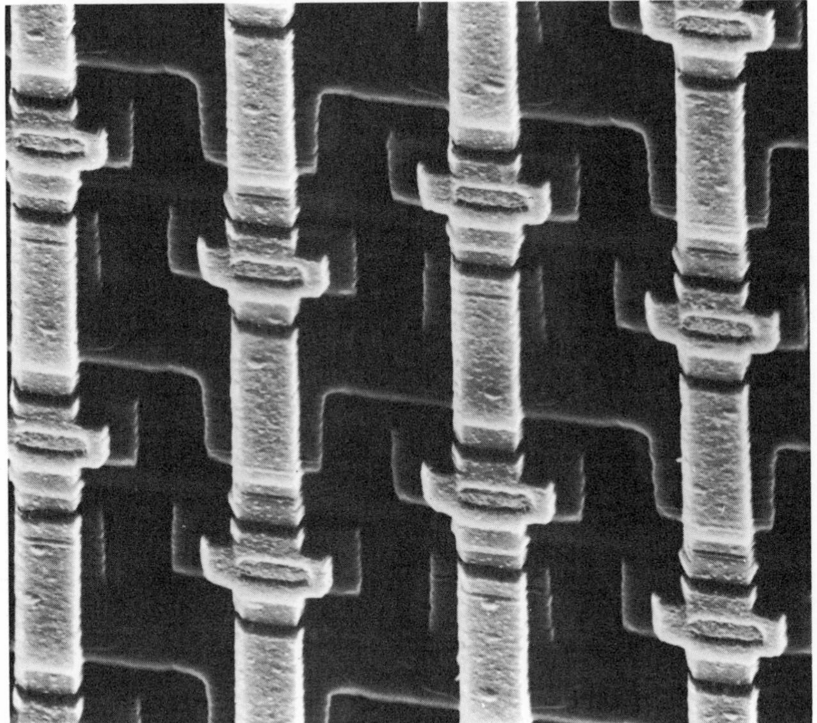

Courtesy, IBM Corp.

Electron micrograph shows small portion of a high-density, 8,192-bit field-effect transistor (FET) computer memory array. Vertical light-colored lines are aluminum interconnections about 2.3 microns wide. The wide, raised, horizontal gray lines and pads are polycrystalline silicon gates, and the thin, lighter gray horizontal lines are doped connectors. The short cross lines on the aluminum are at the locations of the FET gates.

Courtesy, IBM Corp.

A one-ounce "floppy" disk can contain as much information as 3,000 punch cards. It can also be reused and is read much faster than cards by a computer.

might someday achieve densities up to a billion bits per square inch. This corresponds to writing the entire *Encyclopaedia Britannica* on a small postcard. If this density is achieved, bubble memories will probably replace most disks and drums and provide much faster access at a comparable cost. By 1975 densities of more than ten million bits per square inch had already been achieved.

Mass storage. The term mass storage refers to attempts to put together large combinations of devices. One approach is to have a collection of, say, 1,000 magnetic tapes so that any one of them can be automatically moved to a read/write station without human intervention. By 1975 systems holding hundreds of tapes were available commercially. The access times in such a memory system are of the order of many seconds because of the delays in mechanically moving the tapes. Another approach is to use a collection of 100–200 disks, each with 500 million to one billion bits of information. A switching mechanism allows the computer to gain access to a particular disk. A third approach uses magnetic cartridges; IBM Corp. introduced a 2½ trillion-bit model of this type during the latter part of 1975.

This article has emphasized two aspects of memory systems, size and speed, but there are clearly other important ones, such as physical bulk, power requirements, permanency of recorded information, and cost. Costs projected for 1980 ranged from one cent per bit for ultrahigh-speed random-access memory systems to 0.0001 cents per bit for very large auxiliary memory storage. A volume such as the *Yearbook of Science and the Future* has about 100 million bits of information, and thus the memory to store this book could range in price from $100 to $1 million, depending on the system used.

Personal memory systems. Thus far, this discussion has assumed that the memory system is a part of the overall computer system. An important exception to this is the so-called personal memory system, which in 1975 generally consisted of decks of key-punched cards or stacks of program listings stored in offices or other areas. These personal systems had certain advantages. They were secure and private; no computer snooper could get at them. One could examine the information at leisure and make changes on inexpensive and widely available keypunch machines. Punched cards, however, have the disadvantages of being bulky, slow to read, and not reusable. Thus, there was a need to develop a better alternative to them. A number of such alternatives (floppy disks, tape cassettes, magnetic cards) were being explored during the year, but none had yet met all the conditions necessary for them to displace punched cards.

—John R. Rice

Communications systems

Communications developments continued their quickening pace during the past year. Although there was good progress in the basic science and technology of communication, the main activity centered on new uses and deployment of communications systems. Expansion continued in voice and video communication, but even greater progress was made in computer communications systems to serve the business world.

Technical achievements. The most exciting advances occurred once again in the science of transmitting information through hair-thin glass fibers on ultrashort pulses of light. Incredibly, signal losses as small as one decibel per kilometer were demonstrated—half as much as a year earlier and one thousandth of what was achieved ten years ago. Although losses in practical cable structures might be somewhat larger, owing to various imperfections and distortions that could result from packaging, this achievement might permit distances of up to ten miles between amplifier stations.

Courtesy, ITT Corp.

Color television picture is transmitted by optical fibers (foreground).

A companion advance improved the durability of a diminutive source of coherent light pulses, the gallium arsenide laser. This pinhead-sized solid-state device, which in previous years had worn out relatively quickly, was shown during the year to promise a lifetime of close to 100,000 hours. It might well prove to be an ideal transmitter for future optical fiber communications systems, but for more immediate applications the gallium arsenide light-emitting diode (LED) seemed simpler to implement.

These achievements opened the way for using optoelectronic communications links, perhaps in the early 1980s, to carry thousands of telephone conversations or other transmissions, particularly in congested metropolitan areas. Before this could happen, however, practical methods of handling and bundling together the delicate fibers would have to be perfected. (See *1975 Yearbook of Science and the Future* Feature Article: FIBER OPTICS: COMMUNICATIONS SYSTEM OF THE FUTURE?)

Other important progress took place in the field of information storage, useful for both computation and communications systems. Researchers at Bell Telephone Laboratories squeezed 256,000 bits of information (at a density of about four million bits per square inch) together with associated magnets and circuitry into a sealed "bubble pack" measuring only 1 × 2 × 0.6 in. The storage medium was a thin magnetic film capable of sustaining tiny domains of different magnetic orientation, appearing like bubbles when viewed in polarized light through a powerful microscope. Information could be "written in" and "read out" at rates of many hundreds of thousands of bits per second; also, it was nonvolatile in the sense that a power failure need not erase it, as is true with most semiconductor memories.

Technology milestones were also reached in the design, fabrication, and use of so-called microprocessors. These are tiny semiconductor chips containing numerous logic and memory cells so organized that they can be programmed to act as the heart of a microcomputer.

The largest communications research and development organization in the world, Bell Telephone Laboratories celebrated its 50th anniversary in 1975. Many fundamental discoveries, inventions, and pioneering experiments took place there, ranging from the principle of feedback to the transistor, and from the birth of radio astronomy to communications satellites.

Telephones. The world's telephones increased by 23.4 million during the most recent census year, a record gain of 7.5%. The United States retained its lead with 138.3 million, followed by Japan with 38.7 million, the United Kingdom with 19.1 million, West Germany with 17.8 million, and

Courtesy, Bell Laboratories

Scientist examines new magnetic-bubble memory developed by Bell Laboratories. The extremely small memory device can store the information equivalent of 27,000 telephone numbers.

the U.S.S.R. with 14.3 million. The largest percentage growth during the last census year, 24%, occurred in Taiwan, while South Korea led the world with the largest ten-year growth, more than 493%. The number of countries having more than half a million telephones reached 41, with Iran and Venezuela just joining the list.

Among the world's principal cities, Washington, D.C.; Stockholm; and Paris stood out with, respectively, 130, 105, and 82 telephones per 100 persons. An interesting comparison was that of the world's telephone growth, nearly 100% since 1965, despite a population growth during the period of less than 25%. This trend clearly has accelerated in recent years.

Telephone technology made progress on several fronts, including large switching systems, small business communications systems, and hands-free telephony. The 500th central office installation of the pace-setting No. 1 Electronic Switching System (ESS) took place during the year, and a newer, midsized version, No. 2 ESS, its 100th. Completed sooner than the small No. 3 ESS, which was being readied for service in rural areas, the first No. 4 ESS was manufactured and installed in Chicago. A large toll-switching system with enormous capabilities, No. 4 ESS was designed to process 350,000 toll calls per hour on up to 107,000 trunk (channel) terminations; yet it occupies less space and requires less installation time than its electromechanical predecessor, which handles only one-third the call volume.

Like the No. 1, 2, and 3 ESS, the No. 4 was based on stored program control; however, the central processor of No. 4 achieved 30–40 times the circuit density and 4–8 times the speed of its predecessors, besides having been designed to be more dependable and maintainable. Most significantly, No. 4 ESS was the first large switching system to allow many calls to be switched over a single physical path. This was accomplished by "sampling" each call 8,000 times a second and interspersing the very short samples of successive calls less than one millionth of a second apart. Although customers were thus connected intermittently, the quality of voice transmission was as good as on a continuous connection. Furthermore, this approach is consistent with that used in the digital transmission communications links, which, in rapidly growing numbers, were carrying voice signals by means of the same time division method.

The small business communications systems, known as Key Telephone Systems and private branch exchanges (PBX), also took a leap forward during the year with the introduction of new compact electronic systems that required a minimum of space on the subscriber's premises. Typically, tens, hundreds, and sometimes thousands of extensions are served, with many novel capabilities. For example, an outside caller who is put "on hold" may listen to music while waiting; if the called party is not at his desk, the answering party may transfer the call directly to another extension simply by dialing it. A system recently introduced by AT&T, called Dimension PBX, incorporates many conveniences by use of the stored-program approach in conjunction with time-division switching: for example, distinctive ringing to differentiate internal and external incoming calls, and outgoing trunk queuing. The latter means that if a user tries an outside call but finds all lines busy he will be called back automatically rather than having to keep trying.

Hands-free telephony is a convenience not only for a busy executive shuffling papers but also for a busy housewife who has her hands full. For some time the speakerphone, a unit comprising separate microphone and loudspeaker units plus a small housing for the required amplifier, has met this need. During the past year, however, a greatly improved instrument, the Model 4A, with almost everything built into a single speaker enclosure, was introduced. This model also had only a small control unit, which houses the "on" switch (to switch from normal telephone to speakerphone), a volume control, and also the tiny omnidirectional microphone. Quality was remarkably improved, despite the difficult problem of room reverberation when a user speaks at some distance from the microphone.

The problem of reverberation or echoes was the subject of some interesting research published during the past year. Several promising ways of eliminating the "hollow barrel effect" caused by sound waves bouncing off hard walls in a small room were studied and tested. This kind of experimentation seemed certain to bring closer the day when conversation between people, and even groups of people, at different locations would be as easy and of as high quality as conversation across a desk or table.

Mobile telephony marked an important milestone with the granting to common carriers by the U.S. Federal Communications Commission of two radio spectrum allocations totaling 40 MHz in bandwidth in the frequency range between 800 and 900 MHz for development of a high-capacity "cellular" system. Such a system was expected to provide a high-quality, widely accessible mobile telephone service by dividing the landscape and "cityscape" into adjoining small cells, each with its own transmitters and receivers. This would permit hundreds of thousands of mobile units to be served in metropolitan areas, which in 1975 were limited to about 1,000 users.

Transmission technology took a leap forward with the first service on Feb. 27, 1975, of a new high-capacity digital line between Newark, N.J., and Manhattan in New York City. Intended primarily for use in and between large metropolitan areas, the so-called T4M line was expected to be particularly advantageous over distances of ten miles or more when more than 10,000 voice circuits are needed. A ⅜-in. coaxial tube carries a sequence of binary pulses at a rate of 274 million per second, an information flow of 274 megabits/sec. This extremely high-speed stream can comprise up to 4,032 different talkers' speech signal waves, suitably encoded into digital form and time interleaved. Alternatively, the bits could also represent video signal samples, or they could be data from a computer or other business machine, or appropriate combinations of such signals. The T4M system uses 22 coaxial tubes in a single cable (suitable for installation in a four-inch conduit), 10 working tubes for each direction plus a pair of protection standbys. Thus, a maximum of 40,320 two-way voice channels can be handled; however, a somewhat smaller cable for 32,256 channels was also developed.

Courtesy, Bell Laboratories

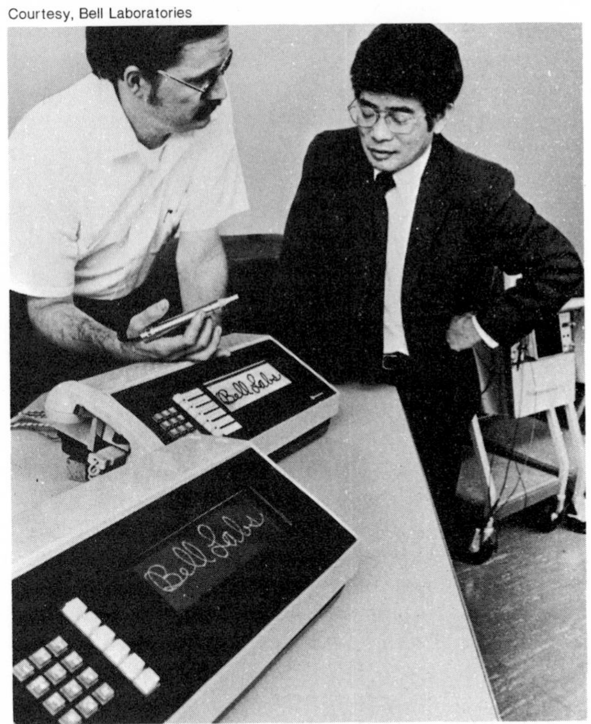

Scientists demonstrate flat-screen video device that can transmit handwriting instantaneously, reproduce pictures, and communicate with a computer.

Data communication. The fastest growing segment of the communications industry continued to be that devoted to interconnecting business machines and computers for electronic data processing. Large-scale applications of this practice were found in industrial, retailing, brokerage, and transportation firms for such purposes as administrative messages, inventory control, price quotations, and reservation service. Other growing applications were for banking and off-track betting. The most prominent growth occurred in the appearance of more point-of-sale (POS) terminals in various establishments, particularly in large stores. These replaced conventional cash registers and sometimes were connected via telephone lines to a central computer for record-keeping and billing in the case of credit sales. Some installations were planned to include also an electronic funds transfer system (EFTS), whereby, after a customer presents a plastic card, the system checks his bank account against the amount of the purchase and then transfers the appropriate amount to the store's account while updating all records.

Along similar lines, the Bell System's Transaction telephone was introduced during the year. In the simplest version of this instrument a sales clerk would insert into it a plastic card bearing a magnetic code; this would automatically call, say, the BankAmericard central computer. Then the customer's credit card would be inserted and the amount of the purchase keyed in on the Touch-Tone dial. A computer-generated voice would reply that the purchaser's credit was satisfactory or that further checking was required.

Still another manifestation of the rapid growth of teleprocessing was the proliferation of cathode-ray terminals whose screens, fed by a distant computer, could fill up with text and numbers faster than the eye can follow. Usually the display is in response to an inquiry that the terminal's user may have "typed" into its keyboard. Many such keyboards may be "clustered" in one location, and many such clusters may all "home" on one central computer via a so-called multipoint circuit, or data network.

After nearly a year of deliberations and of receiving the testimony of various interested parties, the U.S. Federal Communications Commission permitted initial offering of the Bell System's Dataphone Digital Service (DDS) between Boston, New York, Philadelphia, Washington, D.C., and Chicago. Permission was also granted to add 19 more cities. This new private-line service utilized the rapidly growing digital transmission channels, originally conceived for handling digitized speech signals, to transmit business-machine data signals, which are naturally of digital form. Instead of 9,600 bits per second, the highest rate available over analog voice channels, 56,000 bits per second were available from a single voice-equivalent digital channel. By subdividing the bit stream, lower rates—namely 2,400, 4,800 and 9,600 bits per second—were also made available.

The Digital Data System, the network that implemented the service, included not only the existing digital lines but also voice-carrying microwave links augmented to carry digital data signals on the same radio beam. Because the data spectrum was placed just under the existing voice frequency spectrum the arrangement was called DUV, data under voice.

Another data-only network, this one offering switched (dial-up) service, was put into service by Data Transmission Co. Initially linking Chicago with cities in Missouri, Oklahoma, and Texas, this new service also was expected to experience a rapid increase in geographical coverage of the U.S.

Much interest and attention was focused on special data-oriented networks not only in the U.S. but also in Canada and various European countries. Strong emphasis was on message or packet switched systems, in which a block of data—a collection of bits—finds its own way from source to destination, much like a piece of mail. Such networks were increasingly found to be appropriate for data traffic comprising many randomly spaced messages of very short duration. In the U.S., the Arpanet (Advanced Research Projects Agency Network) continued to gain in usage, on a noncommercial basis. Interconnecting 45 government and university computers to fashion a single scientific computation facility of enormous versatility, the network expanded to span Hawaii on its western extremity and London on its eastern border. Traffic between distant points or "nodes" of Arpanet grew from just over one million data packets (of roughly 1,000 bits each) to about three million packets per day during the last year, while the number of nodes rose from 32 to 45. Other packet networks planned in the U.S. included Telenet and Graphnet, while the Canadian Datapac, the French Cyclades, the British EPSS (Experimental Packet Switching System) and the European Economic Community's multinational EIN (European Integrated Network) also registered progress.

—Ernest R. Kretzmer

Satellite systems

Applications satellites in Earth orbit utilize their unique vantage point to provide a number of services of increasing benefit to man. The United States and the Soviet Union continued to launch the great majority of such satellites. Japan, France, West Germany, Canada, and other countries, however, began launching precursors of their own operational satellite systems. Some of these launchings were made by the U.S. National Aeronautics and Space Administration (NASA) on a cost-reimbursable basis.

Communications satellites. In November 1974 the sixth of the Intelsat 4 series of communications satellites was placed in geostationary orbit (remaining over the same portion of the Earth's surface at all times). By 1975 Intelsat (the International Telecommunications Satellite Consortium) comprised 89 member nations, representing about 95% of the telecommunications traffic of the world. Intelsat 4 satellites weigh 3,100 lb (1,400 kg) at launch, stand 17 ft 6 in. (5.3 m) high, and are 7 ft 8 in. (2.4 m) in diameter. Each of these six satellites, in geostationary orbit at an altitude of 22,300 mi (35,900 km), relays up to 5,000 two-way telephone calls, transmits 12 simultaneous color television programs, and handles any combination of telephone, television, digital data, facsimile, and other communications traffic.

In 1975 three Intelsat 4s were stationed over the Atlantic Ocean, two over the Pacific, and two over the Indian Ocean. The last of the Intelsat 4 series was launched in May 1975. Scheduled for launch in mid-1975 was Intelsat 4A, which has double the communication capacity of Intelsat 4. The U.S. Communications Satellite Corp. (Comsat) operates the global network of Intelsat and is a major partner in the consortium.

A number of U.S. firms and industrial consortiums competed during the year for approval of U.S. domestic satellite communications systems. Two Westar satellites were launched in April and October 1974. Operated by Western Union, these satellites were similar to the earlier successful Canadian Telesat (Anik) domestic "comsats." A third Telesat was launched for Canada by NASA in March 1975. Additional new domestic satellites were planned for launch late in 1975. Comsat, American Telephone and Telegraph Co., RCA, IBM, and other major industrial firms were involved in the consortiums producing the satellites.

Similar domestic satellite systems were being considered for many parts of the world, either owned nationally or by a regional consortium. The U.S.S.R. has had a Molniya-satellite relay system for communications and TV since 1965. The Soviet system is not geostationary; rather, a highly elliptical orbit of 300 mi (500 km) perigee and 25,000 mi (40,000 km) apogee is used. Thus, on each orbit the satellites have a relatively long transmission period over the Soviet Union. Approximately 65 tracking and ground communication relay stations are located near major Soviet cities. It was reported recently that the system was to be extended to include Czechoslovakia.

In North Africa, Algeria was building a domestic system of 14 antennas to link the entire nation by satellite. Indonesia was constructing a network of about 50 ground stations to link its several thousand islands by satellite. Two satellites were scheduled to be launched in 1976. Micronesia had plans for a similar system of tracking stations and microwave ground links. The Philippines and Brazil both planned to build much larger systems, each having more than 1,000 antennas. Japan proceeded with plans to launch and operate its own national and regional communications satellite system.

The Middle East, with greatly increased oil revenues, began to plan a pan-Arab network, Arcomsat (Arab Communications Satellite System). In September 1974 telecommunications representatives of the 20-nation Arab League agreed upon a plan to establish a satellite system for all forms of communications, including radio and television, that would serve all cities and populated areas of the Arab world.

In Europe the European Space Research Organization (ESRO), merged into the new European Space Agency in 1975, proceeded with plans for a network of stations. In December 1974 the Franco-West German Symphonie satellite was successfully launched by NASA into geostationary orbit. During the next two years this relatively small developmental satellite is to be tested. If its performance is satisfactory, the two countries planned to use Symphonie satellites for international public and specialized telecommunications from northern Norway to southern Italy.

The launch of NASA's sixth large applications technology satellite (ATS 6) in May 1974 resulted

Using an ordinary walkie-talkie and an antenna built on the frame of an umbrella, a research engineer (left) beams a Morse code message more than 50,000 miles through ATS 3. Satellite orbital locations above the Equator (right) are assigned by the U.S. Federal Communications Commission.

Courtesy, General Electric Research and Development Center

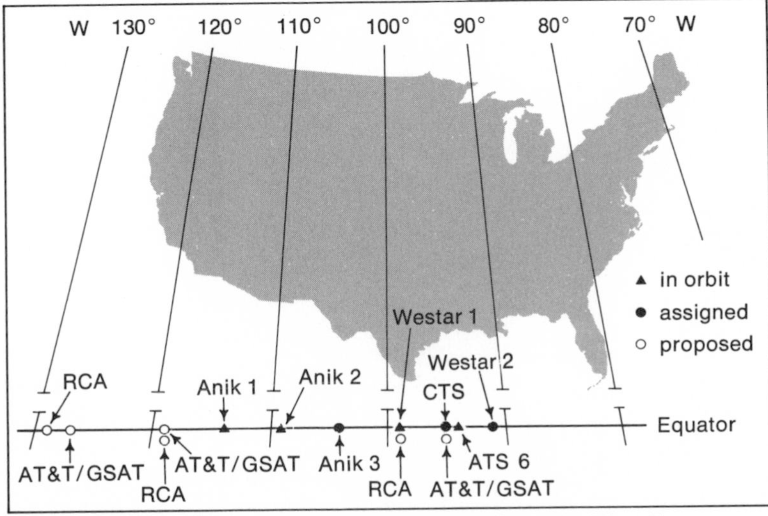

Based on "Users Starting to Hop Aboard U.S. Communications Satellites," by Stephen E. Scrupski, in "Electronics," Oct. 3, 1974, copyright © McGraw-Hill, Inc., 1974

in a new application for communications satellites. Television signals from the 3,100-lb (1,400-kg) craft were relayed to both mobile and small inexpensive receiving antennas, and in this way during the 1974–75 school year educational programs were successfully beamed to schools in Appalachian and Rocky Mountain states not served by conventional television networks. In June 1975 the satellite was repositioned in orbit to assist in an educational service program for thousands of villages in India.

Meanwhile, military satellite communications systems, providing secure strategic and tactical communications from command headquarters to field, ship, aircraft, and outlying posts, continued to be operated and developed. For obvious reasons, little information about these systems was available. In November 1974 NASA launched a second-generation defense communications satellite, Skynet 2B, for the United Kingdom.

Earth observation satellites. This category of satellites has three major forms: weather, Earth resources, and military reconnaissance.

Weather satellites. Current meteorological satellites view the Earth's weather in darkness as well as light by measuring the temperature difference between clouds and the Earth's surface. In addition, sensors aboard the spacecraft obtain a vertical sounding of the temperature of the Earth's atmosphere. Newer techniques not only "see through" clouds but also record rainfall over oceans, map soil moisture, and monitor temperatures and ice conditions in polar regions.

Great Britain's Skynet 2B, the first operational communications satellite to be designed and built outside the U.S. or U.S.S.R., undergoes preflight tests.

The latest addition to the U.S. family of weather satellites is the synchronous meteorological satellite (SMS). Two of these spacecraft were launched in 1974–75, one over the Atlantic Ocean and the other over the Pacific. Located over the Equator in geostationary orbit, the SMSs were able to provide 24-hour coverage of U.S. weather. Using both visible and infrared imaging devices, they transmitted an image of approximately one-quarter of the Earth's surface every 30 minutes.

During the year Japan contracted with Hughes Aircraft Co. for a geostationary meteorological satellite (GMS) to be launched by NASA in late 1976. Weather observation from Hawaii to Pakistan and from eastern Siberia to Antarctica was anticipated.

Earth resources satellites. The first Earth resources technology satellite (ERTS 1) was launched in 1972. Although its performance eventually deteriorated somewhat, ERTS 1 performed better and far longer than originally anticipated. In polar orbit at an altitude of about 560 mi (910 km), it transmitted to the Earth multispectral images that provided data to hundreds of scientific investigators in many disciplines: agriculture, forestry, mineral and land resources, land use, and water and marine resources.

ERTS 1 was renamed Landsat 1, and in January 1975 Landsat 2 was orbited successfully. The two spacecraft are 180° apart in orbit, thereby providing a view of the same local area with the same sun angle every nine days. With only one satellite 18 days would be needed for the same view.

More than 100,000 black-and-white pictures and photographs in strange artificial colors produced from the multispectral scanning system were taken by Landsat 1. Based on those pictures more than 300 separate investigations were conducted; they provided views of the Earth and a better understanding of its surface features. Some of the most valuable uses of Landsat 1 have been: estimating crop acreage; monitoring urban development and planning future land use; locating air and water pollution; mapping strip-mine and forest-fire scars; locating geologic formations that may indicate the presence of minerals and petroleum; updating maps and navigation charts; monitoring the advance of glaciers; and studying flood hazards and managing water resources.

Military reconnaissance satellites. Only the U.S. and the U.S.S.R. have the technological capability and wealth to utilize spacecraft for photographic and electronic monitoring. No official releases are made of such activities, although both countries have been known to take photographs from orbit

Courtesy, Westinghouse Electric Corp.

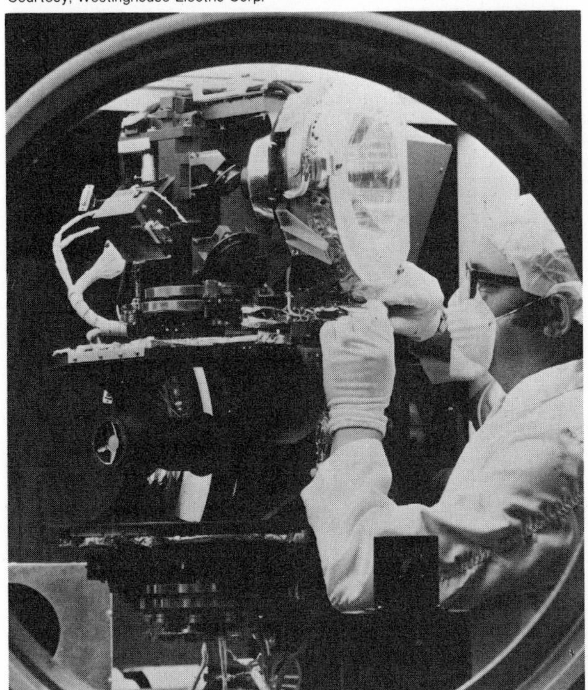

Engineer tests sensor for new U.S. Air Force weather satellite BLOCK 5D. It is expected to return the highest quality weather imagery ever achieved.

and then de-orbit and recover film packages a few days to a week or more later. During 1974 it was reported widely that the Soviet Salyut space stations 2 and 3 were used for military reconnaissance. This belief was based upon the low orbital altitudes of the stations and communications frequencies and modulation techniques.

Navigation satellites. The U.S. Navy Transit navigational satellite system continued to be fully operational and in use by the U.S. fleet throughout the world. Although commercial ships were permitted to use this system, it required a shipboard computer to calculate position based upon the Doppler shift of the satellite's signal. The high cost of such equipment limited civilian and commercial use of the system.

Whereas the Transit system depends upon analysis of Doppler shift, the U.S. Navy began developing a new system based upon precision clocks. Location is obtained by measuring the brief time required for a radio signal to travel from satellite to ship. The first development satellites, known as Timation 1 and 2, carried quartz crystal standards. In July 1974 a navigation technology satellite (NTS 1) was launched carrying a rubidium atomic clock. NTS 2, scheduled for launch in 1976, was to utilize an even more precise cesium-beam tube.

—F. C. Durant III

Information systems

In evaluating technological progress in information science the U.S. National Academy of Engineering reported that the United States leads the world in advanced research but lags behind in the important aspect of applying this knowledge. In order to remedy this situation the National Science Foundation adopted a program called Research Applied to National Needs (RANN). It is an attempt to orient research in science and engineering toward the solution of significant social problems, such as energy, health care, education, and population growth. Successful solutions depend, in part, upon supplying the right information to the right people at the right time.

Information systems and critical national problems. As part of the national commitment to create new energy sources in the U.S. through research and development, a National Energy Information Center was established by the administrator of the Federal Energy Office. Experts at the center were authorized to prepare a comprehensive list of energy data resources in the U.S., to identify and resolve problems associated with the transfer of energy data and information, to create a common energy vocabulary, and to prepare a guide to aid users of computer-processed data bases that contain information on the production, importation, and availability of gasoline and other petroleum products. The center was to be responsible for answering questions about energy matters from Congress, state and local governments, private organizations, and the public. It would facilitate communication among the producers, processors, and users of energy and related information.

A National Fire Data System was designed under contract to the National Bureau of Standards. Fire data were to be collected from fire departments, insurance companies, and hospitals concerning the frequency, severity, and nature of fires. This information would then be stored in a computer and analyzed to determine priorities for programs in fire research and safety. The data would also be used to identify specific problems on which work is required and to assist public and private agencies in developing aids for fire safety and for the management of fire services and resources.

The rapid growth of human population is one of the most critical problems facing humanity. Information systems covering all aspects of population control are needed, for decisions that are based on inadequate information will be inadequate. Fortunately, the information aspect of the population problem has been recognized, and steps are being taken to collect, organize, and disseminate available information on population planning.

The World Population Conference, held in Bucharest, Rom., in 1974, asked that each country make a special effort to obtain data relevant to the development, planning, and formulation of population policies. It was also recommended that research institutions exchange information about ongoing studies. The Association of Population/Family Planning Libraries and Information Centers (APLIC) identified more than 80 organizations with some kind of information, education, or communications component relevant to population studies. At the Bucharest conference the Association organized a population library covering major population subject fields, including demography, population growth, family planning, contraception, statistical methods, and social research techniques. At the conclusion of the conference, the collection of catalogs was turned over to the newly created United Nations Demographic Research and Training Center in Bucharest to serve as a basis for the development of its population library.

Governments of approximately 40 countries were involved during the year in information and education programs in support of population and family planning activities. Many of them needed new informational materials and were not able to produce these locally. In order to provide a source of such materials an International Audiovisual Resource Service was established in London under the joint sponsorship of UNESCO and the International Planned Parenthood Federation. The center collected films, slides, photographs, tapes, and other educational and training aids from family planning programs throughout the world. A prime concern of the service was to adapt locally produced material, such as films and radio scripts, so that it would serve a multilingual audience.

Word list and computer-assigned "index of peculiarity" appear on a cathode-ray screen. They are part of a computer-operated system that can proofread a 100-page book in about three minutes.

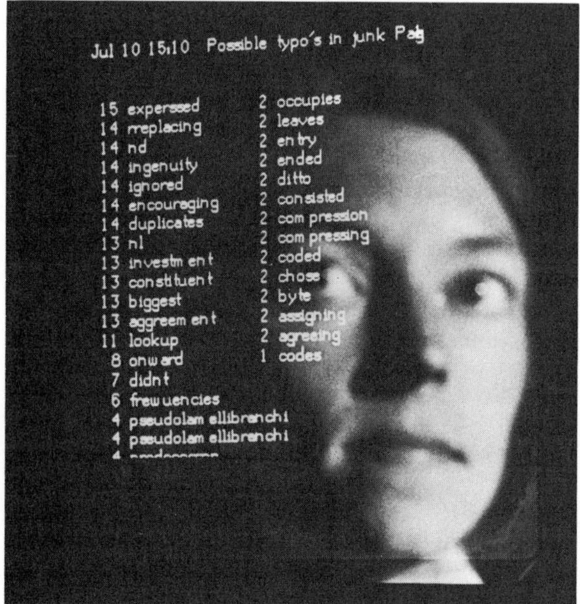

Courtesy, Bell Laboratories

Recent advances in communications technology, specifically communications satellites, combined with on-line computer technology (serial usage of units of a computer without delay or storage between units) opened new possibilities for bringing all the countries of the world into an international information partnership in which every user has equal access to the same information within the same time frame. Popinform, developed by the George Washington University Population Information Program for the storage and retrieval of documents on contraceptive technology, is such a computerized system. Popinform was demonstrated from various cities in the United States, South and Central America, Europe, and Asia by using commercial telephone and satellite communications to gain access to the Washington data base.

Information retrieval. Bioethics is the study of the social and ethical questions resulting from developments in biology and medicine; it embraces such topical issues as euthanasia, experimentation on human subjects, genetic manipulation, and behavior control. Plans were made during the year to establish a Bioethics Information System at Georgetown University, Washington, D.C., to provide systematic access to the literature of bioethics, now widely scattered in the journals of the contributing disciplines. The first projects of the new center were to include the development of an appropriate index language for providing bibliographic control of the rapidly expanding literature, and the design of an automated information retrieval system capable of providing such services as demand searches for relevant literature.

An information system for child placement was designed in recent months, and its feasibility was demonstrated at the North American Conference on Adoptable Children. The problem of child placement is a serious one, for statistics in 1975 indicated that there are almost three-quarters of a million children available for adoption in the U.S. alone. When fully operational, the adoption information system would have available three related files: child profile, family profile, and resource profile. The child profile would contain information on each child available for placement in terms of such factors as age, physical characteristics, handicaps (if any), race, and sex. The family profile would contain information describing the requirements of families wishing to adopt children, and the resource profile would describe the facilities and child-care services available at hospitals, clinics, and other institutions. The system was designed so that caseworkers anywhere in the U.S. could gain access to it by means of a standard terminal connected to a telephone. By dialing the computer, a caseworker would be able to select that group of families whose desires match the characteristics of the child needing placement and also to select that group of children whose characteristics match those of a family desiring to adopt a child.

Disabled individuals recently gained their own computerized retrieval system, called Accent on Information. It consisted of a computer at Illinois State University and advisory personnel from major organizations for the handicapped, and it contained information to help persons with disabilities to help themselves and to live more effectively. Subject and problem areas available from the system ranged from activities of daily living to communications, recreation, and transportation. A phone call set up a search and provided the requested information.

Information systems, in conjunction with computerized typesetting equipment, can be used by authors and publishers to detect typographical errors with only a minimum of human assistance. After the original document is typed into a computer, an "index of peculiarity" is computed for each word in the document. To accomplish this the computer program first divides each word into all possible two- and three-letter segments, known as digrams and trigrams. These are sorted and listed in a frequency table that indicates the number of times each segment appears in the particular document. As each word is processed, the digrams and trigrams are looked up in the previously prepared frequency table. Depending upon the rarity of the letter combinations, the computer assigns to each word a number from 0 to 20 that is equivalent to the word's "index of peculiarity." The words are then listed or displayed on a terminal scope with those having the highest index on top.

In one trial of the proofreading system, a 108-page document of nearly 20,000 words was examined for possible typographical errors in three minutes by the computer. The author of the document then needed fewer than ten minutes to scan the computer-generated word list and locate 30 misspellings, 23 of which occurred in the first 100 words listed by the computer. The system cannot perform the entire proofreading job; it does not find missing or extra words, or detect semantic nonsense, but it does provide important benefits for the typesetting and printing industries.

Future developments. Many countries during the year began efforts to establish national information systems (Natis), and these activities, encouraged and assisted by UNESCO, were expected to have great impact on the future development of information science. A national information system includes all the institutions dealing with knowledge in a country—libraries, information centers, schools, and extension services—that allow the results of research to be brought to the users and decision makers. Thus, planning for Natis included the coordination of all information, documentation, library, and archive services into a system that would ensure the optimum use of available resources and their maximum contribution to the cultural, social, and economic development of each nation.

To help in Natis planning, a UNESCO mission visited Algeria, Morocco, and Tunisia and discussed a variety of topics related to regional planning and the training of information personnel. In East Africa, a mission was undertaken to discuss, with appropriate officials, specific proposals for meeting some of the expressed information needs of the nations in that area. And in Senegal, another UNESCO mission carried out a feasibility study preparatory to the creation of a national scientific and technical information center. The government of Colombia legally ratified the establishment of a National Information System that would be implemented in cooperation with UNESCO and the UN Development Program.

Canada dedicated a new $14.8 million building in Ottawa as the home of the Canada Institute for Scientific and Technical Information. The building contains computerized equipment for the storage, retrieval, and dissemination of information. It houses the National Science Library and the Technical Information Service. The institute coordinates these services and is itself a major node in the evolving Canadian network of scientific and technical information services.

—Harold Borko

Life sciences

A major event in the life sciences occurred when an international group of molecular biologists voted to recommend guidelines for the conduct of certain kinds of experiments involving genetic manipulation. Other developments included discoveries concerning the immune response, a new approach to weed control, fossil evidence for the largest known flying creature, and a vaccine for hepatitis. The human cell continued to be the subject of extensive scrutiny.

Botany

Botanical research continued to support applied botany in the search for solutions to current problems. Many of these relate to environmental concerns such as chemical pollutants and renewable resource depletion. Others relate to plant disease and plant/animal relationships.

Biological weed control. Ecologists have been alarmed over the effects of adding chemicals such as synthetic fertilizers, pesticides, and selective herbicides to natural and agricultural ecosystems. Synthetic fertilizers have been used in large amounts to maximize crop yield in areas under intensive agriculture such as the corn belt of the U.S. In recent years nitrates, presumably from fertilizers, have increased in concentration in waterways and even in well water in some areas. The nitrates have influenced the ecology of streams, and when present in well water it is thought they affect the physiology of infants drinking that water. Pesticides such as DDT were used to control certain damaging insect populations. These pesticides, however, have been known to accumulate in the bodies of animals at the end of food chains. The real and potential damage, not only to the animals but to human beings eating the animals for food, led to rather strict regulation of a number of pesticides.

Selective herbicides have been used to control weeds growing on crop land, along roadways, and along railroad rights-of-way. The secondary effects of using these herbicides are not as well known as those of some pesticides, yet ecologists are learning to be skeptical about adding large amounts of any chemical to an ecosystem. For this reason alternative methods of controlling weeds continue to be considered with limited success. Among these are the selection of crop strains capable of outgrowing weeds and the utilization of natural insect enemies.

Two researchers have stimulated interest in a new approach to biological weed control. Alan R. Putnam of Michigan State University and William B. Duke of Cornell University have demonstrated that the phenomenon of allelopathy may be used to control weeds. Allelopathy is the process by which plants release chemicals that inhibit the growth of other plants. It is much more commonly found in nature than might be expected and thus may be responsible for the competitive abilities of many different kinds of plants.

Putnam and Duke first grew cucumber plants (*Cucumis*) of various species or strains in pots with two economically important weeds, a coarse grass (*Panicum miliaceum*) and a mustard (*Brassica hirta*). When grown with cucumber plants *Panicum* produced as little as 66% and *Brassica* as little as 78% of their growth without the cucumber plants, leading the researchers to reason that one or more chemicals produced by cucumber plants might be causing the inhibition of growth in the weeds. Such chemicals can be collected from the pots in which plants are growing by running a nutrient solution through the potting medium which, in this case, was sterile sand. The nutrient solution provides minerals needed by the plants and collects chemicals excreted into the potting medium by the plant roots. The excess solution, called a leachate, can be collected by a tube connected to a hole in the bottom of the pot. Leachates thus collected inhibited the growth of *Panicum*, and the leachates from some strains of cucumbers were especially effective—verifying that those strains of cucumber have the ability to produce a chemical inhibitory to *Panicum*.

Putnam and Duke suggested that the allelopathy that occurs rather widely in natural plant communities may have been characteristic of the ancestors of some domesticated crop plants only to be lost through the years of artificial selection employed in domestication. It remains to be shown whether the genetic system responsible for allelopathy in laboratory experiments is capable of being bred back into crop strains to the extent that they will give results in the field as promising as those in the laboratory. If so, allelopathy can be expected to contribute to the competitive ability of some crops against weeds.

Forestry developments. Certain forestry practices are being rethought in the light of contemporary ecological and genetic concerns. One of these is the control of fire. For years plant ecologists have collected evidence concerning the function of fire in the formation and preservation of natural ecosystems. At first it was thought that fire was instrumental mainly in the succession of plant communities whereby old community relationships were destroyed, making possible the colonization of new species and the establishment of new combinations of plants in an area. According to this thinking, once a relatively stable community (called climax by some ecologists) became established, fire was an enemy to be avoided because it would destroy the established plants. Such a view supported the policy of U.S. government agencies such as the National Park Service for over 100 years, and the symbol of Smokey Bear enjoined all comers to follow the policy.

As more and more evidence for the function of fire in natural ecosystems accumulated, it became apparent that fire also helps maintain many climax communities by burning over the surface periodically, recycling nutrients by destroying debris,

Sidney Harris

"It sort of made life exciting. I miss DDT."

and killing fire-sensitive seedlings of competing plants. Thus the fire-resistant seedlings of climax plants are allowed to compete successfully and the fuel for devastating forest fires is removed. It is quite certain that many Douglas fir forests in the northwestern U.S. had been maintained this way as were the redwoods of California. The term "fire climax" was coined to identify such communities. Controlled burning of both government preserves and commercial logging areas is being considered. One of the delaying factors, especially in government preserves, is public sentiment against the practice of burning.

Logging practices were also being reviewed, largely because of conservation-group opposition to clear-cutting, the practice of harvesting large blocks of even-aged trees at the same time. Foresters have defended the practice with biological and economic reasons. Biological reasons suggest that only a few kinds of tree seedlings can survive and develop in shade while others must have exposure to sunlight supplied by removal of forest canopy as provided in clear-cutting. Moreover, the rapidly growing trees of importance to logging enterprises are all somewhat shade intolerant. Wildlife is also benefited, it is suggested, many species finding forest openings attractive because the developing forest community provides sources of food. Economic reasons favorable to clear-cutting center on the lower cost of planting and/or management of even-aged stands and of harvesting them at the same time. The net result is an intensive type of forestry emphasizing maximum production.

Opponents to clear-cutting cite its abuses. The unsightliness of a clear-cut area with its debris, denuded landscape, logging roads, and erosion tends to promote conservationists' resistance to the practice. In addition, many feel that natural areas are disappearing at rates too great to be left unchallenged and that these natural areas are necessary to the welfare of man and wildlife. Furthermore, evidence that deforestation interferes with mineral recycling at least temporarily has come from work in experimental forests such as that of the Hubbard Brook Ecosystem Study in New Hampshire. Opposition to clear-cutting has had its impact even to the point of gaining court litigation against new timber sales in certain natural forests until forestry policies are better aligned with environmental concerns.

One example of an area subjected to great public and political scrutiny is the Monongahela National Forest of West Virginia. George R. Trimble, Jr., reported in the periodical *American Forests* that the main objection to clear-cutting in the Monongahela was the alleged failure to consider multiple forest usage instead of the overemphasis on timber production. When timber harvesting is allowed once more, Trimble predicts that changes in policy and performance will give greater consideration to aesthetics, wildlife habitat, and soil and water conservation. Specifically, three measures will be taken. First, improved harvesting techniques such as cable and aerial logging will be designed to reduce logging roads and erosion. Second, the area of intensive timber production will be reduced with logging prohibited in areas of primitive nature, along streams and roads, where other forest uses are planned, and when mature trees are needed to provide wildlife habitat. Third, intensified management will make timber production more efficient where it is allowed. These measures may make clear-cutting more palatable to opponents and still allow for efficient production of needed forest products.

Another concern of foresters involves the genetic future of trees. Plant breeders have learned that genetic variability is extremely important in the maintenance of plant stocks capable of being bred for specific environmental criteria, such as disease resistance and drought tolerance. Unfortunate experiences have taught this lesson, as the recent epidemic of corn blight attests. Now some foresters wonder if the gene pool of trees should not be protected even though forests are not subject to many of the problems of crop plants. For instance, it is unlikely there will be a widespread planting of a single genetic strain of trees since most trees are essentially wild populations and thus genetically diverse and trees live a long time avoiding depletion of the gene pool each year. So for major tree species of wide distribution there is little risk of losing genetic variability. However, those trees that grow in isolated stands may contain genetic diversity not found in other stands or in the general distribution of the species considered. Such stands are extremely necessary to identify and protect against the ravages of man's influence or castastrophe.

Genetic engineering in plants. Three Belgian scientists have succeeded in correcting a genetic deficiency in *Arabidopsis thaliana*, a flowering plant of the mustard family. L. Ledoux, R. Huart, and M. Jacobs secured seeds from strains of *Arabidopsis* which could not synthesize thiamine, an amino acid necessary for normal growth. Because most green plants are able to produce this amino acid, an inability to do so indicates the presence of one or more mutant genes. The attempt to introduce normal genes for thiamine synthesis involved soaking the mutant seeds in solutions of DNA secured from certain bacteria in the hope that normal genes from the bacterial DNA would somehow be taken up by cells in the mutant embryos inside the seeds. The seeds germinated normally, and 0.19% to 0.89% of the resulting plants were no longer thiamine deficient. Evidently the DNA containing genes for thiamine synthesis had been taken up by some embryos, and the newly acquired DNA was functioning along with the resident DNA of the cells.

These researchers investigated the inheritance patterns of the new genes by self-pollinating the corrected plants. The offspring were grown in thiamineless media to demonstrate they had inherited the mutant genes. All offspring up to the F_3 (third generation) and some up to the F_7 were found to breed true. Such breeding true is not expected, but these researchers explained it by invoking meiotic drive, the selection for corrected germ cells during meiosis, about the time that gametes are being formed. This makes sense because these cells would also need thiamine, and it was shown that the addition of thiamine to the media caused some plants to fail to breed true. Evidence from outbreeding with both deficient and normal plants supported the foregoing hypothesis of meiotic drive and demonstrated that the genes were functioning as dominant over the mutant genes for deficiency.

Bacterial diseases in plants. Among bacteria-induced diseases in vascular plants are those that cause tumor formation. One disease, caused by *Agrobacterium tumefaciens*, results when these bacteria enter plant tissues through wounds or are inoculated into the plant experimentally. Once a tumor is established it may continue to grow and give rise to other tumors in the plant even in the absence of the bacteria. A team of French researchers has been able to isolate the tumor-inducing factors from the *A. tumefaciens* which infects *Datura stramonium*, a member of the potato family. These factors are small single-stranded RNA molecules. Interestingly, both cancer-causing and noncancer-causing strains of *A. tumefaciens* contain this RNA, which means that some method yet to be discovered is employed to activate the RNA to initiate tumor production.

In another area of investigation Robert Goodman of the University of Missouri reported that plants have immune systems that are effective against certain bacterial invaders. He found that some plants synthesize proteins that cause certain strains of bacteria to clump soon after entering the plant. Then plant enzymes are able to digest the bacteria. This system bears some similarities to the immune mechanism known to exist in animals. Interest was drawn to such experiments because Goodman found that the plant protein recognized only nonvirulent bacteria. Some characteristics of virulent strains must keep them from being recognized and this may be one of the reasons why they are virulent; the plant cannot defend against them. If plants can be bred to recognize the heretofore virulent bacteria, certain plant diseases could be controlled.

Piñon pines and bird breeding. Investigators continued to be amazed at the intricate ways plants influence the animal world. A way in which pine cone production encourages the breeding of birds was discovered by J. David Ligon of the University of New Mexico. Most temperate bird species seem to breed in the spring, responding to the appropriate photoperiod (length of day and night). However, in areas where spring climatic factors are unpredictable and resulting food availability is affected, some birds may delay breeding until summer or later. Since this uncouples the breeding response from photoperiod, some other cue is necessary to initiate breeding. Experiments with caged piñon jays (*Gymnorhinus cyanocephalus*) showed that the presence of green piñon pine (*Pinus edulis*) cones was accompanied by the development of testes in some males. This suggests that some jays will reproduce in late summer if spring reproduction is lacking and if an abundance of piñon cones is available. The young jays do not eat many pine seeds as long as there is an abundance of insects, but it so happens that the climatic conditions conducive to maturation of abundant cones also ensure abundant insect production. Then later in the winter, when insects are no longer available, piñon seeds will be accessible for food for the young jays. The advantages of this relationship to the pines are not known. Perhaps some degree of seed dispersal is accomplished.

—Albert J. Smith

Microbiology

The most far-reaching event in microbiology during 1974–75 concerned precautionary measures taken by microbial genetic researchers to protect the world from potentially hazardous hybrid bacteria.

Scientists have recently developed a new technique of genetic manipulation that involves constructing hybrid molecules of DNA. The new technique depends upon the use of a newly discovered class of enzymes that cleave DNA into smaller pieces a few genes in length. The separated genes can then be joined with DNA of a different species. The latter is often the DNA of a virus that has the ability to enter into a cell where its hybrid DNA is replicated by the cell's genetic machinery. One common laboratory host for such experimental viruses is *Escherichia coli*, a bacterium that is used extensively by geneticists and molecular biologists. *E. coli* is a common inhabitant of the human intestinal tract. There is thus a remote but possible hazard to society that, owing to man-made arrangements of genetic material that have not occurred in evolution, *E. coli*, unpredictably and unintentionally, may be rendered highly infectious. Such a hybrid organism upon escape from the laboratory could cause a human epidemic of severe proportions.

In July 1974 a group of scientists called for a voluntary moratorium on further research using this technique of genetic manipulation. In February 1975 an international group of biologists held a conference on this issue. The group voted to lift the moratorium, but they established stringent safety precautions and guidelines to be followed. Hybrid DNA molecules resulting from genetic manipulation in the laboratory may solve basic problems in biology that cannot be solved by present techniques. They may also have such practical value as the insertion of bacterial nitrogen-fixing genes into plants or the programming of bacteria to synthesize insulin, to cite only two possibilities.

Medical microbiology. Hepatitis is a debilitating disease of the liver. There are two forms of hepatitis, which are considered to be caused by two viruses. Unfortunately, researchers have been unable to grow the viruses in culture, which is normally a requirement for the production of a vaccine. Scientists now have sidestepped this problem by isolating viral antigens from human carriers of hepatitis B, one of the two forms. For reasons not clearly understood, large amounts of excess viral coat protein are produced by persons harboring the virus. The protein coat material can be separated and purified from blood and used as a vaccine. The new vaccine has been proven successful in chimpanzees. The next step will be testing in humans to ensure that the vaccine is safe, a process that may take as long as five years.

Research is continuing on methods to detect venereal diseases and to develop vaccines. Approximately 17,000 new cases of gonorrhea are reported weekly in the U.S., and more than 84,000 new cases of syphilis are reported yearly. These figures probably represent the proverbial tip of the iceberg because many cases go unreported and many persons are unaware that they are infected with venereal disease. Progress is being made in the development of more efficient detection techniques, but efforts to develop vaccines have been disappointing. The microorganism that causes syphilis cannot be grown in culture, which has impeded efforts to develop an effective vaccine. Despite the fact that the bacterial agent of gonorrhea can be grown in culture, protective immunity does not result when cultured organisms are used as a vaccine, although past claims, later proven to have been premature, have been made for a successful vaccine.

The microorganism that causes leprosy was discovered in 1873, yet scientists have been unable to cultivate it in the test tube. Until recently no satisfactory animal model was available in which to study the disease. It was discovered in 1971, however, that the nine-banded armadillo was susceptible to leprosy and could be used for the study of the disease. During this past year scientists succeeded in cultivating the bacterium that causes leprosy in rodents. The latter bacterium is closely related to the human form, and it had also resisted test-tube cultivation. This discovery may pave the way toward laboratory cultivation of the human form of the leprosy bacillus.

There have been numerous premature announcements of the discovery of human cancer viruses. Now it appears that two viruses have been identified as causative agents of human cancer. In June 1974 Charles McGrath, Marvin Rich, and associates of the Michigan Cancer Foundation announced that they had isolated a virus from human breast cancer cells, while a group from the National Cancer Institute reported in January 1975 that they had isolated a virus from human leukemia cells. Time will tell whether these two reports will withstand scrutiny, but to date the evidence seems firm.

A vaccine has been prepared by Hutton D. Slade and associates of Northwestern University that prevents tooth decay in rabbits under experimental conditions. Tooth decay is considered to begin when certain bacteria, by producing adhesive material, attach to the smooth surfaces of teeth. This adhesive material was isolated and was used as a vaccine in rabbits. The animals formed antibodies that were able to prevent the adherence of bacteria to the tooth surface, thus precluding the onset of tooth decay. Whether this will prove to be practical for humans will require a considerable period of further research and testing.

Microbiology and waste recycling. Traditional solutions to waste disposal such as burning, landfill, and dilution are becoming less acceptable. Instead more sophisticated approaches are being advanced. These approaches are designed not only to make waste acceptable in the environment but also to make waste materials more valuable—for example, by their bioconversion into useful products. Billions of tons of cellulosic waste products result from agricultural crops in the form of stalks, stems, leaves, hulls, etc. Similarly, 40–60% of home and commercial wastes are cellulose. Scientists at the U.S. Army Laboratories in Natick, Mass., have now developed a method of converting cellulose material to sugar by use of an enzyme that is obtained from a common soil fungus. The sugar then can be used for animal feeding, or for further bioconversion to single-cell protein or to such fuels as alcohol or methane. Efforts to convert other wastes, such as animal feedlot wastes, into methane or single-cell protein, also have now reached the pilot-plant scale. (*See* Year in Review: CHEMISTRY: *Applied Chemistry*.)

Microbial nitrogen fixation. There is an increased interest in the study of nitrogen-fixing microorganisms. This is a result of increasing populations and dwindling food reserves and of increasing expense in the production of chemical fertilizers. The latter involves the Haber-Bosch process for ammonia synthesis, which consumes large amounts of fossil-fuel energy. Biological nitrogen fixation is the reduction of atmospheric nitrogen to ammonia by certain bacteria and algae.

The most studied nitrogen-fixing microorganisms are species of *Rhizobium*, the root-nodule bacteria of leguminous plants (*e.g.*, soybeans, clover). These bacteria, however, fix nitrogen only when they are living in symbiotic association (*i.e.*, in root nodules) with the leguminous plants and not when grown in the test tube. Legumes such as soybeans are rich in proteins and, for that reason, are in demand in world markets. Grains, which lack symbiotic nitrogen-fixing bacteria, have fewer proteins of poorer quality.

A promising development was revealed in 1974 when a scientist from Brazil reported the discovery of an association between a nonrhizobial bacterium, *Spirillum lipoferum*, and the roots of a

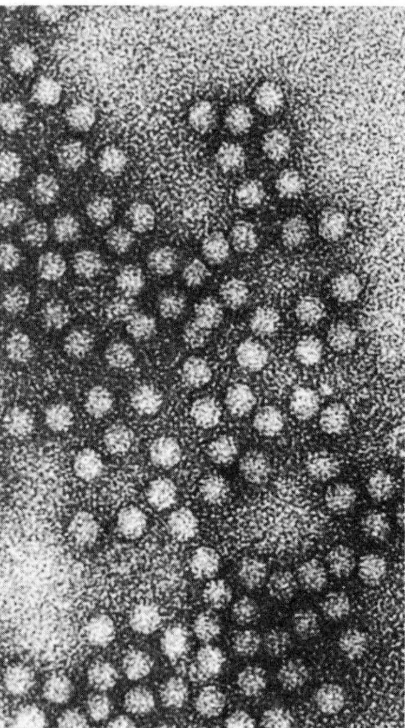

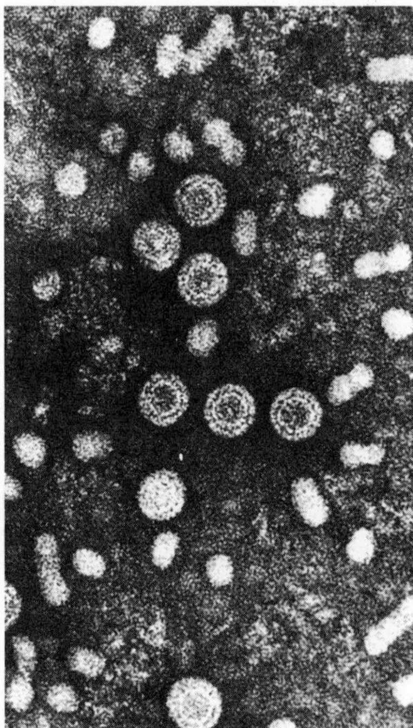

Virus of type A (infectious) hepatitis (left) and purified Australia antigen and viral Dane particles (right) found in patients with type B (serum) hepatitis. The two forms of the disease have been distinguished from one another by new serologic tests using an extract of marmoset liver infected with the type A virus. The tests are expected to simplify the diagnosis of hepatitis A.

Photos, courtesy, Maurice R. Hilleman, Merck Institute for Therapeutic Research tropical grass. The bacteria are located within the inner cells of the root cortex, and their efficiency of nitrogen fixation is almost as high as that of *Rhizobium* species. *S. lipoferum*, however, does not cause root nodules to form and it can fix nitrogen when cultured in the laboratory. It is possible that this discovery may pave the way for man-made associations between nitrogen-fixing bacteria and grasses or grain crops.

Another important development in nitrogen fixation was revealed in reports by both British and U.S. scientists that genes for nitrogen fixation could be transferred to bacteria that do not fix nitrogen naturally. Thus the property of nitrogen fixation was conferred onto nonnitrogen-fixing bacteria. Although highly speculative at the present time, scientists are currently interested in creating new nitrogen-fixing bacterial species via such genetic manipulations with properties that make them attractive to agriculture. There is also considerable interest in transferring bacterial genes for nitrogen fixation to plants in order to endow the recipient plants with the ability to fix their own nitrogen from atmospheric sources. There are several pitfalls to this latter approach, however; *e.g.,* only plants with high capacity to photosynthesize would be suitable because a great amount of energy is needed to fix nitrogen, whether by chemical or biological processes.

Until recently there were no examples of symbiotic associations between *Rhizobium* species and plants other than legumes. This situation was changed when an Australian scientist found a *Rhizobium* species growing in nodules on the roots of a tree in New Guinea. This raises the question whether *Rhizobium* species could be made to form associations with other trees.

Biological nitrogen fixation is not limited to soil systems. A group of researchers from the University of Georgia found that the algal flats on coral reefs at Eniwetok Atoll in the Marshall Islands of the Pacific Ocean fix atmospheric nitrogen at rates comparable to those in managed agricultural systems. The dominant nitrogen fixer appears to be blue-green algae. This provides an answer to a problem that has long puzzled scientists. Coral reef communities are characterized by high rates of biological productivity. These communities obtain their nutrients from the overlying waters, but tropical marine waters are characterized by low levels of dissolved nitrogenous compounds, which should severely limit productivity. The puzzling high rates of biological productivity actually observed can therefore be accounted for by nutrient enrichment owing to biological nitrogen fixation.

Other developments. Microorganisms other than nitrogen fixers also live in close association with the roots of plants and contribute to the well-being of the plants, and recent attention has been given to the association of fungi with roots of trees. In the case of pine trees, the threadlike mycelia of the fungi form a mantle around the cells of the roots. This causes the cells to increase greatly in size with the result that there is an increased uptake of water and nutrients. In fact pine trees will die when set out in poor soil if fungi are not associated with the roots. Efforts are now under way to inoculate artificially the roots of pine seedlings with the proper species of fungi to ensure survival of the trees when transplanted.

In the case of hardwoods and grasses, the fungi actually penetrate the cortex of the root cells. The net result is the same as for pine trees: there is increased water and nutrient uptake. The survival rate of hardwoods when set out is much less than that of pines because the seedling hardwoods do not become infected with the proper fungal species in nurseries. Experiments are now in progress to learn how to inoculate seedling hardwoods with the proper fungi. The fungi involved in association with roots of both pine trees and hardwoods are quite familiar. The fruiting bodies of these fungi include the commonly found puffballs, toadstools, and mushrooms.

It was reported in mid-1974 that bacteria frozen for at least 10,000 years in permanently frozen ground and sediment cores in Antarctica grew when inoculated in culture media. The bacteria were reported to be unlike those occurring in the surrounding environment. This lends credence to the evidence that the growth was really caused by bacteria long frozen in the samples and not by laboratory contamination.

Future trends. The areas in microbiology that appear to be at the forefront for increased development may be discerned by observing where federal funds are being committed for support of research. The U.S. National Science Foundation is seeking proposals for research in five areas of microbiology: (1) development of technology for industrial production and use of food and food proteins by microbial means; (2) engineering research and development of equipment and methods for the computerized simulation design and control processes for microbial technology; (3) molecular biology of industrial microorganisms; (4) development of ways to produce and apply enzymes for industrial and analytical goals; and (5) microbial control of pests in agriculture. There is widespread international interest in these areas and, moreover, the U.S. and the U.S.S.R. have entered into a cooperative agreement for research dealing with these five subjects.

—Robert G. Eagon

Molecular biology

In 1974 *Nature*, a prestigious British weekly magazine that reports on current scientific activities, devoted a major section of one issue to the "coming of age of molecular biology." There was some self-serving in this journalistic celebration since 21 years earlier (on April 25, 1953) *Nature* published the first paper by James D. Watson and Francis H. C. Crick setting forth their proposed model for the structure of deoxyribonucleic acid (DNA), the substance ultimately responsible for hereditary control of life functions. As Robert Olby correctly pointed out in the celebration issue, it is misleading to suggest that molecular biology was born with the Watson-Crick model and came of age with the twenty-first birthday of the model in 1974. There had been, after all, a good reason for Watson and Crick to have spent a hectic year trying to arrive at the probable structure of DNA. Considerable evidence already existed that DNA was the stuff of which genes were made, and by 1953 there was enough known about the properties of DNA to suggest that a search for its structure would not be unproductive. The consequences that almost immediately followed publication of Watson and Crick's model were, however, probably beyond the wildest dreams of any scientist, including Watson and Crick.

The underlying assumption of molecular biology is that by knowing the structure of a biologically significant molecule one gains an understanding of how that molecule functions in the organism. Rarely has that tenet been more convincingly validated than in the research stimulated by the Watson-Crick model. Within two decades after its proposal the basic features of the model were experimentally confirmed: DNA is a long molecule consisting of two chains wound around each other to form a double helix. Each chain is a linear polymer of deoxyribonucleotides, the sugar (deoxyribose) and phosphate portions of which form the backbone, and the bases (the purines, adenine and guanine, and the pyrimidines, thymine and cytosine) of which project almost perpendicularly from the backbone. The two chains in the double helix are of opposite polarity such that the bases of one chain confront the bases of the other chain at regular intervals. Hydrogen bonds form between the base pairs that confront each other, and only four kinds of base pairs are allowed: adenine-thymine; guanine-cytosine; cytosine-guanine; thymine-adenine (where the first-named base is on the same chain).

Not only was the model confirmed by a variety of ingenious experimental investigations but the hoped for understanding of the biological action of DNA also came to pass. The various chemical activities of the cells of which the living organism is composed are brought about by a variety of protein catalysts or enzymes, each enzyme capable of catalyzing only a specific reaction. Prior to Watson's and Crick's contribution, it was already known that each gene governed the production of a specific enzyme. A protein was known to be a polymer of amino acids; there were about 20 different kinds of amino acids in proteins, but it was the particular sequence of amino acids in a protein molecule that conferred upon it a specific pattern of folding and, therefore, a specific three-dimensional configuration upon which the particular catalytic activity of the protein depended.

Watson's and Crick's discovery led to the knowledge of how a gene determines protein specificity: a gene corresponds to a particular segment of a DNA double helix, often several hundred base pairs in length. Proceeding from one end of the gene to the other, each succeeding triplet of base pairs codes for a particular amino acid in the protein governed by that gene. Thus, the specific sequence of base pairs in a gene determines the specific sequence of amino acids in a protein. (In fact, an enzyme is sometimes made up of two to four different kinds of amino-acid chains called polypeptides. Each polypeptide is specified by a particular gene. An enzyme made up of more than one kind of polypeptide is, therefore, determined by more than one gene.) How this knowledge has been acquired is beyond the scope of this article; it must suffice here to summarize the principal features of the mechanism by which genes determine the structures of proteins, so that current work in molecular biology can be appreciated.

Transcription. The action of a gene is indirect. By means of an enzymatically catalyzed process called transcription one of the two strands of a gene becomes a template for the synthesis of a transcript. The transcript is not an exact copy of the transcribed gene strand but is a chain of ribonucleotides, or RNA, that is nevertheless complementary in its base sequence to that of the transcribed gene strand. Thus, the transcript is similar in its sequence of bases to that in the nontranscribed strand, said to be the complement of the transcribed strand. (The only exception is that uracil replaces thymine in the transcript.) The transcript is referred to as the gene's messenger, or mRNA, because it leaves the chromosome where it is formed to enter the cytoplasm. There, in association with a complex organelle (specialized cellular part) called a ribosome, the messenger is said to be "read" or translated.

In the ribosome, amino acids are joined to each other in a sequence specified by the order of base triplets in the mRNA, which was previously determined by the gene. The "reading" of the instructions of the mRNA is facilitated by a class of small RNA molecules called transfer RNA (or tRNA). Different tRNA molecules combine reversibly with different amino acids in the cytoplasm; at one region of each tRNA molecule a specific triplet of bases is exposed. The tRNA triplet complementary to the triplet in the mRNA being read is effectively brought into the ribosome, and its attached amino acid is then joined to the polypeptide chain that has already been formed under the direction of the segment of the mRNA molecule that has made its way through the ribosome.

The tRNA molecule is thus a device for transporting a particular amino acid to the polypeptide chain in the process of formation. Once the amino acid that it carries is joined to that chain, the tRNA molecule leaves the ribosome and is free to attach reversibly to another member of the amino-acid species that its structure recognizes.

By 1975 a large number of the various kinds of tRNA molecules had been purified (freed from extraneous, alien material). In some cases the structures of these molecules had been determined, so that researchers knew where in the overall molecular configuration a specific amino acid is reversibly attached as well as the location of the triplet of bases responsible for recognizing a complementary triplet, or codon, in an mRNA molecule. Because one can know which amino acid a given tRNA carries and because the recognizing triplet (or anticodon) can be distinguished in the configuration of the tRNA molecule, it has become possible to assign unequivocally a specific amino acid to several of the various possible codons. The correspondence between codons and amino acids is called the genetic code, and a major triumph of molecular biology has been the successful "breaking" of this code.

Nucleotide sequence of yeast phenylalanine transfer RNA reveals a cloverleaf pattern. Circled bases indicate constants in all tRNAs; triangles indicate accessibility of various bases to chemical modification.

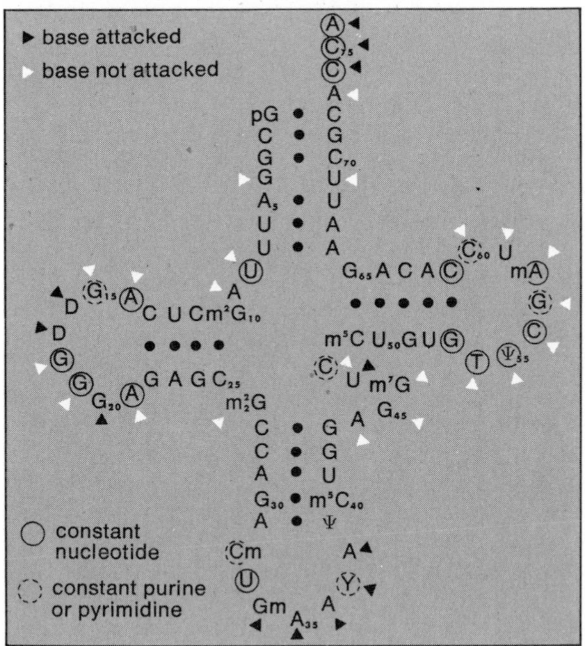

Based on "Three-Dimensional Tertiary Structure of Yeast Phenylalanine Transfer RNA," by S. H. Kim et al., in "Science," vol. 185, pp. 435–440, Aug. 2, 1974, © AAAS

The deciphering of the genetic code did not need to wait upon the purification and structural analysis of tRNA molecules, however. It had been, in fact, accomplished several years earlier when artificial mRNA molecules of known nucleotide sequence were synthesized and used for directing the synthesis of polypeptides in vitro: ribosomes extracted from the cell will construct polypeptide chains when placed in an appropriate medium containing, among other things, mRNA and tRNA molecules charged with amino acids. The amino-acid sequence of the polypeptides that are synthesized is in strict conformity with the sequence of codons in the mRNA.

Polymerases. To be inherited a gene must be duplicated (replicated) and the copies distributed between daughter cells. In molecular terms this means that the sequence of base pairs in a segment of DNA constituting a given gene must somehow be reproduced. The Watson-Crick model suggested how DNA reproduction might occur, and the suggestion has been fully confirmed: the complementary strands of a DNA helix are separated and unwound while each strand serves as a template for synthesizing a complementary copy. Eventually two new DNA helices identical to each other are made. Another mechanism assures that the two helices are then disjoined into daughter cells.

While this basic pattern of gene reproduction undoubtedly occurs, the complete details of the process are not yet fully understood. What is known is that a number of enzymes called DNA polymerases are involved in DNA reproduction. At least three different DNA polymerases have been isolated from bacterial cells; possibly as many will be detected and isolated in the cells of more highly evolved organisms.

It is still too early to indicate the exact role of each polymerase, but the answer may lie in their involvement in other processes besides replication. These other processes are repair and recombination. When DNA interacts with certain physical and chemical agents, such as ultraviolet light, ionizing radiation, and alkylating agents, the consequences are either breaks in one or the other of the two DNA strands or alterations in the bases themselves on either strand. Altered nucleotides, or nucleotides near breaks, are often excised by a specific group of enzymes; the gaps must eventually be filled in with the help of polymerases. Finally, the interruption between one nucleotide and the immediately adjoining one on the same strand must be resealed by an enzyme known as DNA ligase. In the process of excision and repair synthesis, "incorrect" nucleotides (those having bases noncomplementary to the corresponding ones in the undamaged strand) are sometimes substituted for the excised nucleotides. In this way heritable mutations in gene structure are produced.

The other process in which DNA polymerases are probably involved is that of genetic recombination. Occasionally, two similar DNA helices confront each other and exchange parts. Such genetic exchange, or "crossing-over," is known to happen when homologous chromosomes (of maternal and paternal origin) pair in sex cells prior to their disjunction into gametes. This genetic exchange or recombination effectively joins genes of maternal origin to genes of paternal origin and is believed to be of considerable evolutionary significance in increasing the genetic diversity within populations of species. That DNA repair enzymes are also involved in recombination is indicated by the fact that cells defective in repair are often defective in recombination and conversely.

Another class of polymerases uses DNA (or one strand of it) as a template but catalyzes the synthesis of complementary RNA rather than DNA. These are the so-called DNA-dependent RNA polymerases, and they are undoubtedly involved in the in vivo transcription of mRNA. Indeed, such enzymes are essential for the construction of RNA in vitro using DNA as a template.

In recent years an unanticipated type of polymerase was discovered: it uses RNA as a template and catalyzes the synthesis of complementary DNA. It is called, half-jocularly, reverse transcriptase, or, more seriously, RNA-dependent DNA polymerase, and is found in cells infected by certain viruses composed of RNA. The significance of this enzyme is apparently as follows: The infectious viral RNA contains a messenger for the making of reverse transcriptase on host-cell ribosomes; the reverse transcriptase catalyzes the synthesis of cDNA (DNA complementary to the viral RNA); the cDNA can then either reproduce via the cell's battery of DNA polymerases or may be integrated into the host DNA by a process involving recombination enzymes. If only part of the cDNA is integrated, the host cell may express some new property (such as a change in the protein composition of the cell membrane) determined by a viral gene without ever producing virus. If all of the cDNA is integrated, new copies of viral RNA are made every time the cDNA is transcribed by a normal transcriptase.

Procaryotic and eucaryotic organisms. Much of the evidence scientists possess regarding gene action stems from work done with very small and relatively simple organisms, the bacteria. These organisms are said to be procaryotic because of the simple nature of their chromosomes. Bacterial genes are linked to each other in a continuous double helix of DNA having no free ends. Such a group of linked genes, or chromosome, is thus a loop or "circle" of DNA; it is not firmly associated with any proteins and is not separated from the rest of the bacterial cell (the cytoplasm) by a nuclear membrane.

All organisms that have evolved further than bacteria and blue-green algae have a more complex chromosome and are said to be eucaryotic. The eucaryotic chromosome is a linear DNA helix with ends; there may be more than one kind of chromosome in respect to gene content per chromosome set. Moreover, in the eucaryotic chromosome, DNA is regularly imbedded in a protein matrix, and the chromosomes are separated from the cytoplasm by a nuclear membrane except during mitosis, when daughter chromosomes are to be separated to opposite poles of the cell.

In a given eucaryotic species the different members of the chromosomal set are most clearly distinguished when the chromosomes are condensed or contracted in preparation for mitotic separation. Chromosomes, therefore, are best recognized in actively dividing cells, and this is often arranged by removing cells from a multicellular eucaryotic organism and growing them in a suitable culture medium. By appropriate techniques one can accumulate cells arrested in mitosis, and these cells can then be chemically treated and stained to reveal the chromosomes.

Regulation of gene action and chromosome structure. Genes are not expressed in all cells at all times. In procaryotes gene expression is known to be regulated in a number of ways. Procaryotes possess a class of regulator genes, the function of which is to determine the structure of specific regulatory protein molecules. These regulatory proteins, called repressors or activators, combine with specific sites in a procaryotic chromosome and either prevent or activate, respectively, transcription of mRNA from immediately adjacent regions. The interaction of regulatory proteins with specific chromosome sites may in turn be modulated by environmental conditions. In some cases, the specific regulatory proteins have been isolated and purified, and the sites of the chromosome to which they specifically attach have been identified. The site of the region whose transcription is being coordinately regulated, called the "operator," has been found to be sensitive to certain "restriction" enzymes, which cut DNA only at specific sites. The base sequence of the operator has been partially determined in some cases and turns out to have a unique structure characterized by a twofold axis of symmetry that allows the strands of the operator to fold back upon themselves. This structure probably underlies the sensitivity of the operator to restriction enzymes and its power to combine with certain proteins. In any event, regulation of gene expression in procaryotes is visualized as an environmentally modulated system of specific DNA-protein interactions.

Unfortunately, much less is known about the regulation of gene expression in eucaryotes, although there is reason to believe that in a multicellular eucaryote such as a man or a mouse different genes are being expressed in different cells. Regulatory proteins acting upon specific eucaryotic genes had not been discovered by 1975. Investigations were being made, however, of the proteins associated with the chromosomes. Chromatin, the substance of which eucaryotic chromosomes are composed, consists of a continuous chain of DNA to which are attached proteins of two general classes, histones and nonhistones. Histones are rich in such basic amino acids as lysine and arginine and have been thought at various times to regulate gene action. What seems more certain, however, is that a chemical change (phosphorylation) in at least one kind of histone triggers chromosomic contraction and is a signal for mitosis.

Experimenters recently reconstituted chromatin from purified DNA, histones, and nonhistone proteins, and were able to use it for making mRNA in vitro. The kind of mRNA made appeared to depend on the source of the nonhistone proteins rather than the histones, thus indicating the former as controllers of eucaryotic gene expression.

Occasionally, nucleic acid of foreign origin enters the eucaryotic nucleus and may actually become integrated into chromatin. This is true, for example, of a class of viruses that are associated with the induction of cancerous growth in the cells they infect. Depending upon the species of virus, it may contain, outside the host cell, a protein coat and a core of either RNA or DNA. Upon infection only the nucleic acid of the virus is believed to enter the cell. In the case of the RNA viruses, reverse transcriptase is believed to be responsible for the making of DNA complementary to the virus; this cDNA is integrated into a chromosome, possibly a specific one at a specific site, and subsequent transcription makes new viral RNA copies. This is the way in which the virus causing Rous sarcoma in chickens is believed to act. In the case of DNA viruses (such as simian virus 40 or the mammalian adenoviruses) whole virus particles are sometimes found in the cancerous cells, or the virus is absent but a part of the viral DNA is detectable in the malignant cell.

By the use of restriction enzymes, adenoviral DNA was separated into a number of discrete fragments. Then, mRNA was synthesized in vitro from each of the fragments, which acted as templates. The mRNA in turn was used to hybridize with the DNA removed from malignant cells. In this way, it was found that only mRNA transcribed from a specific fragment hybridized to the malignant cell's DNA, indicating that only a specific part of the viral gene constitution has to be integrated into the chromosome of the host cell for a malignant transformation to occur.

Of obviously great interest is the possibility that some human cancers are caused by such viruses. By 1975 the evidence was overwhelming that certain viruses are involved in the induction of cancerous cell growth in humans. Much of the evidence remained circumstantial, but hybridization experiments showed in many cases that the chromosomes of human cancerous cells contain DNA complementary to the nucleic acid of viruses known to induce malignant growth in animals or in human cells in culture.

Cytoplasmic organelles. The ribosome has not only been isolated and used to synthesize proteins in vitro under the direction of natural and synthetic messenger RNA, but also it has been dissected by careful biochemical procedures that preserve the biological activity of individual ribosomal components. Thus, the several different kinds of protein and the special class of RNA that comprise the ribosome have been separated and purified. Fortunately, under appropriate conditions of mixture, the purified components will reassemble to form synthetically active ribosomes. By leaving out a specific component or by using one component from the ribosomes of a different species of organism, researchers were able to determine the functional roles of some of the components as well as the degree of similarity of ribosomal components in different species. Thus, by means of antibodies that can readily detect fine differences in structural detail, two proteins of the ribosomes of a bacterium, *Escherichia coli,* were determined to be structurally similar to two proteins found in rat ribosomes.

Other cytoplasmic organelles were also being biochemically dissected. Mitochondria are bodies containing the energy-producing machinery of the cell, while chloroplasts contain the photosynthetic machinery of the cells of green plants. Both of these organelles underwent analyses in recent years. It is particularly interesting that DNA was found inside them. This DNA is characteristic of the organelle and contains genes that are inherited independently of chromosomal genes. Indeed, organellar DNA often manifests what is called maternal inheritance. For example, because sperm contribute only chromosomal genes to the fertilized egg, which counts upon its own organelles as a source of mitochondrial genes, characters determined by mitochondrial genes show a pattern of maternal transmission from parent to offspring.

The genetics of mitochondria and chloroplasts was beginning to advance after a period of relative neglect caused by the difficulties involved in the subject. For example, both chloroplasts and mitochondria appear to contain proteins determined by nuclear genes and produced in cytoplasmic ribosomes, as well as proteins determined by organellar genes and produced in ribosomes located within the organelles themselves.

Developmental biology. The current direction of molecular biology involves attempts to utilize the knowledge already acquired as a result of the convergence of genetics, biochemistry, and microbiology in order to explain the fundamental phenomena associated with highly evolved forms of life. Molecular research, therefore, was turning increasingly from the study of bacteria and unicellular forms of life to the study of highly organized multicellular organisms.

A major question is the origin of cellular differentiation in the bodies of multicellular organisms. Each of these organisms begins its life as a single cell, yet the mitotic descendants of that cell become differentiated from one another. Underlying this differentiation are molecular changes resulting in specific proteins being made almost exclusively in certain cells. In vertebrates, embryonic muscle cells synthesize actin and myosin primarily, and in reticulocytes hemoglobin is almost the only protein produced. Experimenters were able to insert the nuclei of differentiated cells into enucleated eggs; when the latter were activated to divide, they developed normally, thus indicating that the nuclei of differentiated cells retain the full chromosomal gene complement necessary for normal development. Development must, therefore, involve differential gene expression. Attention during the year was focused upon the problem of how that occurs. While the environmentally modulated repressors and activators of procaryotic genes suggest ways in which eucaryotic genes may be turned "on" or "off," molecular scientists remained a long way from having a clear idea of eucaryotic development.

Nevertheless, experimenters introduced some interesting new methods that bid fair to elucidate the processes of development in the near future. One of these methods consists of injecting biological macromolecules into large eggs, such as those of the African toad, *Xenopus laevis.* When one injects mRNA from rabbit reticulocytes, which makes primarily the protein globin (of hemoglobin), this mRNA is translated by the toad ribosomes so that rabbit globin continues to be made while the toad develops, at least as far as the tadpole stage. This result reveals two things. First, it is evidence that the genetic code is universal: codons are translated the same way regardless of species. Second, the ribosomes appear to be able to read all kinds of messengers, so that the origin of differentiation does not appear to be in the ribosomes themselves.

Another technique consists of transferring whole chromosomes or nuclei from one cell type to another. Thus, cultured hamster cells genetically incapable of making a specific enzyme may be exposed to human chromosomes containing the gene that specifies that enzyme. In ways that are not yet clear, the human chromosomes are absorbed and their genes are expressed so that the host hamster cells soon begin to make the lacking enzyme. Some of the human chromosomes may eventually be lost from the dividing hamster cells, but so long as a particular human chromosome (the sex, or X, chromosome in this case) is retained the enzyme in question is made. In this way, therefore, it is possible to assign a specific human gene to a particular chromosomal location.

More commonly employed is the technique of cell fusion. Under certain conditions cells of different types and from different species fuse, and often the two nuclei of the fused cells combine to produce a common nucleus. One can then determine which, if any, of the specialized molecular products of the two cell lines are expressed in the fused or hybrid cell. One of the cell types used in these somatic cell hybridization experiments is often derived from cancerous tissue because cancer cells characteristically reexpress certain molecular activities that are turned off in normal differentiated cells. Thus, mammalian fibroblasts fail to make dopa oxidase (a pigment-producing enzyme) while melanoma (pigmented cancer) cells do. When mouse fibroblasts were hybridized with the cells of a hamster melanoma, the hybrid cells failed to produce dopa oxidase. These findings indicate that the differentiated state depends upon the production of a diffusible agent, which acts to "turn off" the production of specific proteins.

The aging process (senescence) is a feature of development that seems to have a molecular basis. Human fibroblast cells appear to senesce, in that upon aging they fail to replicate nuclear DNA. A fusion of a senescent with a nonsenescent fibroblast results in the absence of nuclear DNA synthesis, indicating that the senescent phenotype is dominant and is due to the accumulation of diffusible products inhibitory to DNA synthesis. This finding does not exclude the possibility that aging may be caused by other factors, such as the accumulation of environmentally induced damages to DNA and especially to genes controlling DNA repair processes.

As already pointed out, heritable changes in the gene complement are not a normal concomitant of development. An exception exists in the case of certain genes that code for the so-called variable part of immunoglobulins, the proteins that act as antibodies against foreign substances (antigens) in mammals. An individual mammal can make in its lifetime an astonishing variety of antibodies that differ in amino-acid composition. These differences in structure appear in a variable part of the immunoglobulin molecule, which is otherwise constant. The variable part is under the control of one or only a few V genes.

How then does antibody diversity arise during an individual's lifetime? One theory is that the gene undergoes random mutation in the lymphocytes that make immunoglobulins and that a mutant producing a specific immunoglobulin may undergo preferential multiplication when in contact with the corresponding antigen. The antigen selects cells producing specific antibodies but does not "cause" the initial appearance of the latter. According to this theory, the rate of mutation must be relatively high in the immunoglobulin-producing lymphocytes. It is interesting, therefore, that an enzyme affecting DNA structure has been found to exist only in that part of the thymus gland giving rise to such cells. This enzyme, going by the name of terminal deoxynucleotidyl transferase, can catalyze the addition of deoxyribonucleotides to the ends of DNA fragments in vitro. Investigations were under way to check whether this enzyme may act as a mutagen (a substance that tends to increase the frequency of mutations) in lymphocytes so as to diversify the V genes in those cells.

Hybrid DNA and the ethics of science. An impressive demonstration of the universality of the genetic code took place recently with the introduction of eucaryotic genes into bacteria. The bacterium employed was *Escherichia coli,* a common inhabitant of the intestines of mammals including man. This bacterium contains a loop of DNA consisting of the thousands of genes indispensable to its survival and also occasionally becomes infected by small DNA loops containing useful but dispensable genes. These small loops are called plasmids, and some of them are highly infectious, capable of reproducing in their hosts without killing them and capable of being transmitted to uninfected hosts. It is by way of such plasmids that pathogenic bacteria resistant to a large number of antibiotic drugs have increased in frequency in recent times: the plasmids carry genes conferring antibiotic resistance. Unwary therapeutic use of antibiotics in treating diseases of human beings has undoubtedly been responsible for this increase.

Plasmid DNA can be isolated and cut into specific fragments by restriction enzymes. Only a certain fragment may contain the genes necessary for infectivity and reproduction. To this fragment may be spliced a specific fragment of DNA obtained by using restriction enzymes on eucaryotic DNA. The eucaryotic fragment may contain, for example, only those genes determining ribosomal RNA. In this way specific genes of the African toad *Xenopus* and of the fruit fly *Drosophila* have been

Escherichia coli *bacterium became the center of a scientific controversy in 1974 when a group of scientists proposed a voluntary moratorium on certain experiments that could render the common intestinal inhabitant dangerous to human life. In 1975 an international meeting of scientists lifted the ban but recommended guidelines for future experiments.*

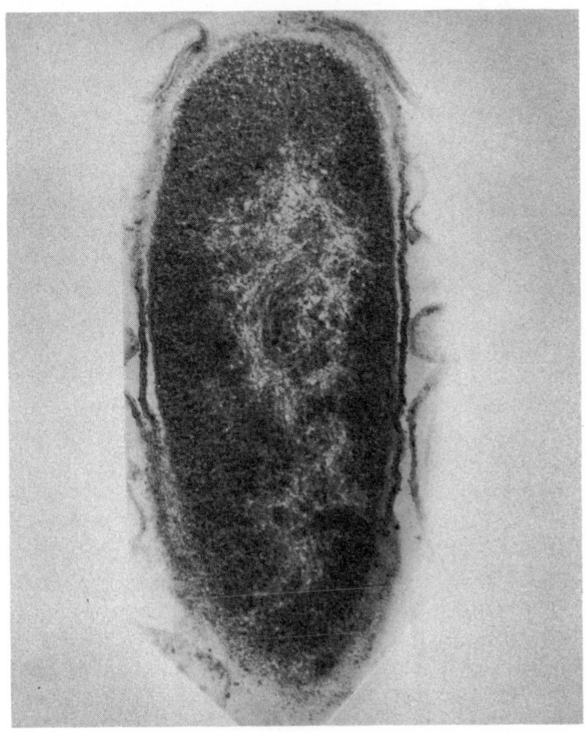

Courtesy, Samuel C. Silverstein, The Rockefeller University

attached to plasmid DNA, and the hybrid, or recombinant, DNA thus formed has been used to reinfect *E. coli.* The infected bacteria proceed to make eucaryotic mRNA, which, upon extraction, can be shown to hybridize specifically with the eucaryotic genes that were used in synthesizing the hybrid plasmids.

From a biological point of view, this neat demonstration of the universality of the genetic code serves to remind us of the unity of living things: their common code attests to a common ancestor, a common origin of life on this planet. The construction of hybrid plasmids has, however, also raised fears that the ingenuity of molecular biology may eventually prove to be a curse. What if human cancer-producing virus DNA were spliced to bacterial plasmids and allowed to multiply in *E. coli,* and then such plasmid-carrying bacteria accidentally reinfected humans where they could then spread in epidemic fashion?

Perhaps for the first time in the history of science, a call was made in 1974 for a voluntary moratorium on certain kinds of experiments: a group of molecular biologists asked that potentially harmful hybrids not be constructed until guidelines could be adopted to assure the security of mankind. In February 1975 an international meeting of scientists was held in Pacific Grove, Calif., to discuss this problem. After much debate a set of guidelines for the conduct of experiments on recombinant DNA molecules was recommended to the community of molecular biologists at large. Because these guidelines are voluntary and are not incorporated into any national laws, it remains to be seen whether scientific research can be regulated for the benefit of mankind by the concerted effort of scientists themselves.

—Arnold W. Ravin

Zoology

That the harvest of dungeness crabs in Oregon was low in 1975 hardly seems worthy of mention. But oceanographer R. Gregory Lough of Oregon State University, Corvallis, reported that the scarcity is related to heavy rainfall in 1971, at the time when this generation of crabs was in its larval state. It appears that the cells of the larvae could not tolerate the temporarily diluted salt concentration in the seawater caused by the excessive rain. This illustrates the close environmental tolerances that affect cells.

Cellular zoology. The cell is the basic unit of all living organisms. Therefore, much work was being done by zoologists throughout the world to learn more about the origin, structure, function, development, heredity, and environment of cells. The cell membrane, the envelope containing the cell, occupied the attention of many investigators. Only 20–30 years ago some zoologists did not believe that such a membrane existed. Others believed that it was there, but many of them thought it was a film not found in live cells but caused by the staining procedures necessary to prepare cells for observation under the microscope. The electron microscope finally proved its existence as, in fact, a two-layered membrane with pores in it. This glimpse of its structure set off such questions as: of what are the layers made, and how does the membrane function with regard to transport of materials into and out of cells?

The cell membrane is known to be composed of molecules of protein and fat (lipid), and their arrangement within the membrane is thought to resemble a slice of bread buttered on both sides; that is, the fat makes up the two outer layers with the protein molecules in between. Certain areas of the membrane are thought to be more reactive than others.

To study membrane transport, oxygen utilization, or enzyme activities in highly localized environments within the cell, nuclear magnetic resonance, fluorescent probes, spin-labeling, X-ray diffraction, gel electrophoresis, and other techniques were used. During the year these methods revealed, for instance, that water has a very high viscosity within cells containing large quantities of membranous material, that proteins may be placed asymmetrically within the lipid bilayers, and that functional proteins are embedded as particles within the lipid matrix of membranes. A group in London (Q. F. Ahking, D. Fisher, W. Tampion, and J. A. Lucy) reported possible ways in which the lipid and protein components of membranes may behave during the process of cell membrane fusion between cells or between organelles (small particles within cells) and their host cells. They suggested that fusion or aggregation of cells occurs when the intramembranous protein particles are temporarily denuded from an area within the lipid bilayer. Future studies were expected to be directed toward assigning functions to protein particles found in more complex membranes and in understanding the significance of particle distribution, an investigation requiring computerized analysis.

A group headed by Roscoe O. Brady at the U.S. National Institute of Neurological Diseases and Stroke reported that certain forms of cancer occur in cells that can no longer produce certain lipid (ganglioside) molecules necessary for their cell membrane. This occurs because the cancer virus may have interfered with the genes that ultimately code for the production of those lipids.

Sam Falk from the "New York Times"

A sea snake can dive to and surface quickly from depths of 120 feet without suffering from the bends. This happens because the snake has a "hole in the heart" which allows the blood to bypass the lungs and thus slows the buildup of nitrogen from the blood.

Finally, the cell membrane may also be the biological clock, that miraculous timing mechanism which tells animals when to perform such activities as molting, migrating, mating, and eating. A group at Harvard University (J. Woodland Hastings, David Njus, and Frank N. Sulzman) reported a model biological clock that consists of ions passing back and forth through a cell membrane as a feedback system. Light is known to affect the rhythmic, cyclical changes in animal (and plant) cells. The group reported that light influences the major phase shifts of ions in light receptor cells such as retinal cells. Hormones might then relay messages to other cells in the body and set off such reactions as molting or migration.

The mapping of chromosomes, the identification of the exact locations of the genes controlling every body trait or function, is an exciting area of zoological research. For example, a group at the University of Minnesota became the first to find that a gene for immune response is on human chromosome number six. This gene is significant because it codes for hay fever and for tissue incompatibility (which causes rejection of organ transplants). Meanwhile, a group at Yale University reported that the chromosomes that make interferon, a glycoprotein produced by the body in response to virus attack, are numbers two and four. This, they believe, is the first assignment of a gene product to more than one chromosome. (*See* Feature Article: THE LIVING CELL.)

General zoology. Geneticists at the California Institute of Technology, Pasadena, and the University of Freiburg in West Germany conditioned fruit flies (*Drosophila melanogaster*) to avoid specific smells and lights. They interpreted their results as "evidence that fruit flies can learn with respect to optical stimuli." The scientists hoped to find and isolate mutant flies with altered learning abilities and then to find the exact cells on which the genes for learning exert their effect.

Another investigator (Jacob Ishay at Tel Aviv University in Israel), knowing that many insects construct elegant nests, webs, and cocoons instinctively and without being trained, wondered how they use parts of their bodies for feedback during nest construction. He reported observations on comb-building activities of intact hornets as compared with those of hornets with variously amputated wings. He found that wing-amputated hornets constructed combs with abnormal spatial orientation, suggesting that the wing tips of hornets are useful in their spatial perception.

To illustrate further the scope of zoological investigation, a group in Australia (C. J. Corben, G. J. Ingram, and M. J. Tyler at the South Australian Museum in Adelaide) described a frog (*Rheobatrachus silus*) found in Queensland in which the female carries its embryos and young in her stomach and ejects them as well-formed juveniles. Major physiological and behavioral adaptation under the control of reproductive hormones must be present in the parent to keep her from eating food or digesting her offspring while carrying them in this unusual manner. These mechanisms were the subject of further investigations.

A West German zoologist (Hans E. Hagenmaier of Aachen Technical University) discovered how

Courtesy, David Bentley, photos by Alma Raymond

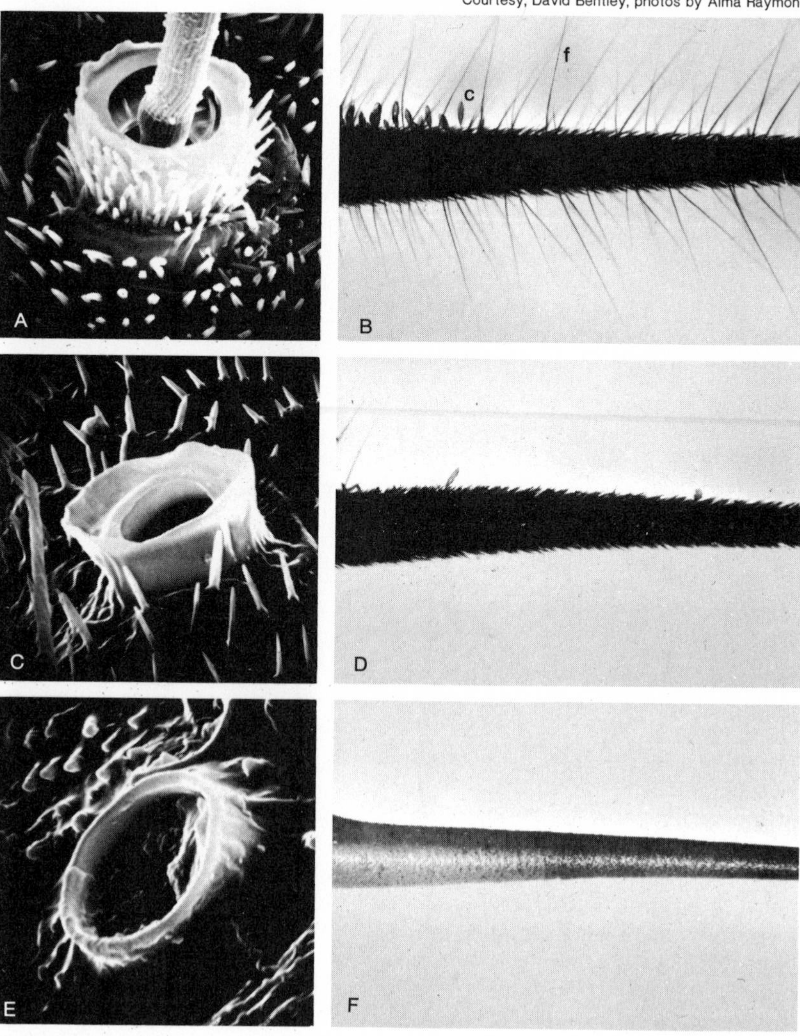

Effects of filiform mutation, a single-gene mutation of a cricket, can be seen on the sensilla, the sense organs that stimulate the evasion response. A and B show the filled filiform hair socket and the filiform hairs (which initiate the evasion response) on the cercus (posterior sensory appendage) of a nonmutant cricket. In C and D the filiform socket of a young adult mutant is empty and the long filiform (f) and clavate (c) hairs on the cercus are missing. E and F show a mature adult mutant; the empty hair socket is flattened, and the cercus lacks all hairs.

embryo fish get free of the gelatinous spawn in which they are laid. He discovered in tiny glands at the head of the embryo a new enzyme that reaches maximum production at the time of hatching. He planned next to find out what triggers the release of the enzyme at just the right time.

Roger Seymour of Monash University, Clayton, Victoria, Australia, reported that sea snakes, which are among the most poisonous reptiles known, can dive to and surface rapidly from depths of 120 ft apparently without suffering the crippling effects of the bends. Seymour said that sea snakes have a "hole in their heart." In humans this is a defective condition that allows the blood to bypass the lungs and causes "blue babies." But in the sea snake this bypass slows the buildup of nitrogen from the blood at a rate that increases as the snake dives to greater depths.

J. C. Hickman of Swarthmore (Pa.) College found an ant that pollinates a certain annual plant (*Polygonum cascadence*) in Oregon. This was an important discovery because all previously accepted pollinating animals fly from plant to plant. Walking, said Hickman, requires less energy than flying and, therefore, these results add further to the growing field of pollination energetics.

Hard, dry-land insects could dry out quickly if they did not have extremely effective waterproofing mechanisms, and they would also be vulnerable to attack, especially after molting, if their new cuticles (shells) did not harden rapidly. Two investigators at the University of Cambridge, Eng., J. F. Terherne and P. G. Wilmer, reported evidence that the waterproofing in cockroaches may be hormonally controlled. Meanwhile, Danish zoologist S. D. Anderson at the University of Copenhagen reported two mechanisms for cuticle hardening whereby amino acids are acted upon by enzymes to cause cross-linking of cuticle proteins and thus hardening of the layers.

Crickets and giraffes continued to be the subject of interesting experiments for a variety of reasons.

Crickets are often agricultural pests, and knowledge of their responses to mating calls could be a way of trapping them so that their numbers could be decreased without using pesticides. In this context, S. M. Ulagaraj and Thomas J. Walker of the University of Florida at Gainesville broadcast synthetic male calling songs of the mole cricket *Scapteriscus acletus* using loudspeakers outdoors. They found females to be responsive to the songs. Evidently, crickets also like loud music because the response increased thirtyfold when the sound level was increased 38 decibels or more above natural intensities.

Confusion would result if sexually responsive females of all the cricket species responded to the same male calling song. Therefore, the song pattern of each species is different and is stored in the crickets' genes. The songs thus become clues to the links in genetic information regarding the development and organization of the nervous system, as David Bentley and Ronald R. Hoy, neurologists at Cornell University, Ithaca, N.Y., pointed out. They continued, "Just as investigation of ... *Escherichia coli* and ... *Drosophila* has been fundamental to current knowledge of molecular biology and genetics, we hope that through the study of cricket singing ... some doors in neurobiology ... will be ... opened."

Colleagues of Bentley and Hoy at Cornell devised an optical technique for the measurement of the mechanical vibration of the live cricket's eardrum without interfering with the motion of the membrane. The procedure involves the focusing of a helium-neon laser beam on the cricket's eardrum and the analyzing of the small shifts of frequency in the scattered light bounced back into a photo-multiplier. The technique, which has broader application to the mechanism of hearing within the human inner ear, can measure vibration amplitudes as small as four ten-trillionths of an inch or one-tenth the diameter of a hydrogen atom on a surface as small as ten square microns.

Some other interesting uses of technological gadgetry included studies on the giraffe described by James W. Warren at the College of Medicine at Ohio State University. August Krogh of Denmark (who won the Nobel Prize for Physiology or Medicine in 1920 for studies on capillaries and blood vessels) once wondered how the giraffe kept from getting filtration edema (swelling) in its legs. Others have wondered what happens when the giraffe bends his head down to drink? How does the animal get blood to its brain? Prone, this animal has a blood pressure of 260/160 as compared with approximately 120/80 for man. Yet the blood pressure in the giraffe's brain is 120/80, an indication that the brain blood pressure control is set about the same for most mammals. To study the blood pressure changes in resting and active giraffes, a pressure gauge was inserted into the animal's carotid artery near its head. At the same time an ultrasonic flowmeter was implanted and a small transmitter and mercury batteries were taped onto its neck.

Some of the year's achievements had medical applications. Miles Laboratories developed an artificial pancreas that can control sugar in the blood of diabetics. The system analyzes minute quantities of blood sugar, computes the insulin dosage or glucose required, and triggers these from an attached delivery system. Laurence R. Pinner and his associates at Stanford (Calif.) University used a computer to produce motor functions by electrically stimulating certain areas in a monkey's brain that are associated with elementary movements. Electrodes were programmed to fire in sequence and produce movements useful to the monkey, such as reaching for food, putting the food in its mouth, scratching, and climbing. Further research in this area was expected to benefit human stroke victims.

Fossil "teeth" of a large squidlike animal were found near Bishop, Calif. The animal was estimated to be 100 million years older than previous predator fossils.

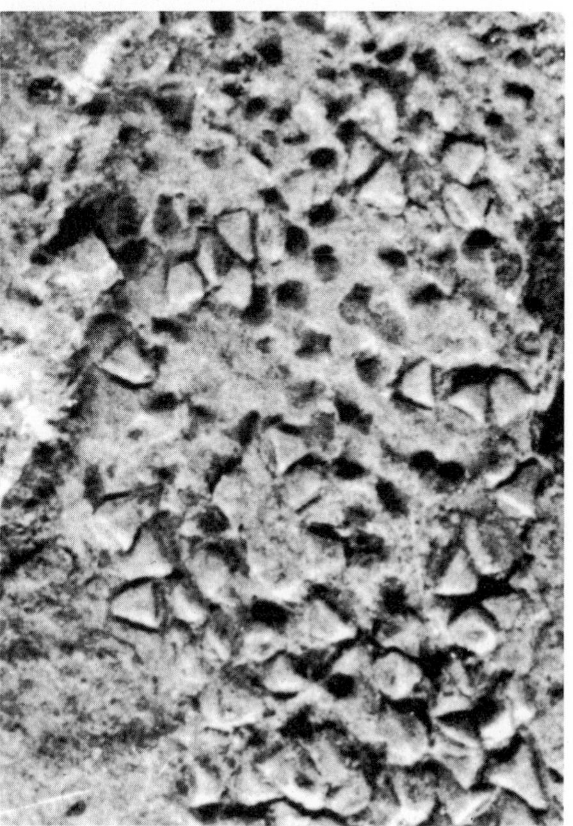

Courtesy, J. Wyatt Durham, University of California, Berkeley

Ecology. For billions of years nature patiently balanced itself through eras of changing environmental conditions. Man's increasing population and impatient technology have, however, increased the rate and number of changes with which nature must cope. In this setting the ecologist has two tasks: to learn what every organism's purpose is in its own niche, in food chains, and in the whole web of nature; and to try to put the brakes on man's attempts to solve environmental problems by expedient, unnatural means. Recently man has been setting off chain reactions in the web of nature, leading to the uncalculated, premature extinction or overgrowth of biological populations and species because action has been taken before the role of every organism in the ecosystem has been fully understood.

Reports on endangered wildlife during the year were both good and bad. In North America 109 species of wildlife, including the timber wolf, Atlantic walrus, whooping crane, polar bear, and southern bald eagle, were in imminent danger of extinction. International whaling treaties were so ineffective that whales, too, face extinction in the near future. Other marine mammals, however, like the Pacific walrus, the sea otter, and the Florida manatee, seemed to be holding their own or even increasing in number. Thirteen national wildlife refuges in the United States were authorized to acquire more than 17,000 ac of prime waterfowl nesting and feeding habitat. This was an effort to restore for migrating ducks, swans, and geese some stopover spots that had been eradicated over the years by building and development.

Environmental scientists at the University of Virginia found abnormally high levels of lead in the bodies of mice, shrews, and meadow voles living along highways. Animals living closest to the road had the highest concentrations, and the lead was therefore presumed to come from automobile exhaust fumes. Meanwhile, autopsy of a 2,100-year-old Egyptian mummy (now on display at the U.S. National Museum of Natural History) was performed at Wayne State University School of Medicine, Detroit, Mich. The mummy, called Pum II, was a 35–40-year-old man at the time of his death, and preliminary reports showed that he had about the same amount of mercury in his bones as modern man but considerably less lead.

Many animals spray, spit, regurgitate, or defecate when attacked. The defensive spray containing ubiquinone-O in the African millipede *Metiche tanganyicense* was isolated for the first time during the year by a group in Kenya. It is effective

Bones of three pterosaurs, possibly the largest flying creatures that ever lived (right), were found in Texas: (a) dorsal view of cervical vertebra of small specimen; (b) ventral and (c) proximal views of left humerus of large specimen; (d) proximal view of left humerus of small specimen.

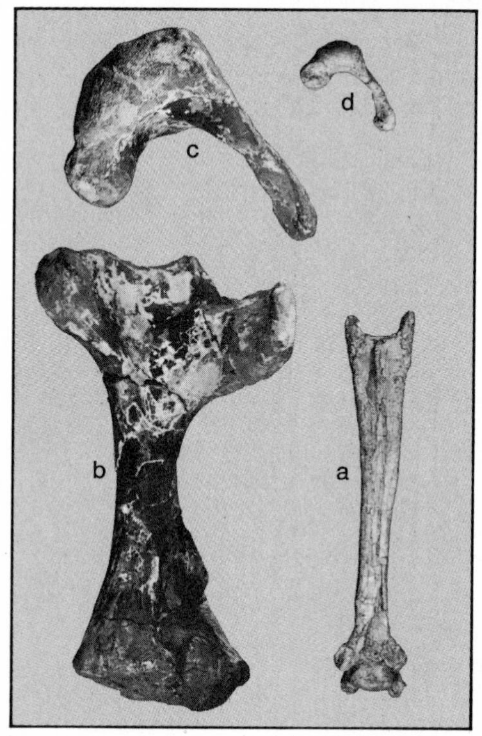

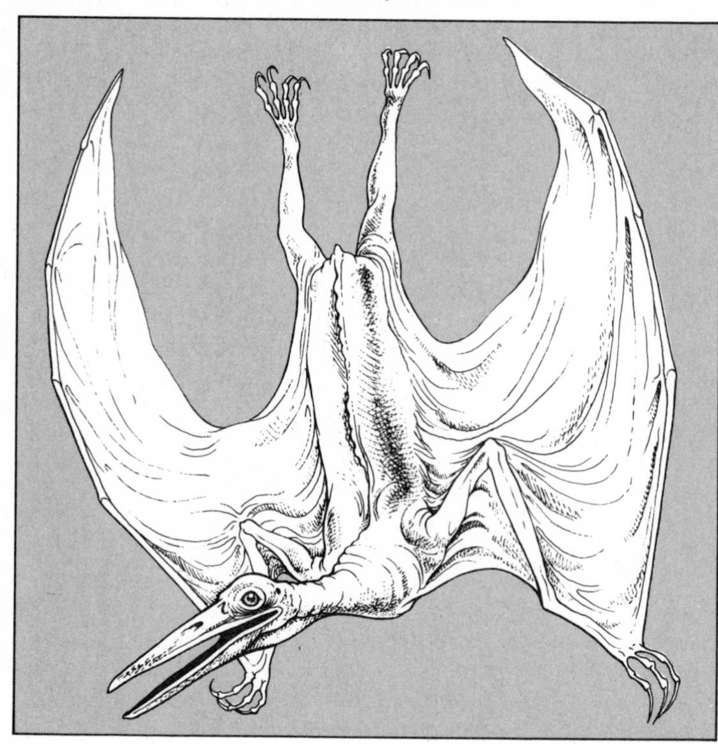

Courtesy, Douglas Lawson

against the mongoose and other predators. A group at Cornell University reported that the larva of the sawfly *Neodiprion sertifer* quickly turns and dabs a droplet of viscous resin from its mouth on the offending object when it is disturbed, causing the obnoxious predator to flee. Upon examination, the fluid was found to contain pine resin, the tree's own defense substance against insects. The investigators concluded that the sawfly has achieved the capabilities of "crashing through" the chemical defenses of the pine, opening the question of the transmission of nonnutritive substances along food chains.

Chemical substances called pheromones are emitted by many animals as signals to attract or warn members of their own species. Among zoologists attracted to the study of pheromones during the year was a group in Ohio (L. R. Nault of the Ohio Agricultural Research and Development Center in Wooster and A. M. Goff of Ashland College), which reported the first known pheromone that an animal makes but cannot release by itself. The species of insect known as the treehopper releases the pheromone only upon rupture of the insect's body and as the result of a wound or combat injury. The substance then released does not repel the predator but repels other treehoppers and, presumably, keeps them away from attacking predators. A group working in the Sudan (D. H. Beach, N. J. Hanscomb, and R. F. G. Ormond) discovered a spawning pheromone released by the crown-of-thorns starfish *Acanthaster planci* L, which synchronizes the spawning in neighboring starfish and induces movement toward the spawning individual. Large groups of these starfish have destroyed much coral in certain Pacific reefs. After this pheromone is identified chemically, an antidote may be produced or the pheromone itself may be used to attract the starfish to gather elsewhere.

Viruses in a form of controlled germ warfare may be the ultimate weapon against pests and the natural replacement for chemical pesticides that tend to overkill. Viruses are specific in terms of target organisms and therefore do not seem to harm other species or pollute as chemicals do. Scientists in Phoenix, Ariz., first produced viruses on a large scale by raising the nuclear polyhedrosis virus (NPV) in tissue culture. These viruses proved as effective against the cotton bollworm, tobacco budworm, alfalfa looper, and beet armyworm as those viruses produced by live hosts.

A group at Cornell placed peregrine falcon chicks in prairie falcon nests, and the chicks were accepted by their foster parents. U.S. Fish and Wildlife workers successfully transplanted bald eagle eggs from Minnesota to nests in Maine. Such transplants may go a long way toward replenishing rare bird and animal populations. (*See* Year in Review: Environment.)

Evolution. Information on the origin and antiquity of many animal species was reported during the year. For example, "teeth," or denticles, of a large squidlike animal, the oldest predator yet found, were discovered in the White Mountains near Bishop, Calif. The find was estimated to be 100 million years older than any previous predator fossils, and its discoverers believe that it fed on trilobites. In Hawaii the first fossils of the flightless ibis, a long-legged wading bird that man finally drove to extinction in the 19th century, were discovered. But the most significant "bird" find occurred in Texas, where a University of California graduate student unearthed a new species of fossil pterodactyl (flying reptile) with a wingspan of 51 ft and a jaw more than 3 ft long. The animal had no teeth and is thought to have been either an insect eater or a feeder on dinosaur carcasses.

In the area of human evolution, James M. Adovasio of the University of Pittsburgh, Pa., and his students reported that they had unearthed remains and artifacts 30 mi from Pittsburgh left by people who lived there both 5,000–6,000 years ago and also 14,000–15,000 years ago. Even more significant were reports by Glyn Isaac and Garniss Curtis of the University of California at Berkeley dating hand axes discovered by Richard Leakey at the Olduvai Gorge in Africa at 1 million–1.5 million years old. But no human remains had been found in that area, and so the question of who or what made the hand axes remained open.

In October 1974 a joint U.S.-French-Ethiopian expedition found in Ethiopia a complete upper jaw and other skull bones, which could extend *Homo* as far back as four million years. If correct, this means an origin several millions of years earlier than even paleontologists had considered for man. Some investigators temporarily suspended prospecting for hominid (manlike) fossils in order to assess and study the hominid specimens that had been collected so far. Apparently both the area in which hominids originated and the time involved are greater than scientists had suspected previously.

—John G. Lepp

Materials sciences

Much of the activity in the materials sciences during the year involved efforts to deal with two of the most important problems of technological society —the production of energy and the protection of the environment.

Courtesy, General Electric Research and Development Center

Scientists use a blowtorch on a pinwheel made of silicon carbide fibers separated by silicon filler to demonstrate its heat resistance (up to 2,500° F).

Ceramics

The past year in ceramic science and technology was not marked by any striking new products or concepts. It did, however, bring advances in many technological areas, particularly those in which ceramics might contribute to energy production and conservation, and to environmental protection. These included new ceramics for high-temperature turbine engines, for the generation and storage of electrical power, for the construction of more efficient structures, and for new applications in electronics.

High-temperature engines. The efficiency of any high-temperature gas turbine increases as the turbine inlet temperature rises. Both commercial and military jet engines use expensive, heavy, high-temperature superalloys and expensive air-cooling schemes for inlet nozzles and turbine blades so that they can operate in the range of 1,000° to 1,100° C. Further advances are becoming increasingly difficult, and many believe that these high-temperature alloys have been pushed nearly to their limits.

To this problem another concern was recently added. A need for small, lightweight, low-cost engines for applications ranging from civilian automobiles to remotely piloted vehicles and cruise missiles was clearly emerging. In such engines, solid ceramic components would eliminate the necessity for costly air-cooling schemes. High-temperature ceramic components would also offer reduced fuel consumption, fewer noxious emissions, and conservation of scarce or strategically limited raw materials.

Under these combined pressures there was a resurgence of interest in the development of stronger, more reliable ceramics for high-temperature structural uses. The characteristic feature of the leading ceramic candidates for these applications is that they are silicon-based, non-oxide materials. Pioneering research in Great Britain several years ago showed that silicon nitride (Si_3N_4) can be fabricated into net or near-net shape very inexpensively by the nitridation of silicon preforms (reaction bonding) and that Si_3N_4 powder can be hot-pressed to near the theoretical density of the material and also to very high strength by the incorporation of certain additives. Both products are much more resistant to thermal shock than are traditional oxide ceramics, and both are leading candidates for engine use. Reaction-bonded Si_3N_4 has proved useful for static components such as combustors and guide vanes, and hot-pressed Si_3N_4 may meet rotating disk and blade requirements.

Research on Si_3N_4 during the past year centered on several problems. While hot-pressed Si_3N_4 has been fabricated to room-temperature strengths as high as about 1 GPa (billion pascals) or 140,000 psi (pounds per square inch), it has been found to be subject to long-term deformation and loss of strength in the range of 1,000° to 1,500° C; both problems appear to result from impurity-induced liquid phases at high temperatures. Recent studies showed that the use of purer materials and of yttrium oxide or rare earth oxides as pressing aids may offer great improvements. In the meantime, recent experiments at Westinghouse Electric Corp. demonstrated that hot-pressed Si_3N_4 inlet vanes, if they are carefully designed for minimum thermal stresses during transients and flameouts, can survive typical operating conditions in commercial electric power turbines intended for high-efficiency, combined-cycle power plants. They and the producer of the blades in the United States, the Norton Co., predict commercial applications within five years.

The other major, current candidate for these applications is silicon carbide (SiC), long known for its high-temperature capabilities. Its potential for engine use increased markedly when dense, strong forms were developed by silicon infiltration of carbon-rich SiC bodies and by hot-pressing. By

1975 hot-pressed SiC could be fabricated to a room temperature strength of about 750–900 MPa (million pascals; 110,000–130,000 psi), and above 1,400° C it was stronger than hot-pressed Si_3N_4. The siliconized, reaction-sintered form is weaker, about 450 MPa (65,000 psi), but is more amenable to inexpensive fabrication.

The most impressive technical advance in SiC, and perhaps in all non-oxide ceramic technology, was the recent development by General Electric Co. of a sintered SiC. Company researchers showed that small amounts of boron and free carbon promote the pressureless sintering of SiC to near the theoretical density of the compound, apparently by providing a greater driving force to achieve the diffusion required in solid-state sintering. Research was under way to understand the effect better and to apply it to other materials of interest, especially to Si_3N_4.

Progress was also made in work on several other experimental materials and processes. Much research was done on silicon nitride-aluminum oxide solid solutions (SiAlON), discovered in recent years by British and Japanese researchers. Sialons appear to have desirable high-temperature mechanical properties and may be the forerunner of a new class of silicatelike materials for structural, metal processing, and electronic applications. A particularly promising announcement made in 1974 was the development of SiC-fiber-reinforced silicon composites at the General Electric Co. According to researchers there, these materials are easily fabricated, have high chemical stability and excellent mechanical properties to about 1,400° C, and demonstrate good resistance to oxidation and to impacts.

Another emerging application of non-oxide ceramics is their use as high-temperature bearings and seals. The strength, hardness, and low coefficient of friction of hot-pressed Si_3N_4, for example, offer great promise for long-lived and fatigue-resistant bearings, pump parts for corrosive environments, and tip seals for rotary engines. Indeed, these may be the first widespread commercial applications for Si_3N_4.

Electrical power storage. The U.S. energy crisis of 1974 had its special impact on ceramic technology. Virtually every process proposed to utilize the nation's coal without further pollution, to harness new energy sources, to generate energy more efficiently, and to store it for optimum use had materials problems. Ceramic research was responding to these problems in several areas. Research on ceramic fuel rods is central to the safe operation of any network of nuclear power plants. More efficient electric power generation may depend on developing turbine materials that can withstand higher temperatures than at present or on the perfection of high-temperature electrodes and insulators for magnetohydrodynamic

Courtesy, Roy R. Whymark

Aluminum sphere 6 mm in diameter is freely suspended by a "levitator," a circular piston that radiates a beam of ultrasonic waves toward a parallel reflecting surface. The levitator makes a container unnecessary, thereby removing a major obstacle in the processing of new materials.

(MHD) generators. (An MHD generator is one that produces electrical energy from an electrically conducting gas flowing through a transverse magnetic field.) Another area of recent interest is ceramic electrolytes for use in high-energy density batteries.

Lightweight, compact batteries for electric cars and for power-plant load leveling require new cell constituents. Sodium/sulfur and lithium/sulfur cells were promising candidates for this purpose, but both imposed harsh electrolyte requirements. An electrolyte for such cells should be a solid with very high sodium or lithium ion conductivity at moderate temperatures, should withstand rapid cell-charging rates, and must resist long-term corrosion or degradation processes. Researchers found that beta-alumina compounds containing varying ratios of sodium oxide and aluminum oxide were promising materials for such electrolytes. Work on these compounds concentrated on their fabrication into dense, reliable shapes and on improved understanding of their degradation during rapid cell-charging.

Other applications. During the past year there has been progress toward producing low-cost reinforcing fibers for composites. The first high-strength carbon fibers cost about $500 per pound. Recent studies, however, showed that carbon fibers with properties adequate for many purposes can be produced from a pitch precursor at an estimated cost of $2 to $3 per pound when production reaches several million pounds per year. Also during the year graphite/epoxy composites, originally emphasized by the military, found many civilian applications. Recent examples ranged from sporting equipment such as fishing poles and golf clubs to large space structures such as communications satellites and large space telescopes.

Recent efforts in fiber optics have been directed mainly at improving transmission. Suitably coated glass fibers can carry light over very long distances, taking advantage of the principle of internal reflection. In fiber optics, light travels along a core of glass that has one index of refraction and is reflected back into the core each time it strikes the surface because the surface is clad with a material that has a lower index. Such fibers, in conjunction with laser light sources, would be especially effective for communications applications. Research in producing purer fibers during the year resulted in losses as low as 1 to 2 decibels per kilometer at wavelengths in the range of 0.8 to 1 micron (one-millionth of a meter).

Analysis of the U.S. National Aeronautics and Space Administration (NASA) Skylab experiments on the processing of materials under zero gravity space conditions was still under way in 1975, but great strides had already been made in understanding solidification from unstirred melts, vapor phase growth mechanisms, and containerless crystal growth. Follow-up experiments in the U.S.-U.S.S.R. Apollo/Soyuz Test Project were to use a new, high-temperature furnace that would permit heating to 1,150° C. NASA also expected to fly 750 to 1,500 lb of space processing experiments each year, using sounding rockets until the Space Shuttle/Spacelab program becomes operational. The shuttle will permit less costly and much larger experiments that will be conducted by a wider range of astronaut-scientists.

—Norman M. Tallan

Metallurgy

Major developments in metallurgy during the year included discoveries concerning the magnetic properties of amorphous alloys and the temper brittleness of alloy steels.

Amorphous alloys. Amorphous alloys, or "glassy metals," have been something of a metallurgical curiosity for a number of years. Within the last year investigation of their magnetic properties has indicated that they have a potential for use as soft magnets.

Amorphous alloys are most conveniently produced by very rapid quenching (so-called splat cooling) from the liquid state, either by directing a stream of liquid metal onto the inner surface of a rapidly rotating drum, or by directing the stream into the gap between two rotating rolls. The material produced is "amorphous" in that X-ray diffraction indicates a lack of the three-dimensional periodicity that is characteristic of the crystalline state.

The mechanical properties of a considerable range of amorphous alloys are remarkably similar: their yield stresses are extremely high, frequently being over 250,000 lb per sq in.; their ductilities are very small; their electrical resistivities are up to three times greater than those of the corresponding crystalline alloys; and their densities are similar to those of the liquid metals. Only a limited range of alloy compositions can be prepared in an amorphous state by the splat cooling technique, and all such alloys are based on the composition $T_{80}M_{20}$ where T is one or more of the transition metals, such as iron (Fe), nickel (Ni), cobalt (Co), and manganese (Mn), and M is one or more of the metalloid elements, such as boron (B), carbon (C), phosphorus (P), and silicon (Si). Among the compositions studied were $Fe_{80}P_{16}B_1C_3$, $Fe_{40}Ni_{40}P_{14}B_6$, and $Fe_{29}Ni_{49}P_{14}B_6Si_2$. Amorphous alloys of the type $T_{80}M_{20}$ crystallize at temperatures near 400° C, although annealing them for only a few minutes at

Courtesy, Allied Chemical Corp.

Ribbon of metallic glass has such desirable physical characteristics as malleability, conductivity, and low acoustic attenuation; it can also be economically manufactured directly from the melt. The nominal composition of the ribbon shown above is $Ni_{36}Fe_{32}Cr_{14}P_{12}B_6$.

temperatures as low as 200° C can cause substantial degradation of both their mechanical and magnetic properties.

Interest in the magnetic applications of these materials was aroused by the reasoning that an amorphous material should have zero magnetic anisotropy; that is, its magnetic properties should not depend on the direction of magnetization. Thus, if zero magnetostriction (change of shape of a ferromagnetic body in a magnetic field) could be achieved, there would be no hindrance to magnetization in vanishingly small magnetic fields and the material would be of infinite permeability. However, measurements of amorphous magnetic ribbons in small (low) magnetic fields indicated that these materials do have an anisotropy, which, necessarily, detracts from their performance as soft magnetic materials. The apparent saturation measured in low magnetic fields was found to be at least a factor of two lower than high-field saturation. (A magnetic material is saturated when a sufficiently intense magnetizing force has been applied to it so that it assumes a state of maximum magnetization.)

The existence of positive magnetostriction, however, suggested that the application of a tensile stress to the material should produce a magnetic anisotropy favoring magnetization parallel to the axis of the field. Experiments showed this to be the case: the application of a tensile stress increased both the level of magnetization produced by the low field and the amount of magnetization persisting in the material when the field was removed; a sufficiently high stress gave, in a field of one oersted, a magnetization essentially equal to the high-field saturation.

In an effort to make these improved magnetic properties permanent, various ribbons were annealed under stress at temperatures that do not lead to excessive embrittlement. Such treatments were found to increase the low-field magnetization, sometimes substantially but never to the full extent that is possible by stress alone. After stress-annealing, the ribbons were found to be more stress-sensitive; that is, full saturation could be achieved with the use of lower fields and lower applied stresses.

In 1975 attempts were being made to correlate the magnetic properties with the degree of amorphousness and the composition of the material. Although no crystallinity was detected by X-ray diffraction, it is possible that annealing produces a crystalline state that has a grain size below the limit of detection.

Temper embrittlement of alloy steels. Although the phenomenon of temper brittleness has been known for decades, only within the last year did the mechanisms of this complex phenomenon begin to be unraveled. Temper brittleness results in loss of impact resistance in low- and medium-alloy steels when they are tempered in the range 350–550° C or are slowly cooled through this range from a higher tempering temperature. (Tempering is the reheating of hardened steel at any temperature below about 700° C in order to increase the toughness.) It has now been firmly established that this type of embrittlement is caused by a buildup of impurity elements such as antimony, tin, phosphorus, and arsenic at grain boundaries. This segregation causes a decrease in grain boundary cohesion and, therefore, a tendency for brittle fracture to occur along grain boundaries. (Grain boundaries are the interfaces between the individual crystals in a metal.) Temper brittleness thus imposes serious limitation on the use of alloy steels in heavy sections, such as in pressure vessels and turbine rotors.

The phenomenon is influenced by a large number of variables, including the type and concentration of impurity metalloid elements, the concentrations of alloying elements, carbon content, thermal history, and microstructural variables such as hardness and grain size. The classical approach to the problem generally involved the measurement of embrittlement as a function of these variables, a method that led to many apparently conflicting explanations of the phenomenon. However, recent application of scanning electron microscopy and Auger electron spectroscopy to the fracture surfaces of embrittled steels began to provide direct information on the mechanism of the embrittlement. Scanning electron microscopy reveals the structure of the fracture surface, and Auger spectroscopy provides quantitative information on the extent of metalloid segregation to the grain boundaries. Using this approach, researchers discovered that embrittlement (as measured by the tough-to-brittle transition temperature) is a unique function of grain boundary concentration of any given metalloid impurity in a steel of fixed hardness and grain size.

In 1975 researchers were working to determine the factors that control the extent of grain boundary segregation in any given steel. Preliminary experiments with an alloy having by weight 3.5% nickel, 1.7% chromium, and 0.008% carbon and also containing 700 parts per million of antimony revealed that both nickel and antimony segregate to the grain boundaries, in spite of the fact that nickel alone is not surface-active in iron. After quenching, tempering at 625° C for one hour, and aging at 560° C, the ratio of segregated nickel to segregated antimony remained fairly constant, indicating that some cooperative process was occurring. Whether or not nickel and antimony exist at the grain boundaries in the form of a stable two-dimensional, surface-active complex was being investigated.

—David R. Gaskell

Mathematics

Geometry was the subject of considerable attention by mathematicians during the year. A revival of interest in the teaching of geometry appeared to be under way, and one of the outstanding developments in research involved the closely related discipline of topology.

Revival of geometry. Under the influence of the proponents of the "new" mathematics, the teaching of geometry almost disappeared from schools and universities after World War II. Even the University of Cambridge, the home of Sir Isaac Newton, who forged the tools for the modern world of science with his *Philosophiae Naturalis Principia Mathematica*, written entirely in geometrical language, abandoned the teaching of projective geometry as a basic course and offered it only as an "enrichment" option. Britain's Open University, in which teaching is mostly carried on over television, a perfect vehicle for geometry, gave no geometry at all in its basic first-year course.

Those who pleaded the continued importance of geometrical teaching were generally denigrated, demoted financially, and denied research support for their work. As a result many schools, particularly those in the newly emergent countries, contained several age groups of teachers who knew no geometry and were, therefore, reluctant to teach it. Fortunately, however, signs of change have begun to appear.

One can now see that, in fact, there have been stabilizing factors favorable to geometry during this period. The Soviets, although they also nearly abandoned the teaching of geometry in their universities, continued to teach the subject to a high standard in their lower schools. The Germans and Austrians continued to teach descriptive geometry in their schools and technical colleges. This geometry discusses methods for representing aspects of three-dimensional objects on a plane, and is essential for engineering and architecture. Yet it is hardly taught anywhere else. Perhaps the influence of Albrecht Dürer of Nürnberg, a great practical and theoretical artist, still lingered in the German-speaking countries. He took the trouble, 450 years ago, to write a textbook on geometry for the instruction of artists, craftsmen, and German youth in general. Of course, there was also the influence of David Hilbert of the University of Göttingen, who started 20th-century mathematics on its impressive axiomatic path with his fundamental investigations into Euclidean geometry. Geometry also continued to flourish in France, the French having been enthusiasts for the subject since the era of Blaise Pascal, Girard Desargues, and René Descartes in the early 17th century.

Because whatever is "new" in education seems to take hold more strongly in the United States than in other parts of the world, the Americans probably abandoned geometry more than other peoples. Recently, however, biochemists and mathematical physicists in the U.S. have expressed their belief in the fundamental importance of geometrical insights for their research. Civil engineers and applied mathematicians, who use mathematics in practical applications, have warmly agreed. Also, the enthusiasm of the young for M. C. Escher's fascinating depictions of impossible spaces, for the psychedelic patterns formed by joining the vertices of many-sided regular polygons inscribed in circles, and for parabolas touched by multicolored threads wound around nails driven into mahogany-stained boards is surely a sign of a yearning for something that has been denied; this is confirmed by reports of the enthusiasm of students exposed to geometrical teaching for the first time.

A well-known French mathematician, René Thom, even went so far as to assert that a training in Euclidean geometry is an essential step in the normal development of man's rational activity, being the first example of the transcription of a two- or three-dimensional procedure into the one-dimensional language of writing. There are heartening signs, therefore, that the brutally severed threads of a mathematical fabric that stretches back for thousands of years are being reunited, to be woven into even more significant and exciting patterns in the future.

Stock market cycle on a manifold illustrates the theory of "catastrophes." Bear market is followed by rising demand and speculation, but when demand begins to fall the bull market meets a catastrophe on the manifold.

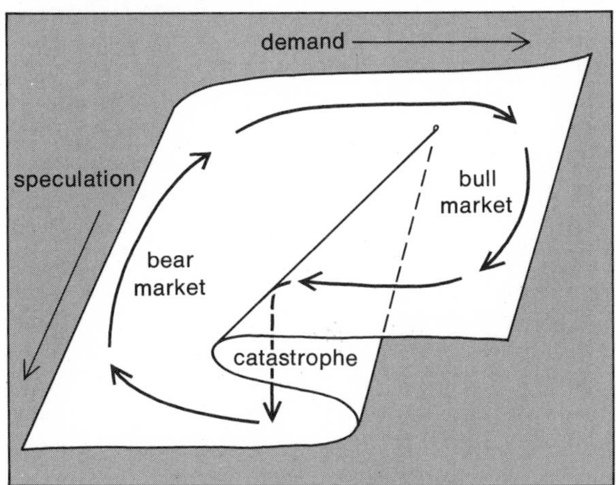

Adapted from "Mathematicians Hail New Theory," by Lynn Arthur Steen, in "Science News," vol. 106, no. 11, p. 166, Sept. 14, 1974

Theory of catastrophes. One of the year's major events in mathematics involved work done by Thom. At the 17th quadrennial international congress of mathematicians at Vancouver, B.C., in August 1974, his theory of "catastrophes" was given considerable attention. An application of topology (which arose from geometry), the theory sought to explain the discontinuous phenomena of nature, such as the boiling of a liquid. Thom wished to show how continuously changing causes may produce results that exhibit sharp and dramatic discontinuities or, in his terminology, catastrophes.

An essential component of Thom's theory is the manifold. This is a surface that is a visual representation of the totality of all possible results when two or more causative agents vary independently; the surface has as many dimensions as there are causative agents. Thom demonstrated how a manifold with an appropriate pleat can explain situations in which continuous causes produce catastrophes. A simple model, as shown at the meeting, is a surface with a smooth pleat. A particular continuous process, such as a bullish stock market, can be envisioned climbing smoothly to the upper part of the pleat. Once there, it may reach a point where it will fall suddenly and catastrophically to the lower part, from which it may once again begin a slow climb.

Thom classified all possible pleats in manifolds of low dimensions and thereby provided scientists with a relatively small number of models for catastrophic phenomena, along with a general proof that these models are the only ones possible. For example, in a two-dimensional manifold (representing two interacting causes) only two types of "elementary" catastrophes are possible, a fold-catastrophe at the boundary of the manifold and a cusp-catastrophe caused by a pleat on the surface of the manifold. According to Thom's classification theorem the number of possible elementary catastrophes becomes infinite when the number of dimensions of the manifold reaches six.

—Dan Pedoe

Mechanical engineering

Sharply higher energy costs in 1974–75 emphasized the need for mechanical systems that were efficient and reliable. In the United States the environmental impacts of new machinery and industrial equipment were analyzed as engineers

Courtesy, Corning Glass Works

Catalytic converters cut down carbon exhaust emissions from gasoline engines by oxidizing unburned hydrocarbons and carbon monoxide. They have the potentially serious drawback, however, of emitting sulfates and sulfur aerosols into the atmosphere.

redesigned pumps, production lines, and power plants to meet the increasingly stringent standards of the Occupational Safety and Health Administration (OSHA), the Environmental Protection Agency (EPA), and the Nuclear Regulatory Commission, the regulatory successor of the Atomic Energy Commission. Mechanical engineers thus found themselves constrained by the laws of man as much as by the traditional limits imposed by nature and economics.

Automotive developments. The current generation of automobile engines consumes fuel 20% more efficiently and has a lower level of exhaust emissions than did previous internal-combustion engines. This was made possible by the use of unleaded gasoline and a catalytic converter to oxidize unburned hydrocarbons and carbon monoxide. However, a question arose concerning the effect of sulfates and sulfur aerosols that are produced in the catalytic converter and emitted into the atmosphere. The annual meeting of the Society of Automotive Engineers (SAE), held in Detroit in February 1975, scheduled a day-long session on the question. EPA Administrator Russell Train postponed for one year the 1977 clean air standards in view of this new pollution problem. New standards to control sulfuric acid emissions by 1979 were proposed.

Another major concern of the SAE meeting was the future supply of liquid fuels for automobile and truck use. It appeared that as the reserves of petroleum are consumed methanol would be a strong candidate to replace gasoline. Methanol, or methyl alcohol, can be made quite readily from natural gas found in areas remote from gas consumption centers. These gas supplies can be transported to their point of ultimate use by converting them to a liquid form, either methanol or cryogenic liquid natural gas (LNG), which must be kept at −160° C. It is far easier to ship, store, and handle methanol than LNG. Although the use of methanol as an automotive fuel is many years away, tests demonstrated that internal-combustion engines can operate satisfactorily with this fuel. Methanol, however, has about half the energy content of gasoline per liter, and vehicles must therefore have larger fuel tanks or stop for refueling more frequently. In Europe small amounts of methyl or ethyl alcohol have been blended with gasoline as a means of extending that continent's imported fuel supplies.

Interest in the Wankel engine continued to decline as General Motors Corp. renegotiated its license agreement with those holding the patent on the engine; under the new agreement GM gained another year in which to decide whether to purchase a royalty-free production contract or to pay a royalty on each engine manufactured. This followed similar action by Ford Motor Co. in 1973. The once bright future of the Wankel, due to its small size and potentially low cost, was clouded by its high fuel consumption and failure to meet increasingly strict emission standards.

Work on alternative power plants such as the stratified charge engine and the gas turbine was slowed during the year because of the uncertainty of future nitrogen oxide emission standards. The major trend in U.S. automotive engineering was weight reduction of the entire passenger vehicle. Industry leaders expected that the weight of full-size cars would be reduced by 1,000 lb by the late 1970s, which was expected to lead to significant improvements in fuel economy.

Energy conservation. The emphasis on energy conservation was not confined to the automotive field. Use of energy in residential and industrial buildings represented 30% of total U.S. consumption, and savings in this area could be significant. The proposed new standard for energy conservation in new building design, American Society of Heating, Refrigerating and Airconditioning Engineers (ASHRAE) Standard 90P, was expected to provide a basis for legislation that would require new buildings to be designed and constructed for efficient utilization losses and efficiencies of heating and cooling equipment. Engineers believed that application of the standard as proposed would result in a substantial reduction in building energy requirements.

The development of solar energy for heating buildings and domestic hot water supplies received widespread attention from federal agencies, universities, and environmental organizations. The principal problem in 1975 was the large initial cost of a solar heating system, approximately 5–20 times the cost of a gas-fired furnace. Another unresolved problem was that of long-term storage of the thermal energy for use during cold, cloudy weather. Tanks containing thousands of gallons of water or antifreeze solution were used on experimental homes. Other units utilized tons of pebbles stored in an insulated room or tank through which air could be passed.

The heat pump received much attention during the year as a home-heating device. Basically it is an air conditioner operated in reverse, extracting heat from cold outside air and delivering it to the warm inside air. In this way approximately two units of energy are delivered inside a home for each unit of electricity consumed. A builder in the Cleveland, Ohio, area, when denied access to natural gas for use in new homes, installed heat pumps instead of furnaces. The thermal insulation of the homes was improved to minimize losses. Overall, the cost of heating these homes was claimed to be quite competitive with that for conventional houses.

Computer applications. Design, manufacture, and operation of mechanical equipment are closely tied to advances in computer technology. During the year minicomputers were being developed to control fuel and braking systems in automobiles. Machine tools produce intricate pieces more quickly and accurately when controlled by computers than by skilled machinists. In the design area large scientific computers were being used to analyze systems in minute detail. One such method of analysis is known as the finite element method. It involves the writing and solution of equations of stress, strain, energy, and geometry of each small subsection of the structure, machine part, or piping system that is under study. As a result, as many as 20,000 equations and unknowns must be solved when dealing with a complex system. The large computers were programmed to handle these immense problems as a routine task.

Despite the use of large high-speed computers in design analysis, mechanical engineers found during the year that the stringent requirements of nuclear power plant design taxed their capacity to the limit. A fossil-fueled power plant has about 20 separate piping systems, which are analyzed to ensure that adequate flexibility for thermal expansion is provided. In addition, these systems must be studied for the loads imposed on their supports. The stress calculations and design of these 20 systems require about 2,500 engineering man-hours. By contrast a nuclear power plant of the same size has more than 300 separate piping systems, which must be analyzed for flexibility, supports, transient fluid pulses or water hammer, the effect of sudden pipe breaks, and seismic shocks. To achieve such a task requires up to 50,000 man-hours and a large computer.

Reliability. The reliability of large steam turbines and engines was under increased study because of the serious economic consequences of failure. The failure of a turbine in a 1,000-megawatt (Mw) power plant may cause a loss of $100,000 per day to the electric utility. Several lawsuits were initiated during the past year by electric utility companies against steam turbine suppliers, alleging millions of dollars of loss caused by faulty mechanical equipment. The size and number of these court actions were expected to have a significant effect on the engineering, manufacture, and inspection of turbines in the future.

On the brighter side, a new record was claimed for large jet-engine reliability when Northwest Airlines, Inc., reported that one of its Pratt & Whitney JT9D engines had 8,312 hours of service on a Boeing 747 transport before being removed from service for maintenance. This is equivalent to almost a year of continuous operation.

Superconductivity. Most developments in mechanical engineering require many years from the time the initial invention is made or the scientific concept advanced until the general public enjoys their benefits. Therefore, an advance in physics reported several years ago is generally a subject of today's engineering work. A case in point is the practical application of the discovery of superconductivity, the complete absence of electrical resistance that certain materials exhibit at temperatures close to absolute zero (−273.15° C).

Since the mid-1960s there have been remarkable discoveries of superconductors that can withstand high-energy magnetic fields and retain their unique characteristic of zero electrical resistance up to temperatures of about 20° K (−253° C). By 1975 the baton of progress had been handed to mechanical and electrical engineers to convert this scientific phenomenon into a means of transmitting large amounts of electrical power without loss. Superconducting transmission line projects at various U.S. government laboratories such as Brookhaven (N.Y.) and Los Alamos (N.M.) demonstrated the feasibility of such zero-loss power transmission. One of the major engineering problems was to provide reliable refrigeration systems that would maintain the cables at temperatures close to absolute zero.

An annular vacuum space with many layers of reflective insulation was employed to minimize the amount of heat flow from the warm surroundings to the cryogenic cable. This type of insulation reduced the heat gain to 200 watts/kilometer, a degree of insulation effectiveness that would enable a seven-room house to be heated by a single 25-watt light bulb in midwinter. Refrigerators that use helium gas as the working fluid were being employed to remove the small amount of energy that penetrates the super-insulation described above. The heart of such a refrigerator is an expansion machine that removes energy from the helium. For such a machine small turbines whirling at 240,000 revolutions per minute and supported on bearings

Seven-strand superconducting cable seen in a mosaic of photomicrographs. Superconducting filaments (light spots) made of a niobium-titanium alloy are embedded in a copper substrate in each strand.

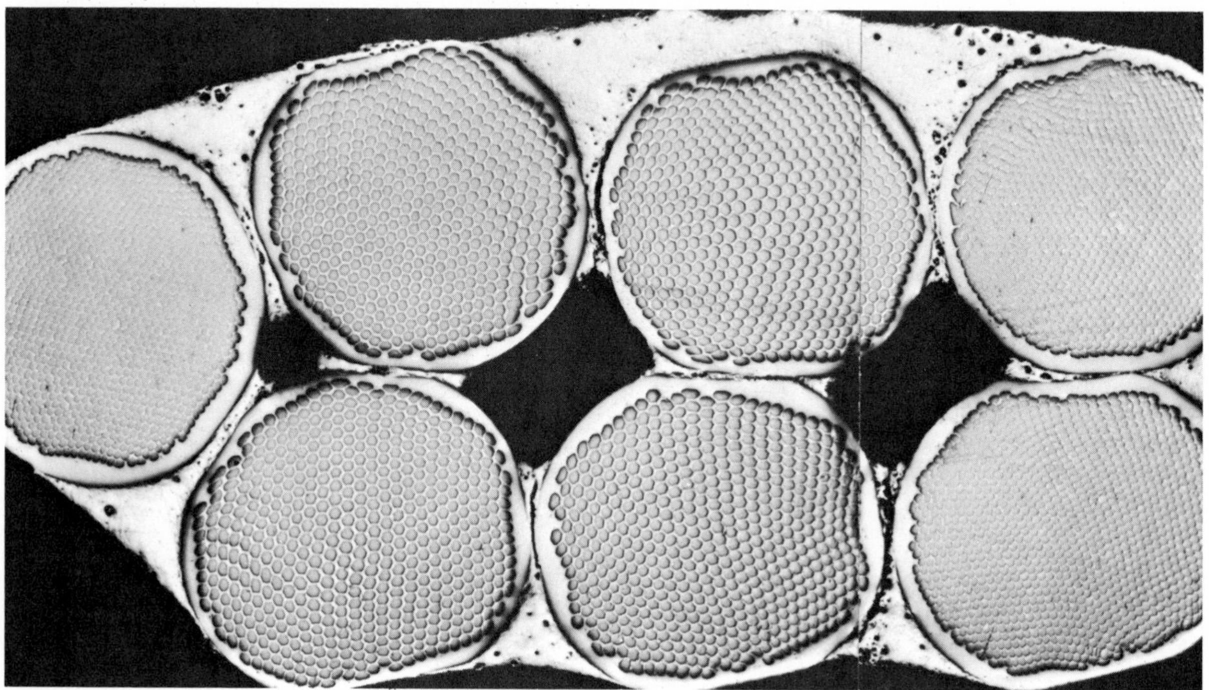

Courtesy, Fermi National Accelerator Laboratory

lubricated by the helium gas itself were first proposed several years ago by H. Sixsmith of the University of Reading in England. By 1975 these had been sufficiently refined so that five years of trouble-free operation had been achieved. Similar developments on the helium compressor should enable systems that provide refrigeration at 4.2° K to operate without difficulty for approximately one year of continuous duty.

—Edward R. Lady

Medical sciences

The range of concerns and accomplishments in the medical sciences during the past year extended from the rapidly rising cost of physicians' malpractice insurance to the implantation of a heart in a cardiac patient without removing the original organ. Cancer researchers continued their hunt for a cancer-producing virus in humans and also developed new methods of diagnosis. In pharmacology some scientists were coming to the belief that a fundamental change in direction was necessary, one in which research would be concentrated primarily on developing drugs that could deal effectively with chemical poisoning from an increasingly polluted environment rather than with infectious diseases.

General medicine

It is thought to be fact that man has only one heart, that doctors and nurses never go on strike, and that Americans have conquered polio, measles, and other infectious diseases. These and other widely accepted "truths" of modern medicine have been proved false by recent events. Some of the new facts mark progress and encouragement in the fight against disease; others definitely do not.

Surgery. In 1967 Christiaan Barnard and his associates at Groote Schuur Hospital in Cape Town, South Africa, performed the first human heart transplantation. Seven years later, the South African surgical team implanted second hearts in each of two patients without removing the original organs. The donor hearts replaced the functions of damaged left ventricles in the patients' original hearts. Theoretically, if the transplanted hearts should suffer the complication of tissue rejection, the original hearts would continue to perform the necessary work alone.

The Barnard cardiac transplants, first tried in animals 70 years ago by Alexis Carrel, were being watched carefully by surgeons throughout the world. Heart transfers have been halted at most surgical centers except Stanford University at Palo Alto, Calif., until scientists learn more about tissue rejection. Despite the rejection problem some

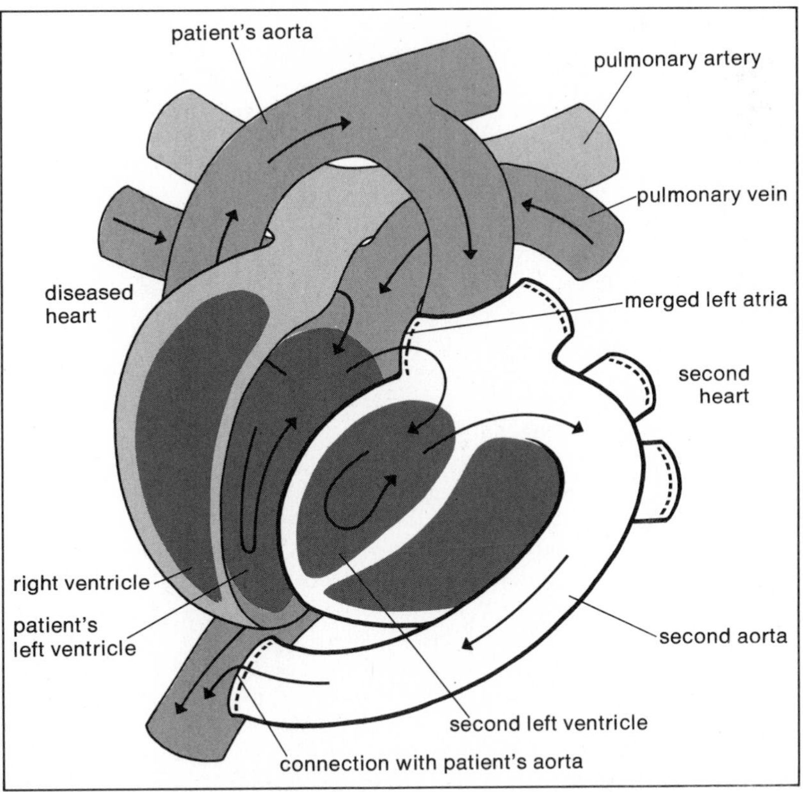

A healthy human heart from a fatally injured donor is transplanted into a patient with cardiac failure and joined to the diseased heart. Much of the patient's scarred left ventricle was removed to make room for the new heart.

Adapted from "Two Hearts that Beat as One," in "Medical World News," p. 6, Dec. 20, 1974

Courtesy, SmithKline Corp.

Live virus influenza vaccine is produced by inoculating the virus into eggs with living embryos.

heart transplant recipients survived for considerable periods of time. Louis Russell, 49, died in November 1974, more than six years after his surgery. The Indiana schoolteacher had held the record as the world's longest-surviving cardiac transplant patient.

Another famous patient, former U.S. Pres. Richard Nixon, underwent an infrequently done operation to prevent passage of blood clots from his phlebitis-racked left leg. A special clip allowing sufficient blood flow was placed on Nixon's iliac vein so that clots could not pass into his heart, brain, or lungs. After requiring five days of intensive care for post-surgical shock, Nixon was eventually able to return to his San Clemente, Calif., home. The anticlotting drugs he continued to take were being monitored by a special computer program, developed at the Medical College of Virginia, Richmond.

Mrs. Betty Ford, wife of U.S. Pres. Gerald Ford, and Mrs. Margaretta ("Happy") Rockefeller, wife of U.S. Vice-Pres. Nelson Rockefeller, each underwent the operation known as radical mastectomy for removal of cancerous breast tissue. (Mrs. Rockefeller also subsequently underwent a simple mastectomy.) Despite continued medical controversy, most surgeons continued to support this extensive procedure, according to the American College of Surgeons. The National Cancer Institute revealed that 89,000 women had breast cancer diagnosed in 1974. If the malignancy had not spread to nearby lymph nodes, the institute estimated that some 75% of these women would survive five years or longer.

In nearly 11 hours of surgery, specialists at the Children's Hospital of Philadelphia separated Siamese twins who were joined at the pelvis. The 14-month-old Rodriguez twins, natives of the Dominican Republic, survived and thrived after the one-of-a-kind procedure, which brought them worldwide attention.

Infectious diseases. The battle against infectious illness recorded some gains and some losses during the year. Smallpox may be virtually eradicated by the end of 1975 if efforts of the World Health Organization go as planned. Endemic to more than 40 countries a decade ago, in 1974 the disease was found only in pockets of Afghanistan, India, Nepal, and Bangladesh. Intense vaccination campaigns were under way to wipe out the illness in those areas.

In the U.S. several diseases previously thought to be eliminated or controlled showed disturbing signs of rebounding. For the first quarter of 1975, cases of rubella (German measles) increased about 15% over the corresponding period of 1974. The outbreaks occurred mainly in unvaccinated adolescents and young adults, who mistakenly believed this type of measles to be an exclusively "kids' disease." During the same period primary and secondary syphilis also increased. Public Health officials had hoped, on the basis of early 1974 trends, that the venereal diseases were on the wane. Instead, an increase of over 5% was observed. Gonorrhea increased 8% nationwide in 1974, with the rate in Florida up 18%.

The Center for Disease Control of the U.S. Public Health Service warned that the level of protection from infectious diseases provided by vaccinations of children was slipping badly. In the mid-1960s approximately 80% of all children in the U.S. under the age of four were immunized against polio. By 1975, the center estimated, only 60% were so protected. Similarly low protection rates existed for red measles and rubella. Only about 75% of U.S. preschoolers were protected against diphtheria, pertussis (whooping cough), and tetanus (by the well-known DPT vaccination), and only one in three in the same age group had received the mumps vaccine. Because of this widespread lack of protection, both physicians and the public were warned against "the substantial risk" that any or all of these now-infrequent infections could blossom into national epidemics.

The nation's only predictable annual epidemic, influenza, was less severe during the winter of 1974–75 than originally predicted. Although influenza/pneumonia-related deaths were above the epidemic threshold from January through mid-March, they occurred considerably less frequently than during the 1973 epidemic. Officials had initially feared that the type A flu, called Port Chalmers strain (for the New Zealand town where it was first isolated), would prove to be virulent and deadly.

A new vaccine for a major type of meningitis (type A) underwent a severe field test when it was flown in to fight an epidemic of the disease in Brazil. It arrived too late for more than 14,000 victims, 10% of whom died. The preparation was also tested, under less than crisis conditions, in Finland and France with encouraging results. Also nearing the time of human use in the U.S. were vaccines against the type of hepatitis often transmitted through blood transfusions. The several types of vaccines, which had been equally successful in animal tests, utilized purifications of the so-called Australia antigen, a substance on the surface of the hepatitis B virus.

No vaccine can be developed for another epidemic pointed out by the American Academy of Pediatrics: teenage alcohol abuse. While the use of such drugs as amphetamines and hallucinogens seems to be declining in this age group, misuse of alcohol appears to be increasing alarmingly. In addition to simple curiosity, the academy reported, most "teen-aholics" turn to alcohol for the same reasons their parents do—to relieve anxieties and tensions.

Therapeutic techniques. New and encouraging therapeutic techniques emerged for a variety of diseases and disorders. In patients with Gaucher's disease, injections of an enzyme missing in their hereditary makeup virtually eliminated the brain-damaging lipid accumulations. Investigators hoped to detect and treat the disease even before an affected fetus is born. Ultraviolet light and a drug that affects nucleic acid synthesis, given in combination, appeared to cure generalized psoriasis, according to clinicians at Massachusetts General Hospital and in Vienna, Austria. The therapy was used in patients whose psoriasis had spread over the entire body.

Excellent contraception was obtained with quarterly implantations of pellets containing a female steroid, estradiol. Medical College of Georgia investigators reported the pellets to be safe and effective in a large group of Latin-American women in whom other methods had failed. Other newly developed medical devices included a portable kidney dialysis unit, weighing only seven pounds, in use by patients at the University of Utah, Salt Lake City. An artificial larynx, which can be used by the patient days after surgery and removed whenever he wants to practice esophageal speech, was developed at Northwestern University, Evanston, Ill. One of the patients wearing this device, who necessarily has to shout occasionally, was an Indiana high-school football coach.

About 900,000 pregnancies were interrupted by planned abortions in 1974, an increase of 150,000 over the previous year according to estimates by Planned Parenthood. But the pace of the procedure slowed pending resolution of the case of Kenneth C. Edelin. The Boston City Hospital staff physician was convicted of manslaughter in connection with the performance of a legal abortion of an 18-to-24-week-old fetus. The courtroom struggle, which featured a lively debate on when life begins, resulted in a guilty verdict and a year's probation for Edelin. He appealed the ruling in higher courts.

Finance and insurance. Controversy also raged within the councils of medicine. The American Medical Association sued the federal government in an effort to prevent it from enforcing new regulations to review the utilization of hospital beds. Meanwhile, physicians in more than 20 states saw their premiums for malpractice insurance skyrocket 100–600% in the face of ever larger judgments against their colleagues. Congressional hearings and suggestions for forming federal insurance consortia to protect physicians resulted.

Rising malpractice premiums contributed to a

visible new militancy among physicians, and this resulted in doctors' strikes in Ohio, Florida, New York, California, and elsewhere. For extended periods, ranging from days to weeks, physicians refused to deal with anything but emergency situations. A strike of 1,500 interns and residents in New York City was resolved after four days by placing a limit on their work hours. But elsewhere, physicians returned to work only when their malpractice insurance was restored at guaranteed or supported rates.

Anesthesiologists in both Ohio and California struck in protest of inadequate insurance coverage. Adding emotional impact to their complaints was a national study showing that kidney and liver disease, cancer, and spontaneous abortions were markedly more prevalent among operating room personnel. The American Society of Anesthesiologists also identified a 25% above normal rate of birth defects among children of male anesthesiologists. As of mid-1975 the causes of these defects, suspected to include anesthetic gases, had not been isolated.

Inflation increased at a greater pace in the health-care field than in any other general category. Caspar Weinberger, the secretary of health, education, and welfare, threatened reinstitution of price controls for physicians and hospitals. Their charges were increasing at annual rates of 17% and 19%, respectively.

The vagaries of the nation's economic structure also caused temporary shortages of such important medical items as opium, imported at controlled rates to make morphine and codeine. When beef production faltered, hospital pharmacies became temporarily short of heparin, the blood-thinning agent made from beef livers.

Not even the American Medical Association itself proved immune from rising costs. In order to help solve its critical financial problems the organization assessed each of its 189,000 members $60 and cut its staff by 10% in addition to restricting other activities.

While most planners had predicted that national health insurance would emerge in the U.S. during the year, the bill establishing such a program remained stalled in Congress. The Watergate scandals, as well as the resignation of Rep. Wilbur Mills (Dem., Ark.) as chairman of the key House Ways and Means Committee, were credited with causing the delays. Medicare and Medicaid, the existing federal health programs, meanwhile passed their tenth anniversaries. They covered an estimated 48 million Americans in 1974, at an annual cost of $26.5 billion. A nationwide survey showed that the majority of physicians favored an insurance plan on a national basis if it could avoid the stresses that Medicare and Medicaid have placed on the Social Security system, which provides funding for both.

Physician shortage? The year also marked a turnabout in the prevailing professional and governmental attitude that the United States is short of physicians. The problem is really one of maldistribution, declared a number of experts, including a special Carnegie Commission panel. As of 1975 there were 114 medical schools with 52,500 students in the U.S. This compared with 32,000 medical students in 87 schools only a dozen years ago. Yet for each doctor who goes to a needy rural or semi-rural area, two go to practice in cities. For every pair of doctors in family (general) practice, three enter surgery.

These patterns of practice caused the introduction of several bills in Congress. Most would support a future doctor's medical education in return for a specified period of service in physician-poor areas. They would be similar to the Berry Program, which allows the U.S. Navy to support medical students in return for which the physicians serve a period of postgraduate military duty.

Historically, physicians have been certified only once to practice permanently. The first recertification examination in U.S. medical history was taken

"Dr. Frankenstein—your monster is suing you for malpractice."

in 1974 by 3,400 specialists in internal medicine. The American Board of Internal Medicine announced that 96% of those taking it passed the test. Although those who failed will not be "decertified," they will lack the diplomate certificates issued. The American Board of Family Practice, which makes recertification mandatory, planned to give its first examination, to 1,700 general practitioners nationwide, in 1976. The trend toward forcing physicians to prove their competence at periodic intervals was expected to increase in the U.S. if and when a national health insurance program is passed.

—Byron T. Scott

Cancer research

In many respects present efforts in cancer research and the history of the National Cancer Institute are closely intertwined. The institute is the principal federal agency in the U.S. for the conduct and funding of cancer research, and it also provides the principal source of financial support for cancer research nationally. Created by an act of Congress in 1937, the institute initially confined its research to its own laboratories and clinics in Bethesda, Md. After World War II the institute began increasingly to support research carried out in universities, research institutes, and hospitals. In 1971 Congress passed new legislation, the National Cancer Act, which increased the rate of funding for cancer research dramatically. Since its creation the institute has received $4.4 billion, of which $2,340,000,000, or 53.1%, was voted in the last five fiscal years.

The pace of research is such that most of the recent effort has not yet been completed nor communicated to the medical profession. Thus, most of a physician's present practice represents work that was begun more than five years ago. As a result of the current increased funding levels much exciting and useful knowledge can be anticipated in the next five years.

Cancer research efforts can be divided into a basic approach, which might be entitled cancer biology, and three applied approaches in the cause and prevention, diagnosis, and treatment of cancer. As a basis for an understanding of cancer both in the experimental animal and in man, it is necessary to define the disease. In general, the cancer cell is abnormal in size and shape as well as function. These changes are transmitted from one cell to the next; that is, they are incorporated into the cell's hereditary mechanism and, as the cells divide, are maintained in the daughter cells. These abnormalities, in ways that are not understood, result in an accumulation of the cells as discrete masses, *i.e.*, tumors, that distort the organ in which they are growing. In addition, these cells invade others by growing beyond their normal confines and thus colonizing in neighboring organs. The colonies are known as metastases and can occur in any organ of the body. In the process of invading and colonizing, tumor masses can alter the functions of the organ. The properties of the cancer cell that permit it to form masses, to invade, and to colonize are not known and constitute a fundamental problem in cancer biology. They have been studied in a number of ways. One explanation is that the new properties that permit individual cells to invade and groups of cells to colonize are caused by the addition of new hereditary information to the cell by a virus. For this reason there has been an intensive effort to identify a virus or a series of viruses that cause cancer in man. Some viruses have been found to cause cancer in animals. Another explanation is that these changes are epigenetic; that is, there is some permanent change in the cell's regulatory mechanism that confers the functions and properties characteristic of the cancer cell. Much less is known about this phenomenon. One of the most interesting of recent observations is that cells grown in test tubes may show changes that are characteristic of malignancy but that when the level of cyclic AMP, a naturally occurring chemical compound, is raised they take on the properties of normal cells. The biological importance of cyclic AMP

Courtesy, Robert E. Gallagher and Robert C. Gallo

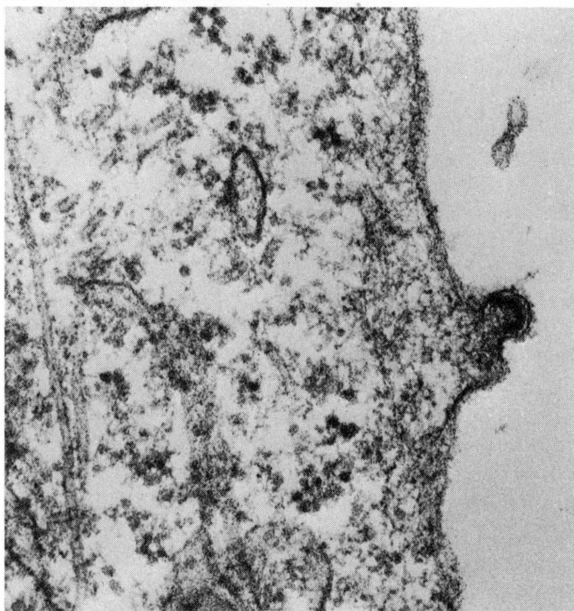

Type C RNA virus buds off a cell membrane of a patient suffering from acute leukemia. This virus resembles a virus that causes cancer in monkeys.

was discovered in 1956 by the late Earl Sutherland, for which he received a Nobel Prize in 1971.

There has been an increasing awareness that basic immunological mechanisms can be important in cancer. This field, tumor immunology, has many important implications, ranging from an understanding of the fundamental nature of the cancer cell through the diagnosis and treatment of cancer. There is evidence that those individuals in whom the capacity to respond to an immunological challenge is decreased, *i.e.*, those who are immunosuppressed, have a high incidence of cancer. This stems from two observations. One is that there are various diseases, often having their onset in childhood, which suppress the immune mechanisms of the body. In such patients there is a striking increase in the incidence of cancer, particularly cancers related to the cells of the immune system. The second observation is that patients given renal transplants coupled with drugs to suppress their immune response and to prevent rejection of the transplanted kidney have a markedly increased incidence of cancer. In fact, some of the tumors in these immunosuppressed patients are structurally similar to those seen in children with a decreased capacity to respond immunologically.

The precise mechanism by which a decrease in immune response results in cancer remains to be determined. A number of research scientists have suggested that if there is a viral cause of cancer in man the failure of immune response leads to a failure to combat the virus. Tumor cells themselves are often antigenic, and it is possible that in certain people the body is unable to respond to antigenic challenge. Although the mechanisms are not well known, it is clear that a considerably increased incidence of tumors occurs in those individuals in whom the capacity to respond immunologically is suppressed.

The existence of tumor-specific antigens has been postulated for some time. It has generally been thought that these are of two types: those occurring in the virally induced cancers in which the antigens are of viral origin and share characteristics of the virus; and those in chemically induced cancers in which most tumors produced by a single chemical are antigenically unique. These lines of distinction are beginning to be blurred. Nevertheless, there does appear to be evidence indicating that tumors are antigenic. There is also opinion that holds that cancer cells arise in all of us but that most of them are rejected by the

Nude mouse (left), a strain lacking immune defenses, is implanted with human large-bowel carcinoma in its right side and then injected with cancer antibodies labeled with radioactive iodine. Three days later a scan (right) shows the antibodies concentrated in the implanted area.

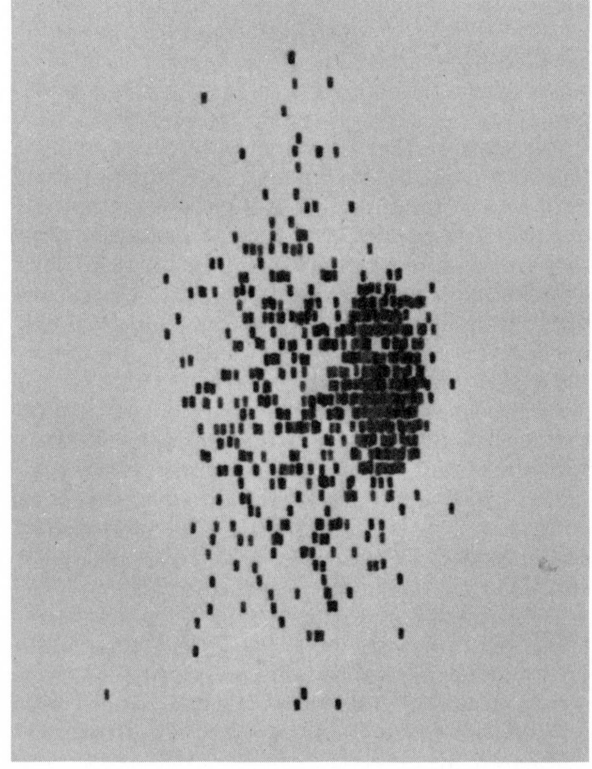

Photos, courtesy, Jean-Pierre Mach, University of Lausanne, Switzerland

Photos, courtesy, Massachusetts General Hospital

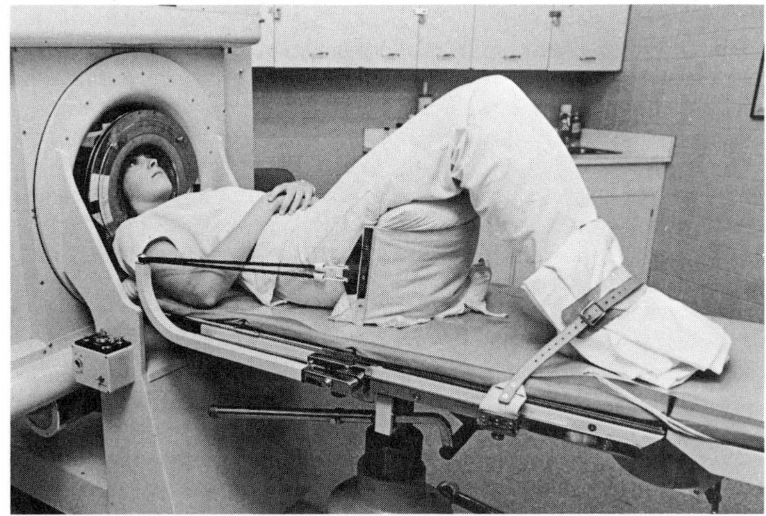

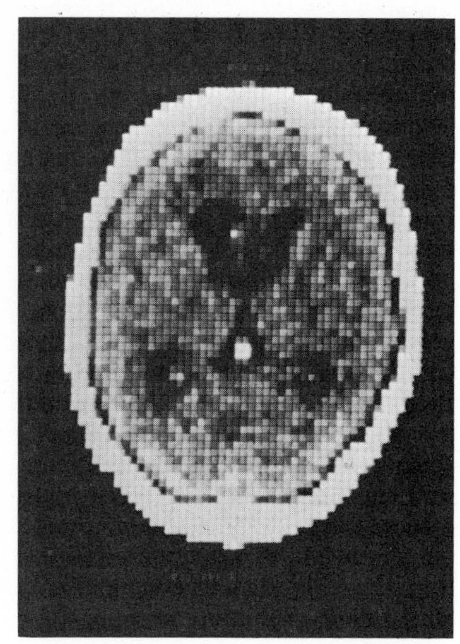

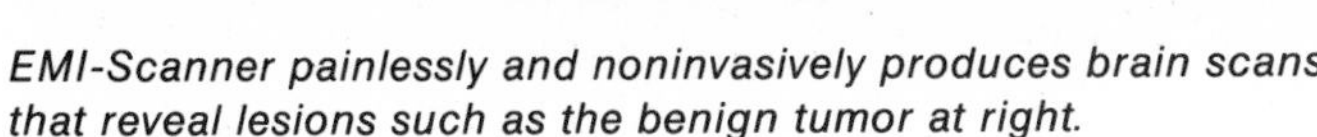

EMI-Scanner painlessly and noninvasively produces brain scans that reveal lesions such as the benign tumor at right.

body's normal defense mechanism before they can grow into detectable masses.

There is a long history of evidence indicating that chemicals can cause cancer. It began 200 years ago when Sir Percival Pott first described scrotal cancer in chimney sweeps. More recently there has been increasing research centering on viruses as the cause of cancer. It has been known since 1911 that a cancer in chickens (known as the Rous sarcoma) is definitely caused by a virus. By 1975 researchers had found a large number (more than 100) of viruses that produce tumors in experimental animals, an observation that has stimulated an intensive search for the identification of viruses causing cancer in man. The search has resulted in a number of claims sometimes accompanied by considerable publicity, little of which has been substantiated. Recently, viruses that may be related to breast cancer have been identified by two independent groups—the Michigan Cancer Foundation in Detroit and the National Cancer Institute. The causal relationship of these viruses to breast cancer in women remains to be proven. As of mid-1975 neither group of investigators had come to a firm conclusion.

Several viruses have been found to cause leukemia in animals. Recently investigators at the National Cancer Institute were able to obtain from the blood of a patient with acute leukemia a virus that had properties very similar to a virus that causes cancer in monkeys. Because this agent has been obtained from only a single patient its designation as a human cancer virus must be withheld. Nevertheless, it is difficult to escape the conclusion that if viruses cause cancer in mammals, including subhuman primates, they may also cause some cancers in man. The mechanism by which a virus induces cancer is the subject of considerable research with a variety of different opinions. It is simplistic to state that the incorporation of the hereditary information of the virus into the DNA of the host cells may provide new information which will cause a normal cell to become a cancer cell (a process known as "malignant transformation").

Other causes of cancer, including a variety of chemicals, are well known. The most common and prominent of these are the chemicals in cigarette smoke. But there are others such as the more recently described bis(chloromethyl) ether, which produced cancer in about 10% of the industrial workers who were exposed to the compound. Another industrial chemical, vinyl chloride, has been shown to induce liver cancer.

Asbestos is another well-known chemical carcinogen. Recent data indicate that the combination of asbestos with cigarette smoking is particularly hazardous. M. F. Stanton, at the National Cancer Institute, has shown that the size and shape of the asbestos particle is an important determinant in producing cancer.

There is an intense effort to develop new methods for the determination of whether any given chemical is carcinogenic. This experimental work must be done in animals. The studies are slow (two to three years to complete), relatively expensive, and have limited capability for studying a large number of compounds. In fact only a small fraction of the new potentially hazardous chemicals introduced into society can actually be studied. What is needed is a simple test-tube method for determining whether a chemical will cause cancer. Studies are under way to determine whether such a method is feasible. Growing cells in the test tube and exposing them to chemicals that may cause cancer has been suggested.

Radiation is another cause of cancer, but in the last few years there has been no new information as to how it does so. The principal radiation problems for man exist in industrial X-ray and medical diagnostic exposure. A principal question to be answered is whether there is a threshold carcinogenic dose; that is, whether there is a dose of radiation below which no cancer is ever produced. General thinking has been that if a threshold dose does exist it would be difficult to determine in an experimental animal and impossible to determine in man. For the protection of the population it will be necessary to minimize hazard by decreasing exposure to radiation.

The diagnosis of cancer. Currently, approximately two-thirds of the patients seeking medical advice because of cancer symptoms already have distant metastases at the time the diagnosis is made. Only occasionally are cancers detected during the course of diagnosis and treatment of other disease, and even less frequently do patients come to medical attention as a result of planned prevention programs through periodic screening. Because most patients have symptoms and most symptomatic patients have metastatic disease, the diagnostic research efforts must focus on the study of asymptomatic individuals. A major research effort is being made to find a simple diagnostic test, preferably on blood or urine, that could be carried out in large populations to separate those who have cancer from those who do not. In 1974 the U.S. Food and Drug Administration licensed the carcinoembryonic antigen test (CEA). As originally described this test was thought to be specific for cancer of the bowel, but it is positive in a substantial number of other diseases and is not positive in all patients with cancer. Nevertheless, this discovery has spurred research to determine whether there are cancer-specific antigens in the blood that can be detected early in the course of the disease. None has yet been identified. Other metabolic products of tumor growth—such as chorionic gonadotropin, the hormone produced by the placenta; alpha fetoprotein, a protein produced by the fetal liver; and one form of lactic dehydrogenase—are also present in the blood of some patients with some cancers. It is likely that the current search for tumor-specific antigens in a variety of tumors will be successful and that a screening test may become available.

Functional studies collectively known as tests of cellular immunity can also be used. These are carried out on cells either in the test tube or in the skin of patients. In the latter instance they are used to determine whether patients will respond to particular fractions of cells that may be associated with cancer, or to an organic chemical, dinitrochlorobenzene. For some patients there is a very good correlation between the ability to respond to dinitrochlorobenzene and their eventual clinical course. Patients who have the capacity to respond, survive; patients who do not have the capacity have a low probability of surviving, that is, a poor prognosis.

One of the most important diagnostic leads is in the field of diagnostic X-rays. Modification of the X-ray technique by EMI Ltd. has opened a new field of X-ray imaging. This technique at present is applicable principally to diagnosis of tumors of the brain. A narrow X-ray beam is passed through the patient's head and the amount of energy absorbed is measured. From a very large series of measurements coupled with extensive computer calculations (tens of thousands of equations), this machine produces sequential images of the head at intervals of approximately two centimeters. This technique is rapidly displacing many of the older techniques for the diagnosis of a variety of diseases of the brain, including brain tumors. Comparable instruments suitable for diagnosis in other parts of the body are being developed, but as of 1975 sufficient data were not yet available to determine their utility.

Ultrasound is another imaging technique, in which sound waves are aimed at a part of the body and their echoes measured. The echoes produced by different tissues from which echo images can be constructed vary. This technique had proved useful in obstetrics because it permitted localization of the placenta. It has been applied extensively to the eye and heart and, now, the brain. A device is currently being developed in the U.S. for ultrasound examination of the breast for cancer.

Another interesting research effort in cancer diagnosis is being carried out at the University of Chicago and at the University of California at Berkeley. In these studies investigators are using charged particles with properties sufficiently different from X-rays to provide an entirely new opportunity for imaging various parts of the body. Exceedingly expensive and elaborate facilities are needed to obtain the charged particles, but on theoretical grounds this technique offers a very exciting possibility.

The last of the newer diagnostic techniques consists of direct visualization of internal portions of the body. Research in Japan has led to development of superb fiberoptic devices that permit direct visual inspection of much of the lung and of

Courtesy, G. Baum, "Fundamentals of Medical Ultrasonography," G. P. Putnam & Sons, New York, 1975

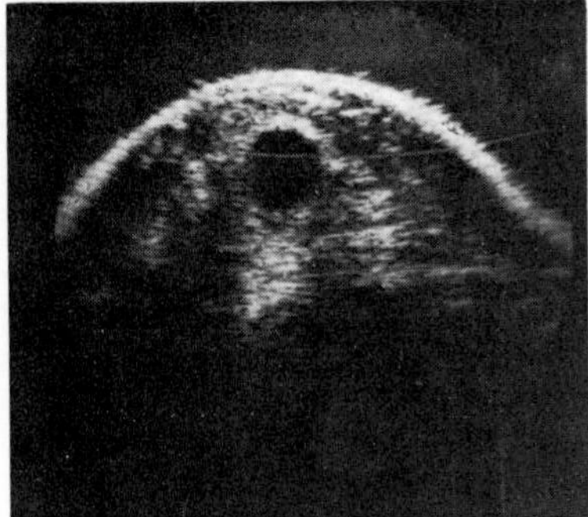

Courtesy, American Cancer Society

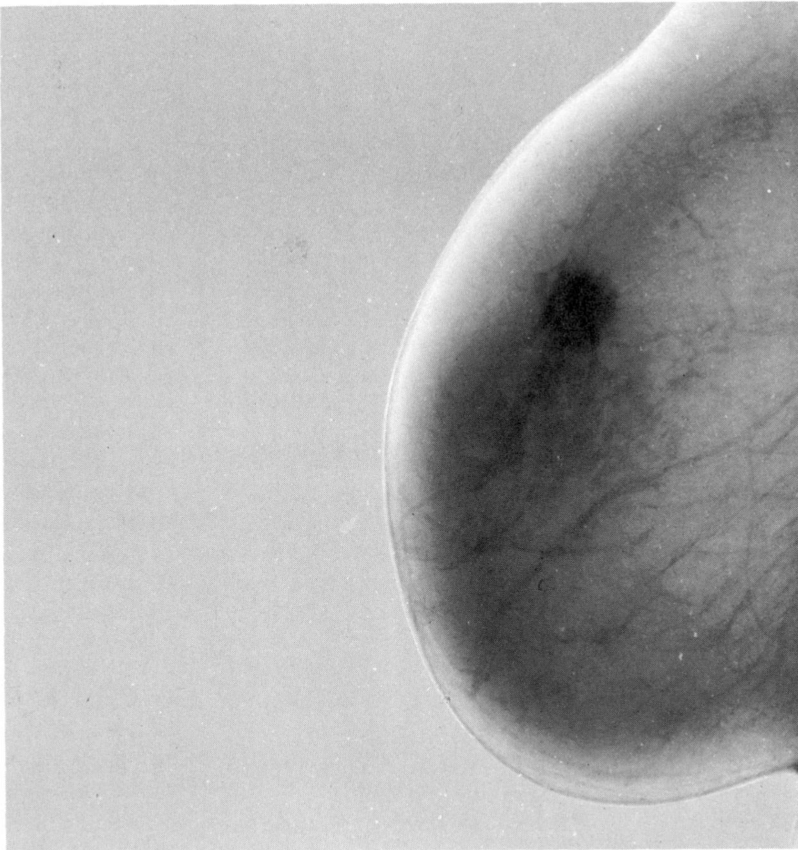

Mammography is the radiological examination of the breast. The X-ray mammogram at the right reveals a malignancy, the dark spot in the upper portion of the breast. The ultrasonic mammogram above shows a cyst as a recognizable black hole.

the entire bowel. These devices also have attachments that record the procedure on videotape, providing a pictorial record that can be stored and used to compare with a later examination. The technique has revolutionized the diagnosis of lung cancer. The impact on the diagnosis of bowel cancer is still to be determined, but it is likely that it will in part replace X-ray examination. The time required for cleansing and inspection of the bowel is similar to that required for X-ray examination. The devices also facilitate the taking of a biopsy.

As a result of research begun in the early 1960s and carried out by the Health Insurance Plan of the state of New York, it was demonstrated that a reduction in breast cancer mortality could be obtained from yearly physical and X-ray examinations of the breast. This encouraged the American Cancer Society and the National Cancer Institute to create breast-cancer-detection demonstration projects throughout the U.S. The initial findings in these demonstration projects have shown that 75% of the women diagnosed as having cancer were at that stage in the development of the disease in which the prospect for a cure was good. This compares with less than 40% in that category in the nation at large.

Also under development are new techniques for the diagnosis of breast cancer. For example, a breast pump developed by Otto Sortorius of Santa Barbara, Calif., has been used to obtain fluid from the breasts of normal women. This fluid often contains cells, and in some women with proven cancer it contains cancer cells. In principle the technique is very much like the familiar Pap test for cancer of the cervix.

In addition to cell examination, breast fluid can be examined chemically. In some of the women there was evidence of an enzyme characteristic of a virus that may be associated with breast cancer. The biological significance of these findings remains to be determined.

Studies begun in the late 1960s show that a screening program may be effective in reducing mortality from cancer of the lung. It has been known for many years that periodic X-ray examination as frequently as every six months will not reduce significantly mortality from lung cancer. It has also been known that some patients with lung cancer have cancer cells in their sputum. The difficulty in the past with the use of sputum cytology has been that frequently cancer cells were found in the sputum but both X-ray examination and visual examination of the bronchi by a bronchoscopist did not visualize the tumor. It was only the development of a superb fiberoptic bronchoscope in Japan and its initial use by S. Ikeda that permitted the localization in the lung of the cancer in each patient who had cancer cells in the sputum. This is a significant advance.

At the Mayo Clinic in Rochester, Minn., male patients over the age of 45 who smoke cigarettes have been routinely examined. Almost one out of every 100 who comes to the clinic for reasons other than lung cancer, whose referring physician did not suspect he had lung cancer, was found to have that malignancy. In each instance the tumor could be localized. The 99% who did not have lung cancer were divided into two groups. One group was asked to have a sputum examination and a chest X-ray every four months; the other group was advised to have a sputum examination and an X-ray yearly. Early results indicate that if a sputum is examined for cancer cells and an X-ray of the chest is done every four months, cancer can be found in what appears to be early stages, at the time that the possibility for cure by surgical removal is greatest.

Time will determine quantitatively how great this improvement will be. Currently it is projected that instead of only 10% of the patients surviving five years (cured) after the diagnosis, as many as 40% may be cured. This is a significant advance, but it is insufficient.

This research recognizes the probability that many people will continue to smoke despite warnings. One of the particularly important research problems is to identify those smokers most likely to develop cancer. Most smokers appear to have some abnormal but not cancerous cells in their sputum. It is likely that those with abnormal cells will have a higher probability of developing cancer. Again, only time will tell. Among smokers, only 15% at the most will develop cancer; how to identify this 15% remains to be discovered. An enzyme present in human cells, aryl-hydrocarbon hydroxylase, converts some chemicals that have the potential for causing cancer into chemicals that do cause cancer. Inhalation of cigarette smoke increases the amount of this enzyme in the cells lining the lung, but the increase is not the same in all people. Preliminary data requiring confirmation indicate that the incidence of lung cancer is greater in the group having the greatest increase in the enzyme.

Treatment of cancer. There are three methods for the treatment of cancer: (1) removal by surgical means; (2) destruction by X-ray; and (3) use of drugs and immunological techniques to destroy the malignancy.

In the field of immunotherapy there is considerable interest in two areas. One is the use of BCG and similar bacterial products to stimulate the body's general ability to respond to immunological challenges. BCG probably does this through nonspecific activation of macrophages, which are a part of the immune system that processes an antigen so that the body cells can respond. Cells respond in two ways: humoral antibodies develop which circulate in the plasma; or lymphocytes of the blood are specifically altered so that they interact with tumor cells. Some have called these "killer cells" or "effector cells." In any event, BCG plays a central role in activation of macrophages, and this may be the mechanism by which

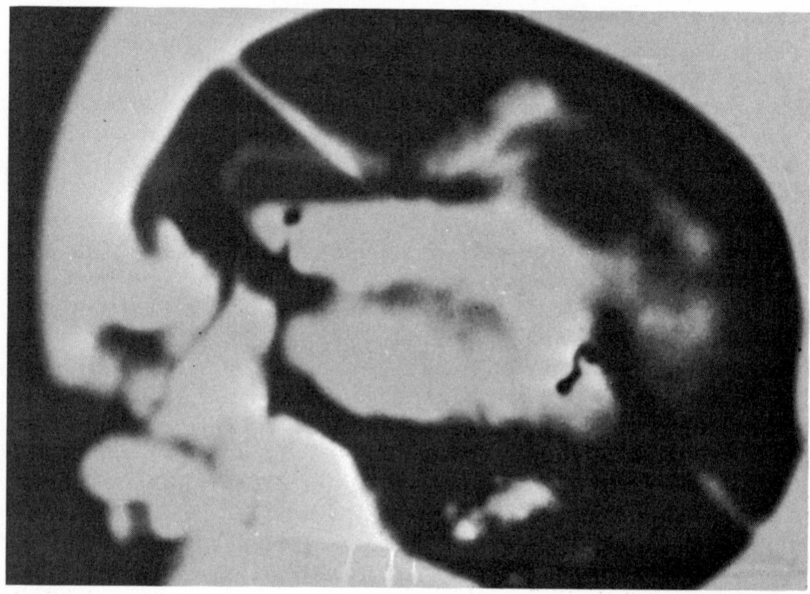

Proton radiograph of a brain reveals a tumor (dark area), an example of the effectiveness of charged-particle radiography for diagnoses.

Courtesy, V. W. Steward

BCG enhances the body's response to an antigenic challenge. The second means of increasing the body's ability to respond immunologically is to use inoculations that consist of tumor cells. The tumor cells may be derived either from the patient's own tumor or from another patient with a similar tumor.

In principle, immunotherapy has much to offer; in practice it is in its infancy. Studies of the immunotherapy of malignant melanoma and of acute myelogenous leukemia have shown some therapeutic promise, but in 1975 they were in a preliminary phase of evaluation and were not ready for widespread application.

To improve cancer therapy new principles of treatment must be found. When a cancer is confined to the organ of origin, then removal of part of the organ or the entire organ will be curative. When the cancer has spread to the other adjacent organs or to the lymph nodes draining that organ, then removal of the cancer, the organ of origin, and its adjacent and regional lymph nodes will be curative. This limits the role of surgery and, for some organs, the role of X-rays in the cure of cancer. However, most patients (two-thirds) have cancer deposits in the liver, lung, brain, and bone when they first come to a physician, and thus the treatment of the cancer by the removal of the organ of origin cannot be effective. Growing recognition of this deficiency has led to several studies in which drug therapy has been initiated at the same time the cancer was removed.

Very significant advances have been made recently in the surgical treatment of cancer of the breast in women who on physical examination alone did not have involvement with cancer of their axillary lymph glands. This is approximately 55% of the breast cancer patients. A large cooperative study group has reported on a comparison of the effectiveness of (1) simply removing the breast; (2) removing the breast and then giving X-radiation therapy; and (3) the conventional (Halsted) radical mastectomy in which the breast, the underlying muscles, and much of the tissues of the armpit are removed. There has long been controversy within the profession as to which of these procedures offers most hope. Data are now becoming available from a national study carried out in more than 30 institutions and involving about 1,000 women. At the present time, with almost three years of follow-up, there is no discernible difference in the number of recurrences in the three groups; there is no significant advantage or disadvantage in terms of recurrence rate to any one of the procedures. The corollary to this is that the simplest procedure, that is, simply removing the breast, appears to be sufficient.

Courtesy, William I. Wolff, Beth Israel Medical Center

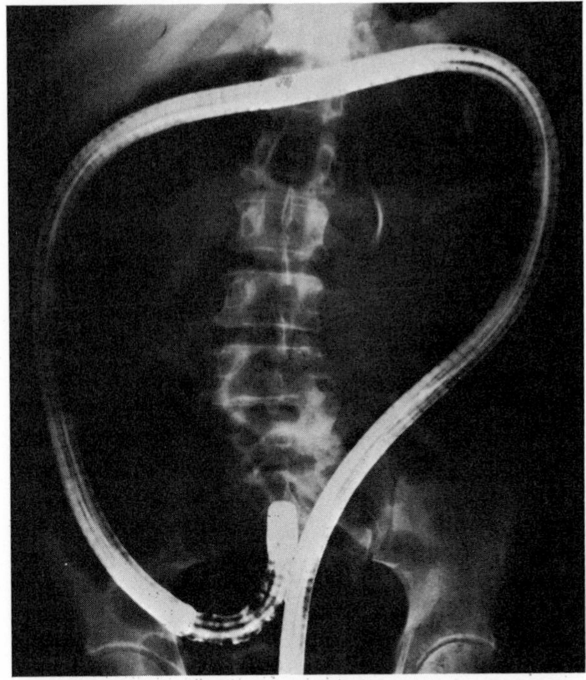

Colonoscope inserted into the large intestine (colon) allows a physician to see six feet up the colon and remove polyps before they become cancerous.

The same group of investigators has begun to design a study to determine whether the entire breast need be removed. The ultimate goal of treatment, of course, would be removal of the smallest portion of the breast that is consistent with good medical care, thereby preserving as much of the breast as is possible.

An indicator of which women at the time of original surgery have a high probability of developing metastatic disease (cancer colonies) and which women do not is whether or not there are cancer cells present in the lymph nodes in the axilla at the time of the surgery. Approximately 75% of the women with cancer cells in axillary lymph nodes will eventually die of breast cancer. In the few women with negative axillary nodes who later develop evidence of cancer in more distant sites, it is possible that the search for cancer cells itself was incomplete; it is also possible that the cancer cells bypassed the axillary nodes and went directly to other parts of the body. Both explanations are likely. Not only is it very difficult to examine each node in the greatest detail, but it is also biologically possible that the cancer cells spread first to more distant sites. Nevertheless, the most likely spread is first to the lymph nodes and then to distant sites. Given this knowledge, a drug study was started in 1972 in which physicians treated women who had breast cancer cells in their axillary lymph nodes. The drug was L-phenylalanine mustard

(L-PAM). The first results of this treatment were reported early in 1975 and are extraordinarily encouraging. In fact this study has shown the greatest therapeutic effect of any that the National Cancer Institute has ever supported. In treated women of all ages, there has been a two and one-half fold reduction, and in the postmenopausal women slightly more than a twofold reduction in recurrences of the cancer.

A similar study using three drugs (cytoxan, fluorouracil, and methotrexate) carried out in the National Cancer Institute in Italy yielded even better therapeutic results. Somewhat greater toxicity was obtained in the Italian study compared with results in the U.S., but none of the women was incapacitated and all were able to continue with their normal occupation. For study purposes the National Cancer Institute physicians and their colleagues have come to the conclusion that in planned studies *no* woman with disease in the axillary nodes at the time of initial surgery will go untreated by drugs.

The common adjuvants of cancer therapy to consider if the cancer recurs are to remove the ovaries in premenopausal women or to remove the adrenal gland or the pituitary gland in women of any age. In addition estrogens (female sex hormones) or androgens (male sex hormones) may be administered. Approximately 30-40% of women who have recurrent breast cancer will respond satisfactorily to one of these methods of treatment by reduction in the size of the cancer deposits. None of the methods is curative, they only delay tumor growth for a time and produce a reduction in the tumor masses. Recent research makes it possible to determine which women will respond to such therapy. The key to determining which women will respond is an estrogen receptor assay done either on the breast cancer tissue itself or on the metastases. In some breast cancers there are specific proteins that have an ability to take up and bind estrogens, a positive estrogen receptor test. Approximately 60% of the women who have a positive estrogen receptor assay will respond satisfactorily to endocrine treatment. No more than 5% of the women who do not possess the estrogen receptor will respond to endocrine therapy. Thus endocrine therapy is indicated only in women who have a positive estrogen receptor assay.

Radiation therapy immediately following removal of the breast is still commonly used despite the fact that several studies have shown that it does not provide significant benefits. Recently data have been compiled from a large number of studies which show that such radiation therapy has a small disadvantage. The women who received X-ray treatment had a slightly shorter life expectancy than the women who were not treated.

It is possible to suggest that for most women with breast cancer the removal of the breast alone is sufficient surgery and that the surgeon should not attempt to remove all the axillary contents but should remove a sufficient number of lymph nodes so that the pathologist can examine them to see if they contain cancer. If the nodes contain cancer, these women should be given long-term treatment with anticancer drugs. The breast cancer tissue should also be examined for the presence or absence of the estrogen receptor and that information recorded for possible future therapeutic use.

There were two reports late in 1974 on the treatment of osteogenic sarcoma, a tumor of bone in which surgical removal of the tumor was combined with drug therapy. Neither of these was a randomized study, and thus is open to some criticism, but both showed significant therapeutic benefit of the drugs. Another bone tumor called Ewing's sarcoma was studied by physicians at the National Cancer Institute. This study also showed that a combination of radiation to the tumor and drug therapy was of great benefit. These studies of chemotherapy shortly after surgery in breast and bone cancer are very encouraging but should not be misinterpreted. They are only beginning to yield results and the data is preliminary.

Epidemiology. In 1974 data from the third National Cancer Survey began to appear. Data on cancer incidence were obtained in nine geographical areas with a population estimated to be 10% of the total U.S. population. Differences in the incidence of cancer in various populations are of considerable significance. A comparison of the incidence of a given cancer in different parts of the world and the change in incidence in cancer in migrating populations has led to the conclusion that perhaps as much as 80% of the cancers in man are in some manner related to environmental influences. The nature of these influences is, however, unknown.

Aside from the possible identification of a virus associated with breast cancer in women, new epidemiological information helps to identify women most likely to get breast cancer. In addition to the well-known higher incidence in Caucasians as compared with Orientals and the affluent as compared with the lower socioeconomic strata, there are now two recent observations. Data from The Netherlands indicate that taller, heavier women have a three- to fourfold increase of breast cancer as compared to short, lean women. The explanation for this is unknown. It is suggested that in the taller, heavier women, greater caloric intake somehow affects hormonal status and that breast

Photos, courtesy, Nicholas R. DiLuzio

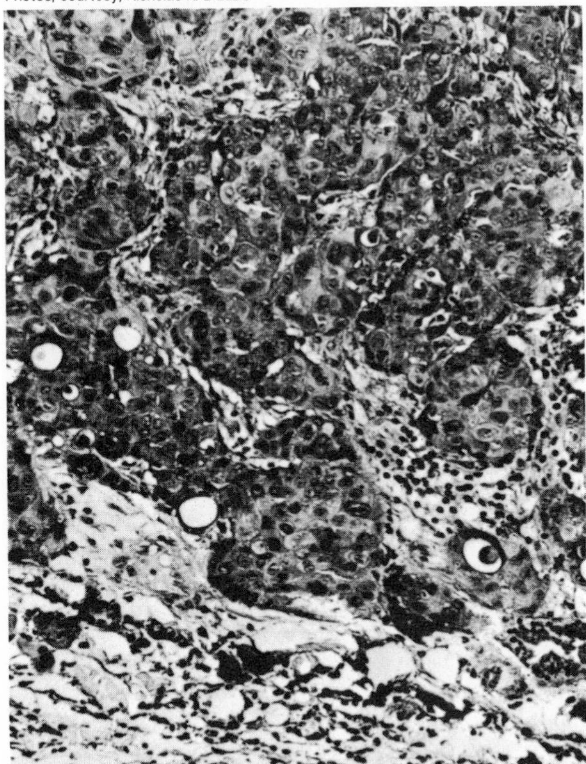

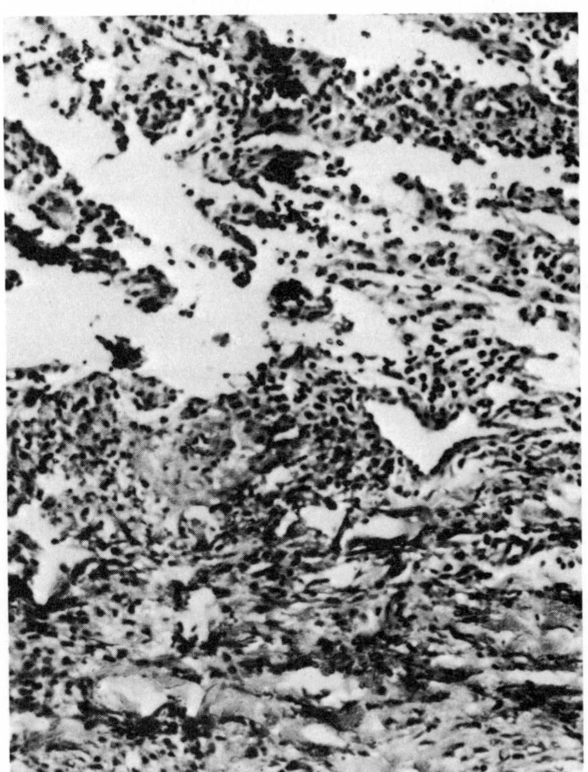

Lung cancer lesion injected with control substance (left) reveals islands of malignant cells. Similar lesion (right) injected with the yeast polysaccharide glucan is free of tumor cells; it contains mostly macrophages, large tissue cells that engulf foreign material.

cancer results from this change. Heavier women have a greater incidence in cancer of the body of the uterus. There is also confirmation of evidence that the incidence of breast cancer is much less in women who have their first child before the age of 20 as compared with women who have their first child at age 35. In fact the incidence of breast cancer increases progressively with the individual's age at the time of her first pregnancy. In Iceland there has been a tenfold increase in the incidence of breast cancer in women born in the decade 1940–49 as compared with women who were born in the decade 1840–49. The incidence of breast cancer is greater in Japanese women who are natives of the U.S. as compared with Japanese women born and raised in Japan.

The most common tumor in the male is lung cancer. The incidence of lung cancer in males in the last 25 years has increased more than fivefold. The evidence that cigarette smoking is a decisive factor in the cause of most lung cancers is now well established. The incidence is rapidly rising in women and again can be attributed to the smoking of cigarettes.

New data of considerable significance are constantly emerging. The virologists are coming close to finding a human cancer virus. The science of immunology is contributing at an enormous rate to our knowledge of cancer and has in theory the capacity to be important diagnostically and therapeutically. New advances in the diagnosis of cancer, particularly new advances in imaging and visualization by inspection of the internal organs, have been reported. If the data from currently available diagnostic and therapeutic studies were put into practice, the mortality from breast cancer could be markedly reduced. A program of sputum examination and chest X-ray at four-month intervals in cigarette smokers may soon become a recommended procedure, though this is much less preferable than the cessation of cigarette smoking. Perhaps most important is the recognition that at the time they are first treated two-thirds of the patients have metastases, and, given this condition, the very encouraging results from instituting long-term drug therapy at the time of surgery may represent a signal advance in the attempt to conquer cancer. Although there have been no dramatic breakthroughs there has been progress, and more progress may be anticipated from current research efforts.

—Nathaniel I. Berlin

Pharmacology for the future

There are two categoric bases for new drug searches. The usual is quantitative and is based on the projection of the future in simple terms of the present. Its predictions and searches deal with drugs and diseases as they are actually known. Development on this basis has been very productive in the discovery of new drugs for treatment of infectious disease, hypertension, heart disease, and mental illness, but with a significantly less substantial inroad on cancer, and these have accounted for the saving of many human lives in the first half of the 20th century. It is not surprising therefore that searches for new drugs continue as if in the future the problems to be faced will be resolved along the same lines. The fact is that such an approach can no longer be expected to make any significant progress. It is a statistical shock to recognize that despite all of the remarkable advances in medical science in recent years the prospect for living longer has not increased at all since about 1950. A cure for cancer, it has been calculated, will increase overall longevity by only a few years; one for heart disease and stroke, a few more; and both would complicate matters of health care delivery, which is already overloaded and for which no expansions are in view.

It is time to look in other directions. The basis for this discussion is qualitative and assumes that the future will not remotely resemble the present and that diseases will be so different that the drugs needed for them are currently without clinical precedent. If the judgments of the doomsayers have credence, there will ultimately be a global crisis that will threaten mankind through overcrowding, undernutrition, and pollution. Mankind has faced two of these many times before; it has survived both famine and pestilence. This time the augury is that pestilence will be a new set of chemical intrusions into the environment.

Today, man's most effective drug therapy is against infectious disease, and, while infectious disease would certainly play an important role and no doubt often deliver the coup de grace, the principal diseases caused by direct environmental intrusions in the doomsayers' world will be chemical poisoning and malnutrition. For chemical poisoning man's present therapies are not very effective and there is the devastating fact that what now exists requires too much technical man-to-man medical attention. The quantities of food needed will not be available in an overpopulated world if progressive soil salination and chemical defilement also prevail.

Can antidotes be developed against environmental poisoning? During World War II the military were successful in a search for drugs to protect against mass exposure to chemical warfare poisons. Recent research in pharmacology has led to drugs that stimulate enzymes which destroy poisons, drugs that increase the rate of excretion of poisonous chemicals, physiological antidotes that completely suppress intolerable physiological upsets by poisons, and chelating drugs that make poisonous chemicals already in the body innocuous.

Studies should be organized on forms of life that can survive in the fumes of volcano and geyser discharges and in soils that drain our polluted waters in the hope of discovering yet unknown biological mechanisms for dealing with poisons. After all, it is not a simple coincidence or even pure serendipity that an important antibiotic-producing bacterium was isolated from water contaminated by sewage outfall. Programs should be initiated to design drugs to enable the human body to deal with those poisons in the environment that man cannot help but ingest in his food and water or inhale when he breathes. If nothing else such drugs could provide time to develop needed but more permanent kinds of help.

Man may have to turn to his genetic abilities. To the extent that there are some of us with pollutant-tolerant genes, interbreeding could spread them about. Unless pharmacogenetic traits that could conceivably be useful in adapting to a polluted environment are deliberately and systematically explored, those now carrying such genes are certain to go unrecognized. If the genes that help man to live in the new environment can be identified, they might be used systematically for breeding. In this regard there are logistic reasons for praying for a boy wunderkind. Such a point of view is not male chauvinism; the fertile life of the male is longer than that of the female, there is no periodicity in sperm production, and unlike eggs, whose number in each female is fixed at puberty, there is apparently no practical limit to the number of effective spermatozoa a male can produce during a lifetime, while by artificial insemination the number of females who can be fertilized by a single ejaculate is very large indeed. And sperm can be frozen and transported, so that the gene can be spread widely and expeditiously. Frozen stores can also be used for repeat trials in cases of failure and by future generations if their sources of supply are limited.

Is the possibility of an adaptive change that will enable man to deal with, even thrive in, what would not be intolerable pollution, too farfetched to discuss seriously? Probably not. After all, there are forms of life already identified that thrive on some of our most flagrant pollutants—the algae, for example, that are now stifling our lakes and still waters. Water hyacinths that consume toxic metals are being used to remove those metals from contaminated lakes.

The possibilities of inducing genetic changes in man are so great, varied, and eerie, and considered by many so likely to be realized if scientists work on it, that already the ethical aspects of experimenting with such problems and perfecting such techniques in the lowest forms of life have been raised. There are many who consider the whole prospect inhumane and who will have nothing to do with it. Even in our rapidly changing society the prospect of deliberately altering genes in man is shocking, even though natural spontaneous genetic change is commonplace and human mutants are certainly no novelty. It should be remembered that man as well as all other forms of life is continuously mutating as part of a natural process. X-rays could be used to increase the rate of mutation in man by directing the stress on the gonads so as to limit adverse effect on vital nongerminal cells. Drugs that sensitize chromosomes so that smaller doses of radiation would be required to induce mutation could also be used. Mutagenic drugs on the whole seem somewhat safer than radiation, but they are so often associated with teratogenicity (production of malformations) and carcinogenicity that their use would pose serious technical problems.

There is also salvation possible in symbiosis. Many animals depend on symbiotic relationships with vastly different forms of life, on microbes that live in their intestinal tracts to supply essential materials such as vitamins or that assist in the digestion of food. It is even conceivable that viable intracellular viruses could be engineered to induce mutation to serve some of our needs. Symbiosis at all physiological levels is not unrealistic, certainly not unnatural, and whether man can be aided by beneficent symbionts should be determined. Symbionts such as bacteria and molds, perhaps worms in the intestinal tract, or even small intracellular microorganisms might supply enzymes that would enable man to live amid our anticipatable repertory of pollutants. The evolutionary process alone certainly does not guarantee that natural selection will select a mutant for survival that is even remotely related to man. Man, as he now behaves, does not have as much biologic importance to the biosphere as do many lower orders.

A dismal pharmacological future for man can follow in the wake of continued extravagant outraging of natural resources and of a chain reaction of vital resources on one another. Every unique accomplishment of man has been for his own use and too often to the detriment of other forms of life. *Homo sapiens* is not essential to evolution: rather evolution is essential to him.

—Walter Modell

W. Eugene Smith from Magnum

Mother in Japan bathes her helpless 16-year-old son, poisoned as a fetus by water polluted with methyl mercury.

Dentistry

Dentistry's involvement in the nascent national health insurance program and proposed legislation to increase health education in school curricula were among prime issues of interest to the dental profession in the U.S. during 1974–75. Although prospects for congressional action on the various bills aiming at the implementation of a national health insurance program appeared slim during 1975, the American Dental Association (ADA) continued to express support for any such program that would include comprehensive dental coverage with a preventive component as a high priority. But ADA officials warned that the profession will continue to oppose any proposals that would utilize public funds to provide health services for nonindigent persons or that would impose a federalized structure on the health-care delivery system. The profession continued to insist that comprehensive dental care for children should receive top priority in any national program eventually adopted by Congress.

Dentists have long been aware that prevention is a key that will unlock the door to a lifetime of better dental health and that prevention should start with children. Thus, the ADA endorsed proposed legislation known as the Comprehensive School Health Education bill, which earmarks federal support for teacher-training grants, pilot and demonstration projects, and comprehensive school health education programs. Dental health is prominently featured within the components of the bill. In congressional hearings, ADA officials stressed that the profession has been working for many years on the implementation of a comprehensive dental health education program for schools, to be conducted by classroom teachers, and that numerous educational materials for the teaching of dental health in elementary and secondary schools are available from the association. Tooth decay is one of the most preventable of all dental problems, yet over 40% of the money spent on dental care each year is for the restoration of decayed teeth.

Mandibular staple, a lightweight titanium bar, is attached to the underside of the jaw. The two long pins at each end protrude into the mouth, where they serve as anchors for a dental bridge.

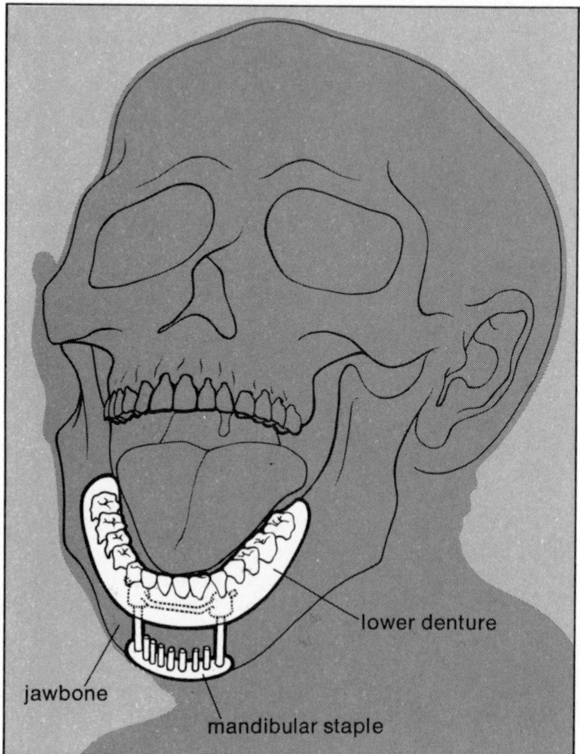

Electrosurgery for oral cancer. Electrosurgery has a definite place in the successful treatment of patients with advanced oral cancer because it may result in a longer survival rate and can also prevent the facial deformities caused by conventional surgical techniques, suggested Maurice J. Oringer of New York City. Electrosurgery, or electrocoagulation, when followed by radiation therapy can be extremely effective in the definitive treatment of advanced oral cancer, including the so-called inoperable cases for which the prognosis appears hopeless even to the radical heroic "commando" surgery. The latter form of surgery usually has a 16% "cure" rate and is frequently so mutilating that it causes severe esthetic and functional postoperative facial defects that require reconstructive plastic surgery or use of facial prostheses. Electrosurgery is cutting of tissues with a radiofrequency current applied to the tissues with electrodes. It cuts like a sharp knife but without the usual bleeding that occurs with conventional surgical techniques. Studies of electrosurgery treatment of advanced oral cancer cases in a Veterans Administration hospital in the United States indicated a cure rate of up to 12 years and longer for patients, Oringer noted.

Oral diagnosis. The dentist, through careful examination, may not only pinpoint oral health problems but may be the first one to spot signs in a patient's mouth signaling trouble in other parts of the body. This suggestion was made by Robert A. Goepp, associate professor of oral pathology at the University of Chicago and radiologist at the university's Walter G. Zoller Memorial Dental Clinic. Abnormal healing of tooth extraction sites or resorption of the jawbone, although most often attributed to dental origin, can in some cases be the first indication of far more serious problems in the rest of the body. Symptoms that can be found in the mouth through visual or X-ray examination can be indications of a whole gamut of other disorders ranging from vitamin deficiencies to rheumatic fever, diabetes, kidney and digestive tract disorders, leukemia, and anemia. One of the primary concerns of a dental examination is to check the patient for signs of oral cancer. But in addition to primary oral cancer, which may occur in the soft tissues of the mouth, the dentist may also be the first one to spot secondary cancers, which have spread from other parts of the body and may become visible in the jawbone, for example.

In another area of oral diagnosis, dental researchers from Columbia University found that changes in the saliva may provide significant clues to a wide variety of systemic disorders such as alcoholic cirrhosis, diabetes, and salivary gland problems. The study of saliva as a diagnostic tool is a relatively new research field and clues derived from it could greatly aid the health professions in speedy diagnosis and treatment if properly interpreted, according to David Abelson and Michael Marder of New York City. Reporting on their research at the general session of the American Association for Dental Research, they pointed out that previous studies have shown that alterations in the quantity or biochemical composition of saliva may reflect many changes in body functions related to such diseases as hypertension and cystic fibrosis. Their recent study of patients suffering from alcoholic cirrhosis indicates a possible relationship between changes induced in salivary glands as a result of the liver disease and similar changes observed in other related systemic disorders such as pancreatitis and hypertension. Diabetic patients appear to be subjected to an increased susceptibility to periodontal, or gum, disease and an alleged decrease in salivary flow. Their investigation revealed that a diabetic person secretes more than the normal amount of calcium into his saliva. And calcium may well be involved, they said, in the formation of dental plaque on teeth and in the mineralization of the plaque to form calculus.

Immunization against gum disease. The possibility of future immunization of adults against periodontal disease, which is the major cause of tooth loss in adults, was advanced by a researcher from the State University of New York at Buffalo's School of Dentistry. Immunoglobulins found in secretions of diseased gums can reduce the disease-producing bacteria or interact with them to cause tissue damage characteristic of gum disease, according to Russell J. Nisengard. When samples of bacteria are taken from the mouths of patients known to have gum disease and are subjected to an immunofluorescent test, the bacteria in some patients are reduced by the immunoglobulins but those in other patients actually react to cause gum damage. Based on these findings,

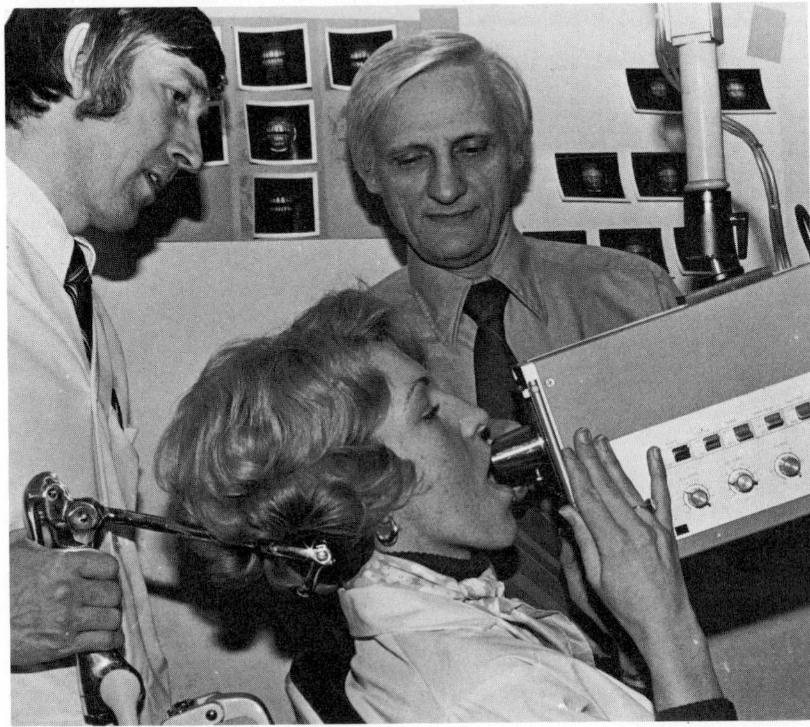

Courtesy, State University of New York at Stony Brook, photo by Antoinette Bosco

Camera photographs interior of patient's mouth using ultraviolet light. Such pictures reveal cavities and plaque on teeth before they become visible in ordinary light. At right is Israel Kleinberg, chairman of the department of oral biology at the State University of New York, Stony Brook. His research led to the development of the camera.

there appears to be some hope for developing a possible vaccination mechanism against periodontal disease.

For patients suffering from gum recession a new grafting technique promises less chair time, better hemostasis (control of bleeding), and a lower incidence of postoperative swelling. Under traditional methods patients with this type of periodontal disease are treated by transplanting gingiva, or gum tissue, from the palate to a surgically prepared area where tissue has receded or was lost. According to Bruce Dordick of the University of Pennsylvania School of Dental Medicine, better results may be obtained by transplanting the soft tissue directly onto the bone. The procedure was successfully tested in 60 patients, none of whom showed any infection during the almost one-year duration of the investigation.

Other developments. Although grinding or clenching of teeth is a habit that is common, an unlucky 20% of the population cannot afford relieving tension in this manner. The reason, according to Nathan A. Shore of New York City, is that these individuals have faulty bite caused by badly positioned teeth or by more extensive dental problems. Tooth grinding, or bruxism, for them results in referred pains in the head and ears and frequently spasms of the neck, arms, and shoulders. Until recently, he noted, the profile of a typical clencher was that of a person between the ages of 30 and 40, short in figure, fair in complexion, frequently hypertensive, and most likely a woman. But now, Shore reported, more and more young people between 16 and 20 years are becoming clenchers, probably because of anxiety stemming from today's troubled times. But there is nothing imaginary about their pain from bad bite, which mimics a variety of ailments ranging from sinusitis to migraine and often baffles the physician from whom the victims seek help. One frequent source of the pain can be dysfunction of the jaw hinge or temporomandibular joint, and symptoms can range from "clicking" or "gravel-like" sounds in the joints to severe pain in the joints and other parts of the head when the mouth is either opened or closed.

In many instances old, discolored crowns can now be remodeled in the patient's mouth during a 30-minute dental appointment. J. Daniel Cox of the College of Medicine and Dentistry of New Jersey said "we have waited for years to have the ability to change color and replace discolored plastic directly in the mouth. It can now be done inexpensively." Through the use of new plastic veneers and of a powder mix, a dentist can quickly remodel discolored crowns including those where the gold is showing. The resins and powder mix are applied directly to the surface of gold or acrylic crowns to produce an attractive tooth.

Swedish dental scientists successfully tested a new freezing technique to keep extracted teeth and their attached root tissues alive for more than a year for possible reimplantation in the same patient or even other individuals. P. Otteskog of the Karolinska Institutet in Stockholm reported at the annual session of the International Association for Dental Research, held in April 1975 in London, that the freezing method might pave the way toward the eventual establishment of "tooth banks" that could supply teeth for rational matching of donor and recipient in what is called allotransplantation. In his study 21 healthy teeth extracted for orthodontic reasons were first treated with a cell cultivation technique and then placed into a liquid nitrogen freezer at temperatures dipping down as low as −197° C. This storage technique may be beneficial to patients for whom an immediate transfer of a tooth to a new position should be delayed until the jaw and gums have healed, especially if the extraction site has become infected.

—Lou Joseph

Veterinary medicine

The organized veterinary profession in the U.S. is little more than a century old, but it is appropriate during this bicentennial celebration to note that the first quasi-official mention of veterinary services in the U.S. dates to 1776. In December of that year Gen. George Washington appointed Col. Elisha Shelden commandant of the Connecticut Regiment of Horse and directed that one farrier be attached to each of his six troops of dragoons. A farrier-veterinary service was later attached to the artillery, and in 1777 Washington himself mentioned by name one Joseph Fox, whom he identified as a light-horse farrier. Veterinary schools had been established in Europe since 1762, and the first mention of a graduate in the U.S. dates to 1778 when a Count Saxe, said to be "a man of great skill and celebrity in his profession," was a member of Baron von Steuben's general staff. In 1792 an act of Congress included farriers in the tables of organization for the Army.

What is probably the first mention of veterinary services in the American colonies appears in the court records of Virginia for 1625, when "an expert cow doctor," William Carter, was involved in a lawsuit. The first veterinary surgeon in the U.S. was John Haslam, a graduate of the London veterinary school (established 1792), who entered practice in Baltimore in 1803 and published veterinary articles in the agricultural press. In 1807 the physician Benjamin Rush advocated establishment of a

UPI Compix

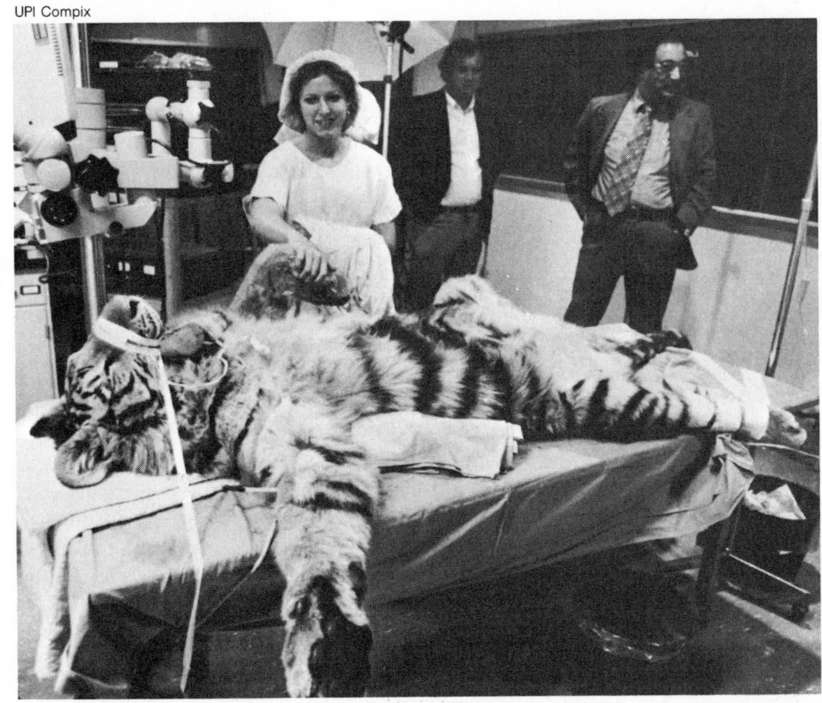

Brutus, a 179-lb Siberian tiger at Chicago's Lincoln Park Zoo, is readied for a cataract operation. The operating team followed a relatively new procedure that required only one tiny incision per eye.

veterinary school at the University of Pennsylvania, but this did not come about until 1884; among the still-extant schools it followed those at Ontario (1862), Montreal (1866), and Iowa (1879).

Only about a dozen veterinary works (identified as "farriers") were published in the U.S. before 1800, the first of consequence being *The Citizen and Countryman's Experienced Farrier* (Wilmington, Del., 1764) by J. Markham, G. Jefferies, and Discreet Indians, which went through four editions to 1839. Like others of its stripe it was primarily a wretched compilation of earlier British writings. The first truly native work was *The New-England Farrier* (Newburyport, Mass., 1795) by Paul Jewett. The first journal was *Farrier's Magazine* (Philadelphia, 1818), published by James Carver, a London graduate, but only two issues appeared. This was followed by George Dadd's *American Veterinary Journal* (Boston, 1851–59), published at a time when there were only about a dozen graduate veterinarians in the U.S. Dadd was a British physician who turned to veterinary practice in about 1845 and was the first veterinarian in the U.S. to make regular use of anesthesia in surgical operations. He also established the first functional, but short-lived (1854–60), veterinary school, known as the Boston Veterinary Institute.

The veterinary profession remained small and had little influence until after the Civil War. In 1863 a group of 40 practitioners founded the United States Veterinary Medical Association, which espoused the development of education, research, and writings that paved the way for the rapid flowering of veterinary medicine to its present high status. During the half century from 1868 to 1918, some 40 public and private veterinary schools were founded, of which only 10 remained after 1933. The private schools in the U.S. produced about 10,000 of the approximately 16,000 graduates before 1935, but the cost of meeting steadily increasing educational requirements gradually forced the schools out of business.

No new schools were founded until the mid-1940s, when attrition of the veterinary work force and increasing demand for services resulted in the establishment of eight additional schools; one new school (in Louisiana) was opened in 1974 and another (in Florida) was opened in 1975, bringing the total to 20. By 1975 plans for new schools in Tennessee, North Carolina, Mississippi, Wisconsin, and New England had been announced. Oregon and Idaho entered into a shared-curriculum agreement with Washington, and South Dakota signed a contract agreement with Iowa, as did New Jersey with Pennsylvania.

Veterinary education and practice had long been oriented toward equine and food-animal medicine, but after World War II the trend toward other aspects of the field rapidly accelerated. This was favored by the great increase in facilities that were comparable in many respects to those in medicine, and also by the numbers of highly qualified science-oriented applicants. As an unintended result many more graduates opted for small-animal

"The Times," London

Eddie, a seven-year-old giraffe, receives acupuncture treatment for arthritis in his fetlock joints. He was the first zoo animal in the United Kingdom to be so treated.

practice, in part because the increasing affluence of pet owners made it feasible to use various sophisticated methods of diagnosis and treatment that formerly had been restricted largely to teaching situations. Despite the substantial increase in the number of graduates since World War II and increased output by both the existing and new schools, a deficit of 5,000 to 7,000 practitioners was anticipated by 1980. This would equal the total current output of the existing schools for five years, and if the recent trend toward small-animal practice were to continue the present shortage of food-animal practitioners would become even more severe.

How acute this shortage actually becomes will depend on several factors. A continued recession would probably affect veterinary medicine less than many other endeavors, but it would slow the rate of expansion. A substantial number of veterinarians would be likely to work a few years longer to bolster retirement income, and this would automatically augment the number of working practitioners. The rapidly increasing availability of well-trained veterinary technicians has made it possible for some practices to hire such personnel, often instead of an additional veterinary graduate and in any event increasing the volume of service delivered by a given number of veterinarians. Within the past two or three years several of the schools have developed specific programs to make food-animal practice more attractive, and the opportunities for practitioners to acquire new skills in this area have been greatly expanded. Considering the enormous cost of providing additional facilities for veterinary education, the current expansion program testifies to the faith professional leaders have in the future of veterinary medicine.

In addition to the problem of how best to meet the increasing demand for veterinary services, the profession was concerned with finding methods for better control of unwanted animals and for disseminating information on the application of birth-control methods. By late 1974 the Los Angeles municipal spay and neuter clinics, instituted in 1971, had been credited with effecting a substantial reduction in the number of unwanted animals in the Los Angeles area.

Reports appeared in the popular media suggesting an association between feline leukemia and leukemia in humans. The reports were based largely on the earlier demonstration that feline leukemia virus (FeLV) would multiply in cultures of human tissue cells and on reports that FeLV had been found in a high proportion of cats in households where cases of human leukemia had been diagnosed. One report appearing in the *American Journal of Public Health* in 1972 indicated the presence of FeLV in leukemic persons, but subsequent investigation failed to confirm any association with the disease in cats.

In some areas veterinarians reported a large increase in the number of cases of leukemia and related diseases in cats during the past few years, but this could be attributed in part to greater awareness of the disease and to the recent development of a fluorescent antibody (FA) test for the virus. The FA test could detect FeLV in many healthy cats, and it was inevitable that some FA-positive animals would be found in households where leukemic persons lived. This put many veterinarians in the uncomfortable position of weighing a recommendation to have such cats destroyed, for public health reasons, against the obvious reluctance of owners to do so.

According to an editorial that appeared in the *New England Journal of Medicine*, despite the prevalence of the virus in cats, "there is yet no definitive evidence for linking cats with human cancer." This confirmed the earlier conclusions reached by a panel of veterinary experts and gave support to the more rational recommendation that cats with clinical signs of leukemia (which would soon die anyway) be humanely destroyed and that FA-positive but otherwise healthy animals be examined at intervals.

—J. Frederick Smithcors

Optical engineering

Optical engineering encompasses much more than such traditional areas as the design of camera lenses, microscopes, and other optical instruments. Much of the exciting and important work of the last year involved laser applications, miniaturization of optical systems, and inclusion of optical components in large complex systems in which the boundary between optical engineering and other specialties is blurred.

Optical communication. Optical fibers the size of a human hair can be used as waveguides to transmit information with beams of light. (See *1975 Yearbook of Science and the Future* Feature Article: FIBER OPTICS: COMMUNICATIONS SYSTEM OF THE FUTURE?) Fiber optical communication has an information-handling capacity far greater than telephone lines or microwave transmission. A major drawback has been the large losses of light in the fibers. Low light loss is necessary to reduce the number of "booster" stations required to make fiber optics systems practical. During recent months scientists at Bell Telephone Laboratories and Corning Glass Works developed new glass fiber waveguides that absorb only one-third of the initial light intensity over a distance of one mile. Such attenuation would occur within 0.1 in. in ordinary glass. The Bell Laboratories' fiber consists of a pure fused-silica core to minimize absorption and a borosilicate glass coating on the core to assure light guidance along the fiber by means of total internal reflection.

To realize the full potential of fiber optics systems, light sources, modulators, detectors, and optical components of dimensions small enough for the efficient coupling of light with the fibers are needed. Considerable progress was made in this direction in the last year with the development of thin-film gallium-aluminum-arsenide ($Ga_{1-x}Al_xAs$) laser diode light sources having an active cross-sectional area of less than one-billionth of a

Courtesy, Bell Laboratories

Optical fibers made from commercial glass of the highest known purity are assembled as orderly strands within a protective sheath. The width of a ribbon consisting of several fibers is about one-eighth of an inch.

square inch. The advantage of these lasers is that they emit in the infrared, a wavelength region of low fiber attenuation, and can be modulated directly by controlling the electrical current through the diode. The optical properties are strongly dependent on *x*, the fraction of aluminum, thereby making it possible to tailor the waveguiding behavior.

Light emitted in a $Ga_{1-x}Al_xAs$ laser can be confined as desired by appropriately varying the composition of the laser and, therefore, the optical refractive index at the desired boundaries. Techniques were also developed for fabricating optical lenses of less than 0.001 in. in diameter directly on optical fiber surfaces; this method resulted in a fourfold increase in the coupling efficiency of the laser beam and optical fibers.

That optical communication is coming of age was demonstrated by the installation in 1974 of a fiber optics communication system on the USS "Little Rock," the U.S. Navy's Sixth Fleet flagship. The U.S. Department of Defense hoped to save more than $50 million a year by introducing fiber and integrated optics.

Laser fusion. Many energy experts believe that nuclear fusion is one of the best hopes in the long run for safe and pollution-free energy, and they believe that laser technology is the best means of achieving the extraordinarily high temperatures required to make fusion practical. This use of laser technology imposes stringent requirements on optical components. The optical problem is to shine laser light onto a pellet only 0.02 in. in diameter that contains deuterium and tritium, heat it to 100 million degrees Kelvin (180 million degrees Fahrenheit), and cause a microscopic thermonuclear explosion. To achieve this, the optical components used for transmitting and focusing the high-intensity laser beam must withstand intensities of 60 billion watts per square inch. However, high-intensity light absorbed in the optical components can severely damage the optical materials. Such absorption also induces a change in the index of refraction of the glass, causing the laser beam to split into filaments that prevent focusing of the beam onto the target.

In an effort to deal with these problems new classes of phosphate glasses have been developed with lower refraction indexes than the traditional silicate glasses. The use of an electron beam to evaporate refractory materials such as titanium dioxide (TiO_2) resulted in achieving close

Experimental fiber optics telephone system consists of central electronic switching exchange (left), terminals (right), coils of optical fibers, and electrical-to-optical converters.

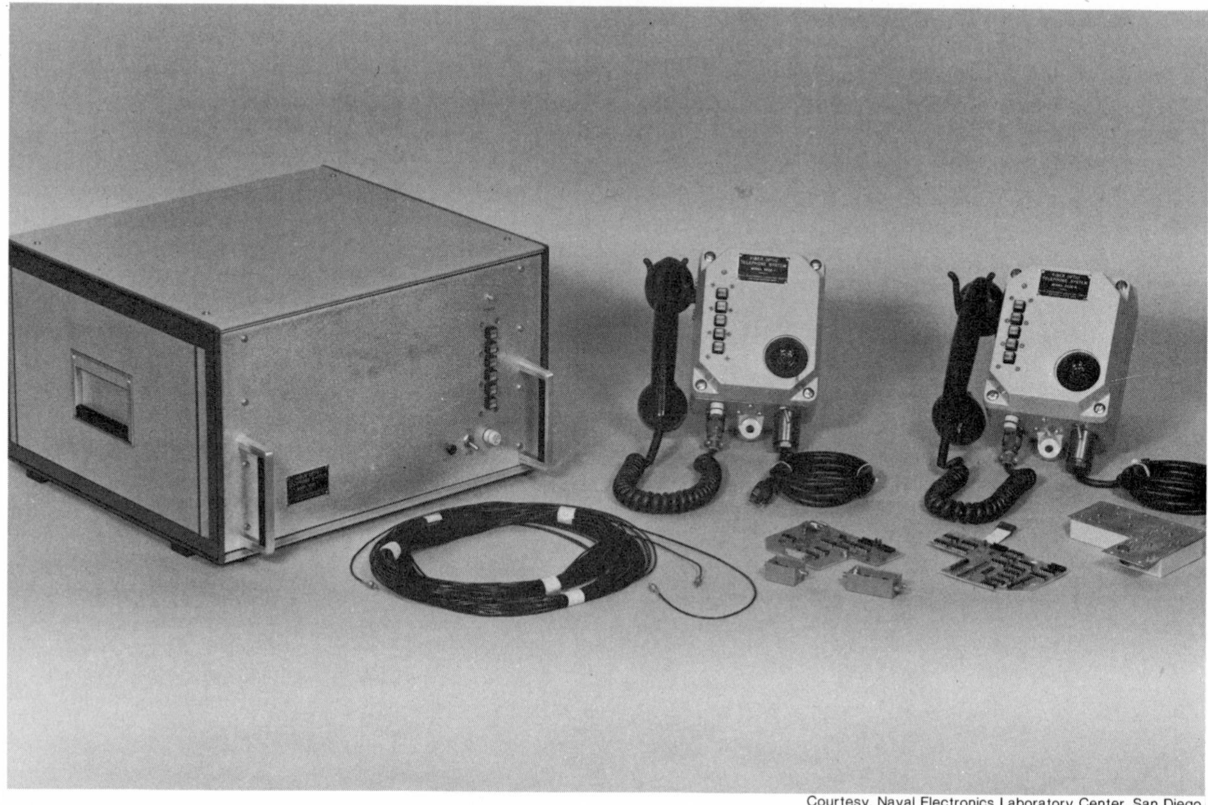

Courtesy, Naval Electronics Laboratory Center, San Diego

to theoretical damage limits in thin-film antireflection coatings. Although progress continued in improving the optical components of such fusion systems, much more powerful lasers remained to be developed.

Isotope separation. In June 1974, at the eighth International Quantum Electronics Conference in San Francisco, five groups reported results on laser-induced isotope separation. Similar work was also reported in Israel and the U.S.S.R. At the San Francisco meeting, Anthony Bernhardt of the University of California at Davis reported the results of barium isotope separation experiments carried out at Lawrence Livermore Laboratory by using a photodeflection technique. Bernhardt described the technique as somewhat like using table tennis balls to deflect a stream of moving bowling balls. The bowling balls in this case were the various isotopic forms of the barium atom, and the table tennis balls were discrete bundles of light energy, called photons, emitted by the laser.

Other methods of laser isotope separation were reported. All had in common the use of the monochromaticity of the laser beam; that is, the characteristic of a laser to emit light at essentially a single wavelength. Because the different isotopes of an atom have small variations in atomic weight, each isotope absorbs light at a wavelength slightly different from that of any other. If the shift in wavelength is large enough, it is possible to tune a laser so that one isotope of a particular element will absorb the laser light while the other isotopes are unaffected. The development of such tunable lasers has thus been the key in making laser isotope separation possible.

The greatest incentive for the commercial development of laser isotope separation is for uranium enrichment. To meet future energy needs, increasing amounts of uranium-235 will be needed for nuclear reactors. Only 0.7% of natural uranium consists of the fissionable isotope uranium-235. Spurred on by the potential market and the estimated savings of as much as $100 billion in capital investment over traditional gaseous diffusion plants, Exxon Nuclear Co. and Avco Everett Research Laboratories planned to build jointly a pilot plant for uranium enrichment that would utilize a laser process.

Automated supermarket checkout. Probably the largest single order for laser equipment ever placed by a civilian organization was that of the National Cash Register Corp. (NCR) for an estimated 3,000 to 4,000 laser scanners purchased from Spectra-Physics, Inc., at a cost of $9.8 million. Delivery of the scanners began in 1975. They will be part of an electronic label reader system designed to automate supermarket checkouts. The complete system includes a computer to be provided by NCR. Trial testing in certain stores began in mid-1974.

As packages pass by a slot window in the checkout counter, a laser beam scans a rectangular array of bars printed on the label. The bars and spaces define a pattern that identifies the manufacturer and the product. The laser scanner identifies the product by its code and sends the information to the computer. The computer then obtains the price from where it is stored in the computer memory, subtracts the number of units sold from inventory, and delivers to the customer a receipt with the name and price of each item.

A number of requirements dictate the use of a laser for scanning. These include the need to illuminate the code with a single color for which reflection varies strongly so that the code can be easily detected as the bars and spaces are scanned. The beam must also be narrow so that the bars in the printed code can be easily resolved even when the package being scanned is passing by at several feet per second.

Integrated circuits. Microscopic integrated electronic circuits consist of as many as 1,000 active elements on a single wafer, or chip, of silicon measuring only a tenth of an inch on a side. An essential step in the manufacturing process is the use of a mask to produce the exact circuit geometry on the silicon chip. The mask allows passage of light through accurately defined windows onto the chip. A chemical photographic process is then used to convert the resulting pattern image into an actual circuit.

Have the practical limits to making even smaller circuits been reached? A limiting factor is the quality of the optical path through which the circuit pattern is transmitted from mask to silicon chip. Contact "printing" has been the mainstay of the industry for years. However, defects can be produced by forced intimate contact of the mask with the chip. This situation is greatly improved if the pattern is projected onto the chip. The pattern can be focused using a system of optical lenses. Potentially, the best technique is to use an all-reflective projection system. The Perkin-Elmer Corp. reported in 1974 the development of such a system, which, compared with a lens-based projection system, substantially reduces the amount of light-scattering and eliminates chromatic aberration and the need for narrow-band optical filters. The new projection printer should be of particular value in the fabrication of solid-state memories and other large-scale integrated circuits. It holds promise of producing up to ten million elements on a single quarter-inch chip.

—Frederick Wooten

Physics

Among major developments in physics were the discovery of several new subatomic particles, possibly including one that had only one magnetic pole, and the first formation of element 106. An unusual achievement was the rediscovery of the method for making Damascus steel. Light, sharp, and flexible weapons of this steel were well known in the Middle Ages, but its formula had been lost for centuries.

High-energy physics

The most exciting and probably the most important advance in high-energy physics during the past year was the discovery of several new particles with rest energies in the range of from three to five billion electron volts (GeV). Two of these particles have extremely narrow widths; that is, extremely well-defined masses. While the ultimate theoretical explanation of these particles and their relationship to previously known particles remains in doubt, there is no doubt that their discovery has opened a substantial new area of investigation in high-energy physics, for both theorists and experimenters. Indeed, the first announcements of their existence, by experimental groups working at Brookhaven (N.Y.) National Laboratory and at the Stanford (Calif.) Linear Accelerator Center, produced an unprecedented surge of theoretical articles and experimental investigations.

Brookhaven and Stanford experiments. The initial discoveries were made in experiments that studied how pairs of particles, consisting of an electron and a positron, produce, and are produced by, particles of the general class known as hadrons. The hadrons are characterized by the property of converting into one another very easily when brought together in collisions. Hadrons include protons, neutrons, and most of the other known subatomic particles, but do not include electrons or positrons.

In the Brookhaven experiments, done by a group of physicists from the Massachusetts Institute of Technology and Brookhaven, protons of about 25-GeV energy were allowed to collide with the protons and neutrons in a beryllium target, and the pairs of an electron and a positron that are sometimes produced were detected. These pairs can be produced with various total energy. It was found, however, that electron-positron pairs with

Physicists at the Brookhaven (N.Y.) National Laboratory (left) display data indicating the presence of a new subatomic particle with a rest energy of 3.1 GeV. Group at the Stanford (Calif.) Linear Accelerator Center (right) also detected the particle.

Courtesy, Brookhaven National Laboratory

Courtesy, Stanford News & Publications Service, photo by Charles Painter

an energy of 3.1 GeV are produced much more often than pairs with nearby energies. The interpretation made of this was that in the proton-beryllium collision an uncharged particle with rest energy of 3.1 GeV—more than three times that of the proton—is produced and that this particle transforms rapidly (decays) into the electron-positron pair that is observed. Because it was set up to detect only the pairs, the Brookhaven experiment could not determine whether this particle had other ways of decaying. The experiment did suggest, however, that under the conditions studied the production of the new particle was quite rare, occurring in only about one collision in every 100 million.

The Stanford experiments took place at an installation called by the acronym SPEAR, from Stanford Positron-Electron Asymmetric Ring. They were carried out by a group from Stanford and from the University of California. In these experiments a beam of electrons and a beam of positrons, with equal energies but traveling in opposite directions, are made to collide. As the energies of the electrons and positrons are varied, the particles produced in the collisions are studied. At total energies of 2–3 GeV (1–1.5 GeV for each particle), hadrons are produced at a roughly constant rate per collision. As the energy reaches 3.1 GeV, the rate of production of hadrons suddenly increases by a factor of 50 or more; at slightly higher energy it drops back to the original value. Such a rapid change in the production rate indicates that a new process is occurring, one that is interpreted as the production of a particle from the colliding electron and positron followed by its decay into the observed hadrons. Both production and decay occur very quickly, in less than 10^{-20} sec, so that the particle that is produced and

Diagrams by the Brookhaven team (left) and the Stanford physicists (right) reveal the difference in the experiments performed by each group. The new particle was detected in the e^+e^- mass spectrum at Brookhaven and in the $e^+ + e^- \rightarrow$ hadrons spectrum at Stanford.

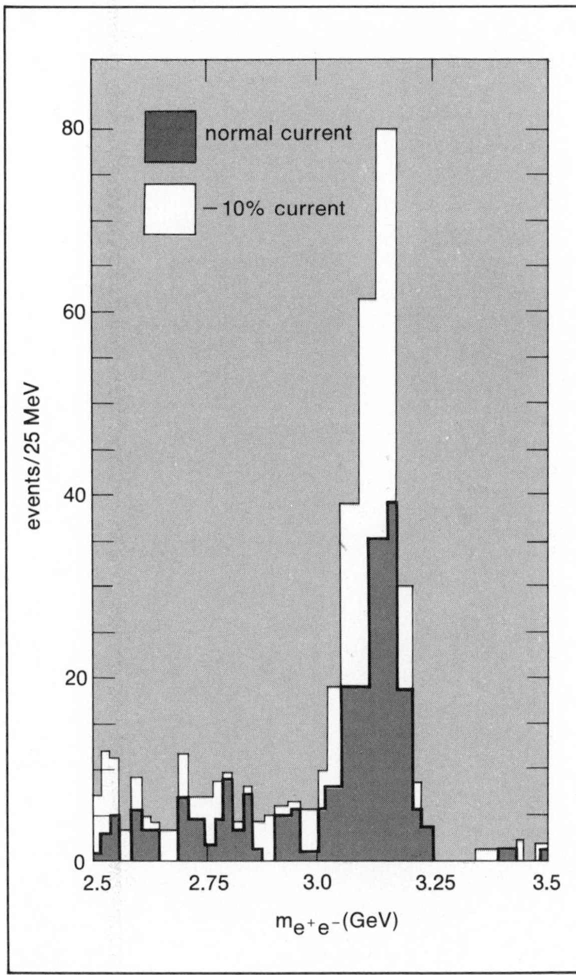

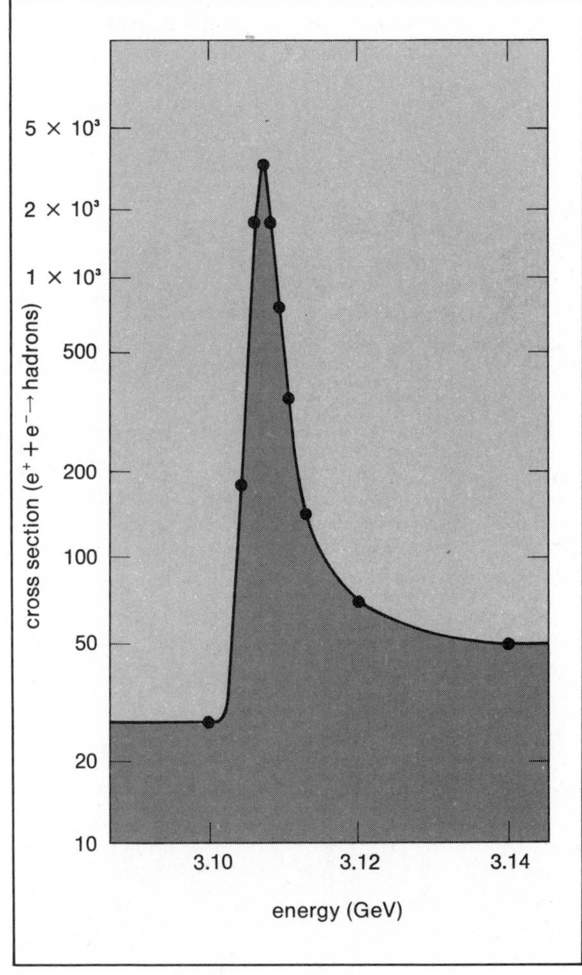

decays is not observed directly. Instead, its properties must be inferred from its decay products.

It is apparent that the chain of events observed at SPEAR is the reverse of that at Brookhaven, in that, in the former, hadrons come from electron-positron pairs while, in the latter, electron-positron pairs come from hadrons. It was the consistency of the properties of the inferred new particle as revealed in both experiments that caused physicists to become convinced that the particle existed.

At SPEAR it was possible to control the energy of the beam very accurately and to determine the precise energy at which the production of the new particle occurs. Also, the range of energies over which the particle is produced could be measured. This range, known as the width of the particle, is an indication of how rapidly the particle decays into other particles. In general, the greater the width, the more rapid the decay. Experiments at SPEAR have shown that the width of the 3.1-GeV particle is about 100,000 eV, which is quite small for such a massive particle. For comparison, a particle called the rho meson, with a mass of about 750 million electron volts (MeV), has a width of 150 MeV, approximately one thousand times greater than that of the new particle. The smallness of its width is one of the major puzzles connected with the 3.1-GeV particle.

Because the new particle can decay into electron-positron pairs, as at Brookhaven, and into hadrons, as at SPEAR, it is important to determine the relative probabilities with which each of these occurs. All decays of individual subatomic particles obey the laws of quantum mechanics and happen randomly within their overall probabilities, and it is therefore necessary to observe many instances of the decay of a particle in order to determine the relative probability of different decay modes. This was done at SPEAR with the result that the predominant decays of the 3.1-GeV particle are those containing hadrons. Further experiments were in progress in 1975 to determine which hadrons, as well as other particles such as photons, are among the decay products. The determination of its precise decay modes is of great importance in making a theory to describe the 3.1-GeV particle.

An important quantity that characterizes any particle is its spin, the angular momentum that it has when at rest. It is possible to determine this for the 3.1-GeV particle by measurements of the angle made between the initial electron that produces it and the electron that it may decay into. Such measurements have shown that the way in which this angle varies, over many examples of the decay, is consistent with a spin of one unit of angular momentum. This value is also suggested by the fact that a single 3.1-GeV particle can be readily produced in electron-positron collisions.

Soon after the discovery of the 3.1-GeV particle a similar phenomenon was detected, again at SPEAR, at an energy of 3.7 GeV. This was interpreted as the production and decay of a second new particle. The 3.7-GeV particle has a somewhat bigger width than the 3.1-GeV particle, about 500,-000 eV, which is still a good deal less than would be expected for such a heavy particle. This suggests that the two particles are part of a new family, distinct in some way from the previously known hadrons. This view was reinforced by the discovery that the dominant decay mode of the 3.7-GeV particle is its transformation into the 3.1-GeV particle plus two pi mesons. It is not yet known what other decay modes the 3.7-GeV particle has, and these are also under investigation.

Finally, it appears that yet another new particle was produced in electron-positron collisions at

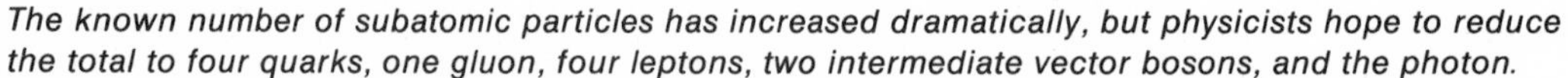

The known number of subatomic particles has increased dramatically, but physicists hope to reduce the total to four quarks, one gluon, four leptons, two intermediate vector bosons, and the photon.

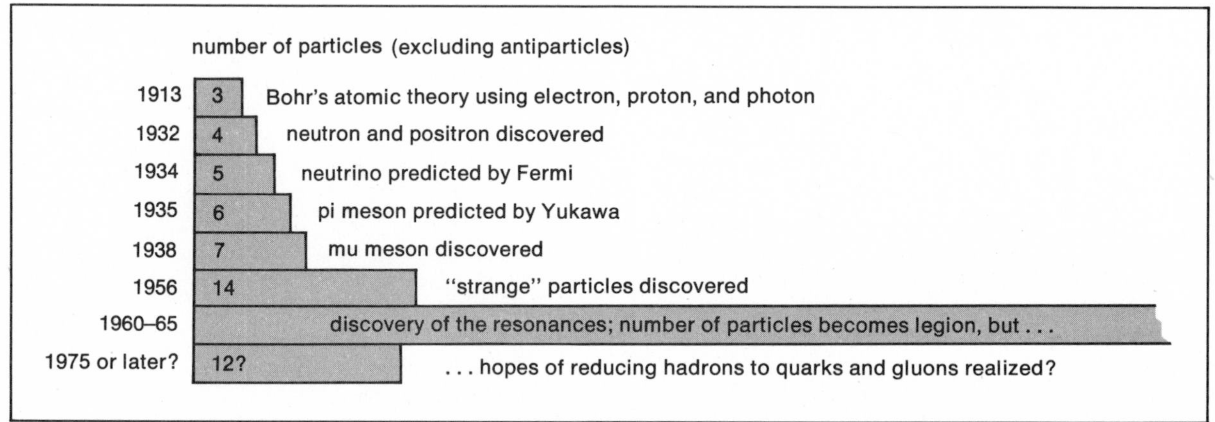

From "Physics Flowers but Funds Wither," by Robert Walgate, in "New Scientist," p. 496, Feb. 27, 1975

SPEAR, one with a rest energy of about 4.2 GeV. This particle has a much greater width than the other two, more than 100 MeV. In mid-1975 it was not yet clear what its decay products are, or whether it is really part of the same family as the 3.1- and 3.7-GeV particles.

The 3.1-GeV particle was also detected at other laboratories, produced in collisions among various particles. The two heavier particles have not yet been detected except in electron-positron collisions, but physicists expected that they would be as soon as experiments of sufficient precision are carried out. No similar neutral particles of still higher rest energy were found at SPEAR or elsewhere, although searches were made for such particles with rest energies up to about 5 GeV.

Particle theories. The response of theoretical physicists to these discoveries was an outpouring of articles that proposed various hypotheses about the nature and properties of the new particles. There was no constant theme to these proposals, and at this writing it is too early to tell which, if any, of the many hypotheses that have been proposed is correct.

Perhaps the most widely studied and widely accepted idea about the new particles is that they are a kind of atom or bound system made of two particles, facetiously called "charmed quarks." According to this hypothesis, the three newly discovered particles would really be analogous to three different states of one atom, such as a hydrogen atom, rather than to three intrinsically different atoms, such as hydrogen, helium, and iron. The particles that are supposedly bound together in the new particles have themselves never been detected as isolated objects, unlike the case of ordinary atoms, made of nuclei and electrons that can occur either bound together or separated. However, many of the properties that the charmed quarks must have if they exist have been inferred indirectly from the properties of known subatomic particles. Thus, they must have a spin of ½ unit and must occur both as a particle and an antiparticle, with electric charges of $+\frac{2}{3}$ and $-\frac{2}{3}$ that of a proton. The bound states corresponding to the new particles consist of a charmed quark and a charmed antiquark, whose total charge adds up to zero and whose total spin is one unit.

The charmed quarks are related in kind to "ordinary" quarks, which are also hypothetical particles that bind together to make the previously known subatomic particles. However, the charmed quarks cannot transform easily into ordinary quarks. This would help account for the small widths of the new particles, because their decay into ordinary hadrons would require just such a transformation of a pair of charmed quarks into ordinary quarks.

The reason why quarks, either charmed or uncharmed, cannot be produced as isolated particles remains somewhat mysterious and has been the subject of much speculation by theoretical physicists. It is possible that the bound-state model will shed some light on this question, since some of the mechanisms proposed to prevent quarks from separating should also have important influences on the properties of the bound states.

The bound-state model of the new particles has several implications that should be testable by experiments. In addition to the bound states with a spin of one that have been discovered, there should also be different bound states of a charmed quark and a charmed antiquark with other values of angular momentum, such as zero. These other states cannot easily be produced in electron-positron collisions nor do they decay into electron-positron pairs, but they should be produced and decay in other ways that can be detected, and searches for them are under way.

Also, it should be possible for charmed quarks and ordinary quarks to bind together, producing "charmed hadrons." These should be similar to ordinary hadrons, although somewhat higher in energy, because the charmed quarks have higher energy than ordinary quarks. Charmed hadrons should have even narrower widths than the 3.1-GeV particle. If particles with these properties exist, it should be relatively easy to discover them. Other models of the new particles do not imply the existence of these many, yet undiscovered, bound states, and the results of searches for them will play an essential role in deciding whether the bound-state model is correct.

—Gerald Feinberg

Nuclear physics

Nuclear science spans a broad range, from the most fundamental and abstruse topics that concern the basic laws and symmetries of nature to the most practical and applied subjects relating to the world's energy, hunger, and health problems. During the past year major advances took place in all these areas; only a few representative examples can be included in this discussion.

Radiation damage. One of the most serious problems facing both fusion and fission breeder reactor designers, on whom the United States inevitably must depend for some substantial fraction of its future energy, is the behavior of structural metals while under prolonged neutron bombardment. Normal stainless steel, for example, swells by about 30% under a neutron bombardment that corresponds to a 20-year lifetime of either of the above types of reactors. If it were

Courtesy, Walter Brown of Bell Laboratories

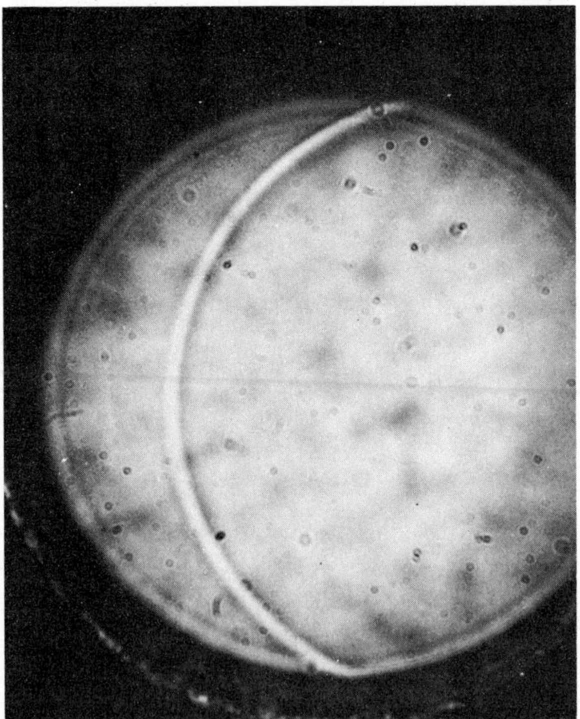

Photomicrographs show a cross section of a quartz optical fiber bombarded with ions from the left. The image at the top, in white light, reveals a crescent that corresponds to the end of the range of ion penetration. The interference photomicrograph below shows changes in the refraction index of the fiber.

necessary to design to allow for such expansion, reactors would necessarily be much less efficient, much more costly, and quite possibly less safe.

Fortunately nuclear scientists had both the equipment and the requisite knowledge to address this problem quickly and effectively. The damage is caused primarily by neutrons bouncing atoms of the metal out of their normal crystal lattices and thereby creating the swelling. In order to test metal samples under conditions to be expected in reactors, using neutrons themselves, it would be necessary to load samples into the nation's largest reactors and leave them there for 10–20 years; clearly, this is too long to wait. Instead, however, a heavy-ion accelerator can be used to fire iron nuclei, for example, into stainless steel (or niobium nuclei into samples of niobium metal, which might well be used as the containment vessel for a fusion reactor). Four or five hours of such bombardment, directly putting the alien nuclei into the metal instead of recoiling them there from neutron collisions, can be expected to duplicate 10–20 years of neutron damage.

This bombardment process has made possible the development of new classes of alloys that are much more resistant to radiation than were previous ones. For example, given conditions under which normal 316 stainless steel swells by 35% in volume, a new alloy developed at the Oak Ridge (Tenn.) National Laboratory swells only 6%. This can have dramatic consequences in the nation's search for more abundant safe energy.

Ion implantation. The field of ion implantation, in which particle accelerators are used to insert nuclei or atoms of any desired impurity into an existing crystal lattice in programmed fashion, has been shown to have far-reaching consequences in fabricating ever smaller, yet ever more powerful, integrated circuit devices. Less than a quarter of an inch square, the small silicon chip that forms the heart of many pocket calculators contains thousands of transistor, resistor, and capacitor elements; all were produced automatically in the crystal itself by using nuclear acceleration and beam-handling techniques.

The power of ion-implantation techniques in producing such devices, in modifying surfaces to obtain better hardness or wear characteristics, in creating chemically impossible alloys, and in producing new surfaces for industrial catalysis and for medical and biological culture media is limited only by the imagination of the scientists and technologists involved. This represents a revolutionary new frontier in materials science.

Medical applications. With increasingly sophisticated understanding of fundamental nuclear structure and dynamics it has become almost rou-

Courtesy, Cornelius Tobias and Eugene Benton

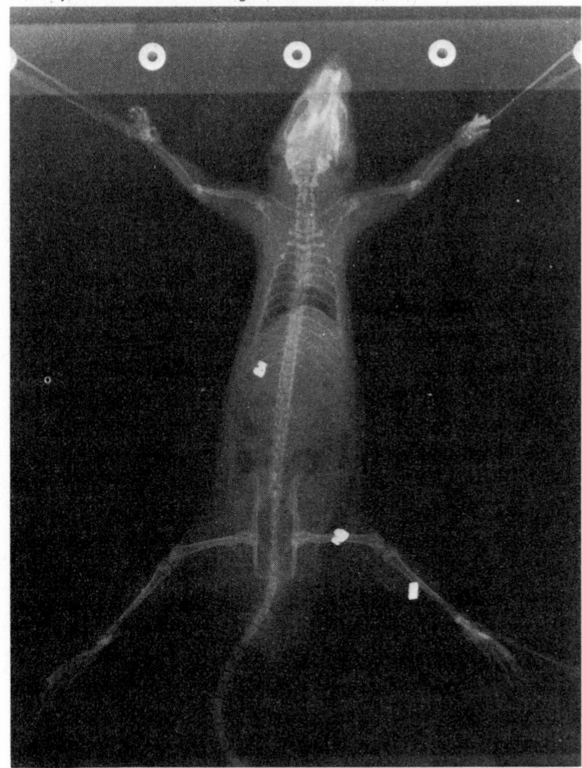

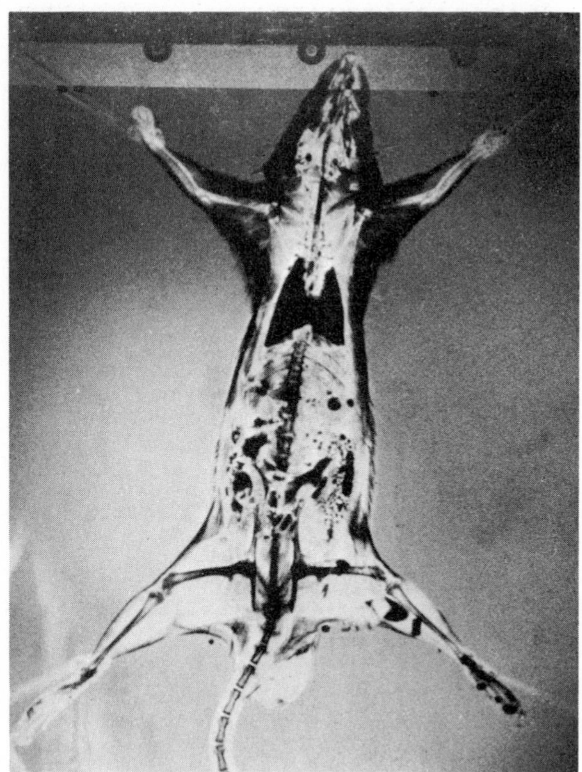

X-radiograph of rat (left) shows bones and hard tissue. Heavy-ion radiograph (right), obtained by suspending the rat in a water tank and exposing it to a beam of 400-MeV-per-nucleon ions, allows scientists to construct in detail the rat's internal organs and muscles.

tine to tailor radioisotopes to the job at hand instead of making do with those that happened to be available. New accelerators and techniques provided access to a vastly greater spectrum of such radioisotopes than had yet been seen. In nature there are only 300 stable nuclear species; since the beginning of nuclear science physicists have produced and studied some 1,300 more. But collisions of high-energy uranium beams with uranium targets can yield as many as 6,000 new species.

A striking example of a tailored radioisotope is the radioiodine used in thyroid cancer therapy. Iodine is chosen because it is selectively concentrated in the thyroid. Initially, iodine-131 was widely used because it was known and available, even though it had a half-life of eight days and much more penetrating radiations than were desirable. The production of iodine-123, however, which has a 13-hour half-life and produces only soft radiations that are absorbed entirely in the thyroid, has provided physicians with an enormously better clinical tool. More than 100,000 U.S. citizens received iodine-123 treatment during the past year.

Dramatic progress was made—as yet on an experimental but soon to be on a routine basis—in the use of high-energy neutron beams from cyclotrons in the clinical treatment of a wide range of human cancers. Certain types of skin cancers (melanomas), cancer of the salivary glands, and a number of other malignancies, which in the past had been characterized by less than 10% remission or recovery rates, were characterized by greater than 80% remission and recovery rates when treated with this neutron therapy. As techniques become perfected, physicians expect that additional types of cancer will yield to such treatment. All large urban hospitals will, as a matter of course, eventually have their own cyclotron installation both for clinical treatment of patients and for production of tailored radioisotopes that cannot be produced any other way.

With the availability of superhigh-energy beams of heavy ions during the past year from the University of California (at Berkeley) Bevalac accelerator, it became possible to evolve an entirely new form of radiography that may have interesting medical applications in the future. In a normal X-radiograph, X-radiation is passed through the body and registered on a photographic plate; bones and heavy material attenuate the X-radiation and thus show up clearly as dark areas in the photographic

Courtesy, Karl Van Bibber, MIT

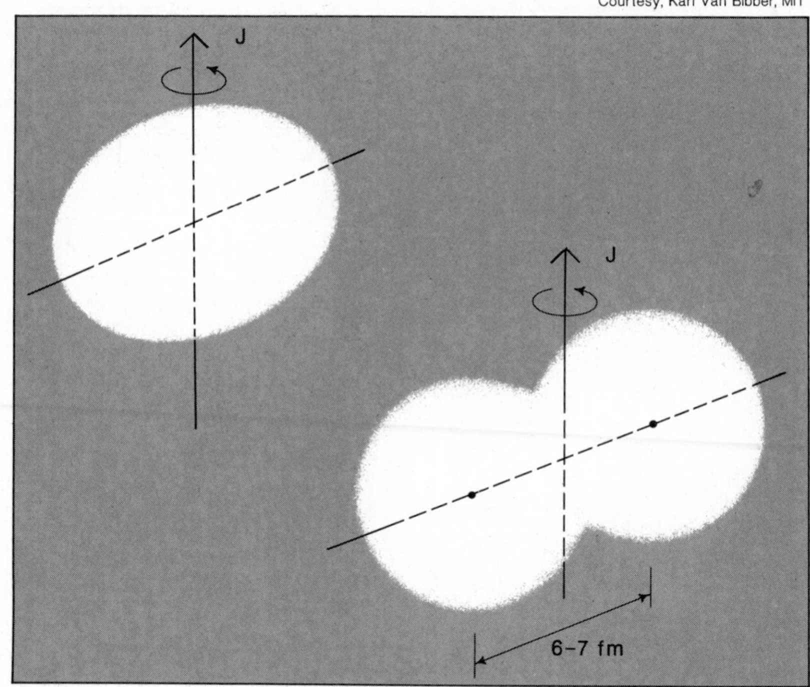

Nucleus of magnesium-24 in its ground state is football-shaped (upper left), but as energy is added it assumes a dumbbell configuration with a nucleus of carbon-12 at each end. The center-to-center distance of the two carbon-12 nuclei is 6–7 fm (fermis; 1 fm = 10^{-13} cm). J is the axis of rotation.

negative. But little information is obtained about the soft tissues. In heavy-ion radiography, accelerator beams of very-high-energy nuclei, such as neon, are passed through the body and registered in a suitable nuclear detector that is sensitive to the amount of energy each beam particle lost in traversing the body. In initial experiments with animals C. A. Tobias and his colleagues at Berkeley obtained remarkably detailed pictures of the soft internal tissues, thus complementing X-radiation. Moreover, the heavy-ion radiographs have the additional intrinsic advantage that they can reproduce the internal structure of the subject in three dimensions rather than the two dimensions characteristic of the X-ray plate or photograph.

Basic nuclear science. In the science of nuclear physics itself, the year was a fruitful one. Emphasis both in the U.S. and abroad continued to shift to studies involving the use of heavy ions—projectiles of nuclei ranging from lithium to uranium—as these became available from accelerators.

Karl A. Erb and his associates at Yale University demonstrated for the first time that it is experimentally possible by using ultrathin targets (millionths of an inch thick), accelerator beams of very high energy homogeneity, and high-resolution nuclear magnetic detectors to resolve and study the individual quantum (energy) states of heavy nuclear species. This opened up an entirely new nuclear spectroscopic field because heavy ions reveal quite different things about nuclear structure than do the light projectiles, such as protons, deuterons, and alpha particles, that have been in common use in the past. Erb's initial measurements provided the first convincing evidence for the exquisitely detailed higher order quantum-mechanical processes in which the incoming particle first excites the target to a higher energy level and then interacts with it, or in which the interaction takes place first and is followed by an excitation of the residual nucleus by the outgoing reaction product. Such higher order processes interfere with the simple direct processes to produce coherent "beat" phenomena reminiscent of those familiar in music.

Eric Cosman and his associates at Massachusetts Institute of Technology, Brookhaven National Laboratory, and Argonne National Laboratory found convincing evidence that in addition to the normal football-shaped structure of a nucleus such as magnesium-24 (which contains 12 neutrons and 12 protons), a new family of dumbbell-like structures emerges as energy is added. In these the ends of the dumbbell are carbon-12 (each containing six neutrons and six protons). As more energy is added the structures spin end over end faster and faster. These new molecular states are important additions to the knowledge of nuclear structure; it is a tribute to recent progress in theoretical nuclear science that they can be fully understood on a microscopic basis.

John R. Huizenga and his associates at the University of Rochester, N.Y., and the University of California at Berkeley, and Vadim V. Volkov and his collaborators at the Joint Institute for Nuclear Research at Dubna, near Moscow, discovered a new kind of nuclear interaction that occurs when two heavy nuclei come together in an accelerator-induced collision. In the past it was believed that the interactions could be considered either as direct, in which a few neutrons or protons were exchanged or a few surface waves excited, or as compound, in which the colliding nuclei simply fused and lost all memory of their prior history before breaking up statistically into various allowed final states. Huizenga and Volkov found an intermediate situation in which surface friction between the colliding nuclei results in the formation of a relatively short-lived molecular complex that exchanges substantial amounts of nuclear matter before the molecule fissions under the mutual electrostatic repulsion of the protons in the end fragments. This new process could have an important impact on the approach to the formation of desired new radioisotopes and on physicists' eventual ability to form superheavy nuclei.

Formation of element 106 was reported for the first time both by Georgi N. Flerov and his associates at Dubna and by Albert Ghiorso and his associates at Berkeley. Different techniques were used and different isotopes of the new element were reported; some controversy concerning the results continued. The Soviet workers chose to exploit their development of very heavy projectiles, such as chromium and molybdenum, and bombarded stable targets such as lead-208; the U.S. scientists exploited the availability of very heavy radioactive targets such as californium-252 and bombarded them with much lighter projectiles such as oxygen. Curtis E. Bemis, Jr., and his collaborators at the Oak Ridge National Laboratory completed arrangements to fly a trailer truck containing a fully equipped nuclear detection laboratory to Moscow, drive to Dubna, and combine Soviet heavy projectiles with U.S. targets in a pioneering example of cooperation that promised to resolve some long-standing disagreements and discrepancies between the Soviet and U.S. groups in the area of heavy elements.

E. W. Schopper and his collaborators at the University of Frankfurt, West Germany, and at Berkeley observed what appear to be shock waves in nuclear matter as a very-high-energy oxygen ion plows through a much heavier nucleus. This is analogous to the shock wave from an aircraft moving at speeds greater than that of sound in air; from the measurements it is possible to extract the speed of sound in pure nuclear matter. This in turn measures the compressibility of this matter, a value crucial to astrophysics and elementary-particle physics as well as to nuclear scientists.

Fundamental physics. As the only natural entity in which all four forces of nature operate simultaneously, the nucleus is a natural microscopic laboratory for the examination of these fundamental forces and of the laws and symmetries that they impose on natural systems. One of the most exciting and far-reaching programs in contemporary physics is that directed toward the demonstration that two, if not indeed three, of these natural forces are not independent but rather are different manifestations of a more fundamental parent force. Major attention focused on the electromagnetic and weak nuclear forces as being most closely related. The weak nuclear analogue of the electric radiation in the electromagnetic field had long been observed in nuclear beta decay; but the so-called "weak magnetism," the corresponding analogue of the magnetic radiation in the electromagnetic field, had never been detected.

R. E. Tribble and Gerald Garvey, working at Princeton University and at Brookhaven National Laboratory, succeeded in finding this weak magnetism for the first time in a detailed study of the nuclear mass 8 system. The preliminary results were consistent with the abovementioned coalescence of the weak and electromagnetic forces, but important questions concerning the number, nature, and interactions of the entities postulated to underlie all matter—the quarks—remained unresolved. This work illustrated the important fact that as knowledge of nuclear structure and dynamics becomes more precise and comprehensive, the ability to use the nucleus as a base for making fundamental observations is greatly enhanced.

—D. Allan Bromley

Solid-state physics

With the standard of living of modern man increasingly threatened by the depletion of once plentiful and inexpensive natural resources such as oil, minerals, and even clean air and water, scientists are challenged to develop new technology to alleviate these problems. Whereas in the two decades following World War II there was considerable financial support from both government and industry in the United States for research aimed principally toward the increase of scientific knowledge, by the 1970s there was an increasing reluctance to support work that was not related to some degree to practical problems.

This led to a large amount of quiet and largely unorganized reexamination of the relationship between science and the technology necessary to solve practical problems. The interaction between the two is particularly important in solid-state

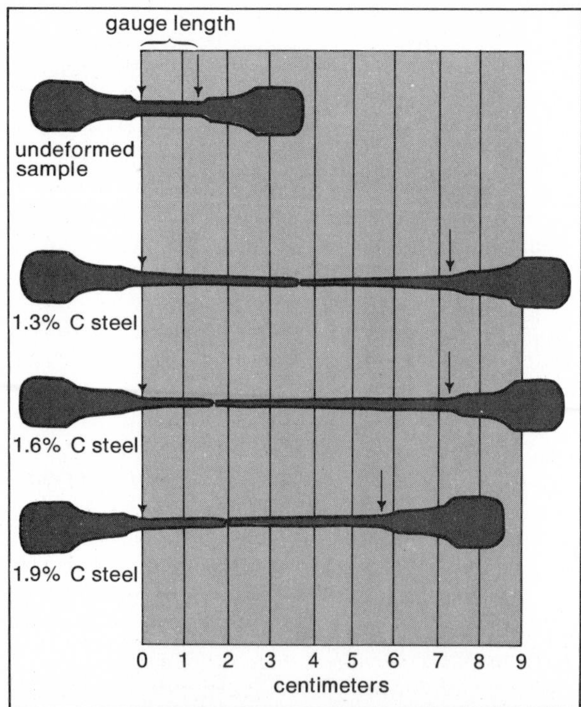

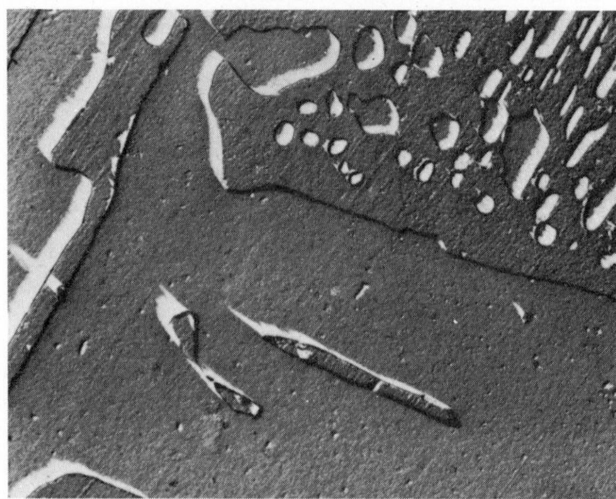

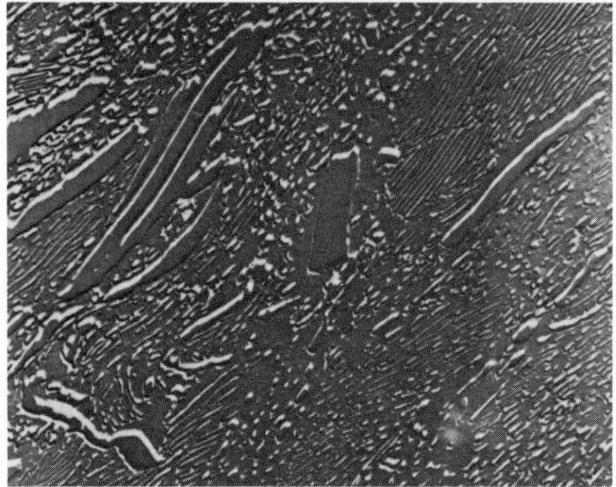

Courtesy, William E. Spicer

physics because of the dependence on solids in almost any practical endeavor one might examine. In this year's review, a few examples of recent advances that illustrate the relationship between the development of technology and scientific knowledge are singled out for detailed examination.

"Damascus" steel. One can distinguish between two different ways in which technology grows. The first is through the systematic application of scientific knowledge. This might be called scientific engineering. Nuclear energy, electronic computers, and space vehicles were all developed by means of scientific engineering. The second type of growth occurs empirically through practical experience. In such instances one does not attempt to understand in depth what is happening; rather, one learns through accumulated practical knowledge how to provide a desired product.

Because it lacks a basis in fundamental knowledge and depends on procedures with applications that may vary appreciably from generation to generation, there is always a danger that empirically based technology will be lost. In particular, if only a few people have mastered a complex process for which there is no general understanding, the process might be lost with their death.

A fascinating case in point is that of the legendary "Damascus" steel. There are many stories from the Middle Ages about the quality of swords made from this material. The swords were light and flexible yet capable of receiving and retaining an extremely sharp edge. Typical is the tale told of a meeting, during a truce in the Crusades, between the two chief adversaries, Richard I the Lion-Heart, king of England, and Saladin, the sultan of Egypt and Syria. Richard demonstrated his strength and the weight of his heavy two-handed sword with a massive overhead stroke, cutting through an iron mace handle. Saladin was politely impressed but pointed out that his Damascus blade could also cut through iron. He then threw a feather into the air and cut it in half before it hit the ground with an easy stroke of his blade. He then asked Richard to repeat this feat with his heavy blade. Richard did not even try.

Although the remarkable properties of Damascus steel are well documented, the ability to make such steel gradually became a lost art. Despite all the advances of scientific technology, modern man did not seem capable of producing it. In

Newly developed high-carbon steel, similar to the Damascus steel of antiquity, can be drawn out over five times its original length at 650° C without breaking (top). This can be done by working the steel at high temperatures so the large grains that cause brittleness (center) become fine (bottom).

early 1975, however, a group at Stanford University succeeded not only in reproducing Damascus steel but also in understanding scientifically the reason for its special properties.

There appeared to be a paradox with Damascus steel. It is believed to contain a relatively large amount of carbon; in modern experience, however, steels with high carbon (1–2%) content are known to be too brittle for wide practical application. Cast iron is an example of a typical high-carbon steel. The Stanford group developed new processing techniques for high-carbon steel and studied microscopically the changes in the structure as well as the changes in the properties of the finished product. In this way, they seemed not only to have reproduced the properties of Damascus steel but also to have developed a scientific understanding as to the reason for its properties.

The success of the Stanford group appeared to result from obtaining steel made up of two very finely dispersed components. One of these consists of grains of iron, and the other of grains of an iron-carbon composite material called cementite. By means of special processing, the size of each type of particle is placed in the range of 1–2 microns (1 micron = 10^{-6} m); thus, each grain is about 10,000 atoms in diameter. In contrast, steel with the same carbon content (typically 1.6%) formed conventionally would have much larger grains. Establishment of grain size was possible only by the use of an electron microscope.

One method of obtaining the fine grain size is first to heat the steel quite close to its melting temperature of 1147° C in order to dissolve the cementite. The steel may also be mechanically worked at that temperature by rolling, forging, extrusion, or any other procedure that will break up any large grains present. In the next step, the material is continuously mechanically worked as it is cooled to about 600° C. The final step is additional mechanical working at a fixed temperature between 550° and 650° C. At all temperatures the mechanical working serves to break up large grains and prevent their reforming.

The steel so formed demonstrates superplastic flow at 650° C. This means that it can be formed easily into desired shapes at that temperature without breaking. Samples can be drawn out over five times their original length before they break. This is associated with the small particle size and the two phases of the material. It appears that the small particles flow rather easily past each other, whereas large particles would impede the flow and lead to early breakage as the metal is elongated or otherwise formed into shape.

The properties of the material at room temperature are desirable. It is strong, tough, and nonbrittle. It may also be relatively inexpensive to produce. One interesting aspect of the knowledge gained in this work is that one can easily see how the conditions of working the steel near the melting point and as it is cooled could also have been achieved by a blacksmith with his anvil, hammer, and hearth. Thus, contact is made between a lost art and modern scientific technology.

Practical superconductors. Since the discovery of superconductivity by Heike Kamerlingh Onnes at the beginning of the 20th century, it has been apparent that this phenomenon has great potential for practical application. Because electric currents flow through superconductors with no loss of energy, the efficiency of generating, transmitting, and utilizing electricity can be increased by the use of such materials. The need for energy conservation has increased the urgency for the practical application of this phenomenon. As discussed in previous years, it is the fact that these materials must be operated at very low temperature that has prevented extensive practical utilization to date. In 1975 strong research efforts were being directed toward overcoming these problems. Throughout the world one could identify at least three different approaches.

In the first approach, engineering programs were underway in the U.S., Japan, and Europe to make practical machinery and electrical transmission lines based on the niobium alloys, which were readily available. These alloys become superconductors at temperatures below 10° K (−263° C); therefore, the cooling must be done by liquid helium, which makes the apparatus expensive and complex. A further complication is the relative scarcity of helium, particularly outside of the U.S.

The other two approaches, much more long-term in nature, were based on the development of new high-temperature superconductors that can operate above 21° K and thus be cooled by liquid hydrogen rather than helium. The two approaches differ in the type of materials utilized in the search for higher temperature superconductors. The more conventional of the two is associated with a class of metallic alloys having the β-tungsten crystal structure, an arrangement of three mutually perpendicular, nonintersecting linear chains that seems to characterize many high-temperature superconductors. Examples are V_3Si, Nb_3Ge, Nb_3Sb, and V_3Ge. Some of these materials exhibited transition temperatures near 21° K; however, they are very difficult to fabricate and the samples that were made only became superconducting well below 21° K. Unfortunately, at present, most of the samples have superconducting transition temperatures too far below 21° K to be useful. Furthermore, the materials are very hard, brittle, and nonductile. As a result, it will be difficult to fabricate them into practical forms without destroying their high-temperature superconductivity.

In response to these problems, work was under way to develop new methods of materials preparation. Rather than following the usual procedure of melting the components of the alloys together in a crucible, these new methods involve forming the alloys by deposition from the vapor. Another method is to use bombardment by protons to "tear up" the material to a sufficient extent so that atoms can move and reach their correct position at lower temperatures than could ordinarily be used.

The third general approach to practical superconductors was the most revolutionary. It envisioned the development of superconductors from materials, such as organic compounds, that are ordinarily not good conductors. Interest in these materials was generated by theoretical considerations which indicated that strong interactions leading to superconductivity might occur in such materials. In 1973 and 1974, researchers demonstrated a considerable increase in conductivity in the vicinity of 60° K in the organic material (TTF) (TCNQ). At first, physicists believed this might indicate high-temperature superconductivity. As of mid-1975, however, this expectation had not been fulfilled and laboratories were having difficulty confirming the original results.

A less spectacular but clearly well-founded development occurred in 1975 that revived much of the interest in materials that are not ordinarily good conductors. A group working at the IBM Research Division Laboratory, San Jose, Calif., and at Stanford University found that a polymeric material, polysulfur nitride $(SN)_x$, becomes a superconductor at about 0.25° K. This is the first superconductor made entirely of elements which themselves are not good conductors. It was of further interest because the material is composed of long, one-dimensional chains of sulfur and nitrogen [the *x* in the formula $(SN)_x$ refers to the number of atoms in the chain]. What makes this so encouraging is that it is in general agreement with the theoretical work which suggested that chain-like materials that to a good approximation are one-dimensional may have special properties that will make them superconductors at reduced temperatures, even though their high-temperature conductivity is much less than that of a metal.

Spin-polarized electron source. Each electron possesses a quantized electron spin that produces a microscopic magnetic field or magnetic moment associated with that electron. Physicists have long desired to develop a relatively easy method for providing a convenient source of electrons with spins all aligned in the same direction. Such a source was in particularly strong demand in high-energy physics, where, for example, electrons with extremely high energy (1–20 GeV) are scattered from positrons or other particles. For solid-state physics such sources would allow the possibility of low-energy (about 100 eV) scattering experiments in which the scattering is affected by the interaction between the aligned magnetic moments of the electrons and the magnetic order—*i.e.*, the geometric ordering of the magnetic moments associated with individual atoms—of the scattering medium (for example, the surface of a ferromagnetic sample). This is important because much needs to be learned about the effects of surfaces on ferromagnetic order.

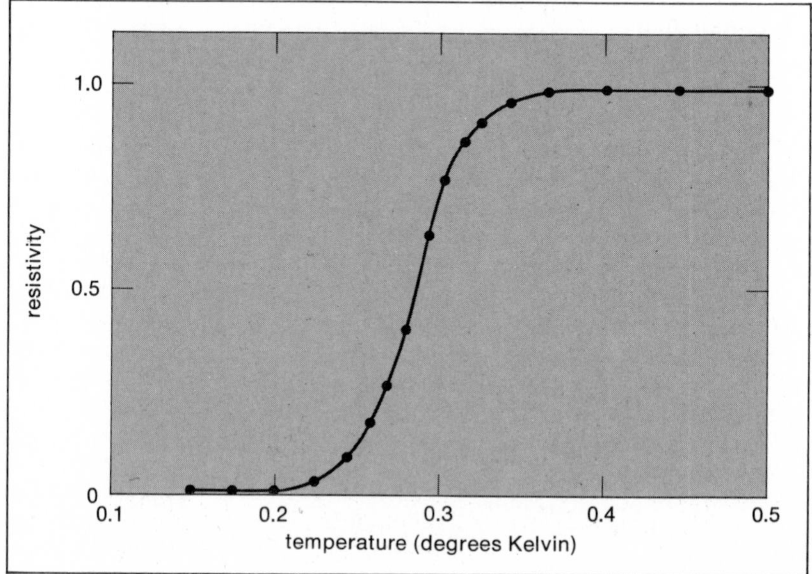

Polysulfur nitride becomes a superconductor at about 0.25° K. It is the first superconductor made entirely of elements that are not themselves good conductors.

Courtesy, IBM Corp.

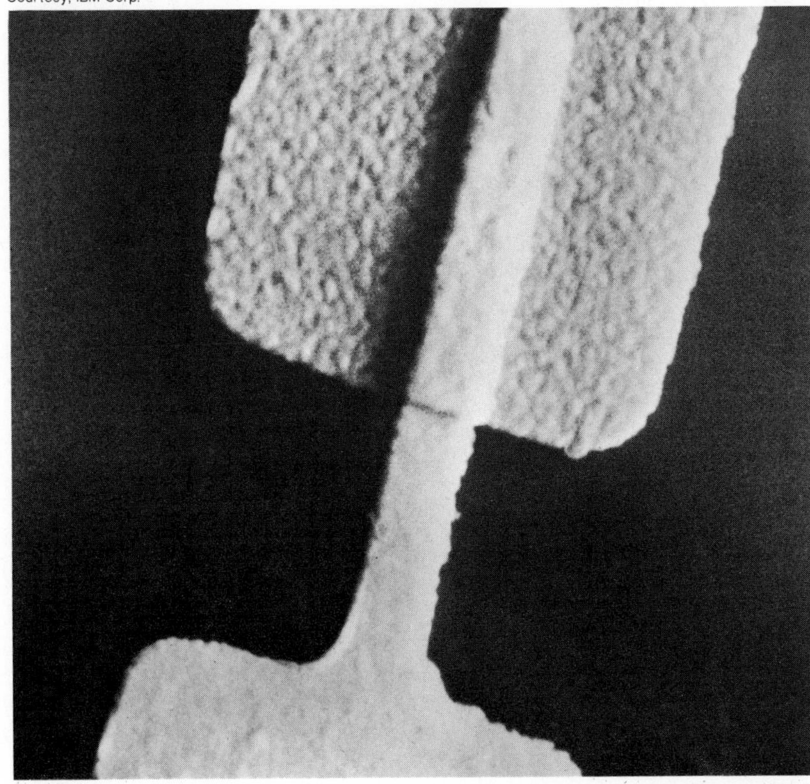

Josephson junction, two superconductors separated by a thin layer of insulating material, measures 1.3 by 7 micrometers. Electric current in the form of pairs of electrons flows from one superconductor to another across the insulator in the absence of an applied voltage.

In 1975 a group at the Laboratory for Solid State Physics in Zürich, Switz., achieved what appeared to be the first conventional source of spin-polarized electrons. Their accomplishment provided a strange example of scientific engineering. Making use of prior scientific knowledge, they achieved success in a relatively short time and in accordance with theoretical predictions. Several different types of knowledge were brought together. The first of these was knowledge of quantum (energy) states that characterize semiconductors, such as gallium arsenide, and the way in which circularly polarized light preferentially excites electrons with one spin polarization. By reversing the direction of polarization, researchers can quickly reverse the direction of spin. The second type of knowledge was an understanding of the process by which excited electrons can escape from such a semiconductor into a vacuum after being optically excited (photoemission). Based on the available fundamental knowledge of these processes plus empirical knowledge about the preparation of gallium arsenide surfaces to maximize the electron yield, it was possible to build and demonstrate this efficient polarized electron source in about a year. A number of other laboratories began working on such sources and their application to different experiments.

—William E. Spicer

Psychology

Events within psychology during the past year demonstrated a continuation of significant recent trends and also sharpened some differences of opinion regarding the future of the discipline. As the following discussion indicates, many of the noteworthy developments were in the area of professional rather than academic psychology. Breakthroughs of a scientific sort are few and far between; nevertheless, the slow and erratic but steady accumulation of knowledge continued, and on an ever widening front.

Social affairs. With respect to the role that psychologists play in social and political affairs, the most significant event of the past year was the official initiation of a new organization established within the framework of the American Psychological Association (APA) but with legally independent organizational status. The new organization, the Association for the Advancement of Psychology (AAP), was designed to offer a more moderate alternative to the vigorous politico-legal activities of the fully independent Council for the Advancement of the Psychological Professions and Sciences (CAPPS). The latter organization, which concentrated on legislation related to private practice, remained in operation, and it was too early to tell how the two would relate to one another or

Floyd Jillson from "Atlanta Journal-Constitution Magazine"

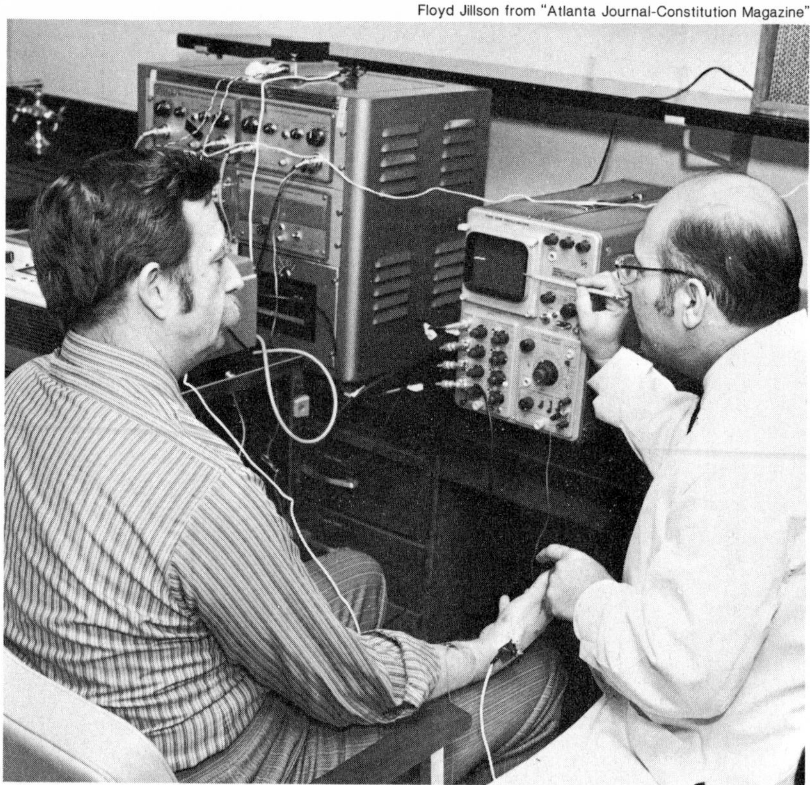

John Basmajian at Emory University, Atlanta, Ga., connects a subject with a biofeedback device that monitors electrical activity in the body.

how each would represent psychologists in their social and political involvements with governmental agencies and the public.

Among the significant developments for psychology during the year was the initiation of a legislative attempt to place biofeedback devices, largely developed by psychologists, under the control of medical agencies. Biofeedback refers to the learning of voluntary control over physiological functions, such as heartbeat, that are not normally controllable; this is accomplished by means of selective reinforcement of responses made to the overt display of measures of these internal processes.

Unfortunately, the application of biofeedback techniques seemed to be running well ahead of solid scientific tests of their efficacy, as often occurs when a striking new technique is developed. Most interested psychologists recognized the need to achieve some kind of control of untried devices and techniques, but they strongly opposed the kind of restrictions on research that would be likely to accompany legislation unless it is carefully prepared.

Professional practice. The accentuation of emphasis on applied problems throughout psychology continued during the past year. This trend could be seen in the type of jobs that new holders of the Ph.D. were being offered; for example, there were more openings in human than in animal learning. Renewed attacks by politicians on projects that showed little possibility of practical application were another reflection of this trend; thus a prominent U.S. senator, William Proxmire (Dem., Wis.), generally regarded as politically "liberal," took issue with various "pure" research projects supported by the National Science Foundation (such as the research of a social psychologist concerned with the determinants of romantic attachment).

The general public suspicion of psychology and psychological procedures surfaced in a number of ways. The American Civil Liberties Union renewed its attack on programs in mental hospitals in which tokens are used as "monetary" rewards for good behavior, claiming them to be violations of individual rights, and the START (Special Treatment and Rehabilitative Training) program of behavior modification in federal prisons was declared unconstitutional.

Attempts to cope with this growing public distrust of psychological efforts were made. Albert Bandura, in his APA presidential address, explained the widespread public antipathy to behavioral theory and practice on the grounds that many people equate the behavioral approach with conditioning, which in turn is identified as mechanical response of the sort that Pavlov's dogs showed

Ernie Hearion from the "New York Times"

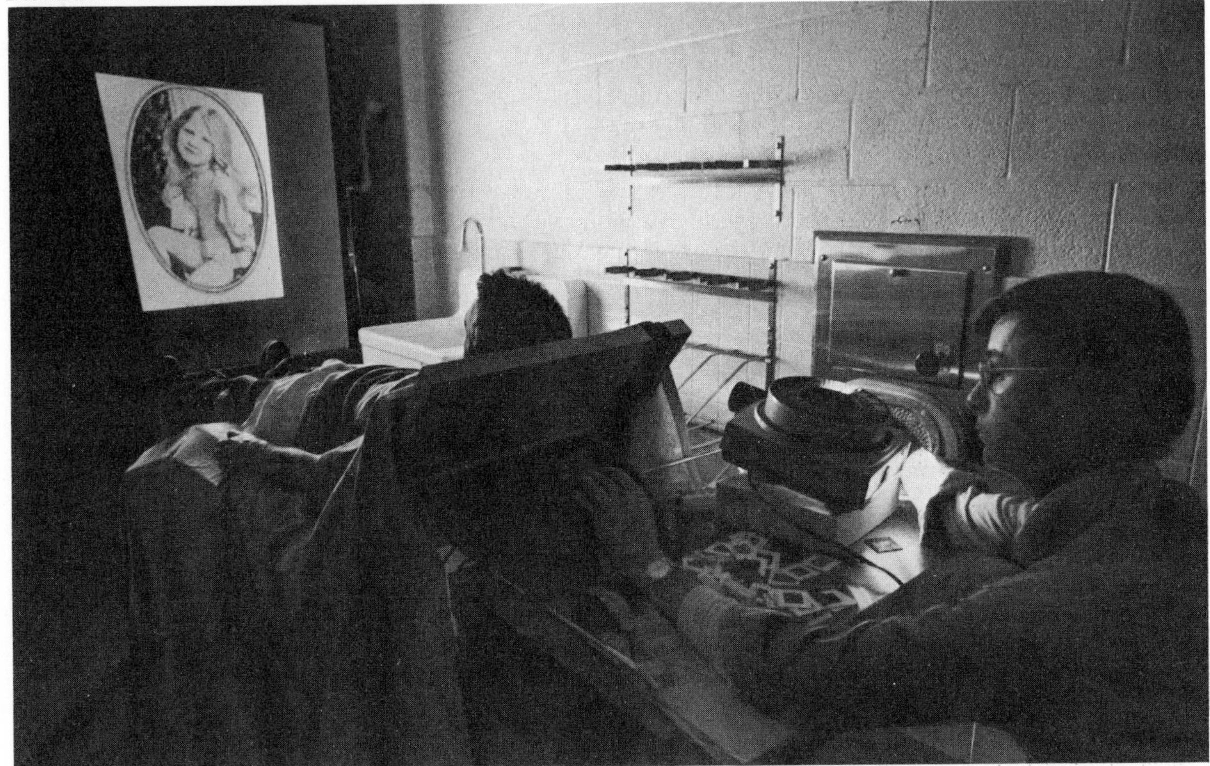

As slide picture of a child is shown, a man imprisoned for child molestation receives an electric shock. Such behavior-modification techniques drew considerable criticism from civil liberties groups.

in the original demonstrations of conditioning. Bandura argued that this narrow view of behavioral practice is misleading.

Professional training. In clinical psychology there was an increasing tendency to relate practice and research. On the one hand, clinical practice moved closer to research results, and on the other hand there was a movement of research designs in the direction of more immediate applicability of data to the requirements of practice. A good example is behavior modification, in which both practical and theoretical aspects of psychology are closely related.

The New Jersey Graduate School of Applied and Professional Psychology opened in September 1974 as an independent but coordinate part of Rutgers University. The pilot class consisted of 25 students in clinical psychology and 15 in school psychology, selected from more than 2,000 applicants. The new school signaled the accentuation of a trend toward independent graduate training in the practice of psychology (California recently developed a state school with a similar purpose). In line with this trend the school offered the practitioner degree, Psy.D., in contrast with the research-oriented Ph.D.

Academic psychology. Within social psychology, the emphasis on cognitive factors continued to grow. The special focus of this interest was on the problem of "attribution of causality." This concept is concerned with how people react to other people's behavior in terms of their perception of what has caused the behavior. For a commonplace example, if one is insulted by a person who is perceived to be drunk, one's reaction will be different than when the insulter is perceived as sober.

The psychology of women has become a topic of unusual interest throughout psychology. This growing concern was reflected in the appearance of a chapter on this topic, for the first time, in the *Annual Review of Psychology* for 1975.

Within theoretical psychology generally, there was a growing tendency to recognize interactions between variables, as opposed to the simpler main effects of the variables, as crucial relationships. For example, Benton J. Underwood, in his Distinguished Scientific Address at the national meetings of the APA, argued for incorporating individual-difference variables into behavior theory. He held that general, universal theories "should be formulated in a way that will allow an immediate individual-differences test" so that the interactions effected by individual differences can become a kind of "crucible in theory construction."

In the area of human learning and memory the most significant development over the past year was the steadily increasing acceptance of a relatively new interpretation, offered as an alternative to the short-term–long-term distinction. This new view is called depth of processing. It holds that all memories vary with regard to the depth or level at which they have been processed, ranging from sensory to semantic (meaningful) levels. The deeper the processing that has occurred, the more closely the semantic level is approached. It is the deeply processed meanings that are more readily retrieved.

Attention within physiological psychology was also focused on the problem of memory. One of the interesting general trends was the questioning of the reality of amnesia. This view asked whether there is any solid evidence for a failure of memory storage. Like so many issues in psychology, the problem is at heart a methodological one. Just as learning cannot be demonstrated without performance, so memory storage cannot be demonstrated without memory retrieval. Psychologists maintained that there are factors other than storage failure that can account for the failure of retrieval. To give an example, the long-accepted demonstration of failure of memory consolidation as a result of electroconvulsive shock—which is based on the failure of organisms so treated to show memory of habits established just prior to the shock—was being questioned; an alternative interpretation of these results is that the shock affects the retrieval mechanism rather than the memory storage itself.

A major setback was suffered during the past year by researchers on and proponents of parapsychology, who had long labored to improve the scientific status of their work and to have it accepted as a respectable part of the scientific community. Their efforts had been rewarded, as evidenced by the formation of a parapsychology division in the American Association for the Advancement of Science. In 1974, however, one of the leading parapsychological researchers was detected tampering with an automated experiment in which the ability of rats to exercise psychokinetic influence (mental effect on physical movement) was under investigation. The investigator subsequently resigned.

The affair was unfortunately made more significant by the fact that the researcher involved had been the "main hope for the Institute," according to 79-year-old J. B. Rhine, founder and director of the Institute for Parapsychology in Durham, North Carolina. It was too early in 1975 to predict the severity of the blow to parapsychology, but no one doubted that it would seriously impede the attempts of the field to achieve full scientific respectability.

An important publishing development within the academic and research wings of psychology was the initiation, in 1975, of four separate sections to replace the previous single section of the prestigious *Journal of Experimental Psychology*. The four subjournals focused on (1) human learning and memory, (2) human perception and performance, (3) animal behavior processes, and (4) general subjects.

While locked in an acoustically shielded room, psychic Uri Geller drew responses to target pictures selected at random in a nearby room. Controversy continued to surround Geller's performances during the year.

(Left and right) Courtesy, Stanford Research Institute; (center) "The Washington Post"

Personalized system of instruction. During the past year the Personalized System of Instruction (PSI), also called the Keller Plan (for its major founding father), continued to expand. The core of this instructional program is the division of the subject matter of a course into units and the allowing of a student to proceed to master each unit in order at his own pace, as determined by his taking of a test whenever he feels ready for it. The tests are given by trained student proctors, and there is no penalty for failure; if a test is failed, another equivalent test is provided, again when the student feels ready for it, after discussing his test performance with the proctor. Ordinarily, full credit (a grade of A) is given for mastery of the course (all units passed successfully). This system was being used at many levels of instruction.

In his review paper marking the tenth anniversary of the introduction of PSI, Fred Keller characterized it by saying that "the teacher of the future [should] no longer be a classroom entertainer, an information vender, a critic, or debater, but . . . an educational engineer—a manager of student learning." This paper appeared, incidentally, in the first issue of a new journal, *Teaching of Psychology*, published by the American Psychological Association and its division on teaching.

—Melvin H. Marx

Space exploration

Highlights of the year in space exploration included the completion of the Apollo/Soyuz Test Project, the first international manned space flight. Also, the unmanned U.S. space probes Pioneer 11 and Mariner 10 achieved successful flybys of Jupiter and Mercury, respectively.

Manned flight

During the last year the United States and the Soviet Union launched one joint manned space mission, and the Soviets also placed two space stations and five manned space flights in orbit. All missions were limited to Earth orbit.

In the U.S. the development of the space shuttle continued with efforts concentrated on completion of the design phase. A conglomerate of nations in Western Europe continued work on the development of a cylindrically shaped manned spacelab that would be carried into orbit with the space shuttle.

Apollo/Soyuz Test Project. The space teams from the U.S. and the Soviet Union devoted much of the year to the necessary development, testing, and operational planning for the joint Apollo/Soyuz Test Project (ASTP). This program, initiated in 1970, was the first international manned space flight project. Its major objective was to demonstrate the space rescue capability of the two nations should a space crew be stranded in Earth orbit. To accomplish this objective, scientists designed a docking module that allowed an Apollo command module and a Soyuz spacecraft to join together in space and permitted crewmen to move from one spacecraft to the other. The U.S. built the docking module, which met the design specifications of both nations. The docking module and Apollo command module were launched by the Saturn IB booster rocket. At launch, the docking module was carried within a shroud in the same

Model of the U.S. space shuttle is assembled at Rockwell International Corp. in California. Part airplane and part spacecraft, the reusable shuttle is expected to become operable in 1979.

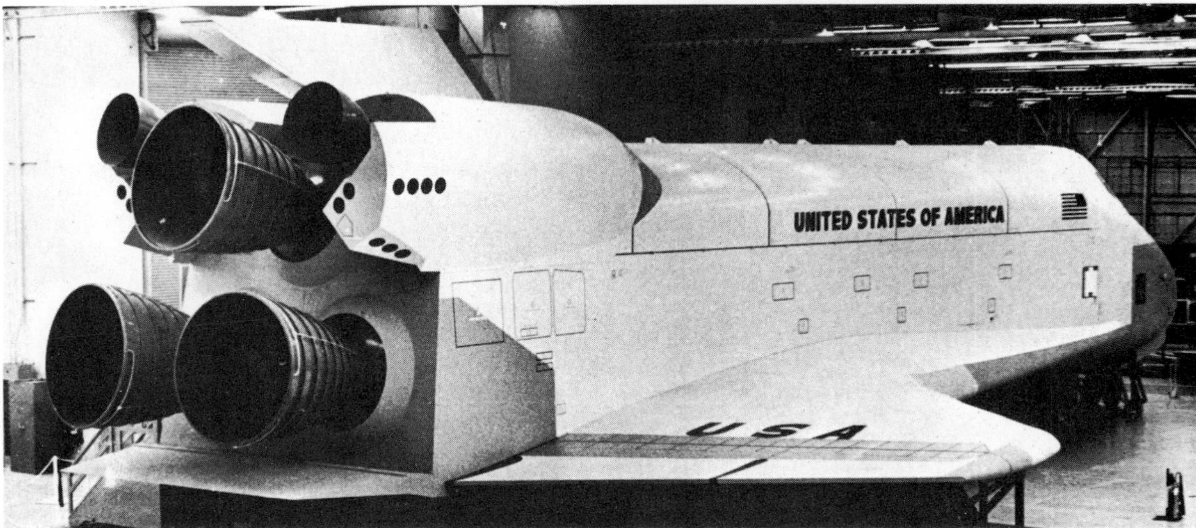

Wide World

location under the command and service module as the lunar module was carried in the Apollo lunar landing program.

Other objectives of the joint test project included scientific studies carried out jointly and, in some instances, solely by either crew. These studies included Earth observations, acquisition of biomedical data to study the effects of space flight on man, biological experiments to study microorganism growth in the weightless environment, and an artificial solar eclipse.

During past years members of both nations' manned space flight teams made many visits to both the U.S. and the Soviet Union. Formal working groups with representatives from both nations discussed and settled technical problems and completed the necessary planning. Crewmen from each nation studied the other nation's language, and by 1975 all crewmen could converse in either English or Russian.

The test project was launched on July 15, 1975. The Soyuz spacecraft was launched first, into an Earth orbit ranging from 115 to 136 mi. This orbit was changed by firing the Soyuz propulsion system in order to achieve a circular orbit of approximately 140 mi. Seven and one-half hours after the Soyuz launch, the Apollo command module and the docking module were launched into an orbit ranging from 95 to 106 mi. Once in orbit, the U.S. astronauts separated the command module from the docking module shroud, turned the command module 180°, docked with the docking module, and withdrew the docking module from the shroud. The astronauts then performed maneuvers to fly to the proper orbit for a rendezvous and docking with the passive Soyuz.

After the docking, the astronauts began such activities as joint science studies, television transmission from each spacecraft, and the exchange of crewmen from one spacecraft to the other. After approximately two days of docked operations, the Soyuz spacecraft separated from the Apollo and then landed on July 21 in the Soviet Union. The U.S. astronauts landed in the Pacific on July 24. During the descent, failure by the astronauts to throw two switches caused the Apollo cabin to fill with potentially harmful nitrogen tetroxide gas. The astronauts were placed under medical observation for several weeks, but none appeared to have suffered any lasting ill effects.

The Soviet crew was commanded by Aleksey A. Leonov, a space flight veteran who flew in 1965 in the Voskhod 2 spacecraft and was the world's first space walker. The flight engineer was Valery N. Kubasov, also a veteran space crewman who flew in Soyuz 6 in 1969. The U.S. crew was commanded by Thomas P. Stafford, who had previously flown on two Gemini missions and on the Apollo 10 circumlunar mission. The docking module pilot was Donald K. Slayton, and the command module pilot Vance D. Brand, both space flight rookies.

Salyut 3. The U.S.S.R. launched Salyut 3, its third space station, into Earth orbit with a Proton rocket on June 25, 1974. The space station weighed approximately 20 tons and had an internal diameter of approximately 13 ft. The Salyut was equipped with a docking module to permit the manned Soyuz spacecraft to dock and to allow the transfer of cosmonauts into the space station. The docking module was designed so that it could be isolated from the spacecraft and space station. It contained the necessary equipment for decompression and recompression, and a hatch to allow the cosmonauts to move into free space to carry out extravehicular activities.

Salyut 3 used solar panels to convert solar energy into electrical power so as to operate the station systems and experimental equipment. The interior of the space station was separated into living quarters, a control center, and a work compartment. Previous Salyut space stations contained only a single large compartment. The spacecraft was maintained at a temperature of about 71° F.

Medical instrumentation used in obtaining biomedical data was provided in the crew living compartment. A treadmill was located in this area to allow the crew to exercise by simulated walking, running, jumping in place, and other exercises. The crewmen attached elastic straps from a special exercise suit to the treadmill to simulate the pull of Earth's gravity and to load the muscles while exercising. The floor of the space station and the bottom of the cosmonauts' footwear were covered with a Velcro-like material to allow the crewmen to anchor their feet to the floor and walk.

The larger cylindrical section of Salyut 3 was used as a laboratory and contained scientific equipment to conduct medical and other types of technical studies. A control compartment was located between the living quarters and laboratory section. It was described as a "captain bridge" where the cosmonauts controlled the flight and monitored and operated the space station systems and equipment.

Soyuz 14. A two-manned cosmonaut team was launched from the Soviet Union in a Soyuz-type spacecraft on July 3, 1974. The crew was commanded by Pavel R. Popovich, a veteran space traveler who flew on Vostok 4 in 1962. The flight engineer was Yuri P. Artyukhin, who made his first flight on Soyuz 14. On the second day of the mission the Soyuz spacecraft was flown by the crew to rendezvous and dock with the orbiting Salyut 3

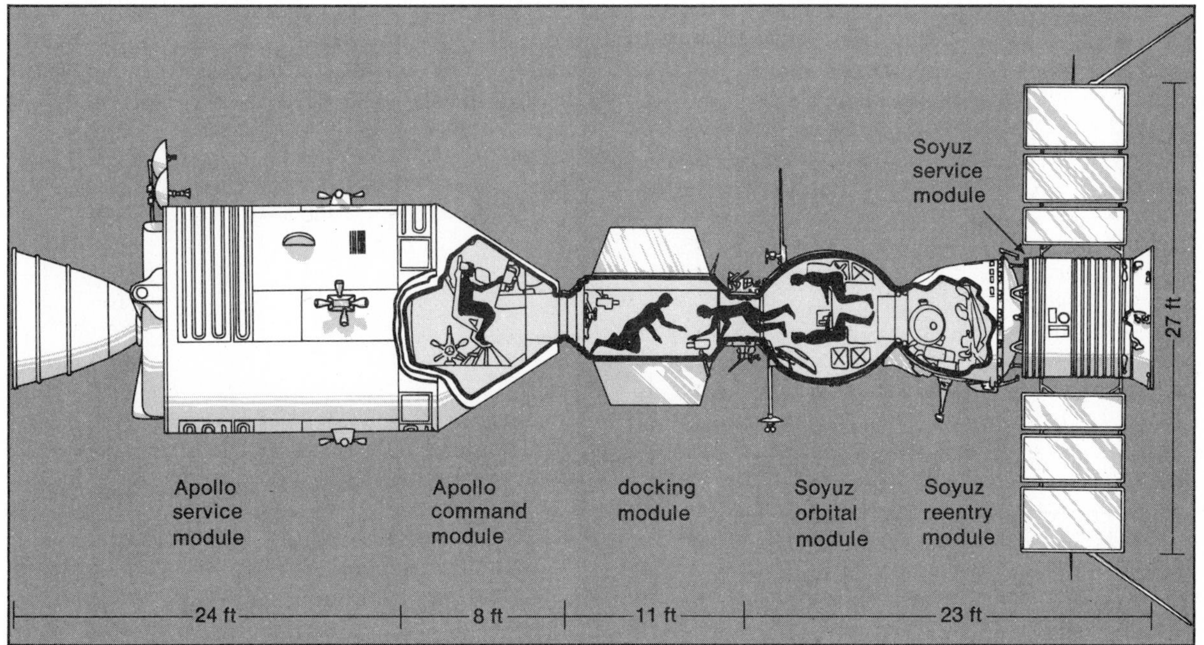

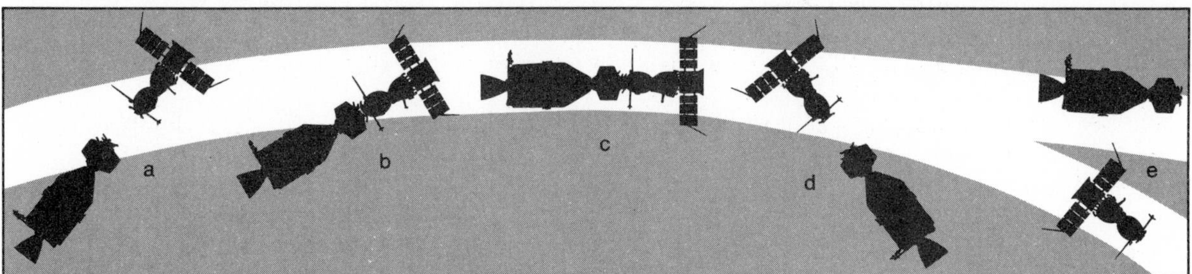

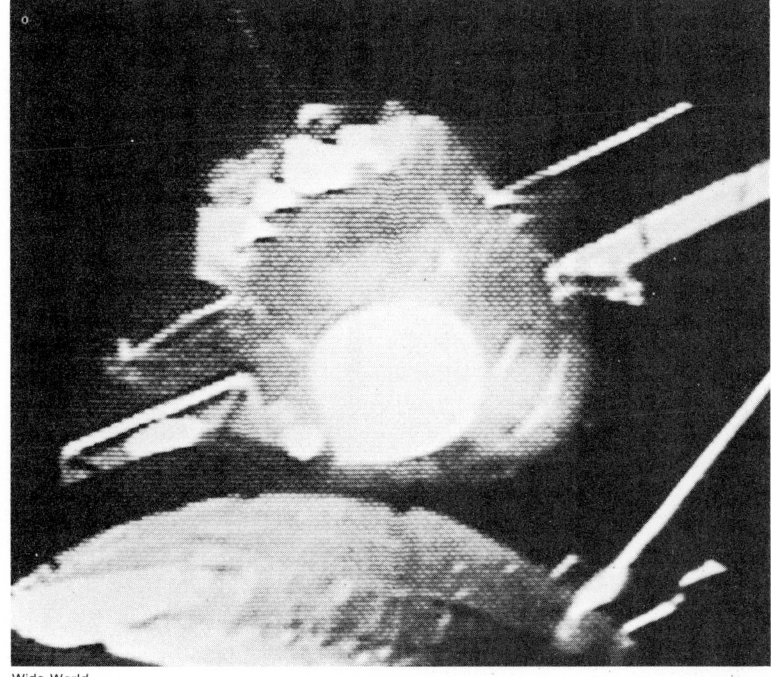
Wide World

U.S. and Soviet astronauts joined one another in space for two days during the Apollo/Soyuz Test Project in July. At the top is a cutaway drawing of the Apollo and Soyuz spacecraft and the docking module. The docking maneuvers (center) began with the rendezvous (a) between the Apollo (lower left) and Soyuz. The two craft docked (b), and U.S. astronauts Thomas Stafford and Donald Slayton entered Soyuz (c). After 1½ days of joint experiments the two spacecraft undocked temporarily for a solar eclipse experiment (d) and then, after redocking for another ten hours, achieved final separation (e). Soyuz is photographed from the Apollo (bottom) during final separation.

Novosti from Sovfoto

Soviet space flight control center near Moscow communicates with the cosmonauts aboard Soyuz 16 in December 1974. The mission tested the center's readiness for the joint U.S.–Soviet flight in July 1975. Photographs are of the two Soyuz 16 cosmonauts, Anatoly V. Filipchenko (left) and Nikolay N. Rukavishnikov (right).

space station. The stated objectives of the Salyut 3/Soyuz 14 mission were astronomical observation, Earth resources studies, medical tests to establish the effects of space flight on man and to establish the proper work regimes for crewmen, and the testing of improved space station structures and systems.

The Soyuz 14 mission lasted approximately 15 days. On July 19, 1974, the crew transferred back from the Salyut space station to the Soyuz spacecraft. The crewmen undocked the spacecraft and landed in the U.S.S.R. The Salyut 3 space station remained in orbit for another possible visit.

Soyuz 15. The Soyuz 15 spacecraft was launched into Earth orbit from the Soviet Union on Aug. 26, 1974. The mission was commanded by Gennady Sarafanov, and the flight engineer was Lev Demin. Both crewmen made their initial space flight after approximately ten years in the cosmonaut training program.

On the second day of the mission, the crew maneuvered the spacecraft to rendezvous with the Salyut orbital station and apparently tried unsuccessfully to dock with it. U.S. space officials had speculated that the Soyuz 15 crewmen had intended to dock with the space station and conduct a one-month mission, similar to the first U.S. Skylab flight. A Soviet space official indicated that the Soyuz 15 crew had encountered a problem with the rendezvous and docking system. The mission was aborted after a flight that lasted two days, and Soyuz 15 landed in the U.S.S.R. on Aug. 28, 1974.

On Dec. 25, 1974, Soviet space officials announced that the remote control of Salyut 3 was to be terminated. On Jan. 24, 1975, the space station was taken out of orbit, after which it fell into the Pacific Ocean.

Soyuz 16. The U.S.S.R. launched the Soyuz 16 spacecraft into Earth orbit on Dec. 2, 1974, as a dress rehearsal for the Apollo/Soyuz Test Project. The crew for this mission were Anatoly V. Filipchenko and Nikolay N. Rukavishnikov. Both men also served as back-up pilots for the ASTP Soviet cosmonauts. Filipchenko had flown earlier on Soyuz 7, and Rukavishnikov on Soyuz 10.

The purpose of the Soyuz 16 mission, which lasted approximately six days, was to flight-test the Soyuz spacecraft modifications required for the joint flight project. The spacecraft was launched into an elliptical orbit, after which the crewmen ignited the spacecraft propulsion system to achieve a circular orbit with an altitude of 120 naut mi. During the mission the crew lowered the spacecraft atmosphere from a pressure of 14.7 psi to 10 psi, the latter being required for the joint docking operations with the Apollo command module. During the mission U.S. operational personnel conducted ground tracking exercises similar to those to be performed in the ASTP.

Salyut 4/Soyuz 17. The fourth U.S.S.R. space station was launched on Dec. 26, 1974, into a circular orbit 214 mi above the Earth. Salyut 4, launched unmanned, was similar in design to Salyut 3. On Jan. 11, 1975, the Soyuz 17 spacecraft was launched into Earth orbit. The next day the crew maneuvered the spacecraft to dock with Salyut 4. The commander for this mission was Aleksey Gubarev, and the flight engineer was Georgy Grechko. Both cosmonauts were making their first space flight.

Sidney Harris

"Say—this offers a valuable insight into the origins of the universe."

The mission objectives for Salyut 4 appeared to be more scientifically oriented than were those for Salyut 3. Extensive medical studies were conducted, similar to those carried out by the U.S. Skylab astronauts. These included cardiovascular system measurements under exercise and other stresses, and the acquisition of blood samples for postflight analysis. A solar telescope was used to study the Sun; X-ray telescopes were employed to make measurements of other celestial bodies; and instruments and cameras were used for Earth resources experiments. On Feb. 9, 1975, 30 days after launch, the Soyuz 17 crew separated their spacecraft from Salyut 4 and landed safely in the Soviet Union.

Soyuz 18. On April 5, 1975, the U.S.S.R. unsuccessfully attempted to launch Soyuz 18 into orbit to dock with the Salyut 4 space station. The commander for the mission was Vasily Lazarev, and the flight engineer was Oleg Makarov. Both crewmen had flown on the Soyuz 12 mission in September 1973. The Soyuz 18 mission was aborted several minutes after launch when a malfunction occurred in a separation device in the launch booster. At the time of the failure the spacecraft was near orbital altitude, and it then traveled in a suborbital flight to land in western Siberia, 1,000 mi from the launch site. The crew was recovered in good condition several hours after landing.

—Richard S. Johnston

Space probes

Despite the outstanding performances and contributions to science of space probes such as Mariner 10 and Pioneers 10 and 11, the future outlook for the U.S. unmanned space program reflected a considerable reduction in activity, mainly because of inflation. Although the budget for the U.S. National Aeronautics and Space Administration (NASA) seemed to have stabilized after a steady decline since 1968, the shrinking value of the dollar and the rising costs for labor and materials left the agency with no "new starts" for fiscal 1976.

Probing Jupiter. Pioneer 11, launched on April 5, 1973, from Kennedy Space Center, first detected Jupiter in June 1974, when its radiation detectors responded to bursts of high-speed electrons that were known to be produced by the planet. However, its cameras, or photometers to be more exact, were not turned on until November 18, when the probe was 9.2 million mi from Jupiter. On November 25, Pioneer 11 crossed Jupiter's bow shock wave, formed where the solar wind meets the planet's magnetosphere. The shock wave was sensed at a distance of about 4.8 million mi, or 109 radii, from the planet. The probe's predecessor, Pioneer 10, had detected the phenomenon at a distance of 108 radii; thus, the probes were in close agreement on the distance. The following day Pioneer 11 crossed the magnetopause, a layer of turbulence behind the shock wave, and a day later the probe was in the magnetosheath, or magnetic field, surrounding Jupiter.

Only six hours before reaching its closest point of approach to the planet, on December 2, Pioneer 11 was in the heart of the planet's intense radiation belts, similar to the Van Allen belts that surround the Earth. At that time, it was approximately 250,000 mi above the top of the clouds that swirl around the planet. The radiation by protons it encountered was more than 100 times greater than that experienced by Pioneer 10. At a speed of 107,630 mph, Pioneer 11 dipped within 26,600 mi of the top of the cloud layer, some 54,000 mi closer than Pioneer 10 had approached. Because of the tremendous speed of the probe, however, it spent less time than did Pioneer 10 in the radiation belt; thus, Pioneer 11 received a total radiation dose less than that of its predecessor.

Pioneer 11 revealed that the peak flux of high-energy protons occurred about 7° north of Jupiter's magnetic equator rather than at the physical equator, as scientists had expected. Thus, observers concluded that the protons are concentrated in bands or shells in the innermost radiation belts.

The Pioneer 11 mission experienced one disappointment. It was hoped that the probe would pass close enough to one of Jupiter's moons to detect a flux of magnetic lines linking the satellite to Jupiter. If such a flux were found, then there would be a mechanism to explain certain radio waves generated by the planet. However, none was discovered. The answer may be that the probe simply did not fly through the flux.

The probe's close approach to Jupiter at the planet's high latitudes also confirmed what scientists had already theorized from the data sent by Pioneer 10. The huge planet is a sphere of liquid hydrogen, except for possibly a small rocklike core, and radiates more heat than it receives from the Sun.

Additionally, Pioneer 11 made possible an accurate estimate of the density of Callisto, one of Jupiter's 13 moons. Using the new information, astronomers should be able to improve their estimates of the densities of Jupiter's other 12 satellites. The probe also took pictures and ultraviolet radiation measurements of the satellites Ganymede, Europa, Io, and Amalthea.

Micrometeoroid detectors aboard the probe proved that there was no swarm of extremely small meteoroids in orbit about Jupiter. The few detected were found to be orbiting the Sun.

A radio-occultation experiment produced data that turned out to be equivocal at best. As Pioneer 11 passed behind the planet, radio signals were transmitted through its thick, turbulent atmosphere to the Earth. Based on the extent to which those signals were refracted, or bent, it is possible to determine the temperature and pressure of the planet's atmosphere. However, anomalies in the transmitted data reduced the confidence scientists placed in them.

Having finished its investigation of Jupiter, Pioneer 11, officially renamed Pioneer Saturn by NASA, assumed a trajectory that would take it out of the ecliptic, or plane in which the Earth orbits the Sun, for a five-year trip to Saturn. In 1977 the probe was to reach a peak height of approximately 100 million mi above the plane. Then it would dip back two years later for a rendezvous with Saturn in September 1979, after a 1.5 billion-mi journey from Jupiter. If the mission goes as planned, Pioneer 11 will travel through the 9,000-mi gap between the innermost ring of Saturn and the planet itself. On its outward-bound trajectory Pioneer Saturn will pass Titan, the largest of the satellites of Saturn. This moon is of particular interest because it is known to have an atmosphere and may possibly have liquid water, making it a possible abode for some sort of life. Telemetry signals from Pioneer Saturn at that time will take 1½ hours to reach the Earth. *See* Feature Article: VISIT TO A LARGE PLANET: THE PIONEER MISSIONS TO JUPITER.

Return to Mercury. On March 16, 1975, Mariner 10 returned to Mercury for the third time since its launch in November 1973. The pictures it returned were not as good as hoped for, being marred by a blurring (the probe was traveling at 25,000 mph relative to the planetary surface) and an electronic failure of the 210-ft-diameter tracking antenna at Canberra, Australia. Nevertheless, enough photographs were obtained to be compared with similar views made during the probe's first two visits. Because Mariner 10 approached within 198 mi of the

U.S. space probe Pioneer 11, being tested before launch, passed within 26,600 miles of the top of Jupiter's cloud layer in December 1974. After flying by Jupiter, the probe continued on into space for a rendezvous with Saturn in 1979.

Courtesy, NASA

planet's surface on the third flyby, even the degraded pictures, once enhanced by computers, were valuable in studying the topography of Mercury. Features as small as 150 to 300 ft in length or diameter were revealed.

On its last visit to Mercury, Mariner 10 proved that the planet has an intrinsic magnetic field but no encircling Van Allen belts of trapped radiation. The probe also provided much more accurate data than previously obtained on the mass of Mercury and its true orbit around the Sun, and it established that there was a very thin atmosphere of helium present on the planet.

The primary mission of the third flyby was to investigate the magnetosphere of Mercury, which trails out behind it on the side shadowed by the Sun. During the first swing past the planet, magnetometers aboard the probe had detected a bow shock wave and a magnetopause, components indicative of the interaction of the solar wind on a magnetosphere. The measurements made at that time, however, were not conclusive enough to permit scientists to determine how the magnetic field around Mercury is generated. They could not tell whether it is the result of an internal dynamo process such as that possessed by the Earth or whether it is the result of the bombardment of Mercury's surface by the solar wind.

In commenting on the findings, Norman F. Ness, of NASA's Goddard Space Flight Center, in Greenbelt, Md., said, "The magnetic field experiment on Mariner 10 dramatically confirmed and extended the results obtained earlier at the Mercury 1 encounter. The magnetic field which is responsible for deflecting the flow of solar wind is unquestionably intrinsic to the planet and not associated with . . . the solar wind interaction with the planet."

Ness added, however, that it was still not possible to determine precisely the means by which the field is formed. It could result from a core of magnetized rock or from the Earthlike dynamo mechanism of a rotating fluid core.

Having left Mercury for the third time, Mariner 10 spent eight days performing tracking and engineering experiments. On March 24, it was turned off and began a permanent, silent orbit about the Sun. The orbit was fixed by the laws of celestial mechanics so that Mariner 10 would never again visit Mercury except at distances of more than several million miles.

Approaching the Sun. On March 15, 1975, Helios 1 became the first man-made object to make a relatively close approach to the Sun. The solar probe, launched from Kennedy Space Center by a Titan Centaur on Dec. 10, 1974, was a joint scientific venture of the U.S. and West Germany. Named for the Greek god of the Sun, it passed within about 28 million mi of the solar sphere. Thus, Helios 1 broke the record of Mariner 10, which went past the Sun in 1974 at a distance of 45 million mi. Aboard Helios 1 were ten instruments to measure magnetic fields, solar wind, cosmic rays, zodiacal light, and micrometeoroids. As the probe sped past the Sun at a speed relative to it of 143,000 mph, the mission control center at Oberpfaffenhofen, located near Munich, reported that all instruments were working as planned.

Helios 1 incorporated several novel devices and

Courtesy, NASA

Cloud tops of Jupiter near the planet's north pole were photographed by Pioneer 11 from a distance of about 3.7 million miles. Resolution of this photograph is several times better than ever obtained by a telescope on the Earth.

Courtesy, Martin Marietta Aerospace

Viking spacecraft is prepared for tests of its ability to withstand a rigorous year-long journey to Mars, where it will soft-land and investigate the planet's surface and atmosphere and the possible existence of life.

techniques in order to protect its delicate scientific instruments from the searing heat of the Sun, estimated to be 700° F at the point of closest approach. These included mirrors, special insulating materials, and continuous spinning of the craft itself. These various means protected the interior of the probe and kept the temperature there about 86° F. Temperature sensors on the high-gain radio antenna of the craft reported a temperature of only 310° F as it neared the Sun. Scientific data produced by the probe were being analyzed.

Probing the Moon. The Moon, neglected by the U.S. since the end of the Apollo program, remained a target for Soviet space probes. Luna 22 became a satellite of the Moon on June 2, 1974. On Nov. 11, 1974, after 1,778 revolutions around the Moon, the probe had its orbit changed to produce an apolune of 913.3 mi and a perilune of 106 mi. Data from its instruments continued to flow to the control center near Moscow.

Launched on Oct. 28, 1974, Luna 23 went into orbit around the Moon on November 2. The probe was programmed to make a soft-landing four days later in the Sea of Crises. There, it was to have drilled approximately eight feet into the lunar soil and, though the Soviets did not say so, return to the Earth with the samples. During its landing, however, the probe was damaged to an extent that prevented the fulfillment of the mission. Whether the craft fell over on impact or whether the drilling mechanism was broken was not revealed by the Soviets. The fact that the U.S.S.R. publicly admitted the failure of the mission on November 10 was unusual in itself and a radical departure from its past practices in publicizing its space program.

A note from Mars. During the year, the Soviets revealed that color photographs from their Mars 4 and 5 probes showed that the plains of the planet are orange but that the color of the highlands is blue and many craters are a kind of bluish green. The blue and green colors had never before been detected on the planet. If true, they would indicate light-absorption capabilities greatly different from those of the predominantly orange properties previously observed for the planet.

Prospects for Viking. Project Viking, the most ambitious space probe ever undertaken by the U.S., encountered problems during the latter part of 1974 and the early months of 1975. Designed to soft-land on Mars and transmit photographs of the surface as well as to perform biochemical analyses of the planet's soil in a search for Earthlike life forms, the probe was beset from several quarters.

In order to save money, estimated at $9 million, NASA deleted the construction of a back-up lander and orbiter from the project. Plans had earlier called for a test vehicle and two flight vehicles. By eliminating one of the flight vehicles and reworking the test vehicle to perform its role, NASA effected some of the savings. Even so, the program was estimated to be costing some $5–10 million more than estimates made only a year previously. Additionally, 120 personnel of the flight control team were eliminated.

With reference to what he called the "crisis in astronautics," Sen. Frank E. Moss (Dem., Utah) looked hopefully to possible assistance from the U.S.S.R. in developing future Mars landers. He said, "Sooner or later, we are going to want to examine that Martian soil more closely. . . . A roving vehicle on Mars is another possibility and here the Soviets' experience with their lunar rover could be quite useful."

Probing the future. With the successes of Mariner 10 and Pioneers 10 and 11 proving the feasibility of the swing-by mode of visiting planets, it seemed that such a technique had been perfected by the end of 1975. Future missions, however, were threatened by the specter of constantly shrinking space budgets and rising inflation.

Despite such gloomy prospects, the Space Research Board of the U.S. National Academy of Science recommended in November 1974 that NASA undertake a Mariner polar-orbiting mission of Mars and a Pioneer hard-landing on the planet sometime between 1978 and 1983. The board further recommended a polar-orbiting lunar probe as a future project for the space agency. Also suggested were probes to rendezvous with comets and a possible Mariner to swing by Jupiter and then visit distant Uranus.

By mid-1975, however, only two future projects had been funded by NASA. They included two Mariner flyby missions to Jupiter and Saturn in 1977. The spacecraft were to be larger and more sophisticated than the Mariner 10, and NASA hoped that they would provide experience in developing probes for future flights to Uranus and Neptune. The other funded program, scheduled for launch in 1978, consisted of a mission to Venus by two small Pioneer probes. One was to orbit the planet, while the other would eject small probes that would descend through the atmosphere of Venus, telemetering data on its structure and dynamics as they did so.

—Mitchell R. Sharpe

Transportation

Continuing inflation and a severe recession in the United States and other industrialized countries took its toll in the transport technology field. Research and development funds were cut sharply for many programs, forcing some to be stretched out and others to be dropped or placed in a study-only status. This proved to be particularly true with projects concerning futuristic types of transport, such as air-cushion and linear-induction vehicles, as well as with various automated personal rapid transit systems.

Aircraft, motor vehicle, rail, and ship manufacturers continued to express strong objections to the steady stream of U.S. government rules and regulations relating to pollution, noise, and safety. These regulations, they claimed, forced them to devote most of their technology efforts to meeting these standards without increasing their equipment costs substantially. One major corporation, Chrysler, announced that it was getting out of the business of making heavy-duty trucks because compliance with the new U.S. standards for such items as noise and brakes had become too difficult to permit profitable operations.

While this admittedly was a debatable issue, actions by the government, in at least two instances, seemed to support claims that the new standards were being imposed too quickly without knowledge of their acceptability and practicability. In one instance, the government-imposed requirement was not acceptable to the general public (the linking of auto ignitions to seat belts), and a new law was passed repealing it. In another instance, the requirement could be creating a problem as

Sidney Harris

"Just what we need—mass transit."

bad as the one intended to be solved. This was the use of catalytic converters in automobile exhaust systems to meet an anti-pollution standard deadline. The deadline had to be extended because it appeared that a different type of emission from the converters could be as harmful as the emissions being eliminated.

The manufacturers also emphasized the strong pressures on them to find ways to reduce fuel consumption, a necessity because of the high cost of fuels and the poor fuel-supply outlook. This required considerable research into the development of more efficient engines and in finding lighter materials that could be used without a sacrifice in performance.

Transport technology was also being strongly influenced by legislation. For example, the U.S. Congress in 1974 passed into law a bill that will permit the construction of offshore, deepwater terminals to handle the huge tankers that cannot enter U.S. ports because of shallow depths. Another new act strengthened federal regulatory control and enforcement over the movement of hazardous materials by all modes of transport. This posed many problems in the areas of packaging, handling, and transporting for shippers of the rapidly expanding number of commodities that fall within this category.

A third new act authorized states in the U.S. to increase by nearly 10% the maximum weight of trucks using the 42,500-mi interstate highway system. This, of course, permitted larger payloads and an increase in productivity, which truckers claimed were necessary to offset the higher costs of fuel and the slower operations forced by the extended national 55 mph highway speed limit. The higher weights should also encourage the introduction of more powerful and efficient engines.

Intermodal transport. The use of legal and/or administrative methods to block advances in transport technology was evidenced by the attempts of labor and port interests on the U.S. Atlantic and Gulf coasts to prevent what has been termed "minibridge" service between Japan and U.S. eastern cities. This service envisaged using container ships from Japan to West Coast ports in the U.S. and then transferring the cargo to railroads for the transcontinental journey. By mid-1975 attempts to get the courts or the Federal Maritime Commission to rule against this innovation failed despite its effect on the ports that would lose the traffic that otherwise would go by all-water routes.

Similar minibridge service was being provided between Europe and U.S. West Coast cities. Of equal significance was the inauguration by Seatrain Lines, Inc. and Japan Line of a full landbridge service between Europe and Japan, using separate container ships in the Atlantic and Pacific that were linked with trans-U.S. rail service. This long-delayed service was made possible when Japan's Fair Trade Commission removed restrictions on its use in that country. The new 10,000-mi weekly service, according to Seatrain, offered greater speed at competitive rates compared with the all-water services now primarily used and with the trans-U.S.S.R. rail landbridge service that handled about 10% of the Japan–Europe container traffic.

Air transport. The high cost of fuel and cutbacks in travel continued to force many airlines to curtail

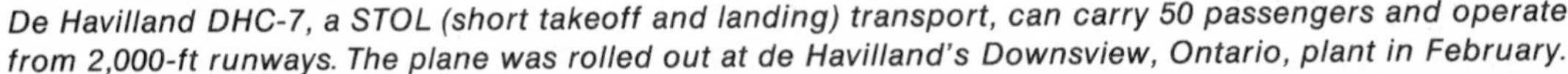

De Havilland DHC-7, a STOL (short takeoff and landing) transport, can carry 50 passengers and operate from 2,000-ft runways. The plane was rolled out at de Havilland's Downsview, Ontario, plant in February.

Courtesy, The de Havilland Aircraft of Canada, Ltd., photo by Ron Nunney

use of Boeing 747s for passenger service, although successful operations by Lufthansa German Airlines and World Airways, Inc., with their 747 all-cargo and convertible (passenger and/or cargo) versions, helped stimulate orders for those models. Seaboard World Airlines, Inc., started 747 all-cargo service across the North Atlantic in competition with Lufthansa. Seaboard's service featured a new intermodal aluminum container especially designed for joint air and truck movements. The new container, measuring 20 ft by 8 ft by 8 ft and capable of being coupled into 40-ft units, was built in West Germany and, according to Seaboard, could meet both U.S. and European highway standards for strength and safety.

While the trend toward ever larger sizes in air passenger transports appeared to have peaked, this did not seem to be the case for air cargo planes. One sign of this was the negotiations that were under way between Lockheed Aircraft Corp. and Iran for possible reopening of the production line for the huge C-5A transport; the line had been closed for almost two years following completion of deliveries to the U.S. Air Force.

Plans for a fleet of even larger air cargo jets were formalized by a group of retired U.S. industrialists, headed by Edward Cole, former president of General Motors Corp. Formed into a corporation named International Husky, they were promoting a $7.5 billion program to operate 300 such aircraft. Each would be capable of hauling intermodal containers in its rectangular-shaped fuselage, which would contain five side-by-side cargo bays 90 ft long. The promoters claimed that the concept could reduce the costs of air cargo by 50% and thus make air freight competitive with trucks for about half the tonnage moved 500 mi or more in the U.S. As of mid-1975, however, no airlines had publicly expressed any formal interest in the plan.

The European-built Airbus A-300B began scheduled service between Paris and London, where it recorded a 98% reliability rating. The twin-engine, wide-body jet was designed for short- and medium-distance markets and carries 251 passengers (26 in first class and 225 in a single, unpartitioned economy section). Of 43 orders placed (26 of them options) for the $18 million transport, 35 were from nationalized airline participants in the A-300B program. No orders from airlines in the United States were announced.

Controversy over the environmental effects of a fleet of supersonic transport (SST) airplanes continued. A U.S. Department of Transportation study concluded that the 16 Concordes and 14 Soviet Tu-144s scheduled for commercial operations would not damage to any significant degree the Earth's ozone protection from the Sun, as was also brought out in a separate report by the U.S. National Academy of Sciences. The latter said that 100 SSTs would cause about 1.4% more instances of skin cancer, and that this figure would rise to 20% for a fleet of 300–400. The serious economic problems of the Concorde, however, will undoubtedly limit its use for some time, its price having risen to $60 million per plane (more than double that of a 747) and its passenger capacity totaling only 108 to 128, depending on the cabin layout. The first commercial flights of the Concorde were scheduled for early 1976, and top priority was being given to transatlantic routes between London-Paris and New York-Washington.

Airtransit Canada, a subsidiary of the nationalized Air Canada, began STOL (short takeoff and landing) shuttle service between Ottawa and Montreal with specially modified de Havilland DHC-6 Twin Otters, each carrying 11 passengers. Operating on 40-minute schedules between STOLports, the small STOLs permitted downtown-to-downtown trips of 75 minutes. The DHC-7 was introduced in February.

Highway transport. As noted previously, the U.S. Department of Transportation officially revoked its ignition-seatbelt interlock requirement. The department also seriously questioned new automobile bumper standards, claiming that the benefits do not offset the higher costs incurred by added weight and reduced fuel mileage. At the same time, however, the department set new brake standards for large interstate trucks, trailers, and buses, which manufacturers claimed would increase operators' costs by as much as $400 million a year. The U.S. Environmental Protection Agency (EPA) added another challenge to motor vehicle technologists by issuing final noise standards for interstate motor carriers, with enforcement scheduled to begin Oct. 15, 1975.

Also mentioned previously was the granting of a one-year delay in compliance with tough automobile exhaust emission standards. This resulted from EPA studies which showed that the catalytic converters developed to meet the standards were emitting potentially harmful sulfuric acid. Thus, plans to use such converters at a cost of several hundred million dollars in research and development funds, plus the unit costs of $100–$150 per automobile, had to be revised. The manufacturers still must eliminate virtually all harmful emissions and develop an engine that will bring about a 40% improvement in fuel economy.

RCA Corp. claimed that it had developed a new microprocessor that could save 40% in automobile fuel use by means of electronic starts, shifting, and cruising—at a cost of about $100 a unit in mass production. Automakers quickly challenged

Chris Ware from Keystone

Dial-a-bus, an innovation of London Transport, operates in Hampstead Garden Suburb. The 16-seat vehicles, which run every 15 minutes, pick up customers at their homes and at several fixed stops, and the driver will also stop when hailed if it is safe to do so.

such claims, however, asserting that results of their own partial testing of the innovation had failed to show such promising results.

The fuel situation, both high cost and questionable supply, continued to play a major part in automakers' growing lack of interest in the rotary engine. General Motors indefinitely postponed its planned "late 1975" introduction of the Wankel rotary engine because of related emission and fuel problems. Ford Motor Co., which had already discontinued its rotary engine research program, announced that it was also discontinuing research and development on the Rankine-cycle steam engine, after spending about $4 million. The federal Urban Mass Transportation Administration (UMTA), on the other hand, awarded AMF Inc. and Steam Power Systems $1 million each to design and build prototype five-passenger, low-pollution, steam-powered, "para-transit" vehicles able to accept a range of fuels and designed to accommodate both handicapped and elderly riders. (Para-transit vehicles are those that fall between large mass transit rail cars and buses, and small taxis and private autos in capacity and convenience.) Delivery of the vehicle prototypes was set for May 1976.

Another UMTA development program took a step forward. The agency's three 40-ft prototype urban buses of the future were being operated in extensive commercial service demonstrations in a number of large U.S. cities. The new buses, developed by three competing companies, are 25% wider than normal city buses and have many safety, economy, and comfort features. The objective of the program is to develop a single performance specification for future urban buses.

As cities indicated a desire to try demonstration services, UMTA continued work on the dial-a-ride bus concept for city and suburban travel needs. Previous efforts had not been successful, the latest casualty being the discontinuance of dial-a-ride at Haddonfield, N.J., despite a three-year trial costing UMTA more than $5 million and state and local jurisdictions $420,000. It obviously is not an easy job to get the vast majority of drivers to desert their cars, even for a customer-tailored small-bus service at reasonable cost.

Rising costs also threatened the success of two major UMTA projects which were to demonstrate the capabilities of an automated, urban transit system utilizing bus-type, rubber-wheeled vehicles operating on a concrete guideway. One, Skybus in Pittsburgh, created so much local controversy and extended delays that UMTA warned city officials they must soon reach an agreement about it or forfeit a development grant that totaled $15 million in U.S. funds.

A similar type of project in Morgantown, W.Va.,

Authenticated News International

High Performance Personal Rapid Transit vehicle, built by Boeing Vertol Co. in a design competition for the U.S. Department of Transportation, seats 12, is powered by an electric motor, and has front and rear radar systems for collision avoidance. A digital computer inside the vehicle operates it automatically by communicating through radio antennas with computers located in the stations and in central control.

likewise was in trouble, both financially and technologically, but UMTA continued providing the necessary financial backing to put it in full-scale use. The Boeing Airplane Co., the builder of the 45 automated cars, or buses, to be used in the system, started deliveries of the vehicles during the year, and the 2.2-mi tracked system began public service in mid-1975.

Intercity truck lines, encouraged by higher highway weight limits authorized by the U.S. Congress, foresaw larger payloads and more economical use of fuel and drivers. They also envisioned major savings in fuel by means of joint efforts to eliminate empty backhauls. With the approval of the Interstate Commerce Commission one group of truck lines reduced sharply such empty backhauls on the Chicago–St. Louis route through computerized matching of inbound and outbound movements; additional truck lines planned to take similar action.

Pipelines and tunnels. Because of rapidly escalating costs that threatened to boost the final price tag of the English Channel tunnel project to $2.4 billion by 1980, the British government officially withdrew from its agreement with France to build it. The plan, which would have initially cost about $1.1 billion, envisioned twin tunnels for train movements over a distance of 32 mi. Project shutdown costs were estimated at $83 million.

Rising costs, along with forced changes for environmental reasons, also pushed the new total cost for the 798-mi trans-Alaska oil pipeline to nearly $6 billion ($900 million originally). Following rapid construction of a 360-mi state service road from the Yukon River to Prudhoe Bay, actual construction on the huge line got under way. The targeted completion date of mid-1977 was expected to be met, even though the initial capacity of the line was increased to 1.2 million barrels per day. This increase was made possible by adding three pumps to the five originally planned.

Another major pipeline about ready for the start of construction is a 510-mi, 30-in.-diameter crude oil line from Freeport, Texas, to Cushing, Okla. A novel feature of this $160 million line, named Seaway Pipeline, is that it was designed to link with Seadock, a deepwater unloading terminal scheduled to be built about 30 mi off the Texas coast.

A major surge in pipeline construction in the U.S. could come quickly, but not for moving oil. If Congress gives its approval to the utilization of eminent domain via federal courts, coal producers and consumers, primarily utilities, are expected to build a number of long-distance coal-slurry lines. One would extend 1,030 mi from Wyoming to southern utilities; another would run 1,100 mi from Colorado to Houston; a third 800 mi from Wyoming to the Northwest; and two more for 180 mi between the Southwest and Nevada-Arizona. The successful operation of the 273-mi Black Mesa coal-slurry line from mines in Arizona to a utility in Nevada apparently proved the economic feasibility of this new mode of transport.

Rail transport. Railroads continued to stress unit-train operations (moving of a single commodity in one-way shuttle-type service), and Canadian National Railways reported developing an ideal two-way unit-train operation. It hauled potash one way and phosphate the other on round trips of 2,500 mi, using 85-car trains composed of 100-ton hoppers. In the U.S. the ICC was urged by the government's National Commission on Productivity and Work Quality not to implement a recommended order that would impose limits on the use of cars in unit-train service during shortages. The commission argued that such service has obvious benefits in generating heavy loads, higher unit revenues, and car purchases, and it also feared that the service would be destroyed by such an order.

A recent innovation took place when the ICC allowed three U.S. railroads to experiment with a computerized car-pooling system in order to increase equipment utilization. The Federal Railroad Administration gave another railroad approval to test a remote-control train stop capability over 552 mi of track in Ohio.

Despite its continued heavy losses, Amtrak obtained additional funds from the U.S. government to order 435 new rail passenger cars, 235 of which were newly designed bi-level cars, which, officials hoped, would stimulate long-distance rail travel. To be built by Pullman-Standard, a division of Pullman Inc., at a cost of $167 million, the cars were to be flexible in that their interiors could be arranged into coaches, diners, sleepers, diner-lounges, and coach-snack cars. This should permit the tailoring of the cars to the needs along different routes. As a coach, each car would be able to carry 86 persons, and all would be capable of being hauled at speeds of up to 120 mph.

Amtrak also placed into service several second-generation, French-built turbotrains, after completing a successful year of service with two first-generation models between Chicago and St. Louis. By mid-1975 it had bought six and planned to buy seven more, which would be built in the U.S. They were to be placed in other U.S. high-density passenger corridors. France's SNCF rail system predicted its third-generation turbotrains would be capable of top speeds of 155–185 mph and provide Paris–Lyons service (262 mi) in less than two hours by 1980.

United Aircraft Corp., builders of seven turbotrains, withdrew during the year from the passenger-train business. The British Railways, on the other hand, approved a step-up in research and development on new diesel-powered High Speed Trains (HSTs) and prototype electric-powered Advanced Passenger Trains (APTs). The HSTs were designed to reach speeds up to 125 mph and were scheduled to be put into service by 1976, while the

Auto-Train carries family automobiles in special cars while the owners ride in lounge cars. The service operates two runs, one between Florida near Disney World and Washington, D.C., and the other between the Florida terminus and Louisville, Kentucky.

Courtesy, Auto-Train Corp.

APTs with a new vehicle suspension unit and body-tilt system were expected to be able to attain speeds up to 155 mph on existing tracks, with first operations predicted in 1977. The well-publicized high-speed passenger trains in Japan encountered progressively worsening track problems and delays during the year, mainly because of excessively high loads, and labor union officials began insisting on stepped-up maintenance.

The successful operations of Auto-Train Corp., which hauls family automobiles in special cars while the riders sit, eat, sleep, and are entertained in lounge cars, were given another boost by the introduction of new tri-level auto carriers, each of which carried 12 cars instead of the 8 on the bi-level carriers. Auto-Train also reached agreement with three railroads to establish, subject to ICC approval, a truck-train service between Florida and points in the northeast and midwest. Trucks would be piggybacked, and their drivers provided with meals and sleeping accommodations in other cars.

The final link in the San Francisco–Oakland Bay Area Rapid Transit network (BART) was completed, a 3.6-mi tube under the San Francisco Bay, and the entire system became operational. Automatic operations were still limited, however, because of unresolved technical problems, and BART sued its engineering consultants and three major suppliers for $238 million in damages, charging breaches of contracts.

Sankei Shimbun

Fearing contamination of their fishing grounds, Japanese fishermen mass their boats to prevent the nuclear freighter "Mutsu" from leaving port.

Water transport. The sudden change in the oil supply situation—from one of a huge demand to one of a forced reduction in demand—caused serious problems in the tanker construction industry, as many orders for jumbo tankers were cancelled and others stretched out. The need for tankers to haul Alaskan oil, starting about mid-1977, did not diminish, however, and U.S. Coast Guard officials announced the readying of eight stations to serve as electronic eyes for monitoring the movements of supertankers along carefully laid out water routes between Valdez, Alaska, and a central California deepwater port. The purpose of the tight surveillance will be to minimize the chances of accidents and spillages. The possibility of spillage is remote, according to shipping experts, who also claimed that spilled oil could now be easily and quickly scooped up without any long-term effect on beaches or marine life.

An oceangoing, integrated catamaran tug-barge vessel neared completion for service on a ten-year lease to Shell Oil Co. Called the "Catug," the 120-ft tug mates with a 409-ft tank barge that can carry 320,000 bbl of oil. A sister ship was also being built by Kelso Marine Inc. for the owner, Seabulk Corp. The coupling feature, utilized in several other instances with standard push-type tows, permits separate use of the power unit during loading/unloading, or when used with two or more barges.

A U.S. National Academy of Sciences study concluded that nuclear merchant ships would not be commercially feasible over the next 20 to 30 years, but a U.S. government shipping official strongly disputed this because of the tripling of fuel costs since the study was made. The noneconomic problems of nuclear ships, which most believed to be of far more significance than the economic ones, were illustrated when the Japanese nuclear merchant ship "Mutsu" made its initial trial voyage and developed a leak in its reactor. Fearing contamination of fishing grounds, fishing fleet operators refused to let the "Mutsu" return to port for 50 days. After heated controversy the ship was finally allowed to return home for repairs, although it reportedly must now find another permanent port.

Lykes Brothers Steamship Line, in order to permit unlimited offshore operations with its new Seabee ships, which carry float-on/off barges on their three decks, purchased special watertight hatch covers. Lykes operates three Seabees and

246 barges, which can service both coastal and inland ports.

An innovation to facilitate LASH (Lighter-Aboard-Ship) service, in which ships contain their own cranes for loading and unloading small barges called lighters, was the use of a floatable V-shaped "false bow" with a diesel-powered thruster for helping tows position the lighters for pick-up and discharge along the Atlantic Coast. The innovation permits the tows used in support of LASH ships to boost the number of lighters being towed from four to seven, to increase their speed from four to nearly seven knots, and to operate more efficiently and safely in strong tides and currents, as well as in generally rough weather conditions.

A new rail-car ferry service was announced, linking Thunder Bay, Ont., and Superior, Wis., on Lake Superior. The 382-ft, 14-knot ferry was able to carry 26 50-ft or 32 40-ft railroad freight cars on five pairs of track over a 200-mi direct water route that took only 14 hours as compared with the three days required by a circuitous overland rail line.

Another innovation, for expediting the handling of containers at ports, was the installation of a twin lift at Norfolk, Va., claimed to be the only one on the market that could handle two unattached 20-ft container units simultaneously and individually. A second such lift was being built for the Port of Melbourne.

—Frank A. Smith

U.S. science policy

The relationship between science and the public in the United States, a love match that had been showing signs of strain in recent years, grew even more tense in 1975. Like many a middle-aged couple, as each party became more dependent upon the other more questions were raised and more suspicions aroused.

Fears about public disaffection with the scientific endeavor had swept through the research community in recent years, centered on the notion that the layman looked upon science as the handmaiden of technology and that it was therefore equally guilty of the evils that technology had spawned—hideous weapons of war, environmental pollution, toxic chemicals in food, and so on. There certainly appeared to be, in some intellectual circles, a tendency to blame science/technology for a national overemphasis on material things as contrasted with spiritual or aesthetic values. But when the National Science Board, the policymaking arm of the National Science Foundation, commissioned a study of public attitudes toward science, it uncovered reassuring data to the effect that most Americans looked upon science and technology as, on the whole, beneficent forces in our society. Although a small fraction did register their concern with unbridled technological advance, the percentage of Americans who credited science and technology with improving the human condition was high.

If the general public was not disaffected with science, from where did the original set of alarming signals come? Perhaps from the press. A survey of science writers published in the fall of 1974 demonstrated that a substantial fraction of them believed that the U.S. public was taking a dim view of science and its works. It was all too likely that their reading of that situation had produced an atmosphere of doubt and distrust that had been misread by the apprehensive leadership of science as the attitude of the public itself.

Or perhaps the reporters were simply prescient. No sooner had the scientific leadership reassured itself than a report appeared concerning the work of two political scientists at the University of California. Their survey indicated that while the public is not hostile toward science and technology, "the current assessment of the public as largely, and somewhat vacantly, enamored with science and technology does not hold." They observed that an identifiable fraction of the population, young and regarding themselves as liberal, constituted a potential opposition to technological development on the basis of conflicting values.

"We can only speculate," the authors said, "whether, as these younger people grow older, they will carry their uneasiness about technology with them. Were they to do so, and were this group to be joined by still younger people who also hold wary attitudes, the context of scientific and technological work could become much more fraught with political controversy. Another point emerging from our interpretation is how very crucial to continued free scientific inquiry is the distinction between scientific work and technological activities apparently now made by a sizable portion of the public. Should this distinction become lost, perhaps through continued merging of science's role with technology's by the popular press, attitudes now mainly associated with technology could spill over to scientific research as well."

Funding. To the extent that the welfare of the U.S. scientific enterprise depends upon public appreciation and support and to the extent that the public derives its good or bad feelings about science from its relationship with technology, the scientific community will have a hard question with which to wrestle over the next few years. As of 1975 there seemed little doubt that the public, insofar as it was represented by Congress, still looked to advances in technology, medicine, and agriculture as the primary reason for the support of science. During the period 1970–74 outlays for *basic* research by the federal government increased by 6% in current dollars, but because of inflation the purchasing power of those dollars actually declined by 15%. During the same period federal expenditures for *applied* research rose by 15%, declining by 8% in constant dollars. The difference in fields other than defense or space was even more marked. Basic research in such areas as health, environment, and natural resources grew by 11% in constant dollars, while applied research in those "civilian" areas increased by 34% in constant dollars, three times as fast.

The proposed federal government budget for research and development for fiscal year 1976, when issued in February 1975, contained a 15% increase over the same budget for 1975, from $18.8 billion to approximately $21 billion. Assuming an inflationary effect of 8%, the administration thus appeared to be taking a positive attitude toward science and technology. Most experienced observers, however, were reserved in their enthusiasm, aware that Congress had yet to make its voice heard. In addition, were the overall budget to threaten the administration's financial game plan, the Office of Management and Budget would very likely suggest severe cutbacks during the course of the year. It had done so during fiscal 1975. At the time that the fiscal 1976 budget was being submitted, Pres. Gerald Ford also proposed to Congress that more than $500 million be stricken from the research and development budget for 1975, most of it coming from the money allocated to the National Institutes of Health.

The support of basic research, especially in universities, was also eroded by federal decisions virtually to abandon institutional grants to universities except for agricultural schools. Direct support of individual graduate students and postdoctoral fellows was also markedly reduced.

The situation with regard to federal funding has been bearish for the scientific community since 1967; the proposed R & D budget for fiscal 1976, after allowance for inflation, is 10–15% less than that for fiscal 1967, depending on how the figures are interpreted. Yet the impact of this reduction on the progress of science has been hard to discern.

Dick Swanson from Woodfin Camp

Relics of an age already past, a rusting gantry stands on a desolate launch pad at Cape Canaveral, Florida, where the first U.S. manned space flights began.

Courtesy, National Science Foundation

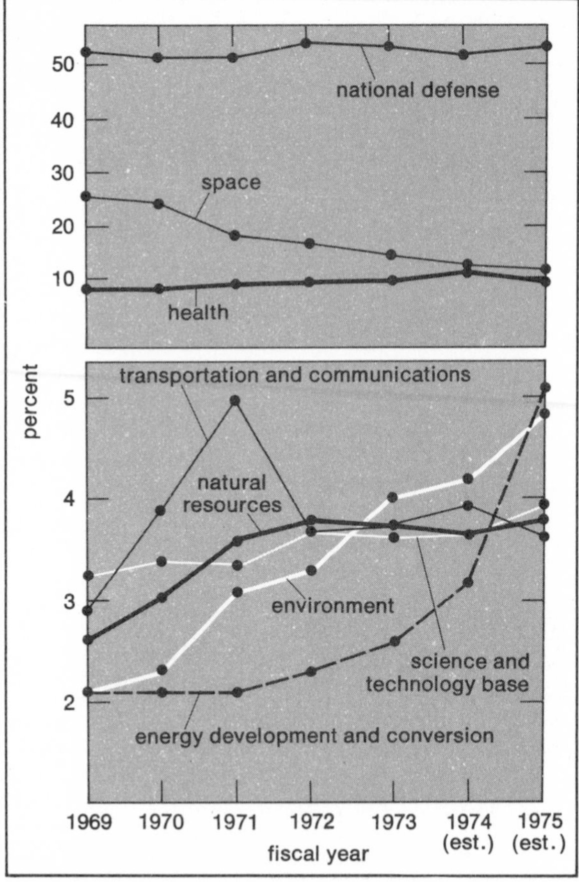

U.S. government spending on scientific research and development is shown on a percentage basis. Primary mission of a program determined its category.

Those individuals in a position to survey the entire front confessed that the austere budgets have had surprisingly little effect. In his annual report to the members of the National Academy of Sciences, their president, Philip Handler, declared that "American science seems to hurtle along with enormous vigor and success."

Employment. The first hard effect may well be on the professional employment of scientists. Although engineers, especially in electronics, had been hard hit by the sudden cutbacks in space and defense research several years ago, academic scientists seemed to escape without severe damage. A number of researchers had found funding difficult to obtain, but the percentage of them who were not employed in their field was considerably smaller than in many other professions. By 1975, however, academic scientists also were beginning to encounter employment problems. A projection of the supply and utilization of scientists and engineers with the doctoral degree, during 1980–85, issued by the National Science Foundation in February, estimated that between 375,000 and 400,000 such individuals would be available to the U.S. economy in 1985 as compared with 295,000 positions that would be available for people with that level of education.

Concern over such bleak projections prompted leaders in specific scientific fields to appoint special committees to report on the prospects. A special report of the American Chemical Society, published in April 1975, found the long-range outlook for Ph.D. chemists as "ominous," with no prospects for a reversal for more than a decade. Academic positions would be hardest hit, the report said, due in part to the fact that student bodies were likely to be smaller and young faculty members would therefore tend to hold on to their positions. The report predicted that the demand for chemists on university faculties in 1990 would be approximately half as large as it was in 1972.

A report on an astronomy manpower committee of the National Academy of Sciences was even more bearish. According to its calculations, 600 new astronomy Ph.D.s would be competing for fewer than 200 openings during the next five years. The authors of the report, all well-known astronomers, advised universities to discourage young people from entering the field by making sure they are aware of the employment problems they would have upon graduation. To those who had participated in the tremendous national drive in the late 1950s and 1960s to persuade youngsters to enter science as a career, it was a momentous turn of events.

Student achievements. Even more depressing was the news from the high schools. In February the National Assessment of Educational Progress, a project responsible to state boards of education and supported by the federal government, issued a survey of the changes in knowledge of science between 1969 and 1973. In the three age groups tested—9, 13, and 17—the average decline in science achievement between 1969–70 and 1972–73 was about two percentage points, which corresponded to a loss of about a half-year of learning experience.

The authors of the survey expressed the hopeful observation that because their survey included all children in those age groups it might be possible that children especially interested in science were scoring higher than before, the lower scores resulting from the fact that fewer children overall were interested in the subject. But even if that were true, it was not good news for those concerned with increasing the number of Americans having an adequate knowledge of science.

Furthermore, the report concluded, the gap between national and group performance in science had not been narrowed for females, black Americans, or students from the inner city. Said the authors: "The facts highlighted in this report should stimulate discussions about national interest in science, school responsibility for science education and the long-term implications of a downward trend in science achievement."

Policy issues. A population growing more ignorant about science was depressing not only because it occurred following the investment of very large amounts of public money into science education but also because an increasingly large proportion of the most worrisome national and international policy issues were heavily dependent on scientific information, or, if not information, scientific intuition.

Several of these issues were of some years' standing, and had become more complicated as the nation found itself threatened by extraordinarily large budgetary deficits. One was the question of environmental pollution and the cost of its abatement. No sooner had Congress passed a series of amendments to the Clean Air Act of 1970 than it was forced to reexamine the standards set for automotive emissions. It was one thing to order a 90% reduction in some of the most noxious elements in automobile exhausts but quite another to justify adding several hundred dollars to the cost of an automobile at a time when automobile sales were nosediving. Again, it was cheering to the environmentalists when opposition mounted to the levels of sulfur oxides permitted in stack gases from power plants, but this became a controversial stand when the price of low-sulfur fuel oil and natural gas rose sharply.

When the National Academy of Sciences was asked by the Senate Committee on Public Works to verify that the benefits to be derived from the legislated levels of automobile emissions were worth the costs of achieving them, it reluctantly agreed that they were but protested that the data necessary to validate the judgments were sadly lacking. Nor was much confidence displayed in the available methodologies used to determine cost/benefit ratios.

Even more uncertain were the best available judgments on the desirability of shifting to a greater dependence on nuclear energy. There, the problem was not only one of costs measured against benefits but also one of actual physical risk to populations in the vicinity of the nuclear power plants. A highly vocal sector of the public

Sidney Harris

"Remember—it's better to light just one little thermonuclear power station than to curse the darkness."

argued that, despite the reassurances contained in the so-called Rasmussen report issued by the Atomic Energy Commission, the risks of a truly catastrophic accident were too high to be acceptable. Not so, said the proponents of nuclear power, pointing out that in any event the amount of petroleum available for power generation would not be sufficient to meet demand through the period of time necessary to change over to coal, of which an estimated two or three centuries' supply was available (if strip-mining of the Western coal lands was permitted).

Because the Rasmussen report had come under such heavy fire from public-interest groups, the Atomic Energy Commission asked the American Physical Society to reexamine the data. According to the APS report, released on April 28, the Rasmussen study had indeed underestimated the risks. According to a press statement: "in the event of a major accident in which containment fails and radioactive material is spread across the countryside, the long-range consequences to human health are greater than had previously been supposed."

Although, said the APS study group, it found no reason for substantial short-range concern regarding the risk of accidents, "the probabilities for their occurrence deserve more attention than so far received within the reactor safety community." Such attention may come from the National Academy of Sciences. In mid-May it was notified by the Energy Research and Development Administration, successor to the AEC, that it would soon be asked to undertake a massive study of the "nuclear option," to determine the risks and benefits inherent in the development of conventional and breeder reactors and the extent to which these power generators might be necessary to meet national needs for energy.

As the National Academy of Sciences prepared itself for this task, it was asked by President Ford to tackle another problem in science and public policy of even greater scope. Acting in response to growing fears that international population growth had already begun to outstrip the global capacity to produce food, Ford asked the Academy to work with a number of government agencies to recommend ways in which agricultural productivity worldwide might be enhanced to forestall, and—if at all possible—to avoid entirely the threatening food crisis.

The pathways between Congress and the scientific establishment, barely discernible a quarter-century ago, were well-trod during the past year. As questions and answers were traded back and forth, many of the latter bristling with qualifications, the problems seemed to grow in number and dimension almost daily. In addition to energy, the environment, and food production, there was the problem of possible damage to the ozone layer in the stratosphere, first through supersonic air traffic and then—incredibly—through the use of halocarbons as propellants in aerosol spray cans. Fairly reliable data indicated that for each 1% decrease in the strength of the ozone layer, there would be a corresponding 2% increase in ultraviolet radiation and in human skin cancer.

Conflicting estimates on reserves of fuels and other mineral resources again brought policymakers to the scientists. One major dispute between the scientists outside and inside the government on the amount of recoverable petroleum and natural gas under the U.S. and its offshore possessions was eventually resolved in favor of the outsiders when the U.S. Geological Survey suddenly lowered its own estimates.

Few could disagree with the statement issued in March by the American Association for the Advancement of Science: "It is clear that most of the driving issues of the last half of the 1970's and into the 1980's will be related to major social problems of which scientific knowledge and technological development are pervasive and critical aspects. . . . While science and technology alone will not have all the answers, they certainly will have important roles in illuminating questions of choice, feasibility, and alternatives."

Clear endorsement came from the chairman of the House of Representatives Committee on Science and Technology, Olin E. Teague (Dem., Tex.): "Science and technology are an element of our contemporary culture as pervasive and important as economics or education or labor or environment. Like them, science and technology are interwoven into all the major missions with which government is involved."

Congressional criticism. Suddenly, in the midst of this warm regard for the social and political value of science and technology, came an exploding mortar shell, lobbed in by Sen. William W. Proxmire (Dem., Wis.). In a statement released to the press on March 11, Proxmire declared: "My choice for the biggest waste of the taxpayer's money for the month of March has to be the National Science Foundation's squandering of $84,000 to try to find out why people fall in love." He proceeded then to tick off a series of studies that he regarded as unnecessary and wasteful: a $15,000 study of hitchhiking, an $81,000 study of the social behavior of the Alaskan brown bear, $25,000 for the study of primate teeth, and $112,000 for a study of the African climate during the last Ice Age.

Questioning the wisdom of federal support of individual grants on the basis of their titles was not invented by Senator Proxmire. Ever since the size of the federal allocation to academic research had reached the $1 billion level, it had attracted the attention of the Department of the Treasury and members of Congress. The attack by Proxmire, however, was of a different order of magnitude: He was the chairman of the Senate subcommittee governing appropriations to the National Science Foundation. And it was not his first attack nor his most serious. The previous November, he had lashed out at the Foundation's management of its expenditures in support of R & D, claiming that NSF had been "outrageously remiss in its stewardship of tax dollars."

That attack, considered intemperate by many friends of the NSF, nevertheless drew some support by members of Congress previously friendly to the Foundation. One of them, Rep. James W. Symington (Dem., Mo.), chairman of the House Subcommittee on Science, Research and Technology, observed that although his committee had been a strong supporter of the Foundation program, "The committee will want to examine the fiscal 1976 program with particular care this year. A number of questions of both policy and practice will require our attention."

The two charges by Proxmire had to be treated separately. As noted above, Representative Symington was willing to associate himself with the questioning of the Foundation's priorities, and he was joined in this by a number of science educators who objected to cutbacks in the Foundation's support of science education. Proxmire's attacks on the system by which the Foundation evaluated individual projects, however, drew strong rebuttal from almost every source. James Reston, in the *New York Times,* noted that if the study on romantic love and marital discord produced some useful information, "it could be the best investment of Federal money since Mr. Jefferson made the Louisiana Purchase."

President Handler of the National Academy of Sciences saw more worrisome possibilities. "For years," he said, "some commentators and members of Congress have sought to make temporary capital by inveighing against federal funding of research projects with seemingly silly titles, usually in the life or social sciences, without examination of the actual subject matter or its potential significance." Although he was grateful for the number of journalists and commentators who had taken Proxmire to task, Handler said that "the seeds of distrust in the judgment-making apparatus of science-supporting agencies is being firmly planted."

Handler's "seeds of distrust" all too soon bore bitter fruit for the National Science Foundation. On April 9, after a stormy three-hour debate over an NSF-sponsored introductory anthropology course, the House of Representatives passed, 212 to 199, an amendment to the NSF authorization bill that would essentially give Congress veto power over all NSF project support. Introduced by Robert E. Bauman (Rep., Md.), the amendment would require the director of the Foundation to provide to Congress each month a list of all proposed grants to be made by the Foundation, together with literature explaining how the national interest would be furthered by the approved projects. Both houses of Congress would then have 30 calendar days of continuous session either to take no action, which would constitute approval, or, by means of a resolution of either House, to veto any or all of the proposed grants.

The Foundation, the House Committee on Science and Technology, which is responsible for the NSF authorizations, and even Senator Proxmire were startled by the unexpected action of the House. For one thing, the offensive anthropology curriculum, "Man, A Course of Study," had already been written off by the Foundation in March after receiving criticism about its controversial contents. For another, no one seemed to be clear on how Congress could carry out its self-imposed assignment should the bill get through the Senate. As Rep. Robert Krueger (Dem., Tex.) declared, in an unsuccessful last-ditch effort to head off the amendment, "We have neither the time nor the talent to make determinations, as a body, on individual scientific projects."

It appeared doubtful that the Bauman amendment would be acceptable to the Senate because both Proxmire, who chaired the appropriations subcommittee, and Sen. Edward M. Kennedy (Dem., Mass.), chairman of the authorization committee, opposed the measure. Nevertheless, no one doubted that the episode would have long-lasting effects. H. Guyford Stever, director of the Foundation, saw it as a watershed, "a signal that all scientists should heed." Its major impact, whether or not it becomes legislation, will most surely be on the sensitivity with which the Foundation will examine all proposals in the life sciences and the social sciences. The most probable immediate effect will be a change in the titles of projects so that their eventual utility will be more apparent. Thus, "A Study of the Sweat Glands of Australian Aborigines" would become "A Possible Biological Control for Heat Prostration Among Military Personnel." Research that could not be so easily targeted by title would suffer most, however, and with it, the sound development of scientific knowledge.

—Howard J. Lewis

Scientists of the Year

Honors and awards

The following is a selective list of recent awards and prizes in the areas of science and technology.

Archaeology

Drexel Medal. Since 1902 the University Museum of the University of Pennsylvania has periodically awarded the Lucy Wharton Drexel Medal "for the best archaeological excavation or publication by an English-speaking scholar." The 1974 gold medal was awarded to John Grahame Douglas Clark, Disney Professor of Archaeology at Cambridge University, author of numerous books and a pioneer in the ecological approach to prehistory.

Architecture and civil engineering

Brown Medal. The Franklin Institute of Philadelphia named Hans Liebherr of West Germany as recipient of the 1974 Frank P. Brown Medal. Liebherr was cited "for his advanced concepts and development of portable, self-erecting, tower climbing, slewing cranes to facilitate the more rapid, efficient, and economical construction of modern, tall skyscraper-type buildings."

Paul Flory

Courtesy, Stanford News & Publications Service, photo by Charles Painter

Honor Awards. A special jury of the American Institute of Architects annually bestows awards for excellence in architectural design. In 1975 only 9 designs from more than 600 entries were deemed worthy of special citation: Hanselmann Residence of Fort Wayne, Ind. (Michael Graves, architect); I.D.S. Center, four buildings in the heart of Minneapolis (Philip Johnson & John Burgee); Kimbell Art Museum of Fort Worth, Texas (Louis I. Kahn); Columbus East High School of Columbus, Ind. (Mitchell/Giurgola Associates); Park Central, an urban renewal complex in Denver, Colo. (Muchow Associates); Herbert F. Johnson Museum of Art in Ithaca, N.Y., and 88 Pine St. in New York City (I. M. Pei & Partners); Cedar Square West, an urban housing project in Minneapolis, Minn. (Ralph Rapson & Associates, Inc.); and The Republic, an office and newspaper plant in Columbus, Ind. (Skidmore, Owings & Merrill).

Astronomy

Cresson Medal. The 1974 Elliott Cresson Medal for astronomy, administered by the Franklin Institute of Philadelphia, was bestowed on Bruno B. Rossi of Cambridge, Mass., for his "many important contributions to our understanding of cosmic rays; and for his pioneering work in space physics, and gamma-ray and X-ray astronomy."

Also honored. The Henryk Arctowski Medal was awarded to Jacques Maurice Beckers of the Sacramento Peak Observatory in New Mexico. The Henry Draper Medal, given for astronomical physics, went to Lyman Spitzer, Jr., of the Princeton University Observatory. The James Craig Watson Medal, awarded approximately every three years, was bestowed posthumously on Gerald Maurice Clemence of Yale University.

Chemistry

Adams Award. The Roger Adams Award in Organic Chemistry, which carries with it an honorarium of $10,000, was awarded in 1975 to Rolf Huisgen of the University of Munich in West Germany. The award acknowledged his discovery of a reaction route, called 1,3-dipolar addition, which led to the synthesis of completely new systems of heterocyclic compounds.

Barnard Medal. The trustees of Columbia University of New York City presented the quinquennial Barnard Medal for Meritorious Service to Science to Louis P. Hammett, who was cited for "helping to found the important borderline field of physical-organic chemistry." Hammett invented a mathematical relationship by which it became possible to predict the effect of substituents on reaction rates and equilibria. This equation and its progeny have become the cornerstone of quantitative studies on reactivity.

Cresson Medal. Theodore L. Cairns of Greensville, Del., was named recipient of the Franklin Institute's Elliott Cresson Medal for chemistry "for his recognition of the possibility of totally unknown classes of percyano and other compounds that might be expected to have very unusual properties, and subsequent creation of methods for their synthesis and exploration of their chemical and physical properties."

Garvan Medal. The American Chemical Society's 1975 Garvan Medal, with its stipend of $2,000, was given to Marjorie C. Caserio of the University of California at Irvine. Caserio's creative research on organic reactions opened new pathways for the synthesis of new and unusual compounds.

Kipping Award. Each year, in recognition of distinguished research in organosilicon chemistry, the Dow Corning Corp. grants a $2,000 award, administered by the American Chemical Society. In 1975 the Frederic Stanley Kipping Award was given to Hans Bock of the University of Frankfurt in West Germany.

New Materials Prize. The International Prize for New Materials, sponsored by the IBM Corp. and awarded by the American Physical Society, was given to Heinrich Welker, director of Siemens AG in Erlangen, West Germany. Welker, first recipient of the new award, was cited for "pioneering investigations on the physical and chemical properties of the binary compounds of the elements in groups III and V of the periodic table."

Nobel Prize. The prestigious Nobel Prize for Chemistry was awarded to Paul J. Flory for work "both theoretical and experimental in the physical chemistry of macromolecules." Flory discovered ways of analyzing polymers so that it became possible to develop new plastics and other synthetics in a systematic manner. Many of the materials have the ability of increasing the length of their chains; he also discovered the means by which one molecule under growth can stop and pass its growth ability on to another. Flory obtained his Ph.D. from Ohio State University in 1934. He served on the faculty of Stanford University in California from 1961 to 1975.

Olney Medal. The American Association of Textile Chemists and Colorists named R. Lee Wayland, Jr., as the 1975 recipient of the Olney gold medal and $1,000 honorarium for technical and scientific contributions to textile chemistry.

Priestley Medal. Each year the American Chemical Society awards a gold medal considered to be the most prestigious honor in U.S. chemistry. Henry Eyring of the University of Utah, "one of the truly great scientists of our age," was chosen to receive the 1975 Priestley Medal. A theoretical chemist with broad interests ranging from molecular biology to sophisticated developments in quantum and statistical mechanics, Eyring has written more than 500 technical papers and eight books.

Earth sciences

Day Lectureship. The Arthur L. Day Prize and Lectureship, administered by the National Academy of Sciences, was inaugurated in 1972 to encourage efforts to synthesize current knowledge about the physics of the Earth. The 1975 honor was bestowed jointly on Drummond H. Matthews of the University of Cambridge and on Fred J. Vine of the University of East Anglia, both in England.

Day Medal. The Geological Society of America chose Alfred Edward Ringwood as recipient of its 1974 Arthur L. Day Medal. Ringwood's activities at the Australian National University at Canberra centered on the development and operation of a laboratory for studying high-pressure phase transformations and their bearing upon the internal structure of the Earth.

Meteorological Prize. The International Meteorological Organization Prize for 1974, which consists of a gold medal and an honorarium of $1,200, was awarded to Joseph Smagorinsky, director of the Geophysical Fluid Dynamics Laboratory at Princeton University. He was cited for "his dedicated work on the general circulation of the atmosphere, his important contribution to international research activities and his services to the cause of international collaboration in meteorology."

Penrose Medal. The Geological Society of America annually awards the gold Penrose Medal to a scientist whose original contributions mark significant advances in the geological sciences. In 1974 W. Maurice Ewing, who directed the Lamont-Doherty Geological Observatory at Columbia University, New York City, from 1949 to 1972, was honored posthumously for his landmark contributions to oceanography.

Prix Scientifique. Le Prix Scientifique du Québec for 1974, together with its honorarium of $10,000, was awarded to Branko Ladanyi of the École Polytechnique in Montreal. In making its choice, the award committee took special note of Ladanyi's studies on rock mechanics and permafrost.

Rossby Medal. Each year the American Meteorological Society awards a gold medal and certificate for outstanding contributions to man's knowledge of the atmosphere. Charles H. B. Priestley of the Environmental Physics Research Laboratories in Australia was named recipient of the Carl-Gustaf Rossby Research Medal for making "fundamental contributions to the understanding of turbulent processes and the links between small-scale and large-scale dynamics in the atmosphere."

Schuchert Award. J. William Schopf of the University of California at Los Angeles was selected to receive the 1974 Schuchert Award of the Paleontological Society for his "research on Precambrian paleobiology and on origin and diversification of early plant life, organic geochemistry of Precambrian rocks, and electron microscopy of microorganisms."

Second Half Century Award. In naming Louis J. Battan recipient of its Second Half Century Award, the American Meteorological Society cited "his many contributions to cloud physics and his objective scientific evaluation of weather modification efforts, his writings fostering public understanding of meteorology, and his great service to the profession in many capacities." Battan is an adviser and contributor to the *Yearbook of Science and the Future*.

Vetlesen Prize. The 1974 Vetlesen Prize, which consists of a gold medal and an honorarium of $25,000, was awarded to Chaim Leib Pekeris of the Weizmann Institute of Science in Israel. The award was based on Pekeris' "application of advanced methods of applied mathematics to the solution of a wide range of fundamental geological and geophysical problems." He is well known for studies on convection within the Earth, propagation of sound in layered media, tides computed on a global scale, and modes of free oscillation of the Earth.

Also honored. The 1975 John J. Carty Medal and Award for the Advancement of Science was given to J. Tuzo Wilson of the University of Toronto. The Paleontological Society Medal Award was bestowed on John W. Wells of Cornell University in New York.

Energy

Lucas Award. Michel T. Halbouty, a geologist and petroleum engineer, received the 1975 Anthony F. Lucas Award of the American Institute of Mining, Metallurgical and Petroleum Engineers (AIME). He was cited "for his contributions toward creative geology and petroleum engineering, new frontiers, total petroleum conservation, scientific literature, inter-disciplinary communication between earth scientists and engineers, and public understanding of the petroleum industry."

Ramsey Medal. The 1975 Erskine Ramsey Medal of the AIME was given to Charles J. Potter, chairman of the Rochester & Pittsburgh Coal Co., "for his leadership and distinguished service to the coal industry."

Courtesy, The Academy of Natural Sciences of Philadelphia

Ruth Patrick

Environment

Environmental Award. Claude E. ZoBell, formerly of the Scripps Institution of Oceanography, was given the 1975 Environmental Conservation Distinguished Service Award of the American Institute of Mining, Metallurgical and Petroleum Engineers for his extensive contributions to "the knowledge of marine microbiology and protection of the ecology of the oceans."

Environmental Quality Award. Recipient of the 1975 National Academy of Sciences Award for Environmental Quality was John T. Middleton of the U.S. Environmental Protection Agency. The $5,000 award, first bestowed in 1972, recognizes "outstanding contributions based in science or technology to improve the quality of the environment or in the control of its pollution by man."

Nuclear Society Award. Edward G. Struxness of the Oak Ridge National Laboratory in Tennessee was chosen to receive the 1975 American Nuclear Society Special Award. The award certificate noted "his experimental and analytical contribution to the science and technology of radioactive waste management, and... his ability to advance the knowledge base by constructive cooperation with other institutions and individuals both domestic and foreign."

Pollution Control Award. In 1975 the American Chemical Society's $3,000 award for Pollution Control was given to Aubrey P. Altshuller of the

National Environmental Research Center in North Carolina. The award called attention to Altshuller's efforts to build a scientific information base for the development of rational photochemical air-pollution abatement methods.

Potts Medal. In awarding the 1974 Howard N. Potts Medal to Jay W. Forrester of the Massachusetts Institute of Technology, the Franklin Institute of Philadelphia noted that his innovative modeling of urban, regional, and global problems created widespread awareness of the possible consequences of efforts to solve them.

Tyler Prize. The $150,000 John and Alice Tyler Ecology Prize, the largest monetary award ever bestowed for scientific achievements, was established in 1973. Pepperdine University in California, which administers the prize, named Ruth Patrick of the Academy of Natural Sciences in Philadelphia as the latest recipient. She is an authority on water pollution and is considered to have had crucial influence in the government's current clean water policy.

Food and agriculture

Babcock-Hart Award. Each year an award for nutrition research is sponsored by the Nutrition Foundation and administered by the Institute of Food Technologists. In 1975 the award was conferred on Donald K. Tressler, president of AVI Publishing Co., for his "distinguished career of over half a century in all areas of food science and technology, including industrial, academic, governmental, and consulting assignments."

Goldberger Award. Ananda S. Prasad, professor of medicine at Wayne State University and chief of hematology at Detroit General Hospital, received the 1975 Joseph Goldberger Award from the Council on Foods and Nutrition of the AMA. Prasad was responsible for the first description of human zinc-deficiency syndrome.

Osborne and Mendel Award. The Osborne and Mendel Award, sponsored by the Nutrition Foundation and administered by the American Institute of Nutrition, was presented in 1974 to DeWitt S. Goodman of Columbia University's School of Medicine. He was cited "for his contributions toward understanding the transport of retinol (vitamin A) by specific retinol binding protein and the alterations in retinol metabolism and retinol binding protein that occur in health and disease, particularly in severe protein calorie malnutrition."

Information sciences

Medal of Honor. The Institute of Electrical and Electronics Engineers, Inc., honored John R. Pierce of the California Institute of Technology with its Medal of Honor. Pierce was cited "for his pioneering proposals and the leadership of communication satellite experiments, and for contributions in theory and design of electron beam devices essential to their success." Since 1917 the award has generally been an annual event.

Life sciences

Coblentz Award. Bernard J. Bulkin of Hunter College of the City University of New York received the Coblentz Society Award, given annually since 1964 to a scientist not yet 35 years old. Bulkin focused on chemical liquid crystals and biological cellular membranes in carrying out his molecular spectroscopy research.

Cresson Medal. An Elliott Cresson Medal was bestowed on Arie J. Haagen-Smit of Pasadena, Calif., "for his work on plant hormones and in the chemistry of natural products; his contributions to techniques used in quantitative microanalysis; his recognition of the chemical and photochemical nature of smog, and for his leadership on the local, state, and national level in programs of air pollution abatement."

Distinguished Service Medal. Willi Hennig, an entomologist with the Staatliches Museum für Naturkunde in West Germany, was chosen to receive the American Museum of Natural History's Gold Medal for Distinguished Service in Science. Hennig developed a new approach to systematics called cladistic analysis that has received worldwide attention. Challenging the universal validity of certain generally accepted principles of evolution, Hennig contends that not all species with numerous similar characteristics have a common recent ancestor, and that some groups (such as crocodilians and birds), despite striking dissimilarities, have in fact had a common ancestor in recent times.

Horwitz Prize. Since 1966 Columbia University has annually conferred the Louisa Gross Horwitz Prize for outstanding research in biology. The $25,000 prize for 1974 was bestowed on Boris Ephrussi, a French geneticist, for "three momentous contributions to modern biology." His research on yeast led to the development of the genetics of mitochondria; he provided the first experimental data to support the idea that normal genes sometimes determine the presence of an enzyme and that abnormal ones may cause its absence; and, after confirming George Barski's suggestion that cells from two different tissue culture lines can fuse to form a hybrid, he demonstrated how somatic cell hybridization could be used to study the basis for cell differentiation and for cancer, and for the analysis of mammalian chromosomes.

Wildlife Conservation Prize. Felipe Benavides, a Peruvian wildlife expert, was named 1975 recipient of the $50,000 J. Paul Getty Wildlife Conservation Prize. The award is administered by the World Wildlife Fund in Washington, D.C. Benavides was cited for his efforts to save the vicuña and other animal species in South America.

Also honored. For studies on metabolism and on the activity of cholecalciferol in animal tissues, E. Kodicek was given the CIBA Medal and Prize by the Biochemical Society. D. R. Trentham received the Biochemical Society's Colworth Medal for research on the kinetics of enzyme action and on the transient kinetics of myosin-ATPase. The 1974 Selman A. Waksman Award in Microbiology went to Renato Dulbecco.

Materials sciences

Douglas Medal. The American Institute of Mining, Metallurgical and Petroleum Engineers (AIME) announced that Petri B. Bryk, former managing director and president of Outokumpu Oy in Finland, was winner of its 1975 James Douglas Gold Medal "for metallurgical leadership resulting in widespread application of many new, environmentally acceptable, nonferrous processes including flash smelting."

Greaves-Walker Award. The National Institute of Ceramic Engineers each year recognizes "outstanding service to the ceramic engineering profession" by presenting one of its members with the Arthur Frederick Greaves-Walker Award. The 1975 recipient was Wayne A. Deringer, former director of materials research and development for A. O. Smith Corp. in Wisconsin.

Jeppson Medal. D. P. H. Hasselman, director of ceramic research at Lehigh University in Pennsylvania, was named winner of the 1975 John Jeppson Medal and Award by the American Ceramic Society. His work included important research on thermal stress fracture.

Mathewson Medal. The Metallurgical Society of the AIME bestowed its 1975 Champion H. Mathewson Gold Medal on E. T. Turkdogan, whose "numerous outstanding publications . . . have made a major impact in the fields of thermodynamics and kinetics of phenomena related to metallurgical processes."

McConnell Award. The 1975 Robert Earll McConnell Award of the AIME went to Norwood B. Melcher, formerly with the U.S. Bureau of Mines, for research that "led to methods of beneficiating low grade iron ores and to increased blast furnace efficiency and production."

Purdy Award. The American Ceramic Society conferred its 1975 Ross Coffin Purdy Award on Anthony G. Evans of Rockwell International in California and on Melvin Linzer of the National Bureau of Standards. The two were chosen for their paper "Failure Prediction in Structural Ceramics Using Acoustic Emission," published in the *Journal of the American Ceramic Society* (November 1973).

Also honored. The AIME bestowed its Extractive Metallurgy Division Technology Award on Robert W. Bartlett and Hsin-hsiung Huang. Meguru Nagamori received the AIME's Extractive Metallurgy Division Science Award. Co-winners of the Robert W. Hunt Silver Medal, conferred by the Iron and Steel Society of the AIME, were Ramchandra K. Iyengar and Frank C. Petrilli. Joint recipients of the John Chipman Award were Loucas Gourtsoyannis, Hani Henein, and R. I. L. Guthrie. Reinhardt Schuhmann, Jr., an expert in nonferrous extractive metallurgy, received the 1975 Mineral Industry Education Award.

Mathematics

Fields Medal. During the quadrennial meeting of the International Congress of Mathematicians, David B. Mumford of Harvard University and Enrico Bombieri of the University of Pisa were each presented with a Fields Medal "in recognition of work already done and as an encouragement for further achievements." Mumford was honored for many fundamental contributions to algebraic geometry; Bombieri, who excels in many fields of mathematics, was cited for his work in number theory and minimal surfaces.

Franklin Medal. Each year since 1914 the Franklin Medal has been awarded for outstanding work in the physical sciences or its technology. Nikolai N. Bogoliubov of the Soviet Union was named winner of the 1974 medal "for his powerful mathematical methods in nonlinear mechanics."

Levy Medal. The 1974 Louis E. Levy Medal for mathematics was shared by J. N. Damoulakis and Angelo Miele, co-authors of an outstanding paper, "Modifications and Extensions of the Sequential Gradient-Restoration Algorithm for Optimal Control Theory," which appeared in the *Journal of The Franklin Institute* (July 1972).

Mechanical engineering

Holley Medal. The American Society of Mechanical Engineers (ASME) awarded its 1975 Holley Medal to George M. Grover of the Q-dot Corp. in Texas for his invention of the heat pipe, a heat-transfer device now widely used in energy utilization, conservation, and conversion systems.

Timoshenko Medal. Albert E. Green of the University of Oxford received the 1974 Timoshenko Medal of the ASME "for his research contributions to several branches of theoretical and applied mechanics, especially in linear and nonlinear elasticity and in the foundations of mechanics of continua."

Westinghouse Medal. In conferring its George Westinghouse Gold Medal on Charles W. Elston of the General Electric Co., the American Society of Mechanical Engineers cited "his outstanding technical contribution and leadership in the areas of development and application of power generation cycles and in the design, development, production and application of gas and steam turbines and electric generators."

Medical sciences

Brookdale Award. Thomas E. Starzl of the University of Colorado Medical Center was named second recipient of the $5,000 Brookdale Award in Medicine. Starzl accomplished the first successful human liver transplant in 1967 and did important work in neurophysiology, in cardiac physiology, and on hepatotropic substances.

Kittay Award. The International Kittay Award of $25,000, presented annually to an outstanding researcher in the area of mental health, was conferred in 1975 on Harry F. Harlow, director of the Primate Research Center at the University of Wisconsin. The International Advisory Board of the Kittay Scientific Foundation pointed out that Harlow's research on the "mother-infant attachment bond" that exists between primates has "extreme significance for understanding those aspects of human behavior related to depression, aggression, or sexual dysfunction, which originated in the formative years of mother-infant interaction."

Langer Award. A University of Chicago virologist, Bernard Roizman, was chosen to receive the $1,000 Esther Langer Award for his studies of the molecular biology of herpesvirus infections as they relate to human cancer.

Lasker Awards. John Charnley, an orthopedic surgeon at the University of Manchester, England, received the $10,000 Albert Lasker Clinical Medical Research Award for 1974. Using artificial materials as replacements for the hip socket and the top of the thigh bone, he devised surgical procedures that enabled thousands of crippled patients to walk normally again. The Albert Lasker Basic Medical Research Award was shared by four persons, each of whom received $5,000. Ludwig Gross of the Bronx Veterans Administration Hospital, Sol Spiegelman of Columbia University College of Physicians and Surgeons, and Howard Temin of the University of Wisconsin did important research on the role of viruses in cancer; Howard Skipper of the Southern Research Institute in Alabama focused on mouse research that proved valuable in cancer chemotherapy.

Nobel Prize. The highly coveted Nobel Prize for Physiology or Medicine was shared in 1974 by Albert Claude, George E. Palade, and Christian de Duve. The three biologists, forming a loosely knit group, were the first to determine the inner structure of cells and the functions of their component parts. Claude, the Belgian-born leader of the group, used centrifuges to separate cells into their component parts and then studied them with an electron microscope. He discovered mitochondria, which store the cell's energy, and the endoplasmic reticulum, a fiber structure that supports organelles. Palade, a Romanian by birth, improved centrifuge techniques and ways to prepare specimens, and discovered ribosomes, the structures that synthesize protein within the cell. De Duve, a Belgian, discovered lysosomes, the organelles that ingest a cell's nutrients.

Sheen Award. A plaque and a $10,000 stipend was awarded to R. Lee Clark, recipient of the 1974 Dr. Rodman E. Sheen and Thomas G. Sheen Award. As director of the University of Texas M. D. Anderson Hospital and Tumor Institute, Clark pioneered the research and development of the prototype of the cobalt-60 irradiator, protective

Harry Harlow

Courtesy, Kittay Scientific Foundation

Wide World

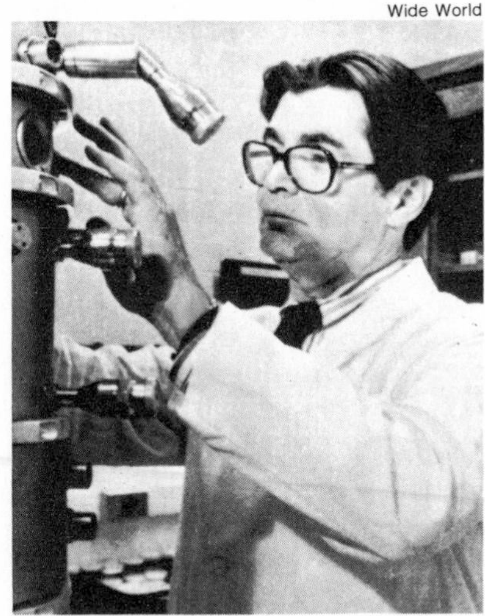

George Palade

Keystone

Albert Claude

Wide World

Christian de Duve

environments for the control of infection in leukemia patients, the IBM Blood Cell Separator for hemotherapy, and the use of the Xenon-133 system for the evaluation of cardiopulmonary function.

Also honored. The American Society for Pharmacology and Experimental Therapeutics gave its $2,000 ASPET Award for Experimental Therapeutics for 1975 to Mackenzie Walser of the Johns Hopkins University School of Medicine. The Paul Ehrlich and Ludwig Darmstaedter Prize for 1975 was shared by G. B. Mackaness of New York, N. A. Mitchison of London, and M. Simonsen of Copenhagen. In 1974 the American Cancer Society bestowed its highest honor, the Annual National Award, on Benno Schmidt, Karl and Ingegerd Hellstrom, and Charles Heidelberger. The Remington Honor Medal of the American Pharmaceutical Association for 1974 was given to Lloyd M. Parks of Ohio State University.

Optical engineering

Ives Medal. Ali Javan of the Massachusetts Institute of Technology was awarded the 1974 Frederic Ives Medal "in recognition of his pioneering contributions to optics and quantum electronics, especially the invention of the gas laser . . . an optical device of unparalleled applicability to scientific research."

Michelson Medal. The Franklin Institute presented its 1974 Albert A. Michelson Medal to Peter Sorokin of the IBM Thomas J. Watson Research Center in New York. He was cited for "outstanding contributions to quantum optics and, in particular, for his discovery of the organic dye laser."

Physics

Bonner Prize. The American Physical Society named Chien-shiung Wu of Columbia University as recipient of its 1975 Tom W. Bonner Prize in Nuclear Physics. The citation noted her contributions "to our understanding of the shapes of beta spectra, to the discovery of the failure of parity conservation in weak interactions and to the evidence for lepton conservation and the conserved vector current."

Buckley Prize. Albert W. Overhauser of Purdue University was named winner of the Oliver E. Buckley Solid State Physics Prize for his "invention of dynamic nuclear polarization and for the stimulation provided by his studies of instabilities of the metallic state." The award is sponsored by Bell Telephone Laboratories and administered by the American Physical Society.

Cresson Medal. In 1974 Robert H. Dicke of Princeton University received an Elliott Cresson Medal "for his many theoretical and practical contributions to modern physics, especially for his direct and seminal role in gravitational experiment and theory."

Langmuir Prize. The Irving Langmuir Award in Chemical Physics, which carries with it a $5,000 honorarium from the General Electric Foundation, was presented to Robert H. Cole of Brown University for his "masterly contributions to the understanding of molecular interactions through study of dielectric behavior."

Lorentz Medal. The Royal Netherlands Academy of Arts and Sciences presented the 1974 Lorentz Medal to John H. Van Vleck, emeritus professor of mathematics and natural philosophy at Harvard

John Deverill from Camera Press, London/Pictorial Parade

Antony Hewish (third from left) and Sir Martin Ryle (far right), co-winners of the 1974 Nobel Prize for Physics, meet with Chinese scientists under one of the large antennas at the Mullard Radio Astronomy Laboratory near Cambridge, England.

University. Van Vleck received the quadrennial award for work that centered mainly on the quantum theory of magnetism and molecular spectra.

Nobel Prize. Two English astronomers, Sir Martin Ryle and Antony Hewish, were selected to receive the Nobel Prize for Physics in 1974. Ryle, an early investigator of extraterrestrial radio sources, developed advanced radio telescopes and a technique called aperture synthesis. It involves two radio telescopes that can be moved on rails to vary the distance between them. The result is equivalent to a single huge telescope, the dimensions of which would have to correspond to the distance between the two separated units. Antony Hewish used this telescope system in 1967 when he became the first to discover pulsars (pulsating radio stars), believed to be the dark remains of once-brilliant stars.

Oppenheimer Prize. The Center for Theoretical Studies at the University of Miami, Fla., named Nicholas Kemmer of the University of Edinburgh recipient of the 1975 J. Robert Oppenheimer Memorial Prize. Kemmer was honored for discovering certain regularities and symmetries in the classification of elementary particles.

Wetherill Medal. The Franklin Institute chose Aage Niels Bohr and Ben Roy Mottelson, research colleagues at the Niels Bohr Institute in Copenhagen, as co-equal recipients of John Price Wetherill Medals. Together they developed the theory of collective states of atomic nuclei.

Also honored. The Faraday Medal was awarded to John Miller Meek for research into electrical discharges in gases, especially the mechanism of the electric spark. The Guthrie Medal and Prize, with its £250 honorarium, went to David Tabor, of the University of Cambridge. The American Physical Society High Polymer Physics Prize went to Walter H. Stockmayer of Dartmouth College.

Psychology

Applications in Psychology Award. Two associates with the Children's Television Workshop shared the 1974 Distinguished Contribution for Applications in Psychology Award. Gerald S. Lesser and Edward L. Palmer were singled out by the American Psychological Association for their sophisticated and imaginative use of psychological research in developing a new kind of human significance for television. Working closely with production personnel for "Sesame Street" and "The Electric Company," they influenced every aspect of these shows with sound child psychology.

Scientific Contribution Awards. The American Psychological Association announced three winners of the 1974 Distinguished Scientific Contribution Awards, each of whom received $1,000. Angus Campbell was honored for utilizing social psychology in surveying attitudes, perceptions, and complex behavioral patterns of various population groups. Lorrin A. Riggs was cited for his many basic and incisive physiological and psychological studies of human vision. Richard F. Thompson received his award for imaginative and extensive research on the neural bases of behavior.

Space exploration

Collier Trophy. John F. Clark, director of the Goddard Space Flight Center, and Daniel J. Fink, vice-president of the Space Division of General

Electric, were named co-winners of the 1974 Robert J. Collier Trophy. The National Aeronautics Association, the organization that bestows the award, acknowledged the vital role the two men played in the success of the first Earth Resources Technology Satellite (ERTS).

Guggenheim Award. The 1974 Daniel and Florence Guggenheim International Astronautics Award, given for an outstanding contribution to space research and exploration during the preceding five years, was bestowed on Hilding A. Bjurstedt, head of the department of aviation medicine of the Karolinska Institutet in Stockholm.

Hill Award. The American Institute of Aeronautics and Astronautics awarded the 1974 Louis W. Hill Space Transportation Award, and its honorarium of $5,000, to Rocco A. Petrone of NASA "for distinguished contributions to space flight technology through his engineering excellence in launch operations and outstanding managerial skills in directing the last six highly successful Apollo missions for the scientific exploration of the moon and lunar environment."

Transportation

Goddard Award. The annual recipient of the Goddard Award and its $10,000 honorarium supplied by the United Aircraft Corp. is selected by the American Institute of Aeronautics and Astronautics. Gordon E. Holbrook of General Motors Corp. and George Rosen of United Aircraft Corp. were named co-winners of the 1975 award "for joint leadership and technical contributions in pioneering the development and production of widely-used turbopropellor propulsion systems."

Gold Medal Award. F. W. Page, recipient of the British Gold Medal for Aeronautics in 1962, was named recipient of the Royal Aeronautical Society's Gold Medal in 1974. The first recipients, in 1909, were Orville and Wilbur Wright. Page performed research on many facets of aeronautics but was especially involved in the development of military aircraft.

Science journalism

AAAS-Westinghouse Awards. Each year the American Association for the Advancement of Science administers science journalism awards sponsored by Westinghouse Corp. A $1,000 prize is assigned to each of three categories: articles appearing in U.S. newspapers having a daily circulation exceeding 100,000; articles appearing in newspapers having more limited circulation; and articles published in magazines. In 1974 the winners were George Alexander, author of three articles in the *Los Angeles Times* that dealt with the universe (Jan. 2, 1974), behavior (March 31, 1974), and fossil-dating techniques (Aug. 18, 1974); Judith M. Roales, for ten articles in the *Delaware State News* on the oil and gas industry (Feb. 17–27, 1974); and Michael Rogers, whose article "Totality, A Report" was published in *Rolling Stone* (Oct. 11, 1973).

Bernard Award. The National Society for Medical Research has since 1967 bestowed the Claude Bernard Science Journalism Award for newspaper and magazine writing that has furthered public understanding of basic research in the life sciences. In 1975 Lawrence K. Altman was honored for his articles on medicine in the *New York Times* and Gene Bylinsky for his science writing in *Fortune* magazine. Each received an honorarium of $1,000.

Grady Award. Jon Franklin, science writer for the *Evening Sun* of Baltimore, Md., was named 1975 recipient of the American Chemical Society's $2,000 James T. Grady Award for Interpreting Chemistry for the Public. Franklin was chosen "for his deft accounts of significant trends in science, many of which reflect the pervasive involvement of chemistry in various fields of scientific research."

Washburn Award. The Bradford Washburn Award, which consists of a gold medal and an honorarium of $5,000, has been annually awarded since 1964 by the Museum of Science in Boston. It is given for "an outstanding contribution toward public understanding of science, appreciation of its fascination, and the vital role it plays in all our lives." The 1974 award was given to ethologists Baron Hugo van Lawick and Jane Goodall.

Miscellaneous

Gibbs Medal. At irregular intervals the National Academy of Sciences presents the Gibbs Brothers Medal and an honorarium of $1,000 for outstanding contributions in the field of naval architecture and marine engineering. In 1974 the recipient was Phillip Eisenberg of Hydronautics, Inc.

Hubbard Medal. The National Geographic Society's Hubbard Medal has been awarded only 26 times since its establishment in 1906. In conferring the award on Alexander Wetmore, a research associate with the Smithsonian Institution, the 1975 award committee recalled Wetmore's "lifetime of outstanding contributions to geography through pioneering explorations and biological studies in the jungles of Central and South America, islands of the Central Pacific Ocean, and worldwide advancement of the science of ornithology."

Kalinga Prize. In 1952 UNESCO awarded the first annual Kalinga Prize for the popularization of science. The £1,000 prize is supplied by Bijoyanand Patnaik, an Indian industrialist, and is given for a distinguished career of public service in the interpretation of science as a writer, editor, speaker, or radio program director. The recipient is expected to spend a month or more in India studying its life and culture. In 1974 José Reis of Brazil and Luis Estrada of Mexico were each given a Kalinga Prize.

Mullard Award. The Royal Society each year rewards outstanding contributions to the advance of science, engineering, or technology that in the preceding ten years led directly to national prosperity in the United Kingdom. In 1974 the gold medal and an honorarium of £1,000 were given to F. B. Mercer, sole inventor of an ingenious yet simple extrusion process for the manufacture of integral or knotless plastic net. Because the Netlon process has an extremely wide range of application, it was able to generate a viable and entirely new industry.

Science Talent Awards. The 34th annual Science Talent Search, sponsored by the Westinghouse Educational Foundation and administered by Science Service, produced the following winners in 1975. The first place award, a $10,000 scholarship, was given to Paul A. Zeitz of Stuyvesant High School in New York City for a project in mathematics involving gamma functions. The second place $8,000 scholarship award went to Alan S. Geller of Ridgewood, N.J., for work in theoretical physics. Winner of the third place $8,000 scholarship was Daniel R. Marshak of La Jolla, Calif., for a project in biochemistry. Three other winners each received a $6,000 scholarship: Byron Bong Siu of New York City; Richard J. Foch of Titusville, Fla.; and Robert M. Claudson of Richland, Wash. Four others received scholarships worth $4,000: Charlene G. Sanders of Narberth, Pa.; Lorraine A. Pillus of Cocoa, Fla.; Craig F. Miller of New York, N.Y.; and H. Britton Sanderford of Metairie, La.

Alexander Wetmore

Courtesy, National Geographic Society

Scientific Achievement Award. On special occasions the American Medical Association presents a medallion to a physician or nonphysician for outstanding work. Philip H. Abelson, editor of *Science* magazine, was given the award in 1974. He is president of the Carnegie Institution of Washington and has devoted many years to research in chemistry, geophysics, and biophysics.

Obituaries

The following persons, all of whom died in recent months, were widely noted for their scientific accomplishments.

Alexanderson, Ernst Frederik Werner (Jan. 25, 1878–May 14, 1975), Swedish-born electrical engineer, revolutionized wireless communications by inventing (1906) a high-frequency alternator that made possible the first radio transmission of speech and music. In addition to repeatedly upgrading the quality of his alternator, he significantly improved transmitting and receiving antennas, ship-propulsion and electric train systems, and telephone relays. His selective tuning device still forms part of modern radio systems. He also developed the amplidyne, an automatic control system; it was first used in factories to automate delicate manufacturing processes but its applications are virtually limitless. In 1927 he demonstrated television in his home and gave the first public exhibition in 1930. Alexanderson, who went to the U.S. in 1901 and retired from the General Electric Co. in 1948 after 46 years of service, renewed an earlier association with RCA in 1952 and developed a color television receiver that became his 321st patent.

Apgar, Virginia (June 7, 1909–Aug. 7, 1974), U.S. physician, won international recognition for developing the Apgar Score, a five-part examination of newborn babies to determine whether they need special medical assistance to sustain life. Immediately after delivery, and again after a five-minute interval, each infant's condition is evaluated on the basis of its skin color, respiration, heartbeat, reflexes, and muscle tone. Apgar

AP, London

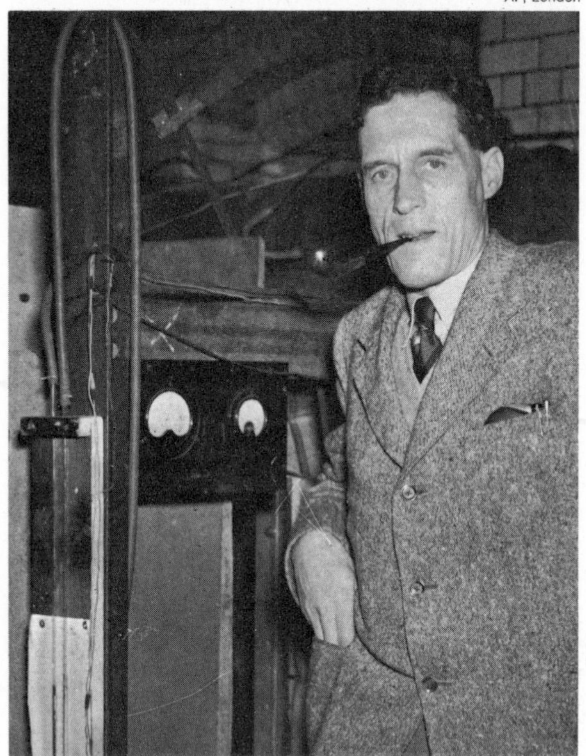

Baron Blackett

Courtesy, Time-Life Films and BBC-TV

J(acob) Bronowski

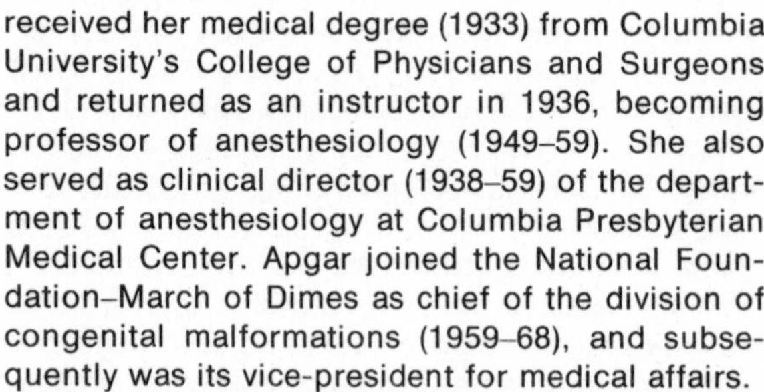

received her medical degree (1933) from Columbia University's College of Physicians and Surgeons and returned as an instructor in 1936, becoming professor of anesthesiology (1949–59). She also served as clinical director (1938–59) of the department of anesthesiology at Columbia Presbyterian Medical Center. Apgar joined the National Foundation–March of Dimes as chief of the division of congenital malformations (1959–68), and subsequently was its vice-president for medical affairs.

Blackett, Baron (Patrick Maynard Stuart Blackett) (Nov. 18, 1897–July 13, 1974), British experimental physicist, received the Nobel Prize for Physics in 1948 for his discoveries in cosmic radiation. As a fellow of King's College, Cambridge (1923–33), he worked under Ernest Rutherford at the Cavendish Laboratory, where he developed the cloud chamber for the study of the collision of nuclear particles.

In 1933 he discovered the positive electron (simultaneously with C. D. Anderson in the U.S.). Blackett was professor of physics, Birkbeck College, University of London (1933–37), and Langworthy professor of physics, University of Manchester (1937–53). During World War II he became director of operations research at the Admiralty. At the University of Manchester he created a school of cosmic-ray research and stimulated developments that led to the building of the Jodrell Bank Experimental Station for Radio Astronomy. From 1953 to 1965 Blackett was professor of physics at the Imperial College of Science and Technology, University of London. From 1965 to 1970 he served as president of the Royal Society. Blackett was awarded the Order of Merit in 1967 and was made a life peer in 1969.

Bronowski, J(acob) (Jan. 18, 1908–Aug. 22, 1974), British research scientist and author, who produced "The Ascent of Man" (1973), a 13-part BBC television series which was widely acclaimed as an outstanding presentation of the evolution of civilization and the place of science in human history. He obtained a Ph.D. in mathematics from Cambridge University (1933) before going to the University of Hull as senior lecturer in mathematics (1934–42). Bronowski's government wartime research included a compilation of statistics on the effects of bombing on industry and economics. Following a visit to Japan in 1945, he wrote a report called "The Effects of the Atomic Bombs at Hiroshima and Nagasaki." After statistical research for the Ministry of Works (1946–50), he directed the Coal Research Establishment at the National Coal Board (1950–59) and served as director general of the board's Process Development Department (1959–63), developing smokeless fuels. From 1964 he was a senior fellow at the Salk Institute for Biological Studies at San Diego, Calif., and was made director of its Council for Biology in Human Affairs in 1970. He broadcast on science and culture, and published *The Poet's Defence* (1939), *William Blake, 1757–1827: A Man Without a Mask* (1944), and *William Blake and the Age of Revolution* (1965). His radio play "The Face of Violence" won the Italia Prize in 1951.

Chadwick, Sir James (Oct. 20, 1891–July 24, 1974), British experimental physicist, received the Nobel Prize for Physics in 1935 for his discovery of the neutron, which led to the development of atomic energy. He investigated gamma-ray emissions from radioactive materials under Ernest Rutherford at the University of Manchester, and then went to Berlin to work with Hans Geiger. Though interned at the outbreak of World War I, Chadwick was able to continue research with aid from German scientists. After the war he followed Rutherford to the Cavendish Laboratory at Cambridge University, where during the 1920s they studied the transmutation of elements under alpha-particle bombardment and investigated the nature of the atomic nucleus. In 1932, as a result of direct experiment, Chadwick discovered the neutron, which had been predicted by Rutherford. In 1935 he was appointed Lyon Jones professor of physics at the University of Liverpool, where he set up a cyclotron. Chadwick traveled to the U.S. in 1943 as head of a British team working on the development of the atomic bomb. After the war he returned to Liverpool to establish a school of nuclear physics. From 1948 to 1958 he served as master of Gonville and Caius College, Cambridge. He was elected to the Royal Society in 1927, was knighted in 1945, and was made a Companion of Honour in 1970.

Coolidge, William David (Oct. 23, 1873–Feb. 3, 1975), U.S. physical chemist and inventor, graduated from Massachusetts Institute of Technology in 1896 before obtaining a Ph.D. (1899) from the University of Leipzig in Germany. In 1908, three years after joining General Electric Co. as a research scientist, he discovered a process that rendered tungsten ductile and therefore eminently suitable for use in incandescent light bulbs (earlier tungsten filaments were exceedingly brittle and short-lived). In 1916 he patented a revolutionary X-ray tube. Using a hot tungsten cathode in place of the traditional cold aluminum cathode, he made a high-vacuum X-ray tube that produced highly predictable and accurate amounts of radiation. The "Coolidge tube" is still the prototype of tubes used in modern X-ray units.

Coolidge later developed portable X-ray units and devised techniques used in the construction

Courtesy, General Electric Research and Development Center

William Coolidge

of 1,000,000- and 2,000,000-volt X-ray machines for treatment of cancer and for quality control in industry. With Nobel laureate Irving Langmuir he also developed the first successful submarine-detection system. During World War II Coolidge extended his research to radar, rockets, antisubmarine devices, and the atomic bomb. He became director of GE's Research Laboratory in 1932 and vice-president and director of research in 1940. He retired in 1944 but remained as a consultant.

Craig, Lyman C. (June 12, 1906–July 7, 1974), U.S. chemist, devised a technique for purifying drugs or isolating certain labile chemical compounds from complex mixtures and thus paved the way for many important biochemical advances. His "countercurrent" technique relies on small differences in the solubility of individual components of a mixture to separate them. It was widely adopted because the time-saving glassware mechanism that Craig created not only permits up to a thousand simultaneous extractions but also eliminates the possibility of degrading the original compounds during the process. Craig studied at Iowa State University (Ph.D., 1931) and at Johns Hopkins University before joining (1933) the faculty of the Rockefeller Institute for Medical Research (now Rockefeller University in New York City).

Gaudin, A(ntoine) M(arc) (Aug. 8, 1900–Aug. 23, 1974), U.S. mineral engineer, was largely responsible for developing ore-processing techniques that yielded uranium for the earliest atomic bombs. He and his colleagues discovered new methods of leaching and ion exchange that were effectively used to extract uranium, which was then purified and used in reactors for plutonium production or, after conversion into uranium hexafluoride gas, was separated into its isotopes. Gaudin also analyzed the principles underlying flotation, a widely used process for extracting specific metals (*e.g.,* copper) comprising only a small portion of the raw ore. Gaudin taught at Columbia University, his alma mater, before successively joining the faculties of the University of Utah, Montana School of Mines, and the Massachusetts Institute of Technology.

Griggs, David T(ressel) (Oct. 6, 1911–Dec. 31, 1974), U.S. geophysicist, undertook experimental research to determine stress-strain relations in Earth materials, and studied plastic flow and fracture in rocks. While at Harvard University (1934–41), he theorized that convection currents within the Earth had created the Himalayan, Rocky, and Andes mountains, and may have produced the world's great landmasses. Beginning in World War II he became consultant to numerous military, governmental, and private agencies. One such association was with the Lawrence Radiation Laboratory in Livermore, Calif., where studies were undertaken to determine the feasibility of using nuclear explosions in engineering projects. But for nearly 27 years (1948–74) his principal loyalty was to the Institute of Geophysics and Planetary Physics at the University of California at Los Angeles. Among his many honors, Griggs received the Medal for Merit in 1946.

Huxley, Sir Julian Sorell (June 22, 1887–Feb. 14, 1975), English biologist and philosopher of science, was the grandson of Thomas Henry Huxley, a noted biologist, and brother of Aldous Huxley, a widely read novelist. Julian's personal fame, both before and after he served as the first director general (1946–48) of UNESCO, was due to a restlessly inquiring mind and a compelling desire to share his ideas with others. To this end he used radio, television, the classroom, the lecture platform, and the press. For the scientific community he produced highly regarded studies on such topics as the courtship of herons and grebes, the growth of animals, and the metamorphosis of the axolotl of Mexico. He was also biology editor of the 14th edition of the *Encyclopaedia Britannica* and won an Academy Award for a film entitled *The Private Life of the Gannets*.

Huxley's most widely discussed and perhaps most challenged theory, however, involved eugenic, behavioral, ethical, and religious principles. Called "evolutionary humanism," it reserved no

Horst Tappe—EB Inc.

Sir Julian Huxley

place for God and postulated man as the sole arbiter of future evolutionary progress. To secure the fullest possible development for the greatest number of individuals, human misery had to be wiped out by selective breeding and population control. Huxley, who studied at Oxford University and received a long list of distinguished honors during his lifetime, was knighted in 1958. Among his writings were: *Religion Without Revelation* (1927), *Problems of Relative Growth* (1932), *Evolution: The Modern Synthesis* (1942), *New Bottles for New Wine* (1957), and *Memories* (1970, 1972).

Julian, Percy Lavon (April 11, 1899–April 19, 1975), U.S. chemist, taught at several universities and did graduate study at Harvard University (1922–26) before receiving his Ph.D. (1931) from the University of Vienna. His synthesis (1935) of physostigmine, used to treat glaucoma, led to his appointment as director of research in the soya products division of Glidden Co. in Chicago. He soon developed an inexpensive soya protein useful in paper manufacturing and in fire extinguishers. With his colleagues, Julian developed scores of soya derivatives. Steroids, derived or synthesized from a soya base, dramatically lowered the cost of treating arthritis and other diseases. In 1945 Julian was named director of research and management of fine chemicals at Glidden but left in 1953 to establish Julian Laboratories, Inc., with a branch in Mexico. From 1964 he served as president of Julian Associates and director of the Julian Research Institute.

Wide World

Percy Julian

Levy, Hyman (March 7, 1889–Feb. 27, 1975), British mathematician, was educated at the universities of Edinburgh, Oxford, and Göttingen. After four years (1916–20) with the Aerodynamics Research Staff of the National Physical Laboratory, he joined the Imperial College of Science and Technology at the University of London, where he remained (1920–54) until his retirement. During this period he also served as dean of the Royal College of Science (1946–52). Though Levy's main interests were numerical methods and statistics, the broad range of topics that engrossed his mind is evident from such publications as *Numerical Studies in Differential Equations* (1934), *Science, Curse or Blessing?* (1940), *Social Thinking* (1945), *Literature in an Age of Science* (1953), and *Finite Difference Equations* (1958). As a prominent member of the British Communist Party, he visited the U.S.S.R. in 1956 and was so appalled at the persecution of Jewish intellectuals and artists that he published *Jews and the National Question* (1957), signaling an end to his Marxist sympathies.

Lewis, Sir Aubrey Julian (Nov. 8, 1900–Jan. 21, 1975), Australian-born psychiatrist, was an independent-minded scholar who preferred eclecticism to dogmatism in regard to competing schools of psychiatry. He was deeply committed to advancing the academic frontiers of psychiatry, and, though not a prolific writer, he produced studies of melancholia and obsessional illness that were widely read. Lewis has also been given generous credit for the rapid development of psychiatric epidemiology in Britain. In 1928 he began an enduring association with Maudsley Hospital in London, serving as clinical director from 1936 to 1948. At the University of London, he was professor of psychiatry from 1946 until his retirement in 1966. He was knighted in 1959.

Maufe, Sir Edward (Dec. 12, 1883–Dec. 12, 1974), British architect, was awarded the Royal Gold Medal for Architecture in 1944 and was knighted in 1954. After studying at St. John's College, Oxford, and at the Architectural Association, he became interested in country houses and gardens. In time, however, his list of major accomplishments included the Anglican cathedral at Guildford, Surrey; St. Saviour's and St. Bede's, two London churches; new buildings for St. John's and Balliol colleges at Oxford, and the Playhouse Theatre; and new structures for Trinity and St. John's colleges at Cambridge, and the Festival Theatre. Among the numerous restorations and renovations that Maufe also undertook were Middle Temple, Gray's Inn, and St. Martin-in-the-Fields, all in London.

Petterssen, Sverre (Feb. 19, 1898–Dec. 31, 1974), Norwegian-born meteorologist, was a renowned expert in both dynamic meteorology, which involves atmospheric motions and the forces creating them, and in synoptic meteorology, which employs empirical data to identify, study, and forecast weather. After many years (1924–39) with the Norwegian Meteorological Service, he accepted a position in the U.S. at the Massachusetts Institute of Technology. In 1942, however, he moved to England to assist the war effort at the British Air Ministry in Dunstable, where he developed new techniques for upper air analysis that subsequently received worldwide acceptance. After the war he returned home to the Norwegian Meteorological Institute, but subsequently returned to the U.S. as director of Scientific Services of the U.S. Air Force Weather Service (1948–52). His last association before retiring was with the University of Chicago (1952–63). For highly original contributions to studies of frontogenesis and convection, the kinematics of weather systems, and the physics of fog, he received many prestigious awards. Among his best-known works are *Weather Analysis and Forecasting* (1940) and *Introduction to Meteorology* (1941).

Rideal, Sir Eric Keightley (April 11, 1890–Sept. 25, 1974), British scientist and distinguished physical chemist, was noted for his work in surface chemistry, particularly on interfaces and in polymer chemistry. In 1930 he became Cambridge University's first professor of colloidal physics and in 1946 was appointed director of the Davy Faraday Research Laboratory and Fullerian professor at the Royal Institution. He completed five years as professor of physical chemistry at King's College, London, before retiring in 1955. Rideal, who was a fellow of the Royal Society and recipient of its Davy Medal, was knighted in 1951.

Robinson, Sir Robert (Sept. 13, 1886–Feb. 8, 1975), British organic chemist, was awarded the Nobel Prize for Chemistry in 1947 for research into the chemical constituents of plants. Long fascinated with the challenge of determining the sequence of chemical reactions through which alkaloids are formed in nature, he eventually synthesized the alkaloid tropinone by the interaction, in dilute aqueous solution at room temperature, of three simple compounds now known to be present in plants. His success in rationalizing the assembly of these organic compounds in terms of the electronic processes occurring during formation and disruption of chemical bonds was a vital step toward understanding all biosynthetic mechanisms. Robinson also contributed important research on anthocyanin pigments of flowers and other vegetable coloring matters; he then proceeded to study the genetics of flower color variations. Still other research centered on the synthesis of penicillin and of female hormones.

Robinson was educated at the Victoria University of Manchester and held the Waynflete chair of chemistry at the University of Oxford from 1930 until he retired in 1955. He was knighted in 1939 and received numerous awards from scholarly institutions throughout the world.

Seversky, Alexander P. de (June 7, 1894–Aug. 24, 1974), U.S. aeronautic engineer, flew more than 50 missions as a Russian aviator despite the loss of a leg during his first combat mission (1915) in World War I. He was assistant naval attaché in Washington, D.C., in 1918 when his country closed its embassy. Seversky thereupon elected to stay in the U.S. During the next half century and more he gained prominence as a persistent and effective advocate of strategic air power. While directing several enterprising corporations that bore his name, Seversky patented a wide variety of inventions. He developed techniques for in-flight fueling, invented landing gear suitable for hazardous surfaces, designed the first fully automatic synchronous bombsight, helped devise the automatic pilot, and received the International Harmon Trophy three times for innovative plane designs. Seversky, whose contributions hastened the arrival of modern aircraft technology, was given the Medal of Merit by the U.S. government.

Stirling, Matthew Williams (Aug. 28, 1896–Jan. 23, 1975), U.S. anthropologist, was chief of the Bureau of American Ethnology at the Smithsonian Institution in Washington, D.C., from 1928 to 1947; thereafter he served as director until his retirement in 1958. Though Stirling led several important expeditions to Europe and the East Indies, Meso-America and South America provided him with his most rewarding discoveries. In southern Mexico a team under his leadership found remains of the long-extinct Olmec civilization. In Panama he identified the site of the region's oldest known village. And in Costa Rica he excavated 11 huge, perfectly formed granite spheres, the function of which still baffles scholars. The Leyden Plate, which he uncovered in Tabasco, Mexico, is a fragmented stone monument bearing a date corresponding to 291 B.C.; at the time, it was the oldest dated relic ever uncovered in the Americas.

Wide World

Alexander de Seversky

Wright, George Ernest (Sept. 5, 1909–Aug. 29, 1974), U.S. archaeologist, became an acknowledged expert on Near Eastern archaeology with the publication of *Pottery of Palestine from the Earliest Times to the End of the Early Bronze Age* (1937). He undertook major excavations in Palestine at Shechem (1956–74) and Gezer (1964–65), and in Cyprus at Idalion (1971–74). At Shechem a team under his direction uncovered artifacts dating as far back as 4000 B.C. In 1966 Wright was named head of American Schools of Oriental Research, which represents some 150 scholarly institutions. His name appears on such diverse books as *Shechem: The Biography of a Biblical City* (1965) and *Ecumenical Dialogue at Harvard: The Roman Catholic-Protestant Colloquium* (1964).

Index

Index entries to feature and review articles in this and previous editions of the *Yearbook of Science and the Future* are set in boldface type, *e.g.*, **Astronomy.** Entries to other subjects are set in lightface type, *e.g.*, Radiation. Additional information on any of these subjects is identified with a subheading and indented under the entry heading. The numbers following headings and subheadings indicate the year (boldface) of the edition and the page number (lightface) on which the information appears.

Astronomy 76–264; **75**–181; **74**–408, 103
aurora polaris **74**–144
Center for Short-Lived Phenomena **75**–151
climatology **76**–184
employment problems **76**–414
honors **76**–418; **75**–247; **74**–249
Natural Satellites, The **76**–224
planetary research **75**–178
red shifts **76**–221

All entry headings, whether consisting of a single word or more, are treated for the purpose of alphabetization as single complete headings and are alphabetized letter by letter up to the punctuation. The abbreviation "il." indicates an illustration.

A

B

E

H

I

Q

R

S

T

U

V

W

X

Y

Z

Acknowledgments

4 Photographs by (top) courtesy NASA; (center) NASA from Black Star; (bottom) David Moore from Photo Researchers

6 Photographs and illustrations by (left to right, top to bottom) David Muench; John Craig; Dennis Magdich; Jay Simon; courtesy, Brookhaven National Laboratory; courtesy, Anglo American Corporation of South Africa Ltd.

38–39 Illustrations adapted from Thomas R. McDonough, "Jupiter After Pioneer: A Progress Report," in "Nature," Sept. 6, 1974

45 Photographs by (left to right, top to bottom) Stephen Deutch; Russell D. Lamb from Photo Researchers; John S. Flannery from Bruce Coleman Inc.; W. H. Hodge from Peter Arnold; Tomas D. W. Friedmann from Photo Researchers; John Padour; John Padour; John Padour; Gerald Brimacombe from Black Star

93, 97–101 Illustrations by Dave Beckes

151 Illustration adapted from Karl von Frisch, "Bees: Their Vision, Chemical Senses, and Language," (1950) Cornell University Press

168 Photographs by (left to right, top to bottom) courtesy, Field Museum of Natural History, Chicago, photograph, John H. Gerard—EB Inc.; Joseph and Helen Guetterman Collection, photograph, John H. Gerard—EB Inc.; Floyd R. Getsinger; courtesy, Field Museum of Natural History, Chicago, photograph, John H. Gerard—EB Inc.; Mac Fall Collection, photograph, Mary A. Root—EB Inc.

169 Photographs by (left to right, top to bottom) Emil Javorsky—EB Inc.; John H. Gerard—EB Inc.; Lee Boltin; Lee Boltin; courtesy, General Electric Specialty Materials Department; Lee Boltin

201, 203 Illustrations by John Craig

213–220 Illustrations by Dennis Magdich

250 Photographs (top to bottom) courtesy, Bell Laboratories; London Daily Express from Pictorial Parade; courtesy, Corning Glass Works; London Daily Express from Pictorial Parade